Discrete Data Analysis with R
Visualization and Modeling Techniques for Categorical and Count Data

CHAPMAN & HALL/CRC
Texts in Statistical Science Series

Series Editors

Francesca Dominici, *Harvard School of Public Health, USA*
Julian J. Faraway, *University of Bath, UK*
Martin Tanner, *Northwestern University, USA*
Jim Zidek, *University of British Columbia, Canada*

Statistical Methods in Agriculture and Experimental Biology, Second Edition
R. Mead, R.N. Curnow, and A.M. Hasted

Statistics in Engineering: A Practical Approach
A.V. Metcalfe

Statistical Inference: An Integrated Approach, Second Edition
H. S. Migon, D. Gamerman, and F. Louzada

Beyond ANOVA: Basics of Applied Statistics
R.G. Miller, Jr.

A Primer on Linear Models
J.F. Monahan

Applied Stochastic Modelling, Second Edition
B.J.T. Morgan

Elements of Simulation
B.J.T. Morgan

Probability: Methods and Measurement
A. O'Hagan

Introduction to Statistical Limit Theory
A.M. Polansky

Applied Bayesian Forecasting and Time Series Analysis
A. Pole, M. West, and J. Harrison

Statistics in Research and Development, Time Series: Modeling, Computation, and Inference
R. Prado and M. West

Introduction to Statistical Process Control
P. Qiu

Sampling Methodologies with Applications
P.S.R.S. Rao

A First Course in Linear Model Theory
N. Ravishanker and D.K. Dey

Essential Statistics, Fourth Edition
D.A.G. Rees

Stochastic Modeling and Mathematical Statistics: A Text for Statisticians and Quantitative Scientists
F.J. Samaniego

Statistical Methods for Spatial Data Analysis
O. Schabenberger and C.A. Gotway

Bayesian Networks: With Examples in R
M. Scutari and J.-B. Denis

Large Sample Methods in Statistics
P.K. Sen and J. da Motta Singer

Spatio-Temporal Methods in Environmental Epidemiology
G. Shaddick and J.V. Zidek

Decision Analysis: A Bayesian Approach
J.Q. Smith

Analysis of Failure and Survival Data
P. J. Smith

Applied Statistics: Handbook of GENSTAT Analyses
E.J. Snell and H. Simpson

Applied Nonparametric Statistical Methods, Fourth Edition
P. Sprent and N.C. Smeeton

Data Driven Statistical Methods
P. Sprent

Generalized Linear Mixed Models: Modern Concepts, Methods and Applications
W. W. Stroup

Survival Analysis Using S: Analysis of Time-to-Event Data
M. Tableman and J.S. Kim

Applied Categorical and Count Data Analysis
W. Tang, H. He, and X.M. Tu

Elementary Applications of Probability Theory, Second Edition
H.C. Tuckwell

Introduction to Statistical Inference and Its Applications with R
M.W. Trosset

Understanding Advanced Statistical Methods
P.H. Westfall and K.S.S. Henning

Statistical Process Control: Theory and Practice, Third Edition
G.B. Wetherill and D.W. Brown

Generalized Additive Models: An Introduction with R
S. Wood

Epidemiology: Study Design and Data Analysis, Third Edition
M. Woodward

Practical Data Analysis for Designed Experiments
B.S. Yandell

Texts in Statistical Science

Discrete Data Analysis with R
Visualization and Modeling Techniques for Categorical and Count Data

Michael Friendly
York University
Toronto, Canada

David Meyer
UAS Technikum Wien
Vienna, Austria

with contributions by
Achim Zeileis
University of Innsbruck
Innsbruck, Austria

1 Introduction

2 Working with Categorical Data

3 Discrete Distributions

I. Getting Started

7 Logistic Regression Models

8 Polytomous Responses

9 Loglinear and Logit Models

III. Model-building Methods

10 Extending Loglinear Models

11 Generalized Linear Models

Discrete Data Analysis with R

II. Exploratory Methods

4 Two-way Contingency Tables

6 Correspondence Analysis

5 Mosaic Displays

CRC Press
Taylor & Francis Group
Boca Raton London New York

CRC Press is an imprint of the
Taylor & Francis Group, an **Informa** business

A CHAPMAN & HALL BOOK

CRC Press
Taylor & Francis Group
6000 Broken Sound Parkway NW, Suite 300
Boca Raton, FL 33487-2742

© 2016 by Taylor & Francis Group, LLC
CRC Press is an imprint of Taylor & Francis Group, an Informa business

Printed on acid-free paper
Version Date: 20151113

International Standard Book Number-13: 978-1-4987-2583-5 (Pack - Book and Ebook)

Library of Congress Cataloging-in-Publication Data

Names: Friendly, Michael. | Meyer, David, 1973-
Title: Discrete data analysis with R : visualization and modeling techniques
for categorical and count data / Michael Friendly and David Meyer.
Description: Boca Raton : Taylor & Francis, 2016. | Series: Chapman &
hall/CRC texts in statistical science series ; 120 | "A CRC title." |
Includes bibliographical references and index.
Identifiers: LCCN 2015033842 | ISBN 9781498725835 (alk. paper)
Subjects: LCSH: Mathematics--Data processing. | R (Computer program language)
Classification: LCC QA300 .F744 2016 | DDC 519.50285/5133--dc23
LC record available at http://lccn.loc.gov/2015033842

Printed and bound in Great Britain by
CPI Group (UK) Ltd, Croydon, CR0 4YY

Visit the Taylor & Francis Web site at
http://www.taylorandfrancis.com

and the CRC Press Web site at
http://www.crcpress.com

Contents

Preface

The greatest possibilities of visual display lie in vividness and inescapability of the intended message. A visual display can stop your mental flow in its tracks and make you think. A visual display can force you to notice what you never expected to see.

John W. Tukey (1990)

Data analysis and graphics

This book stems from the conviction that data analysis and statistical graphics should go hand-in-hand in the process of understanding and communicating statistical data. Statistical summaries compress a data set into a few numbers, the result of an hypothesis test, or coefficients in a fitted statistical model, while graphical methods help us to explore patterns and trends, see the unexpected, identify problems in an analysis, and communicate results and conclusions in principled and effective ways.

This interplay between analysis and visualization has long been a part of statistical practice for *quantitative data*. Indeed, the origin of correlation, regression, and linear models (regression, ANOVA) can arguably be traced to Francis Galton's (1886) visual insight from a scatterplot of heights of children and their parents on which he overlaid smoothed contour curves of roughly equal bivariate frequencies and lines for the means of $Y \mid X$ and $X \mid Y$ (described in Friendly and Denis (2005), Friendly et al. (2013)).

The analysis of discrete data is a much more recent arrival, beginning in the 1960s and giving rise to a few seminal books in the 1970s (Bishop et al., 1975, Haberman, 1974, Goodman, 1978, Fienberg, 1980). Agresti (2013, Chapter 17) presents a brief historical overview of the development of these methods from their early roots around the beginning of the 20th century.

Yet curiously, associated graphical methods for categorical data were much slower to develop. This began to change as it was recognized that counts, frequencies, and discrete variables required different schemes for mapping numbers into useful visual representations (Friendly, 1995, 1997), some quite novel. The special nature of discrete variables and frequency data vis-a-vis statistical graphics is now more widely accepted, and many of these new graphical methods (e.g., mosaic displays, fourfold plots, diagnostic plots for generalized linear models) have become, if not mainstream, then at least more widely used in research, teaching, and communication.

Much of what had been developed through the 1990s for graphical methods for discrete data was

described in the book *Visualizing Categorical Data* (Friendly, 2000) and was implemented in SAS®
software. Since that time, there has been considerable growth in both statistical methods for the
analysis of categorical data (e.g., generalized linear models, zero-inflation models, mixed models for
hierarchical and longitudinal data with discrete outcomes), along with some new graphical methods
for visualizing and interpreting the results (3D mosaic plots, effect plots, diagnostic plots, etc.).
The bulk of these developments have been implemented in R, and the time is right for an in-depth
treatment of modern graphical methods for the analysis of categorical data, to which you are now
invited.

Goals

This book aims to provide an applied, practically oriented treatment of modern methods for the analysis of categorical data—discrete response data and frequency data—with a main focus on graphical
methods for exploring data, spotting unusual features, visualizing fitted models, and presenting or
explaining results.

   We describe the necessary statistical theory (sometimes in abbreviated form) and illustrate the
practical application of these techniques to a large number of substantive problems: how to organize
the data, conduct an analysis, produce informative graphs, and understand what they have to say
about the data at hand.

Overview and organization of this book

This book is divided into three parts. Part I, Chapters 1–3, contains introductory material on graphical methods for discrete data, basic R skills needed for the book, and methods for fitting and
visualizing one-way discrete distributions.

   Part II, Chapters 4–6, is concerned largely with simple, traditional non-parametric tests and
exploratory methods for visualizing patterns of association in two-way and larger frequency tables.
Some of the discussion here introduces ideas and notation for loglinear models that are treated more
generally in Part III.

   Part III, Chapters 7–11, discusses model-based methods for the analysis of discrete data. These
are all examples of generalized linear models. However, for our purposes, it has proved more
convenient to develop this topic from the specific cases (logistic regression, loglinear models) to the
general rather than the reverse.

Chapter 1: *Introduction.* Categorical data require different statistical and graphical methods than
   commonly used for quantitative data. This chapter outlines the basic orientation of the book
   toward visualization methods and some key distinctions regarding the analysis and visualization
   of categorical data.

Chapter 2: *Working with Categorical Data.* Categorical data can be represented in various forms:
   case form, frequency form, and table form. This chapter describes and illustrates the skills and
   techniques in R needed to input, create, and manipulate R data objects to represent categorical
   data, and convert these from one form to another for the purposes of statistical analysis and
   visualization, which are the subject of the remainder of the book.

Chapter 3: *Fitting and Graphing Discrete Distributions.* Understanding and visualizing discrete
   data distributions provides a building block for model-based methods discussed in Part III.
   This chapter introduces the well-known discrete distributions—the binomial, Poisson, negative-
   binomial, and others—in the simplest case of a one-way frequency table.

Chapter 4: *Two-Way Contingency Tables*. The analysis of two-way frequency tables concerns the association between two variables. A variety of specialized graphical displays help to visualize the pattern of association, using area of some region to represent the frequency in a cell. Some of these methods are focused on visualizing an odds ratio (for 2×2 tables), or the general pattern of association, or the agreement between row and column categories in square tables.

Chapter 5: *Mosaic Displays for n-Way Tables*. This chapter introduces mosaic displays, designed to help to visualize the pattern of associations among variables in two-way and larger tables. Extensions of this technique can reveal partial associations and marginal associations, and shed light on the structure of loglinear models themselves.

Chapter 6: *Correspondence Analysis*. Correspondence analysis provides visualizations of associations in a two-way contingency table in a small number of dimensions. Multiple correspondence analysis extends this technique to n-way tables. Other graphical methods, including mosaic matrices and biplots, provide complementary views of loglinear models for two-way and n-way contingency tables.

Chapter 7: *Logistic Regression Models*. This chapter introduces the modeling framework for categorical data in the simple situation where we have a categorical response variable, often binary, and one or more explanatory variables. A fitted model provides both statistical inference and prediction, accompanied by measures of uncertainty. Data visualization methods for discrete response data must often rely on smoothing techniques, including both direct, non-parametric smoothing and the implicit smoothing that results from a fitted parametric model. Diagnostic plots help us to detect influential observations that may distort our results.

Chapter 8: *Models for Polytomous Responses*. This chapter generalizes logistic regression models for a binary response to handle a multi-category (polytomous) response. Different models are available depending on whether the response categories are nominal or ordinal. Visualization methods for such models are mostly straightforward extensions of those used for binary responses presented in Chapter 7.

Chapter 9: *Loglinear and Logit Models for Contingency Tables*. This chapter extends the model-building approach to loglinear and logit models. These comprise another special case of generalized linear models designed for contingency tables of frequencies. They are most easily interpreted through visualizations, including mosaic displays and effect plots of associated logit models.

Chapter 10: *Extending Loglinear Models*. Loglinear models have special forms to represent additional structure in the variables in contingency tables. Models for ordinal factors allow a more parsimonious description of associations. Models for square tables allow a wide range of specific models for the relationship between variables with the same categories. Another extended class of models arise when there are two or more response variables.

Chapter 11: *Generalized Linear Models*. Generalized linear models extend the familiar linear models of regression and ANOVA to include counted data, frequencies, and other data for which the assumptions of independent, normal errors are not reasonable. We rely on the analogies between ordinary and generalized linear models (GLMs) to develop visualization methods to explore the data, display the fitted relationships, and check model assumptions. The main focus of this chapter is on models for count data.

Audience

This book has been written to appeal to two broad audiences wishing to learn to apply methods for discrete data analysis:

- Advanced undergraduate and graduate students in the social and health sciences, epidemiology, economics, business, and (bio)statistics
- Substantive researchers, methodologists, and consultants in various disciplines wanting to be able to use these methods with their own data and analyses.

It assumes the reader has a basic understanding of statistical concepts at least at an intermediate undergraduate level including regression and analysis of variance (for example, at the level of Neter et al. (1990) or Mendenhall and Sincich (2003)). It is less technically demanding than other modern texts covering categorical data analysis at a graduate level, such as Agresti (2013), *Categorical Data Analysis*, Powers and Xie (2008), *Statistical Methods for Categorical Data Analysis*, and Christensen (1997), *Log-Linear Models and Logistic Regression*. Nevertheless, there are some topics that are a bit more advanced or technical, and these are marked as * or ** sections.

As well, there are a number of mathematical or statistical topics that we use in passing, but do not describe in these pages (some matrix notation, basic probability theory, maximum likelihood estimation, etc.). Most of these are described in Fox (2015), which is available online and serves well as a supplement to this book.

In addition, it is not possible to include *all* details of using R effectively for data analysis. It is assumed that the reader has at least basic knowledge of the R language and environment, including interacting with the R console (RGui for Windows, R.app for Mac OS X) or other graphical user interface (e.g., RStudio), using R functions in packages, getting help for these from R, etc. One introductory chapter (Chapter 2) is devoted to covering the particular topics most important to categorical data analysis, beyond such basic skills needed in the book.

Textbook use

This book is most directly suitable for a one-semester applied advanced undergraduate or graduate course on categorical data analysis with a strong emphasis on the use of graphical methods to understand and explain data and results of analysis. A detailed outline of such a course, together with lecture notes and assignments, is available at the first author's web page, `http://euclid.psych.yorku.ca/www/psy6136/`, using this book as the main text. This course also uses Agresti (2007), *An Introduction to Categorical Data Analysis* for additional readings.

For instructors teaching a more traditional course using one of the books mentioned above as the main text, this book would be a welcome supplement, because almost all other texts treat graphical methods only perfunctorily, if at all. A few of these contain a brief appendix mentioning software, or have a related web site with some data sets and software examples. Moreover, none actually describe how to do these analyses and graphics with R.

Features

- Provides an accessible introduction to the major methods of categorical data analysis for data exploration, statistical testing, and statistical models.
- The emphasis throughout is on computing, visualizing, understanding, and communicating the results of these analyses.
- As opposed to more theoretical books, the goal here is to help the reader to translate theory into practical application, by providing skills and software tools for carrying out these methods.
- Includes many examples using real data, often treated from several perspectives.
- The book is supported directly by R packages vcd (Meyer et al., 2015) and vcdExtra (Friendly, 2015), along with numerous other R packages.
- All materials (data sets, R code) will be available online on the web site for the book, `http://datavis.ca/books/DDAR`.

- Each chapter contains a collection of lab exercises, which work through applications of some of the methods presented in that chapter. This makes the book more suitable for both self-study and classroom use.

Acknowledgments

We are grateful to many colleagues, friends, students, and Internet acquaintances who have contributed to to this book, directly or indirectly.

We thank those who read and commented on various drafts of the book or chapters. In particular, John Fox, Michael Greenacre, and several anonymous reviewers gave insightful comments on the organization of the book and made many helpful suggestions. Matthew Sigal used his wizardly skills to turn sketches of conceptual diagrams into final figures. Phil Chalmers contributed greatly with technical and stylistic editing of a number of chapters.

At a technical level, we were aided by the cooperation of a number of R package authors, who helped to enhance the graphic displays: Achim Zeilleis who served as a guiding hand in the development of the vcd and vcdExtra packages; John Fox and Sandy Weisberg for enhancements to the car (Fox and Weisberg, 2015a) and effects (Fox et al., 2015) packages; Milan Bouchet-Valat for incorporating suggestions dealing with plotting rc() solutions into the logmult (Bouchet-Valat, 2015) package; Michael Greenacre and Oleg Nenadic for help to enhance plotting in the ca (Greenacre and Nenadic, 2014) package; Heather Turner for advice and help with plotting models fit using the gnm (Turner and Firth, 2014) package; Jay Emerson for improvements to the gpairs (Emerson and Green, 2014) package.

There were also many contributors from the R-Help email list (r-help@r-project.org), too many to name them all. Special thanks for generous assistance go to: David Carlson, William Dunlap, Bert Gunter, Jim Lemon, Duncan Murdoch, Denis Murphy, Jeff Newmiller, Richard Heiberger, Thierry Onkelinx, Marc Schwartz, David Winsemius, and Ista Zahn.

The book was written using the knitr (Xie, 2015) package, allowing a relatively seamless integration of LaTeX text, R code, and R output and graphs, so that any changes in the code were automatically incorporated in the book. Thanks are due to Yihui Xie and all the contributors to the knitr project for making this possible. We are also grateful to Phil Chalmers and Derek Harnanansingh for assistance in using GitHub to manage our collaboration. Pere Millán-Martínez and Marcus Fontaine helped considerably with LaTeX formatting issues.

The first author's work on this project was supported by grants from the National Science and Engineering Research Council of Canada (Grant 8150) and a sabbatical leave from York University in 2013–14, during which most of this book was written.

Part I

Getting Started

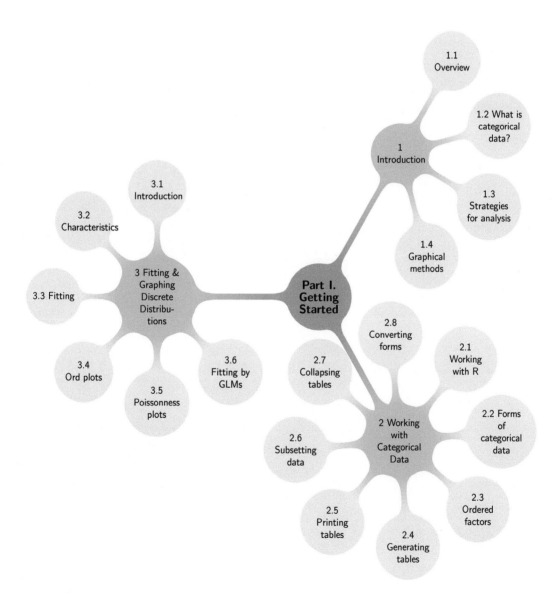

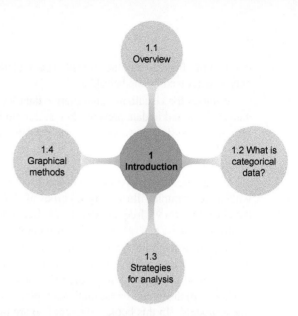

1

Introduction

Categorical data consist of variables whose values comprise a set of discrete categories. Such data require different statistical and graphical methods than commonly used for quantitative data. The focus of this book is on visualization techniques and graphical methods designed to reveal patterns of relationships among categorical variables. This chapter outlines the basic orientation of the book and some key distinctions regarding the analysis and visualization of categorical data.

1.1 Data visualization and categorical data: Overview

> Graphs carry the message home. A universal language, graphs convey information directly to the mind. Without complexity there is imaged to the eye a magnitude to be remembered. Words have wings, but graphs interpret. Graphs are pure quantity, stripped of verbal sham, reduced to dimension, vivid, unescapable.
>
> ---
>
> Henry D. Hubbard, in Foreword to Brinton (1939), *Graphic Presentation*

"Data visualization" can mean many things, from popular press infographics, to maps of voter turnout or party choice. Here we use this term in the narrower context of statistical analysis. As such, we refer to an approach to data analysis that focuses on *insightful* graphical display in the service of both *understanding* our data and *communicating* our results to others.

We may display the raw data, some summary statistics, or some indicators of the quality or adequacy of a fitted model. The word "insightful" suggests that the goal is (hopefully) to reveal some aspects of the data that might not be perceived, appreciated, or absorbed by other means. As in the quote from Keats, the overall aims include both beauty and truth, though each of these are only as perceived by the beholder.

Methods for visualizing quantitative data have a long history and are now widely used in both data analysis and in data presentation, and in both popular and scientific media. Graphical methods for categorical data, however, have only a more recent history, and are consequently not as widely used. The goal of this book is to show concretely how data visualization may be usefully applied to categorical data.

"Categorical" means different things in different contexts. We introduce the topic in Section 1.2 with some examples illustrating (a) types of categorical variables: binary, nominal, and ordinal, (b) data in case form vs. frequency form, (c) frequency data vs. count data, (d) univariate, bivariate, and multivariate data, and (e) the distinction between explanatory and response variables.

Statistical methods for the analysis of categorical data also fall into two quite different categories, described and illustrated in Section 1.3: (a) the simple randomization-based methods typified by the classical Pearson chi-squared (χ^2) test, Fisher's exact test, and Cochran–Mantel–Haenszel tests, and (b) the model-based methods represented by logistic regression, loglinear, and generalized linear models. In this book, Chapters 3–6 are mostly related to the randomization-based methods; Chapters 7–9 illustrate the model-based methods.

In Section 1.4 we describe some important similarities and differences between categorical data and quantitative data, and discuss the implications of these differences for visualization techniques. Section 1.4.5 outlines a strategy of data analysis focused on visualization.

In a few cases we show R code or results as illustrations here, but the fuller discussion of using R for categorical data analysis is postponed to Chapter 2.

1.2   What is categorical data?

A ***categorical variable*** is one for which the possible measured or assigned values consist of a discrete set of categories, which may be *ordered* or *unordered*. Some typical examples are:

- `Gender`, with categories "Male," "Female."
- `Marital status`, with categories "Never married," "Married," "Separated," "Divorced," "Widowed."
- `Fielding position` (in baseball), with categories "Pitcher," "Catcher," "1st base," "2nd base," ..., "Left field."
- `Side effects` (in a pharmacological study), with categories "None," "Skin rash," "Sleep disorder," "Anxiety," ....
- `Political attitude`, with categories "Left," "Center," "Right."
- `Party preference` (in Canada), with categories "NDP," "Liberal," "Conservative," "Green."
- `Treatment outcome`, with categories "no improvement," "some improvement," or "marked improvement."
- `Age`, with categories "0–9," "10–19," "20–29," "30–39," ....
- `Number of children`, with categories $0, 1, 2, \ldots$.

As these examples suggest, categorical variables differ in the number of categories: we often distinguish ***binary variables*** (or ***dichotomous variables***) such as `Gender` from those with more than two categories (called ***polytomous variables***). For example, Table 1.1 gives data on $4,526$ applicants to graduate departments at the University of California at Berkeley in 1973, classified by two binary variables, gender and admission status.

Some categorical variables (`Political attitude`, `Treatment outcome`) may have ordered categories (and are called ***ordinal variables***), while others (***nominal variables***) like `Marital`

Table 1.1: Admissions to Berkeley graduate programs

	Admitted	Rejected	Total
Males	1198	1493	2691
Females	557	1278	1835
Total	1755	2771	4526

Table 1.2: Arthritis treatment data

		Improvement			
Treatment	Sex	None	Some	Marked	Total
Active	Female	6	5	16	27
	Male	7	2	5	14
Placebo	Female	19	7	6	32
	Male	10	0	1	11
Total		42	14	28	84

`status` have unordered categories.[1] For example, Table 1.2 shows a $2 \times 2 \times 3$ table of ordered outcomes ("none," "some," or "marked" improvement) to an active treatment for rheumatoid arthritis compared to a placebo for men and women.

Finally, such variables differ in the fineness or level to which some underlying observation has been categorized for a particular purpose. From one point of view, *all* data may be considered categorical because the precision of measurement is necessarily finite, or an inherently continuous variable may be recorded only to limited precision.

But this view is not helpful for the applied researcher because it neglects the phrase "for a particular purpose." Age, for example, might be treated as a quantitative variable in a study of native language vocabulary, or as an ordered categorical variable with decade groups (0–10, 11–20, 20–30, ...) in terms of the efficacy or side-effects of treatment for depression, or even as a binary variable ("child" vs. "adult") in an analysis of survival following an epidemic or natural disaster. In the analysis of data using categorical methods, continuous variables are often recoded into ordered categories with a small set of categories for some purpose.[2]

1.2.1 Case form vs. frequency form

In many circumstances, data is recorded on each individual or experimental unit. Data in this form is called case data, or data in *case form*. The data in Table 1.2, for example, were derived from the individual data listed in the data set `Arthritis` from the **vcd** package. The following lines show the first five of $N = 84$ cases in the `Arthritis` data,

```
   ID Treatment  Sex Age Improved
1  57   Treated Male  27     Some
```

---

[1] An ordinal variable may be defined as one whose categories are *unambiguously* ordered along a *single* underlying dimension. Both marital status and fielding position may be weakly ordered, but not on a single dimension, and not unambiguously.

[2] This may be a waste of information available in the original variable, and should be done for substantive reasons, not mere convenience. For example, some researchers unfamiliar with regression methods often perform a "median-split" on quantitative predictors so they can use ANOVA methods. Doing this precludes the possibility of determining if those variables have nonlinear relations with the outcome while also decreasing statistical power.

```
2 46    Treated Male   29      None
3 77    Treated Male   30      None
4 17    Treated Male   32      Marked
5 36    Treated Male   46      Marked
```

Whether or not the data variables, and the questions we ask, call for categorical or quantitative data analysis, when the data are in case form, we can always trace any observation back to its individual identifier or data record (for example, if the case with ID equal to 57 turns out to be unusual or noteworthy).

Data in *frequency form* has already been tabulated, by counting over the categories of the table variables. The same data shown as a table in Table 1.2 appear in frequency form as shown below.

```
     Treatment    Sex Improved Freq
1     Placebo Female     None   19
2     Treated Female     None    6
3     Placebo   Male     None   10
4     Treated   Male     None    7
5     Placebo Female     Some    7
6     Treated Female     Some    5
7     Placebo   Male     Some    0
8     Treated   Male     Some    2
9     Placebo Female   Marked    6
10    Treated Female   Marked   16
11    Placebo   Male   Marked    1
12    Treated   Male   Marked    5
```

Data in frequency form may be analyzed by methods for quantitative data if there is a quantitative response variable (weighting each group by the cell frequency, with a weight variable). Otherwise, such data are generally best analyzed by methods for categorical data, where statistical models are often expressed as models for the frequency variable, in the form of an R formula like `Freq ~ ..`.

In any case, an observation in a data set in frequency form refers to all cases in the cell collectively, and these cannot be identified individually. Data in case form can always be reduced to frequency form, but the reverse is rarely possible. In Chapter 2, we identify a third format, *table form*, which is the R representation of a table like Table 1.2.

1.2.2 Frequency data vs. count data

In many cases the observations representing the classifications of events (or variables) are recorded from *operationally independent* experimental units or individuals, typically a sample from some population. The tabulated data may be called *frequency data*. The data in Table 1.1 and Table 1.2 are both examples of frequency data because each tabulated observation comes from a different person.

However, if several events or variables are observed for the same units or individuals, those events are not operationally independent, and it is useful to use the term *count data* in this situation. These terms (following Lindsey (1995)) are by no means standard, but the distinction is often important, particularly in statistical models for categorical data.

For example, in a tabulation of the number of male children within families (Table 1.3, described in Section 1.2.3 below), the number of male children in a given family would be a *count* variable, taking values $0, 1, 2, \ldots$. The number of independent families with a given number of male children is a *frequency* variable. Count data also arise when we tabulate a sequence of events over time or under different circumstances in a number of individuals.

Table 1.3: Number of Males in 6115 Saxony Families of Size 12

Males	0	1	2	3	4	5	6	7	8	9	10	11	12
Families	3	24	104	286	670	1,033	1,343	1,112	829	478	181	45	7

1.2.3 Univariate, bivariate, and multivariate data

Another distinction concerns the number of variables: one, two, or (potentially) many shown in a data set or table, or used in some analysis. Table 1.1 is an example of a bivariate (two-way) contingency table and Table 1.2 classifies the observations by three variables. Yet, we will see later that the Berkeley admissions data also recorded the department to which potential students applied (giving a three-way table), and in the arthritis data, the age of subjects was also recorded.

Any contingency table (in frequency or table form) therefore records the *marginal totals*, summed over all variables not represented in the table. For data in case form, this means simply ignoring (or not recording) one or more variables; the "observations" remain the same. Data in frequency form, however, result in smaller tables when any variable is ignored; the "observations" are the cells of the contingency table. For example, in the `Arthritis` data, ignoring `Sex` gives the smaller 2×3 table for `Treatment` and `Improved`.

```
  Treatment Improved Freq
1   Placebo     None   29
2   Treated     None   13
3   Placebo     Some    7
4   Treated     Some    7
5   Placebo   Marked    7
6   Treated   Marked   21
```

In the limiting case, only one table variable may be recorded or available, giving the categorical equivalent of univariate data. For example, Table 1.3 gives data on the distribution of the number of male children in families with 12 children (discussed further in Example 3.2). These data were part of a large tabulation of the sex distribution of families in Saxony in the 19th century, but the data in Table 1.3 have only one discrete classification variable, number of males. Without further information, the only statistical questions concern the form of the distribution. We discuss methods for fitting and graphing such discrete distributions in Chapter 3. The remaining chapters relate to bivariate and multivariate data.

1.2.4 Explanatory vs. response variables

Most statistical models make a distinction between **response variables** (or *dependent*, or *criterion* variables) and **explanatory variables** (or *independent*, or *predictor* variables).

In the standard (classical) linear models for regression and analysis of variance (ANOVA), for instance, we treat one (or more) variables as responses, to be explained by the other, explanatory variables. The explanatory variables may be quantitative or categorical (e.g., factors in R). This affects only the details of how the model is specified or how coefficients are interpreted for `lm()` or `glm()`. In these classical models, the response variable ("treatment outcome," for example), must be considered quantitative, and the model attempts to describe how the *mean* of the distribution of responses changes with the values or levels of the explanatory variables, such as age or gender.

When the response variable is categorical, however, the standard linear models do not apply, because they assume a normal (Gaussian) distribution for the model residuals. For example, in Table 1.2 the response variable is `Improvement`, and even if numerical scores were assigned to the categories "none," "some," "marked," it may be unlikely that the assumptions of the classical linear models could be met.

Hence, a categorical *response* variable generally requires analysis using methods for categorical data, but categorical *explanatory* variables may be readily handled by either method.

The distinction between response and explanatory variables also becomes important in the use of loglinear models for frequency tables (described in Chapter 9), where models can be specified in a simpler way (as equivalent logit models) by focusing on the response variable.

1.3   Strategies for categorical data analysis

Data analysis typically begins with exploratory and graphical methods designed to expose features of the data, followed by statistical analysis designed to summarize results, answer questions, and draw conclusions. Statistical methods for the analysis of categorical data can be classified into two broad categories: those concerned with *hypothesis testing* per se versus those concerned with *model building*.

1.3.1   Hypothesis testing approaches

In many studies, the questions of substantive interest translate readily into questions concerning hypotheses about **association** between variables, a more general idea than that of correlation (*linear* association) for quantitative variables. If a non-zero association exists, we may wish to characterize the strength of the association numerically and understand the pattern or nature of the association.

For example, in Table 1.1, a main question is: "Is there evidence of gender-bias in admission to graduate school?" Another way to frame this: "Are males more likely to be admitted?" These questions can be expressed in terms of an association between gender and admission status in a 2×2 contingency table of applicants classified by these two variables. If there is evidence for an association, we can assess its strength by a variety of measures, including the difference in proportions admitted for men and women or the ratio of the odds of admission for men compared to women, as described in Section 4.2.2.

Similarly, in Table 1.2, questions about the efficacy of the treatment for rheumatoid arthritis can be answered in terms of hypotheses about the associations among the table variables: `Treatment`, `Sex`, and the `Improvement` categories. Although the main concern might be focused on the overall association between `Treatment` and `Improvement`, one would also wish to know if this association is the same for men and women. A **stratified analysis** (Section 4.3) controls for the effects of background variables like Sex, tests for **homogeneity of association**, and helps to determine if these associations are equal.

Questions involving tests of such hypotheses are answered most easily using a large variety of specific statistical tests, often based on randomization arguments. These include the familiar Pearson chi-squared test for two-way tables, the Cochran–Mantel–Haenszel test statistics, Fisher's exact test, and a wide range of measures of strength of association. These tests make minimal assumptions, principally requiring that subjects or experimental units have been randomly assigned to the categories of experimental factors. The hypothesis testing approach is illustrated in Chapters 4–6, though the emphasis is on graphical methods that help us to understand the nature of association between variables.

EXAMPLE 1.1: Hair color and eye color
    The data set *HairEye* below records data on the relationship between hair color and eye color in a sample of nearly 600 students.

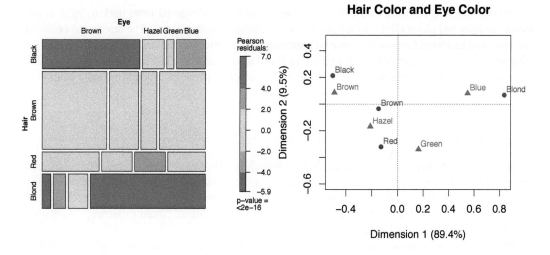

Figure 1.1: Graphical displays for the hair color and eye color data. Left: mosaic display; right: correspondence analysis plot.

```
          Eye
Hair     Brown Blue Hazel Green
  Black     68   20    15     5
  Brown    119   84    54    29
  Red       26   17    14    14
  Blond      7   94    10    16
```

The standard analysis (with `chisq.test()` or `assocstats()`) gives a Pearson χ^2 of 138.3 with nine degrees of freedom, indicating substantial departure from independence. Among the measures of strength of association, ***Cramer's V***, $V = \sqrt{\chi^2/N \min(r-1, c-1)} = 0.279$, indicates a substantial relationship between hair and eye color.[3]

```
                     X^2 df P(> X^2)
Likelihood Ratio 146.44  9         0
Pearson          138.29  9         0

Phi-Coefficient    : NA
Contingency Coeff.: 0.435
Cramer's V         : 0.279
```

The further (and perhaps more interesting question) is how do we understand the *nature* of this association between hair and eye color? Two graphical methods related to the hypothesis testing approach are shown in Figure 1.1.

The left panel of Figure 1.1 is a ***mosaic display*** (Chapter 5), constructed so that the size of each rectangle is proportional to the observed cell frequency. The shading reflects the cell contribution to the χ^2 statistic—shades of blue when the observed frequency is substantially greater than the expected frequency under independence, shades of red when the observed frequency is substantially less, as shown in the legend.

The right panel of this figure shows the results of a correspondence analysis (Chapter 6), where the deviations of the hair color and eye color points from the origin accounts for as much of the χ^2 as possible in two dimensions.

---

[3]Cramer's V varies from 0 (no association) to 1 (perfect association).

We observe that both the hair colors and the eye colors are ordered from dark to light in the mosaic display and along Dimension 1 in the correspondence analysis plot. The deviations between observed and expected frequencies have an opposite-corner pattern in the mosaic display, except for the combination of red hair and green eyes, which also stand out as the largest values on Dimension 2 in the Correspondence analysis plot. Displays such as these provide a means to understand *how* the variables are related.   △

1.3.2 Model building approaches

Model-based methods provide tests of equivalent hypotheses about associations, but offer additional advantages (at the cost of additional assumptions) not provided by the simpler hypotheses-testing approaches. Among these advantages, model-based methods provide estimates, standard errors and confidence intervals for parameters, and the ability to obtain predicted (fitted/expected) values with associated measures of precision.

We illustrate this approach here for a dichotomous response variable, where it is often convenient to construct a model relating a function of the probability, π, of one event to a linear combination of the explanatory variables. Logistic regression uses the ***logit function***,

$$\text{logit}(\pi) \equiv \log_e \left(\frac{\pi}{1 - \pi} \right) ,$$

which may be interpreted as the ***log odds*** of the given event. A linear logistic model can then be expressed as

$$\text{logit}(\pi) = \beta_0 + \beta_1 x_1 + \beta_2 x_2 + \dots$$

Statistical inferences from model-based methods provide tests of hypotheses for the effects of the predictors, $x_1, x_2, \dots$, but they also provide estimates of parameters in the model, $\beta_1, \beta_2, \dots$ and associated confidence intervals. Standard modeling tools allow us to graphically display the fitted response surface (with confidence or prediction intervals) and even to extrapolate these predictions beyond the given data. A particular advantage of the logit representation in the logistic regression model is that estimates of odds ratios (Section 4.2.2) may be obtained directly from the parameter estimates.

EXAMPLE 1.2: Space shuttle disaster

To illustrate the model-based approach, the graph in Figure 1.2 is based on a logistic regression model predicting the probability of a failure in one of the O-ring seals used in the 24 NASA space shuttles prior to the disastrous launch of the *Challenger* in January, 1986. The explanatory variable is the ambient temperature (in Fahrenheit) at the time of the flight. The sad story behind these data, and the lessons to be learned for graphical data display, are related in Example 1.10.

Here, we simply note that the fitted model, shown by the solid line in Figure 1.2, corresponds to the prediction equation (with standard errors shown in parentheses),

$$\text{logit}(\text{Failure}) = \underset{(3.06)}{5.09} - \underset{(0.047)}{0.116} \text{ Temperature}$$

A hypothesis test that failure probability is unassociated with temperature is equivalent to the test that the coefficient for temperature in this model equals 0; this test has a p-value of 0.014, convincing evidence for rejection.

The parameter estimate for temperature, -0.116, however, gives more information. Each $1°$ increase in temperature decreases the log odds of failure by 0.116, with 95% confidence interval $[-0.208, -0.0235]$. The equivalent odds ratio is $\exp(-0.116) = 0.891 \; [0.812, 0.977]$. Equivalently, a $10°$ *decrease* in temperature corresponds to an odds ratio of a failure of $\exp(10 \times 0.116) = 3.18$, more than tripling the odds of a failure.

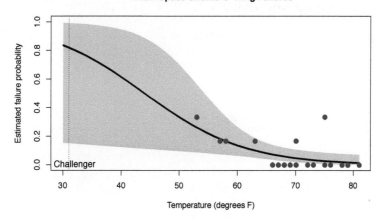

NASA Space Shuttle O–Ring Failures

Figure 1.2: Space shuttle O-ring failure, observed and predicted probabilities. The dotted vertical line at 31° shows the prediction for the launch of the *Challenger*.

When the *Challenger* was launched, the temperature was only 31°. The shaded region in Figure 1.2 shows 95% prediction intervals for failure probability. All previous shuttles (shown by the points in the figure) had been launched at much warmer temperatures, so the prediction interval (the dashed vertical line) at 31° represents a considerable extrapolation beyond the available data. Nonetheless, the model building approach does provide such predictions along with measures of their uncertainty. Figure 1.2 is a graph that might have saved lives.

$\triangle$

EXAMPLE 1.3: Donner Party

In April–May of 1846 (three years before the California gold rush), the Donner and Reed families set out for California from the American Mid-west in a wagon train to seek a new life and perhaps their fortune in the new American frontier. By mid-July, a large group had reached a site in present-day Wyoming; George Donner was elected to lead what was to be called the "Donner Party," which eventually numbered 87 people in 23 wagons, along with their oxen, cattle, horses, and worldly possessions.

They were determined to reach California as quickly as possible. Lansford Hastings, a self-proclaimed trailblazer (retrospectively, of dubious distinction), proposed that the party follow him through a shorter path through the Wasatch Mountains. Their choice of "Hastings's Cutoff" proved disastrous: Hastings had never actually crossed that route himself, and the winter of of 1846 was to be one of the worst on record.

In October, 1846, heavy snow stranded them in the eastern Sierra Nevada, just to the east of a pass that bears their name today. The party made numerous attempts to seek rescue, most turned back by blizzard conditions. Relief parties in March–April 1847 rescued 40, but discovered grisly evidence that those who survived had cannibalized those who died.

Here we briefly examine how statistical models and graphical evidence can shed light on the question of who survived in the Donner party.

Figure 1.3 is an example of what we call a *data-centric, model-based* graph of a discrete (binary) outcome: lived (1) versus died (0). That is, it shows both the data and a statistical summary based on a fitted statistical model. The statistical model provides a smoothing of the discrete data.

The jittered points at the top and bottom of the graph show survival in relation to age of the person. You can see that there were more people who survived among the young, and more who died among the old. The blue curve in the plot shows the fitted probability of survival from a

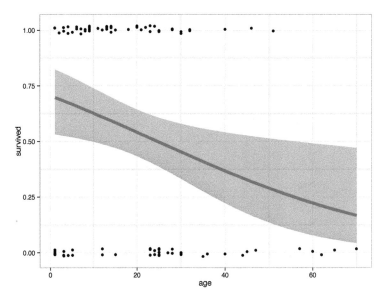

Figure 1.3: Donner party data, showing the relationship between age and survival. The blue curve and confidence band give the predicted probability of survival from a linear logistic regression model.

linear logistic regression model for these data with a 95% confidence band for the predictions. The prediction equation for this model can be given as:

$$\text{logit}(\text{survived}) = \underset{(0.372)}{0.868} - \underset{(0.015)}{0.0353}\,\text{age}$$

The equation above implies that the log odds of survival decreases by 0.0352 with each additional year of age or by $10 \times 0.0352 = 0.352$ for an additional decade. Another way to say this is that the odds of survival is multiplied by $\exp(0.353) = .702$ with each 10 years of age, a 30% decrease.

Of course, these visual and statistical summaries depend on the validity of the fitted model. For

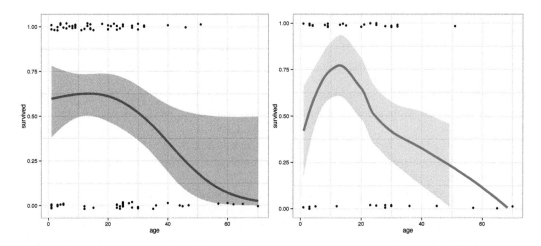

Figure 1.4: Donner party data, showing other model-based smoothers for the relationship between age and survival. Left: using a natural spline; right: using a non-parametric loess smoother.

contrast, Figure 1.4 shows two other model-based smoothers that relax the assumption of the linear logistic regression model. The left panel shows the result of fitting a semi-parametric model with a natural cubic spline with one more degree of freedom than the linear logistic model. The right panel shows the fitted curve for a non-parametric, locally weighted scatterplot smoothing (loess) model. Both of these hint that the relationship of survival to age is more complex than what is captured in the linear logistic regression model. We return to these data in Chapter 7.

△

1.4 Graphical methods for categorical data

You can see a lot, just by looking

Yogi Berra

The graphical methods for categorical data described in this book are in some cases straightforward adaptations of more familiar visualization techniques developed for quantitative data. Graphical principles and strategies, and the relations between the visualization approach and traditional statistical methods, are described in a number of sources, including Chambers et al. (1983), Cleveland (1993b), and several influential books by Tufte (Tufte, 1983, 1990, 1997, 2006).

The fundamental idea of statistical graphics as a comprehensive system of visual signs and symbols with a grammar and semantics was first proposed in Jacques Bertin's *Semiology of Graphics* (1983). These ideas were later extended to a computational theory in Wilkinson's *Grammar of Graphics* (2005), and implemented in R in Hadley Wickham's ggplot2 (Wickham and Chang, 2015) package (Wickham, 2009, Wickham and Chang, 2015).

Another perspective on visual data display is presented in Section 1.4.1 focusing on the communication goals of statistical graphics. However, the discrete nature of categorical data implies that some familiar graphic methods need to be adapted, while in other cases we require a new graphic metaphor for data display. These issues are illustrated in Section 1.4.2. Section 1.4.3 discusses the principle of effect ordering for categorical variables in graphs and tables.

1.4.1 Goals and design principles for visual data display

Designing good graphics is surely an art, but as surely, it is one that ought to be informed by science. In constructing a graph, quantitative and qualitative information is encoded by visual features, such as position, size, texture, symbols, and color. This translation is reversed when a person studies a graph. The representation of numerical magnitude and categorical grouping, and the apperception of patterns and their *meaning*, must be extracted from the visual display.

There are many views of graphs, of graphical perception, and of the roles of data visualization in discovering and communicating information. On the one hand, one may regard a graphical display as a *stimulus*—a package of information to be conveyed to an idealized observer. From this perspective certain questions are of interest: which form or graphic aspect promotes greater accuracy or speed of judgment (for a particular task or question)? What aspects lead to greatest memorability or impact? Cleveland (Cleveland and McGill, 1984, 1985, Cleveland, 1993a), Spence and Lewandowsky (Lewandowsky and Spence, 1989, Spence, 1990, Spence and Lewandowsky, 1990) have made important contributions to our understanding of these aspects of graphical display.

An alternative view regards a graphical display as an act of *communication*—like a narrative, or even a poetic text or work of art. This perspective places the greatest emphasis on the desired communication goal, and judges the effectiveness of a graphical display in how well that goal is achieved (Friendly and Kwan, 2011). Kosslyn (1985, 1989) and Tufte (1983, 1990, 1997) have articulated this perspective most clearly.

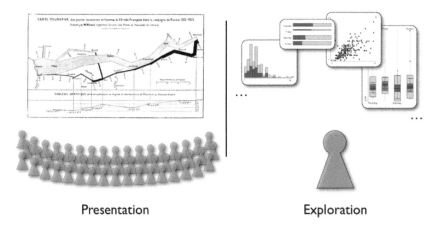

Presentation   Exploration

Figure 1.5: Different communication purposes require different graphs. For presentations, a single, carefully crafted graph may appeal best to a large audience; for exploratory analysis, many related images from different perspectives for a narrow audience (often you!). *Source*: Adapted from a blog entry by Martin Theus, `http://www.theusrus.de/blog/presentation-vs-exploration/`.

In this view, an effective graphical display, like good writing, requires an understanding of its *purpose*—what aspects of the data are to be communicated to the viewer. In writing we communicate most effectively when we know our audience and tailor the message appropriately. So too, we may construct a graph in different ways to: (a) use ourselves, (b) present at a conference or meeting of our colleagues, (c) publish in a research report, or (d) communicate to a general audience (Friendly (1991, Ch. 1), Friendly and Kwan (2011)). Figure 1.5 illustrates a basic contrast between graphs for presentation purposes, designed to appeal persuasively to a large audience (one-to-many) and the use of perhaps many graphs we might make for ourselves for exploratory data analysis (many-to-one).

Figure 1.6 shows one organization of visualization methods in terms of the *primary* use or

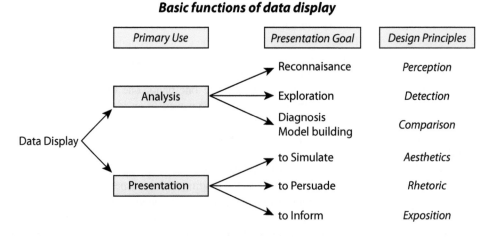

Figure 1.6: A taxonomy of the basic functions of data display by intended use, presentation goal, and design principles.

intended communication goal, the functional *presentation goal*, and suggested corresponding *design principles*.

We illustrate these ideas and distinctions in the examples below, most of which are treated again in later chapters.

EXAMPLE 1.4: Racial profiling: Arrests for marijuana possession

In a case study that will be examined in detail in Chapter 7 (Example 7.10), the *Toronto Star* newspaper studied a huge data base of arrest records by Toronto police for indications of possible racial profiling, i.e., differential treatment of those arrested on the basis of skin color. They focused on the charge of simple possession of a small amount of marijuana, for which enforcement procedures allowed police discretion. An officer could release an arrestee with a summons ("Form 9") to appear in court, or take the person to a police station for questioning ("Form 10") or booking ("Form 11.1"), or order the person held in jail for a bail hearing ("Show cause").

The statistical issue was whether the data on these arrests showed evidence of differential treatment in relation to skin color, particularly in the treatment of blacks vs. whites, controlling, of course, for other factors. Statistical tests on these data (χ^2 tests, loglinear models, logistic regression) showed overwhelming evidence of differential treatment of blacks and whites. However, tables of these results do not reveal the nature of this association.

Figure 1.7 is an example of a graph designed for *analysis*—a mosaic display (Chapter 5) showing the frequencies of those arrested on this charge by skin color and release type. The size of each rectangle shows the frequency and these are shaded in relation to the asociation between skin color and release—blue for positive associations (more than expected if they were independent) to red for negative associations.

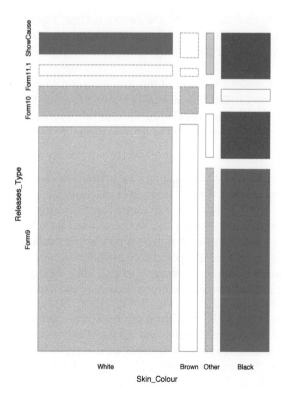

Figure 1.7: Analysis graph: A mosaic display showing the relationship between skin color and release type for those arrested on a charge of simple possession of marijuana in Toronto, 1996–2002.

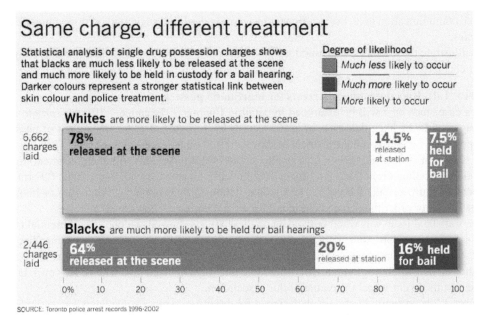

Figure 1.8: Redesign of Figure 1.7 as a presentation graphic. *Source*: Graphics department, *The Toronto Star*, December 11, 2002. Used by permission.

Once you know how to read such graphs, the pattern is clear: blacks were indeed more likely to be held for more severe treatment, whites were more likely to be released with a summons. But this is hardly a graph that would be clear to a general audience, and would require a good deal of explanation.

In contrast, Figure 1.8 shows a redesign of this as a *presentation graphic* prepared by the *Star* and published on December 11, 2002 in conjunction with a meeting between the newspaper and the Toronto Police Services Board to consider the issue of racial profiling. The police vehemently denied that racial profiling was taking place. The revision makes the point immediately obvious and compelling in the following ways:

- It announces the conclusion in the figure title: "Same charge, different treatment."
- The text box at the top provides the context for this conclusion.
- Skin colors "Brown" and "Other," which appeared less frequently, were removed, and the release categories "Form 10" and "Form 11.1" were combined as "released at station."
- The graphic is still a mosaic display, however, it now shows explicitly the number of charges laid against whites and blacks and the percentage of each treatment.
- The labels for whites and blacks were enhanced by indicating what a reader should see for each.
- The legend for color is titled non-technically as "degree of likelihood."

Clear communication is not achieved without effort. The revised graph required several iterations and emails between the graphic designer and the statistical consultant (the first author of this book) in the few hours available before the newspaper went to press. The main question was, "what are we trying to show here?" Starting with the original Figure 1.7 mosaic, we asked, "what can we remove?" and "what can we add?" to make the message clearer.

△

1.4.2 Categorical data require different graphical methods

We mentioned earlier, and will see in greater detail in Chapter 7 and Chapter 9, that statistical models for discrete response data and for frequency data are close analogs of the linear regression and ANOVA models used for quantitative data. These analogies suggest that the graphical methods commonly used for quantitative data may be adapted directly to categorical data.

Happily, it turns out that many of the analysis graphs and diagnostic displays (e.g., effect plots, influence plots, added variable and partial residual plots, etc.) that have become common adjuncts in the analysis of quantitative data have been extended to generalized linear models including logistic regression (Section 7.5) and loglinear models (Section 11.6).

Unhappily, the familiar techniques for displaying raw data are often disappointing when applied to categorical data. The simple scatterplot, for example, widely used to show the relation between quantitative response and predictors, when applied to discrete variables, gives a display of the category combinations, with all identical values overplotted, and no representation of their frequency.

Instead, frequencies of categorical variables are often best represented graphically using *areas* rather than as position along a scale. Friendly (1995) describes conceptual and statistical models that give a rationale for this graphic representation. Figure 1.7 does this in the form of a modified bar chart (mosaic plot), where the widths of the horizontal bars show the proportions of whites and blacks in the data, and the divisions of each group give the percents of each release type. Consequently, the areas of each bar are proportional to the frequency in the cells of this 2×3 table.

As we describe later in this book, using the visual attribute

$$\text{area} \sim \text{frequency}$$

also allows creating novel graphical displays of frequency data for special circumstances.

Figure 1.9 shows two examples. The left panel gives a ***fourfold display*** of the frequencies of admission and gender in the Berkeley data shown in Table 1.1. What should be seen at a glance is that males are more often admitted and females more often rejected (shaded blue); see Section 4.4 for details.

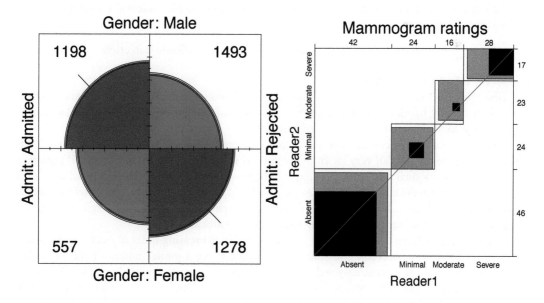

Figure 1.9: Frequencies of categorical variables shown as areas. Left: fourfold display of the relation between gender and admission in the Berkeley data; right: agreement plot for two raters assessing mammograms.

The right panel shows another specialized display, an ***agreement chart*** designed to show the strength of agreement in a square table for two raters (see Section 4.7.2). The example here (Example 4.18) concerns agreement of ratings of breast cancer from mammograms by two raters. The dark squares along the diagonal show exact agreement; the lighter diagonal rectangles allow 1-off agreement, and both are shown in relation to chance agreement (diagonal enclosing rectangles). What should be seen at a glance is that exact agreement is moderately strong and extremely strong if you allow the raters to differ by one rating category.

1.4.3   Effect ordering and rendering for data display

In plots of quantitative variables, standard methods (histograms, scatterplots) automatically position values along ordered scales, facilitating comparison ("which is less/more?") and detection of patterns, trends, and anomalies. However, by its nature, categorical data involves discrete variables such as education level, hair color, geographic region (state or province), or preference for a political party. With alphabetic labels for ordered categories (e.g., education: Low, Medium, High), it is unfortunately all too easy to end up with a nonsensical display with the categories ordered High, Low, Medium. Geographic regions (U.S. states) are often ordered alphabetically by default as are the names of political parties and other categorical variables. This may be useful for lookup, but for the purposes of comparison and detection, this is almost always a bad idea.

Instead, Friendly and Kwan (2003) proposed the principle of ***effect-order sorting*** for visual displays (tables as well as graphs):

sort the data by the effects to be seen to facilitate comparison

For quantitative data, this is often achieved by sorting the data according to means or medians of row and column factors, called ***main-effect ordering***. For categorical data, graphs and tables are often most effective when the categories are arranged in an order reflecting their association, called ***association ordering***.

Another important principle concerns the ***rendering*** of visual attributes of elements in graphical displays (Friendly, 2002). For example, categorical variables in plots (and tables) can be distinguished by any one or more of color, size, shape, or font. The examples below show the use of color to illustrate the precept:

render the data by the effects to be seen to facilitate detection

EXAMPLE 1.5: British social mobility

Bishop et al. (1975, p. 100) analyzed data on the occupations of 3500 British fathers and their sons from a study by Glass (1954), with five occupational categories: Professional, Managerial, Supervisory, Skilled manual, and Unskilled manual.

One would expect, of course, a strong association between a son's occupation and that of his father—the apple doesn't fall very far from the tree. Mosaic plots (detailed in Chapter 5) provide a natural way to show such relationships. Figure 1.10 shows two such plots. The left panel shows the result obtained when the table variables `father` and `son` are read as factors, and therefore ordered alphabetically by default. It is difficult to see any overall pattern, except for the large values in the diagonal cells (shaded blue) corresponding to equal occupational status.

In the right panel, the categories have been arranged in decreasing order of occupational status to show the association according to status. Now you can see a global pattern of shading color, where the tiles become increasingly red as one moves away from the main diagonal, reflecting a greater difference between the occupation of the father and son. The interpretation here is that most sons remain in their father's occupational class, but when they differ, there is little mobility across large steps.

In this example, `father` and `son` are clearly ordinal variables and should be treated as such in both graphs and statistical models. Correspondence analysis (Chapter 6) provides a natural way

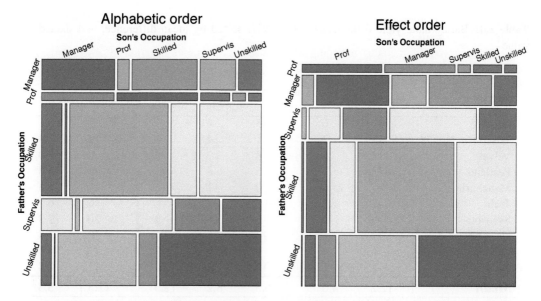

Figure 1.10: Mosaic plots for Glass' mobility table of occupational status. In these displays the area of each tile is proportional to frequency and shading color shows the departure from independence, using blue for positive, red for negative association. Left: default alphabetic ordering of categories; right: occupational categories ordered by status.

to depict association by assigning scores to the categories to optimally represent their relationships. Loglinear models provide special methods for ordinal variables (Section 10.1) and square frequency tables (Section 10.2).

△

The ideas of effect ordering and rendering with color shading to enhance perception can also be used in tabular displays, as illustrated in the next example.

EXAMPLE 1.6: Barley data

The classic `barley` dataset (in lattice (Sarkar, 2015)) from Immer et al. (1934) gives a $10 \times 2 \times 6$ table of yields of 10 varieties of barley in two years (1931, 1932) planted at 6 different sites in Minnesota. Cleveland (1993b) and many others have used this data to illustrate graphical methods, and one surprising finding not revealed in standard tabular displays is that the data for one site (Morris) may have had the values for 1931 and 1932 switched.[4]

To focus attention on this suspicious effect in a tabular display, you can calculate the *yield difference* $\Delta y_{ij} = y_{ij,1931} - y_{ij,1932}$. Table 1.4 shows these values in a 10×6 table with the rows and columns sorted by their means (main-effect ordering). In addition, the table cells have been colored according to the sign and magnitude of the year difference. The shading scheme uses blue for large positive values and red for large negative values, with a white background for intermediate values. The shading intensity values were determined as $|\Delta y_{ij}| > \{2, 3\} \times \hat{\sigma}(\Delta y_{ij})$.

Effect ordering and color rendering have the result of revealing a new effect, shown as a regular progression in the body of the table. The negative values for Morris now immediately stand out. In addition, the largely positive other values show a lower-triangular pattern, with the size of the yield

---

[4]This canonical story, like many others in statistics and graphics lore, turns out to be apocryphal on closer examination. Wright (2013) recently took a closer look at the original data and gives an expanded data set as `minnesota.barley.yield` in the agridat (Wright, 2015) package. With a wider range of years (1927–1936), other local effects like weather had a greater impact than the overall year effects seen in 1931–1932, and the results for the Morris site no longer stand out as surprising.

Table 1.4: Barley data, yield differences, 1931-1932, sorted by mean difference, and shaded by value

Variety	Morris	Duluth	University Farm	Grand Rapids	Waseca	Crookston	*Mean*
No. 475	-22	6	-5	4	6	12	0.1
Wisconsin No. 38	-18	2	1	14	1	14	2.4
Velvet	-13	4	13	-9	13	9	2.9
Peatland	-13	1	5	8	13	16	4.8
Manchuria	-7	6	0	11	15	7	5.5
Trebi	-3	3	7	9	15	5	6.1
Svansota	-9	3	8	13	9	20	7.3
No. 462	-17	6	11	5	21	18	7.4
Glabron	-6	4	6	15	17	12	8.0
No. 457	-15	11	17	13	16	11	8.8
Mean	-12.2	4.6	6.3	8.2	12.5	12.5	5.3

difference increasing with both row and column means. Against this background, one other cell, for Velvet grown at Grand Rapids, stands out with an anomalous negative value.

Although the use of color for graphs is now more common in some journals, color and other rendering details in tables are still difficult. The published version of Table 1.4 (Friendly and Kwan, 2003, Table 3) was forced to use only font shape (normal, italics) to distinguish positive and negative values.

△

Finally, effect ordering is also usefully applied to the variables in multivariate data sets, which by default, are often ordered in data displays according to their position in a data frame or alphabetically.

EXAMPLE 1.7: Iris data

The classic `iris` data set (Anderson, 1935, Fisher, 1936b) gives the measurements in centimeters of the variables sepal length and width and petal length and width, respectively, for 50 flowers from each of 3 species of iris, *Iris setosa*, *versicolor*, and *virginica*. Such multivariate data are often displayed in ***parallel coordinate plot***s, using a separate vertical axis for each variable, scaled from its minimum to maximum.

The default plot, with variables shown in their data frame order, is shown in the left panel of Figure 1.11, and gives rise to the epithet *spaghetti plot* for such displays because of the large number of line crossings. This feature arises because one variable, sepal width, has negative relations in the species means with the other variables. Simple rearrangement of the variables to put sepal width last (or first) makes the relations among the species and the variables more apparent, as shown in the right panel of Figure 1.11. This plot has also been enhanced by using ***alpha-blending*** (partial transparency) of thicker lines, so that the density of lines is more apparent.

Parallel coordinate plots for categorical data are discussed in an online supplement on the web site for the book. A general method for reordering variables in multivariate data visualizations based on cluster analysis was proposed by Hurley (2004).

△

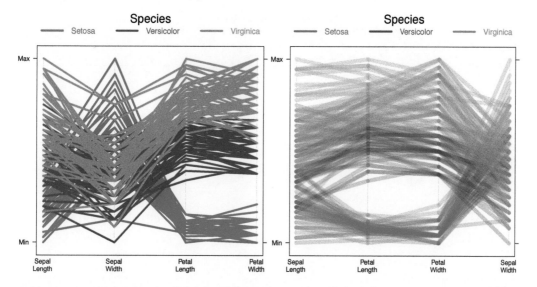

Figure 1.11: Parallel coordinates plots of the Iris data. Left: Default variable order; right: Variables ordered to make the pattern of correlations more coherent.

1.4.4 Interactive and dynamic graphics

Graphics displayed in print form, such as this book, are necessarily static and fixed at the time they are designed and rendered as an image. Yet, recent developments in software, web technology and media alternative to print have created the possibility to extend graphics in far more useful and interesting ways, for both presentation and analysis purposes.

Interactive graphics allow the viewer to directly manipulate the statistical and visual components of graphical display. These range from

- graphical controls (sliders, selection boxes, and other widgets) to control details of an analysis (e.g., a smoothing parameter) or graph (colors and other graphic details), to
- higher-level interaction including zooming in or out, drilling down to a data subset, linking multiple displays, selecting terms in a model, and so forth.

The important effect is that the analysis and/or display is immediately re-computed and updated visually.

In addition, ***dynamic graphics*** use animation to show a series of views, as frames in a movie. Adding time as an additional dimension allows far more possibilities, for example showing a rotating view of a 3D graph or showing smooth transitions or interpolations from one view to another.

There are now many packages in R providing interactive and dynamic plots (e.g., rggobi (Temple Lang et al., 2014), iplots (Urbanek and Wichtrey, 2013)) as well as capabilities to incorporate these into interactive documents, presentations, and web pages (e.g., rCharts (Vaidyanathan, 2013), googleVis (Gesmann and de Castillo, 2015), ggvis (Chang and Wickham, 2015)). The animation (Xie, 2014) package facilitates creating animated graphics and movies in a variety of formats. The RStudio editor and development environment[5] provides its own manipulate (RStudio, Inc., 2011) package, as well as the shiny (RStudio, Inc., 2015) framework for developing interactive R web applications.

EXAMPLE 1.8: 512 paths to the White House

Shortly before the 2012 U.S. presidential election (November 2, 2012) *The New York Times*

---

[5]http://www.rstudio.com.

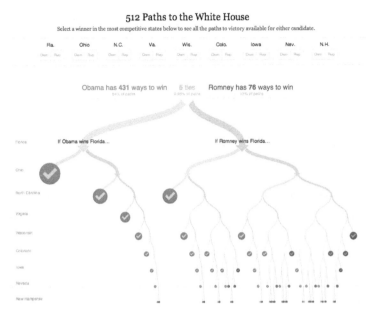

Figure 1.12: 512 paths to the White House. This interactive graphic allows the viewer to select a winner in any one or more of the nine most highly contested U.S. states and highlights the number of paths leading to a win by Obama or Romney, sorted and weighted by the number of Electoral College votes. *Source*: Mike Bostock & Shan Carter, *New York Times* interactive, November 2, 2012. Used by permission.

published an interactive graphic,[6] designed by Mike Bostock and Shan Carter,[7] showing the effect that a win for Barack Obama or Mitt Romney in the nine most highly contested states would have on the chances that either candidate would win the presidency.

With these nine states in play there are $2^9 = 512$ possible outcomes, each with a different number of votes in the Electoral College. In Figure 1.12, a win for Obama in Florida and Virginia was selected, with wins for Romney in Ohio and North Carolina. Most other selections also lead to a win by Obama, but those with the most votes are made most visible at the top. An R version of this chart was created using the rCharts package.[8] The design of this graphic as a ***binary tree*** was chosen here, but another possibility would be a ***treemap*** graphic (Shneiderman, 1992) or a mosaic plot.

△

1.4.5 Visualization = Graphing + Fitting + Graphing . . .

> Look here, upon this picture, and on this.
>
> ———————————————————————————————
>
> Shakespeare, Hamlet

Statistical summaries, hypothesis tests, and the numerical parameters derived in fitted models

---

[6]http://www.nytimes.com/interactive/2012/11/02/us/politics/
paths-to-the-white-house.html.

[7]see: https://source.opennews.org/en-US/articles/nyts-512-paths-white-house/. for a description of their design process.

[8]http://timelyportfolio.github.io/rCharts_512paths/.

are designed to capture a particular feature of the data. A quick analysis of the data from Table 1.1, for example, shows that $1198/2691 = 44.5\%$ of male applicants were admitted, compared to $557/1835 = 30.4\%$ of female applicants.

Statistical tests give a Pearson χ^2 of 92.2 with 1 degree of freedom for association between admission and gender ($p < 0.001$), and various measures for the strength of association. Expressed in terms of the **odds ratio**, males were apparently 1.84 times as likely to be admitted as females, with 99% confidence bounds $(1.56, 2.17)$. Each of these numbers expresses some part of the relationship between gender and admission in the Berkeley data. Numerical summaries such as these are each designed to compress the information in the data, focusing on some particular feature.

In contrast, the visualization approach to data analysis is designed to (a) expose information and structure in the data, (b) supplement the information available from numerical summaries, and (c) suggest more adequate models. In general, the visualization approach seeks to serve the needs of both summarization and exposure.

This approach recognizes that both data analysis and graphing are *iterative* processes. You should not expect that any one model captures all features of the data, any more than we should expect that a single graph shows all that may be seen. In most cases, your initial steps should include some graphical display guided by understanding of the subject matter of the data. What you learn from a graph may then help suggest features of the data to be incorporated into a fitted model. Your desire to ensure that the fitted model is an adequate summary may then lead to additional graphs.

The precept here is that

$$\textbf{Visualization = Graphing + Fitting + Graphing} \ldots$$

where the ellipsis indicates the often iterative nature of this process. Even for descriptive purposes, an initial fit of salient features can be removed from the data, giving residuals (departures from a model). Displaying the residuals may then suggest additional features to account for.

Simple examples of this idea include detrending time series graphs to remove overall and seasonal effects and plots of residuals from main-effect models for ANOVA designs. For categorical data, mosaic plots (Chapter 5) display the unaccounted-for association between variables by shading, as in Figure 1.10. Additional models and plots considered in Section 10.2 can reveal additional structure in square tables beyond the obvious effect that sons tend most often to follow in their fathers' footsteps.

EXAMPLE 1.9: Donner Party

The graphs in Figure 1.3 and Figure 1.4 suggest three different initial descriptions for survival in the Donner party. Yet they ignore all other influences, of which gender and family structure might also be important. A more complete understanding of this data can be achieved by taking these effects into account, both in fitted models and graphs. See Example 7.9 for a continuation of this story.   △

EXAMPLE 1.10: Space shuttle disaster

The space shuttle *Challenger* mentioned in Example 1.2 exploded 73 seconds after take-off on January 28, 1986. Subsequent investigation presented to the presidential commission headed by William Rogers determined that the cause was failure of the O-ring seals used to isolate the fuel supply from burning gases. The story behind the *Challenger* disaster is perhaps the most poignant missed opportunity in the history of statistical graphics. See Tufte (1997) for a complete exposition. It may be heartbreaking to find out that some important information was there, but the graphmaker missed it.

Engineers from Morton Thiokol, manufacturers of the rocket motors, had been worried about the effects of unseasonably cold weather on the O-ring seals and recommended aborting the flight.

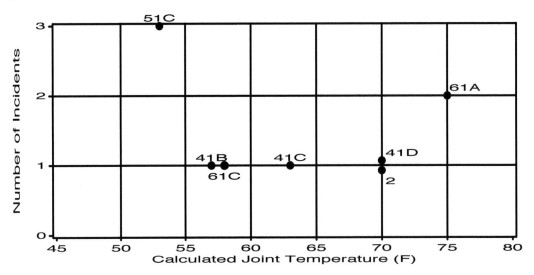

Figure 1.13: NASA Space Shuttle pre-launch graph prepared by the engineers at Morton Thiokol.

NASA staff analyzed the data, tables, and charts submitted by the engineers and concluded that there was insufficient evidence to cancel the flight.

The data relating O-ring failures to temperature were depicted as in Figure 1.13, our candidate for the most misleading graph in history. There had been 23 previous launches of these rockets giving data on the number of O-rings (out of 6) that were seen to have suffered some damage or failure. However, the engineers omitted the observations where no O-rings failed or showed signs of damage, believing that they were uninformative.

Examination of this graph seemed to indicate that there was no relation between ambient temperature and failure. Thus, the decision to launch the *Challenger* was made, in spite of the initial concerns of the Morton Thiokol engineers. Unfortunately, those observations had occurred when the launch temperature was relatively warm ($65 - -80°$F.) and were indeed informative. The coldest temperature at any previous launch was $53°$; when *Challenger* was launched on January 28, the temperature was a frigid $31°$.

These data have been analyzed extensively (Dalal et al., 1989, Lavine, 1991). Tufte (1997) gives a thorough and convincing visual analysis of the evidence available prior to the launch. We consider statistical analysis of these data in Chapter 7, Example 7.4.

But, what if the engineers had simply made a better graph? At the very least, that would entail (a) drawing a smoothed curve to fit the points (to show the trend), and (b) removing the background grid lines (which obscure the data). Figure 1.14 shows a revised version of the same graph, highlighting the non-zero observations and adding a simple quadratic curve to allow for a possible nonlinear relationship. For comparison, the excluded zero observations are also shown in grey. This plot, even showing only the non-zero points, should have caused any engineer to conclude that either: (a) the data were wrong, or (b) there were excessive risks associated with both high and low temperatures. But it is well-known that brittleness of the rubber used in the O-rings is inversely proportional to temperature cubed, so prudent interest might have focussed on the first possibility.[9]

---

[9] A coda to this story shows the role of visual explanation in practice as well (Tufte, 1997, pp. 50–53). The Rogers Commission contracted the reknowned theoretical physicist Richard Feynman to contribute to their investigation. He determined that the most probable cause of the shuttle failure was the lack of resiliancy of the rubber O-rings at low temperature. But how could he make this point convincingly? At a televised public hearing, he took a piece of the O-ring material, squeezed it in a C-clamp, and plunged it into a glass of ice water. After a few minutes, he released the clamp, and the rubber did not spring back to shape. He mildly said, "... there is no resilience in this particular material when it is at a temperature of 32 degrees. I believe this has some significance for our problem" (Feynman, 1988).

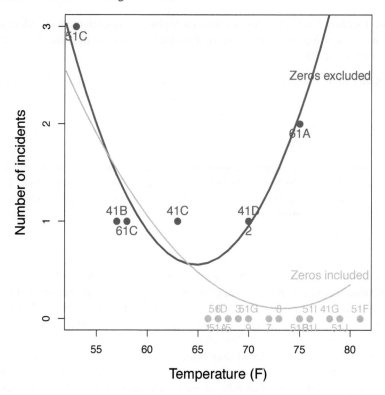

Figure 1.14: Re-drawn version of the NASA pre-launch graph, showing the locations of the excluded observations and with fitted quadratics for both sets of observations.

△

1.4.6 Data plots, model plots, and data+model plots

In this book, we use hundreds of graphs to illustrate aspects of discrete data, methods of analysis, and plots for understanding and explaining results. In addition to the overview of goals and design principles (Section 1.4.1) shown in Figure 1.6, another classification of such graphs is useful to bear in mind as you read this book. We distinguish three kinds of plots:

- **Data plots**: these are well-known. They help answer questions like: (a) What do the data look like? (b) Are there unusual features? (c) What kinds of summaries would be useful?

  An immediate (but bad) example is the plot of failures of O-rings against temperature in Figure 1.13. Many other examples appear throughout Chapter 3, using barplots for discrete distributions, and Chapter 4, using various graphic forms to display frequencies in two-way tables.

- **Model plots**: these are less well-known as such, but also help answer important questions: (a) What does the model "look" like? (plot predicted values); (b) How does the model change when its *parameters* change? (plot competing models); (c) How does the model change when the *data* is changed? (e.g., influence plots).

  Models are simplified descriptions of data. In Section 5.8 we use mosaic plots to show what loglinear models "look" like (e.g., Figure 5.31). Plots for correspondence analysis methods (Chapter 6) show the relationships among table variables as fitted points in a two-dimensional space.

Effect plots (Section 7.3.3) show fitted values for logistic regression models. Models for ordinal variables in terms of log odds ratios (Section 10.1.2) can also be illustrated in terms of simple models plots (Figure 10.4).

- **Data + Model plots** combine these features, and lead to other questions: (a) How well does a model fit the data? (b) Does a model fit uniformly good or bad, or just good/bad in some regions? (c) How can a model be improved? (d) Model *uncertainty*: show confidence/prediction intervals or regions. (e) Data *support*: where is data too "thin" to make a difference in competing models?

Figure 1.2 and Figure 1.3, Figure 1.4 show several data+model plots for the space shuttle and Donner data respectively, both showing confidence bands for predicted values. Figure 1.14 is an another example, comparing two models for the space shuttle data. The model-building methods described in Chapter 7–Chapter 11 make frequent use of data+model plots.

1.4.7   The 80–20 rule

The Italian economist Vilfredo Pareto observed in 1906 that 80% of the land in Italy was owned by 20% of the population and this ratio also applied in other countries. It also applied to the yield of peas from peapods in his garden (Pareto, 1971). This idea became known as the *Pareto principle* or the *80–20 rule*. The particular 80/20 ratio is not as important as the more general idea of the uneven distribution of results and causes in a variety of areas.

Common applications are the rules of thumb that: (a) in business 80% of sales come from 20% of clients; (b) in criminology 80% of crimes are said to be committed by 20% of the population. (c) In software development, it is said that 80% of errors and (d) crashes can be eliminated by fixing the top 20% most-reported bugs or that 80% of errors reside in 20% of the code.

The *Pareto chart* was designed to display the frequency distribution of a variable with a histogram or bar chart together with a cumulative line graph to highlight the most frequent category, and the *Pareto distribution* gives a mathematical form to such distributions with a parameter α (the *Pareto index*) reflecting the degree of inequality.

Applied to statistical graphics, the precept is that

20% of your effort can generate 80% of your desired result in producing a given plot.

This is good news for exploratory graphs you produce for yourself. Very often, the default settings will give a reasonable result, or you will see immediately something simple to add or change to make the plot easier to understand.

The bad news is the corollary of this rule:

80% of your effort may be required to produce the remaining 20% of a finished graph.

This is particularly important for presentation graphs, where several iterations may be necessary to get it right (or right enough) for your communication purposes. Some important details are:

graph title A presentation graphic can be more effective when it announces the main point or conclusion in the graphic title, as in Figure 1.8.

axis and value labels Axes should be labelled with meaningful variable descriptions (and perhaps the data units) rather than just plot defaults (e.g., "Temperature (degrees F)" in Figure 1.2, not `temp`). Axis values are often more of a challenge for categorical variables, where their text labels often overlap, requiring abbreviation, a smaller font, or text rotation.

grouping attributes Meaningfully different subsets of the data should be rendered with distinct visual attributes such as color, shape, and line style, and sometimes with more than one.

legends and direct labels Different data groups in a graphic display shown by color, shape, etc., usually need at least a graphic legend defining the symbols and group labels. Sometimes you can do better by applying the labels directly to the graphical elements,[10] as was done in Figure 1.14.

legibility A common failure in presentation graphs in journals and lectures is the use of text fonts too small to be read easily. One rule of thumb is to hold the graph at arms length for a journal and put it on the floor for a lecture slide. If you can't read the labels, the font is too small.

plot annotations Beyond the basic graphic data display, additional annotations can add considerable information to interpret the context or uncertainty, as in the use of plot envelopes to show confidence bands or regions (see Figure 1.3 and Figure 1.4).

aspect ratio Line graphs (such as Figure 3.1) are often easiest to understand when the ratio of height to width is such that line segments have an average slope near 1.0 (Cleveland et al., 1988). In R, you can easily manipulate a graph window manually with a mouse to observe this effect and find an aspect ratio that looks right. Moreover, in graphs for biplots and correspondence analysis (Chapter 6), interpretation involves distances between points and angles between line segments. This requires an aspect ratio that equates the units on the axes. Careful software will do this for you,[11] and you should resist the temptation to re-shape the plot.

colors Whereas a good choice of colors can greatly enhance a graphical display, badly chosen colors, ignoring principles of human perception, can actually spoil it. First, considering that graphs are often reproduced in black and white and a significant percentage of the human population is affected by color deficiencies, important information should not be coded by color alone without careful thought.

Second, color palettes should be chosen carefully to put the desired emphasis on the information visualized. For example, consider Figure 1.15 showing qualitative color palettes (appropriate for unordered categories) taken from two different color spaces: Hue-Saturation-Value (HSV) and Hue-Chroma-Luminance (HCL), where only the hue is varied. Whereas one would expect such a palette to be balanced with respect to colorfulness and brightness, the red colors in the left (HSV) color wheel are generally perceived to be more more intense and flashy than the corresponding blue colors, and the highly saturated dark blue dominates the wheel. Consequently, areas shaded with these colors may appear more important than others in an uncontrolled way, distracting from the information to be conveyed. In contrast, the colors from the right (HCL) wheel are all balanced to the same gray level and in "harmony." These clearly should be preferred whenever categories of the same importance shall be compared.

Another related perception rule prescribes that lighter and darker colors should not be mixed in a display where areas should be compared since lighter colors look larger than darker ones. More background information on the choice of "good" colors for statistical graphics can be found in Zeileis et al. (2009).

visual impact Somewhat related, important features of a display should be visually distinguished from the less important. This may be achieved by different color or gray shading levels, or simply by contrasting filled with non-filled geometric shapes, or a different density of shading lines. One useful test for visual impact is to put a printed copy of a graph on the floor, rise up, and see what stands out.

---

[10]For example, the `identify()` function allows points in a plot to be labeled interactively with a mouse. The **directla-bels** (Hocking, 2013) package provides a general method for a variety of plots.

[11]For example using the graphics parameter `asp=1`, `eqsplot()` in MASS (Ripley, 2015a), or the equivalents in lattice (`aspect="iso"`) and **ggplot2** (`coord_equal`).

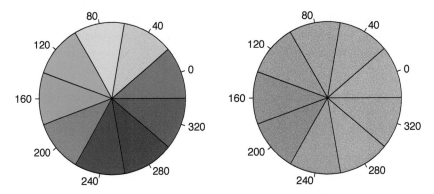

Figure 1.15: Qualitative color palette for the HSV (left) and HCL (right) spaces. The HSV colors are $(H, 100, 100)$ and the HCL colors $(H, 50, 70)$ for the same hues H. Note that in a monochrome version of this page, all pie sectors in the right wheel will be shaded with the same gray, i.e., they will appear to be virtually identical.

Nearly all of the graphs in this book were produced using R code in scripts saved as files and embedded in the text (via the knitr package). This has the advantages of reproducibility and enhancement: just re-run the code, or tweak it to improve a graph. If this is too hard, you can always use an external graphics editor (Gimp, Inkscape, Adobe Illustrator, etc.) to make improvements manually.

1.5 Chapter summary

- Categorical data differs from quantitative data because the variables take on discrete values (ordered or unordered, character or numeric) rather than continuous numerical values. Consequently, such data often appear in aggregated form representing category frequencies or in tables.

- Data analysis methods for categorical data are comprised of those concerned mainly with testing particular hypotheses versus those that fit statistical models. Model building methods have the advantages of providing parameter estimates and model-predicted values, along with measures of uncertainty (standard errors).

- Graphical methods can serve different purposes for different goals (data analysis versus presentation), and these suggest different design principles that a graphic should respect to achieve a given communication goal.

- For categorical data, some graphic forms (bar charts, line graphs, scatterplots) used for quantitative data can be readily adapted to discrete variables. However, frequency data often requires novel graphics using area and other visual attributes.

- Graphics can be far more effective when categorical variables are ordered to facilitate comparison of the effects to be seen and rendered to facilitate detection of patterns, trends or anomalies.

- The visualization approach to data analysis often entails a sequence of intertwined steps involving graphing and model fitting.

- Producing effective graphs for presentation is often hard work, requiring attention to details that support or detract from your communication goal.

1.6   Lab exercises

Exercise 1.1   A web page, "The top ten worst graphs, " `http://www.biostat.wisc.edu/`
`~kbroman/topten_worstgraphs/` by Karl Broman lists his picks for the worst graphs (and
a table) that have appeared in the statistical and scientific literature. Each entry links to graph(s) and
a brief discussion of what is wrong and how it could be improved.

(a) Examine a number of recent issues of a scientific or statistical journal in which you have some
interest. Find one or more examples of a graph or table that is a particularly bad use of display
material to summarize and communicate research findings. Write a few sentences indicating
how or why the display fails and how it could be improved.

(b) Do the same task for some popular magazine or newspaper that uses data displays to supplement the text for some story. Again, write a few sentences describing why the display is bad
and how it could be improved.

Exercise 1.2   As in the previous exercise, examine the literature in recent issues of some journal
of interest to you. Find one or more examples of a graph or table that you feel does a *good* job of
summarizing and communicating research findings.

(a) Write a few sentences describing why you chose these displays.

(b) Now take the role of a tough journal reviewer. Are there any features of the display that could
be modified to make them more effective?

Exercise 1.3   Infographics are another form of visual displays, quite different from the data graphics featured in this book, but often based on some data or analysis. Do a Google image search for
the topic "Global warming" to see a rich collection.

(a) Find and study one or two images that attempt some visual explanation of causes and/or effects
of global warming. Describe the main message in a sentence or two.

(b) What visual and graphic features are used in these to convey the message?

Exercise 1.4   The Wikipedia web page `en.wikipedia.org/wiki/Portal:Global_warming`
gives a few data-based graphics on the topic of global warming. Read the text and study the graphs.

(a) Write a short figure title for each that would announce the conclusion to be drawn in a presentation graphic.

(b) Write a figure caption for each that would explain what is shown and the important graphical
details for a reader to understand.

Exercise 1.5   The R Graph Gallery, `http://rgraphgallery.blogspot.com/`, contains a
large collection of examples of graphs in R, tagged by type or content, together with the R code to
produce them. Explore this collection for the terms (a) association plot (b) bar chart (c) categorical
data (d) fluctuation diagram (e) mosaic plot Find one or two you particularly like and write a few
sentences saying why you do.

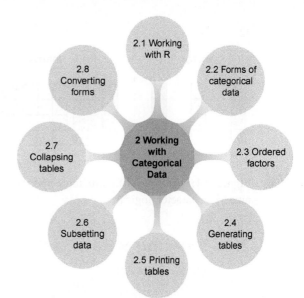

2

Working with Categorical Data

Creating and manipulating categorical data sets requires some skills and techniques in R beyond those ordinarily used for quantitative data. This chapter illustrates these for the main formats for categorical data: case form, frequency form and table form.

I'm a tidy sort of bloke. I don't like chaos. I kept records in the record rack, tea in the tea caddy, and pot in the pot box

---

George Harrison, from
http://www.brainyquote.com/quotes/keywords/tidy.html

Categorical data can be represented as data sets in various formats: case form, frequency form, and table form. This chapter describes and illustrates the skills and techniques in R needed to input, create, and manipulate R data objects to represent categorical data. More importantly, you also need to be able to convert these from one form to another for the purposes of statistical analysis and visualization, which are the subject of the remainder of the book.

As mentioned earlier, this book assumes that you have at least a basic knowledge of the R language and environment, including interacting with the R console (Rgui for Windows, R.app for Mac OS X) or some other editor/environment (e.g., R Studio), loading and using R functions in packages (e.g., `library(vcd)`) getting help for these from R (e.g., `help(matrix)`), etc. This chapter is therefore devoted to covering those topics needed in the book beyond such basic skills.[1]

---

[1] Some excellent introductory treatments of R are: Fox and Weisberg (2011a, Chapter 2), Maindonald and Braun (2007), and Dalgaard (2008). Tom Short's *R Reference Card*, http://cran.us.r-project.org/doc/contrib/Short-refcard.pdf, is a handy 4-page summary of the main functions. The web sites Quick-R http://www.statmethods.net/ and Cookbook for R http://www.cookbook-r.com/ provide very helpful examples, organized by topics and tasks.

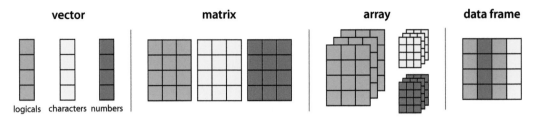

Figure 2.1: Principal data structures and data types in R. Colors represent different data types: numeric, character, logical.

2.1 Working with R data: vectors, matrices, arrays, and data frames

R has a wide variety of data structures for storing, manipulating, and calculating with data. Among these, vectors, matrices, arrays, and data frames are most important for the material in this book.

In R, a *vector* is a collection of values, like numbers, character strings, or logicals (TRUE, FALSE), and often correspond to a variable in some analysis. *Matrices* are rectangular arrays like a traditional table, composed of vectors in their columns or rows. *Arrays* add additional dimensions, so that, for example, a 3-way table can be represented as composed of rows, columns, and layers. An important consideration is that the values in vectors, matrices, and arrays must all be of the same *mode*, e.g., numbers or character strings. A *data frame* is a rectangular table, like a traditional data set in other statistical environments, and composed of rows and columns like a matrix, but allowing variables (columns) of different types. These data structures and the types of data they can contain are illustrated in Figure 2.1. A more general data structure is a *list*, a generic vector that can contain any other types of objects (including lists, allowing for *recursive* data structures). A data frame is basically a list of equally sized vectors, each representing a column of the data frame.

2.1.1 Vectors

The simplest data structure in R is a *vector*, a one-dimensional collection of elements of the same type. An easy way to create a vector is with the c() function, which combines its arguments. The following examples create and print vectors of length 4, containing numbers, character strings, and logical values, respectively:

```
> c(17, 20, 15, 40)

[1] 17 20 15 40

> c("female", "male", "female", "male")

[1] "female" "male"   "female" "male"

> c(TRUE, TRUE, FALSE, FALSE)

[1]  TRUE  TRUE FALSE FALSE
```

To store these values in variables, R uses the assignment operator (<−) or equals sign (=). This creates a variable named on the left-hand side. An assignment doesn't print the result, but a bare expression does, so you can assign and print by surrounding the assignment with ().

```
> count <- c(17, 20, 15, 40)                    # assign
> count                                          # print

[1] 17 20 15 40

> (sex <- c("female", "male", "female", "male"))    # both

[1] "female" "male"   "female" "male"

> (passed <- c(TRUE, TRUE, FALSE, FALSE))

[1]  TRUE  TRUE FALSE FALSE
```

Other useful functions for creating vectors are:

- The : operator for generating consecutive integer sequences, e.g., 1:10 gives the integers 1 to 10. The seq() function is more general, taking the forms seq(from, to), seq(from, to, by=), and seq(from, to, length.out=) where the optional argument by specifies the interval between adjacent values and length.out gives the desired length of the result.
- The rep() function generates repeated sequences, replicating its first argument (which may be a vector) a given number of times, and individual elements can be repeated with each until an optional length.out is obtained.

```
> seq(10, 100, by = 10)              # give interval

 [1]  10  20  30  40  50  60  70  80  90 100

> seq(0, 1, length.out = 11)         # give length

 [1] 0.0 0.1 0.2 0.3 0.4 0.5 0.6 0.7 0.8 0.9 1.0

> (sex <- rep(c("female", "male"), times = 2))

[1] "female" "male"   "female" "male"

> (sex <- rep(c("female", "male"), length.out = 4))   # same

[1] "female" "male"   "female" "male"

> (passed <- rep(c(TRUE, FALSE), each = 2))

[1]  TRUE  TRUE FALSE FALSE
```

2.1.2 Matrices

A **matrix** is a two-dimensional array of elements of the same type composed in a rectangular array of rows and columns. Matrices can be created by the function matrix(values, nrow, ncol), which reshapes the elements in the first argument (values) to a matrix with nrow rows and ncol columns. By default, the elements are filled in columnwise, unless the optional argument byrow = TRUE is given.

```
> (matA <- matrix(1:8, nrow = 2, ncol = 4))

     [,1] [,2] [,3] [,4]
[1,]    1    3    5    7
[2,]    2    4    6    8
```

```
> (matB <- matrix(1:8, nrow = 2, ncol = 4, byrow = TRUE))

     [,1] [,2] [,3] [,4]
[1,]    1    2    3    4
[2,]    5    6    7    8

> (matC <- matrix(1:4, nrow = 2, ncol = 4))

     [,1] [,2] [,3] [,4]
[1,]    1    3    1    3
[2,]    2    4    2    4
```

The last example illustrates that the values in the first argument are recycled as necessary to fill the given number of rows and columns.

All matrices have a dimensions attribute, a vector of length two giving the number of rows and columns, retrieved with the function dim(). Labels for the rows and columns can be assigned using dimnames(),[2] which takes a list of two vectors for the row names and column names, respectively. To see the structure of a matrix (or any other R object) and its attributes, you can use the str() function, as shown in the example below.

```
> dim(matA)

[1] 2 4

> str(matA)

 int [1:2, 1:4] 1 2 3 4 5 6 7 8

> dimnames(matA) <- list(c("M", "F"), LETTERS[1:4])
> matA

  A B C D
M 1 3 5 7
F 2 4 6 8

> str(matA)

 int [1:2, 1:4] 1 2 3 4 5 6 7 8
 - attr(*, "dimnames")=List of 2
  ..$ : chr [1:2] "M" "F"
  ..$ : chr [1:4] "A" "B" "C" "D"
```

Additionally, names for the row and column *variables* themselves can also be assigned in the dimnames call by giving each dimension vector a name.

```
> dimnames(matA) <- list(sex = c("M", "F"), group = LETTERS[1:4])
> ## or: names(dimnames(matA)) <- c("Sex", "Group")
> matA

    group
sex A B C D
  M 1 3 5 7
  F 2 4 6 8

> str(matA)

 int [1:2, 1:4] 1 2 3 4 5 6 7 8
 - attr(*, "dimnames")=List of 2
  ..$ sex  : chr [1:2] "M" "F"
  ..$ group: chr [1:4] "A" "B" "C" "D"
```

---

[2]The dimnames can also be specified as an optional argument to matrix().

(LETTERS is a predefined character vector of the 26 uppercase letters). Matrices can also be created or enlarged by "binding" vectors or matrices together by rows or columns:

- rbind(a, b, c) creates a matrix with the vectors a, b, and c as its rows, recycling the elements as necessary to the length of the longest one.
- cbind(a, b, c) creates a matrix with the vectors a, b, and c as its columns.
- rbind(mat, a, b, ...) and cbind(mat, a, b, ...) add additional rows (columns) to a matrix mat, recycling or subsetting the elements in the vectors to conform with the size of the matrix.

```
> rbind(matA, c(10, 20))

   A  B  C  D
M  1  3  5  7
F  2  4  6  8
  10 20 10 20

> cbind(matA, c(10, 20))

  A B C D
M 1 3 5 7 10
F 2 4 6 8 20
```

Rows and columns can be swapped (transposed) using t():

```
> t(matA)

      sex
group M F
    A 1 2
    B 3 4
    C 5 6
    D 7 8
```

Finally, we note that basic computations involving matrices are performed *element-wise*:

```
> 2 * matA / 100

    group
sex    A    B    C    D
  M 0.02 0.06 0.10 0.14
  F 0.04 0.08 0.12 0.16
```

Special operators and functions do exist for matrix operations, such as %*% for the matrix product.

2.1.3 Arrays

Higher-dimensional arrays are less frequently encountered in traditional data analysis, but they are of great use for categorical data, where frequency tables of three or more variables can be naturally represented as arrays, with one dimension for each table variable.

The function array(values, dim) takes the elements in values and reshapes these into an array whose dimensions are given in the vector dim. The number of dimensions is the length of dim. As with matrices, the elements are filled in with the first dimension (rows) varying most rapidly, then by the second dimension (columns) and so on for all further dimensions, which can be considered as layers. A matrix is just the special case of an array with two dimensions.

```
> dims <-  c(2, 4, 2)
> (arrayA <- array(1:16, dim = dims))      # 2 rows, 4 columns, 2 layers

, , 1

     [,1] [,2] [,3] [,4]
[1,]    1    3    5    7
[2,]    2    4    6    8

, , 2

     [,1] [,2] [,3] [,4]
[1,]    9   11   13   15
[2,]   10   12   14   16

> str(arrayA)

 int [1:2, 1:4, 1:2] 1 2 3 4 5 6 7 8 9 10 ...

> (arrayB <- array(1:16, dim = c(2, 8)))     # 2 rows, 8 columns

     [,1] [,2] [,3] [,4] [,5] [,6] [,7] [,8]
[1,]    1    3    5    7    9   11   13   15
[2,]    2    4    6    8   10   12   14   16

> str(arrayB)

 int [1:2, 1:8] 1 2 3 4 5 6 7 8 9 10 ...
```

In the same way that we can assign labels to the rows and columns in matrices, we can assign these attributes to dimnames(arrayA), or include this information in a dimnames= argument to array().

```
> dimnames(arrayA) <- list(sex = c("M", "F"),
+                          group = letters[1:4],
+                          time = c("Pre", "Post"))
> arrayA

, , time = Pre

   group
sex a b c d
  M 1 3 5 7
  F 2 4 6 8

, , time = Post

   group
sex  a  b  c  d
  M  9 11 13 15
  F 10 12 14 16

> str(arrayA)

 int [1:2, 1:4, 1:2] 1 2 3 4 5 6 7 8 9 10 ...
 - attr(*, "dimnames")=List of 3
 ..$ sex  : chr [1:2] "M" "F"
 ..$ group: chr [1:4] "a" "b" "c" "d"
 ..$ time : chr [1:2] "Pre" "Post"
```

Arrays in R can contain any single type of elements— numbers, character strings, logicals. R also has a variety of functions (e.g., table(), xtabs()) for creating and manipulating "table" objects, which are specialized forms of matrices and arrays containing integer frequencies in a contingency table. These are discussed in more detail below (Section 2.4).

2.1.4 Data frames

Data frames are the most commonly used form of data in R and more general than matrices in that they can contain columns of different types. For statistical modeling, data frames play a special role, in that many modeling functions are designed to take a data frame as a `data=` argument, and then find the variables mentioned within that data frame. Another distinguishing feature is that discrete variables (columns) like character strings (`"M"`, `"F"`) or integers (`1, 2, 3`) in data frames can be represented as *factors*, which simplifies many statistical and graphical methods.

A data frame can be created using keyboard input with the `data.frame()` function, applied to a list of objects, `data.frame(a, b, c, ...)`, each of which can be a vector, matrix, or another data frame, but typically all containing the same number of rows. This works roughly like `cbind()`, collecting the arguments as columns in the result.

The following example generates n = 100 random observations on three discrete factor variables, A, B, sex, and a numeric variable, age. As constructed, all of these are statistically independent, since none depends on any of the others. The function `sample()` is used here to generate n random samples from the first argument allowing replacement (`replace = TRUE`). The `rnorm()` function produces a vector of n normally distributed values with mean 30 and standard deviation 5. The call to `set.seed()` guarantees the reproducibility of the resulting data. Finally, all four variables are combined into the data frame `mydata`.

```
> set.seed(12345)    # reproducibility
> n <- 100
> A <- factor(sample(c("a1", "a2"), n, replace = TRUE))
> B <- factor(sample(c("b1", "b2"), n, replace = TRUE))
> sex <- factor(sample(c("M", "F"), n, replace = TRUE))
> age <- round(rnorm(n, mean = 30, sd = 5))
> mydata <- data.frame(A, B, sex, age)
> head(mydata, 5)

  A  B sex age
1 a2 b1   F  22
2 a2 b2   F  33
3 a2 b2   M  31
4 a2 b2   F  26
5 a1 b2   F  29

> str(mydata)

'data.frame':  100 obs. of  4 variables:
 $ A  : Factor w/ 2 levels "a1","a2": 2 2 2 2 1 1 1 2 2 2 ...
 $ B  : Factor w/ 2 levels "b1","b2": 1 2 2 2 2 2 2 2 1 1 ...
 $ sex: Factor w/ 2 levels "F","M": 1 1 2 1 1 1 2 2 1 1 ...
 $ age: num  22 33 31 26 29 29 38 28 30 27 ...
```

Rows, columns, and individual values in a data frame can be manipulated in the same way as a matrix, using subscripting (`[,]`). Additionally, variables can be extracted using the `$` operator:

```
> mydata[1,2]

[1] b1
Levels: b1 b2

> mydata$sex
```

```
 [1] F F M F F F M M F F M F M M F M M F F M M M M F F F F M M F
[31] M F M F F F F F F M M F F F F F F F F M F M F M M F F M M M M
[61] F F F F F M M F F F M M M F F M F M M F M F M M M M M F F F
[91] F F M M F M F M F M
Levels: F M
```

```
> ##same as: mydata[,"sex"] or mydata[,3]
```

Values in data frames can also be edited conveniently using, e.g., `fix(mydata)`, opening a simple, spreadsheet-like editor.

For real data sets, it is usually most convenient to read these into R from external files, and this is easiest using plain text (ASCII) files with one line per observation and fields separated by commas (or tabs), and with a first header line giving the variable names—called *comma-separated* or CSV format. If your data is in the form of Excel, SAS, SPSS, or other file format, you can almost always export that data to CSV format first.[3]

The function `read.table()` has many options to control the details of how the data are read and converted to variables in the data frame. Among these some important options are:

`header` indicates whether the first line contains variable names. The default is `FALSE` unless the first line contains one fewer field than the number of columns;

`sep` (default: `""`, meaning white space, i.e., one or more spaces, tabs or newlines) specifies the separator character between fields;

`stringsAsFactors` (default: `TRUE`) determines whether character string variables should be converted to factors;

`na.strings` (default: `"NA"`) refers to one or more strings that are interpreted as missing data values (`NA`);

For delimited files, `read.csv()` and `read.delim()` are convenient wrappers to `read.table()`, with default values sep=`","` and sep=`"\t"` respectively, and header=`TRUE`.

EXAMPLE 2.1: Arthritis treatment

The file `Arthritis.csv` contains data in CSV format from Koch and Edwards (1988), representing a double-blind clinical trial investigating a new treatment for rheumatoid arthritis with 84 patients.[4] The first ("header") line gives the variable names. Some of the lines in the file are shown below, with `...` representing omitted lines:

```
ID,Treatment,Sex,Age,Improved
57,Treated,Male,27,Some
46,Treated,Male,29,None
77,Treated,Male,30,None
17,Treated,Male,32,Marked
  ...
42,Placebo,Female,66,None
15,Placebo,Female,66,Some
71,Placebo,Female,68,Some
1,Placebo,Female,74,Marked
```

We read this into R using `read.table()` as shown below:

---

[3]The foreign (R Core Team, 2015) package contains specialized functions to *directly* read data stored by Minitab, SAS, SPSS, Stata, Systat, and other software. There are also a number of packages for reading (and writing) Excel spreadsheets directly (gdata (Warnes et al., 2014), XLConnect (Mirai Solutions GmbH, 2015), xlsx (Dragulescu, 2014)). The R manual, *R Data Import/Export* covers many other variations, including data in relational data bases.

[4]This data set can be created using: `library(vcd); write.table(Arthritis, file = "Arthritis.csv", quote = FALSE, sep = ",")`.

```
> path <- "ch02/Arthritis.csv"  ## set path
> ## for convenience, use path <- file.choose() to retrieve a path
> ## then, use file.show(path) to inspect the data format
> Arthritis <- read.table(path, header = TRUE, sep = ",")
> str(Arthritis)

'data.frame': 84 obs. of  5 variables:
 $ ID       : int  57 46 77 17 36 23 75 39 33 55 ...
 $ Treatment: Factor w/ 2 levels "Placebo","Treated": 2 2 2 2 2 2 2 2 2 2 ...
 $ Sex      : Factor w/ 2 levels "Female","Male": 2 2 2 2 2 2 2 2 2 2 ...
 $ Age      : int  27 29 30 32 46 58 59 59 63 63 ...
 $ Improved : Factor w/ 3 levels "Marked","None",..: 3 2 2 1 1 1 2 1 2 2 ...
```

Note that the character variables Treatment, Sex, and Improved were converted to factors, and the levels of those variables were ordered *alphabetically*. This often doesn't matter much for binary variables, but here, the response variable Improved has levels that should be considered *ordered*, as c("None", "Some", "Marked"). We can correct this here by re-assigning Arthritis$Improved using ordered(). The topic of re-ordering variables and levels in categorical data is considered in more detail in Section 2.3.

```
> levels(Arthritis$Improved)

[1] "Marked" "None"    "Some"

> Arthritis$Improved <- ordered(Arthritis$Improved,
+                               levels = c("None", "Some", "Marked"))
```

△

2.2 Forms of categorical data: case form, frequency form, and table form

As we saw in Chapter 1, categorical data can be represented as ordinary data sets in case form, but the discrete nature of factors or stratifying variables allows the same information to be represented more compactly in summarized form with a frequency variable for each cell of factor combinations, or in tables. Consequently, we sometimes find data created or presented in one form (e.g., a spreadsheet data set, a two-way table of frequencies) and want to input that into R. Once we have the data in R, it is often necessary to manipulate the data into some other form for the purposes of statistical analysis, visualizing results, and our own presentation. It is useful to understand the three main forms of categorical data in R and how to work with them for our purposes.

2.2.1 Case form

Categorical data in case form are simply data frames, with one or more discrete classifying variables or response variables, most conveniently represented as factors or ordered factors. In case form, the data set can also contain numeric variables (covariates or other response variables) that cannot be accommodated in other forms.

As with any data frame, X, you can access or compute with its attributes using nrow(X) for the number of observations, ncol(X) for the number of variables, names(X) or colnames(X) for the variable names, and so forth.

EXAMPLE 2.2: Arthritis treatment
    The *Arthritis* data is available in case form in the vcd package. There are two explanatory factors: Treatment and Sex. Age is a numeric covariate, and Improved is the response—an ordered factor, with levels "None" < "Some" < "Marked". Excluding Age, we would have a $2 \times 2 \times 3$ contingency table for Treatment, Sex, and Improved.

```
> data("Arthritis", package = "vcd")   # load the data
> names(Arthritis)        # show the variables

[1] "ID"         "Treatment" "Sex"        "Age"        "Improved"

> str(Arthritis)          # show the structure

'data.frame': 84 obs. of  5 variables:
 $ ID       : int  57 46 77 17 36 23 75 39 33 55 ...
 $ Treatment: Factor w/ 2 levels "Placebo","Treated": 2 2 2 2 2 2 2 2 2 2 ...
 $ Sex      : Factor w/ 2 levels "Female","Male": 2 2 2 2 2 2 2 2 2 2 ...
 $ Age      : int  27 29 30 32 46 58 59 59 63 63 ...
 $ Improved : Ord.factor w/ 3 levels "None"<"Some"<..: 2 1 1 3 3 3 1 3 1 1 ...

> head(Arthritis, 5)      # first 5 observations, same as Arthritis[1:5,]

  ID Treatment  Sex Age Improved
1 57   Treated Male  27     Some
2 46   Treated Male  29     None
3 77   Treated Male  30     None
4 17   Treated Male  32   Marked
5 36   Treated Male  46   Marked
```

△

2.2.2 Frequency form

Data in frequency form is also a data frame, containing one or more discrete factor variables and a frequency variable (often called `Freq` or `count`) representing the number of basic observations in that cell.

This is an alternative representation of a table form data set considered below. In frequency form, the number of cells in the equivalent table is `nrow(X)`, and the total number of observations is the sum of the frequency variable, `sum(X$Freq)`, `sum(X[,"Freq"])` or a similar expression.

EXAMPLE 2.3: General social survey

For small frequency tables, it is often convenient to enter them in frequency form using `expand.grid()` for the factors and `c()` to list the counts in a vector. The example below, from Agresti (2002), gives results for the 1991 General Social Survey, with respondents classified by sex and party identification. As a table, the data look like this:

		party	
sex	dem	indep	rep
female	279	73	225
male	165	47	191

We use `expand.grid()` to create a 6 × 2 matrix containing the combinations of `sex` and `party` with the levels for `sex` given first, so that this varies most rapidly. Then, input the frequencies in the table by columns from left to right, and combine these two results with `data.frame()`.

```
> # Agresti (2002), table 3.11, p. 106
> tmp <- expand.grid(sex = c("female", "male"),
+                    party = c("dem", "indep", "rep"))
> tmp

     sex party
1 female   dem
2   male   dem
3 female indep
```

```
4    male indep
5  female   rep
6    male   rep

> GSS <- data.frame(tmp, count = c(279, 165, 73, 47, 225, 191))
> GSS

     sex party count
1 female   dem   279
2   male   dem   165
3 female indep    73
4   male indep    47
5 female   rep   225
6   male   rep   191

> names(GSS)

[1] "sex"   "party" "count"

> str(GSS)

'data.frame': 6 obs. of  3 variables:
 $ sex  : Factor w/ 2 levels "female","male": 1 2 1 2 1 2
 $ party: Factor w/ 3 levels "dem","indep",..: 1 1 2 2 3 3
 $ count: num   279 165 73 47 225 191

> sum(GSS$count)

[1] 980
```

The last line above shows that there are 980 cases represented in the frequency table.   △

2.2.3   Table form

Table form data is represented as a matrix, array, or table object whose elements are the frequencies in an n-way table. The number of dimensions of the table is the length, `length(dim(X))`, of its `dim` (or `dimnames`) attribute, and the sizes of the dimensions in the table are the elements of `dim(X)`. The total number of observations represented is the sum of all the frequencies, `sum(X)`.

EXAMPLE 2.4:   Hair color and eye color

A classic data set on frequencies of hair color, eye color, and sex is given in table form in *HairEyeColor* in the **datasets** package, reporting the frequencies of these categories for 592 students in a statistics course.

```
> data("HairEyeColor", package = "datasets")     # load the data
> str(HairEyeColor)                    # show the structure

 table [1:4, 1:4, 1:2] 32 53 10 3 11 50 10 30 10 25 ...
 - attr(*, "dimnames")=List of 3
   ..$ Hair: chr [1:4] "Black" "Brown" "Red" "Blond"
   ..$ Eye : chr [1:4] "Brown" "Blue" "Hazel" "Green"
   ..$ Sex : chr [1:2] "Male" "Female"

> dim(HairEyeColor)                    # table dimension sizes

[1] 4 4 2

> dimnames(HairEyeColor)               # variable and level names
```

```
$Hair
[1] "Black" "Brown" "Red"    "Blond"

$Eye
[1] "Brown" "Blue"  "Hazel" "Green"

$Sex
[1] "Male"    "Female"

> sum(HairEyeColor)                    # number of cases

[1] 592
```

Three-way (and higher-way) tables can be printed in a more convenient form using `structable()` and `ftable()` as described below in Section 2.5.   △

Tables are often created from raw data in case form or frequency form using the functions `table()` and `xtabs()` described in Section 2.4. For smallish frequency tables that are already in tabular form, you can enter the frequencies in a matrix, and then assign `dimnames` and other attributes.

To illustrate, we create the GSS data as a table below, entering the values in the table by rows (`byrow=TRUE`), as they appear in printed form.

```
> GSS.tab <- matrix(c(279, 73,  225,
+                      165, 47, 191),
+                    nrow = 2, ncol = 3, byrow = TRUE)
> dimnames(GSS.tab) <- list(sex = c("female", "male"),
+                           party = c("dem", "indep", "rep"))
> GSS.tab

        party
sex       dem indep rep
  female 279    73 225
  male   165    47 191
```

`GSS.tab` is a matrix, not an object of `class("table")`, and some functions are happier with tables than matrices.[5] You should therefore coerce it to a table with `as.table()`,

```
> GSS.tab <- as.table(GSS.tab)
> str(GSS.tab)

 table [1:2, 1:3] 279 165 73 47 225 191
 - attr(*, "dimnames")=List of 2
  ..$ sex  : chr [1:2] "female" "male"
  ..$ party: chr [1:3] "dem" "indep" "rep"
```

EXAMPLE 2.5: Job satisfaction

Here is another similar example, entering data on job satisfaction classified by `income` and level of `satisfaction` from a 4×4 table given by Agresti (2002, Table 2.8, p. 57).

```
> ## A 4 x 4 table  Agresti (2002, Table 2.8, p. 57) Job Satisfaction
> JobSat <- matrix(c(1, 2, 1, 0,
+                    3, 3, 6, 1,
+                    10, 10, 14, 9,
+                    6, 7, 12, 11),
```

---

[5]There are quite a few functions in R with specialized methods for "table" objects. For example, `plot(GSS.tab)` gives a mosaic plot and `barchart(GSS.tab)` gives a divided bar chart.

```
+                      nrow = 4, ncol = 4)
> dimnames(JobSat) <-
+    list(income = c("< 15k", "15-25k", "25-40k", "> 40k"),
+         satisfaction = c("VeryD", "LittleD", "ModerateS", "VeryS"))
> JobSat <- as.table(JobSat)
> JobSat

         satisfaction
income    VeryD LittleD ModerateS VeryS
  < 15k      1       3        10     6
  15-25k     2       3        10     7
  25-40k     1       6        14    12
  > 40k      0       1         9    11
```

△

2.3 Ordered factors and reordered tables

As we saw above (Example 2.1), the levels of factor variables in data frames (case form or frequency form) can be re-ordered (and the variables declared as ordered factors) using ordered(). As well, the order of the factor values themselves can be rearranged by sorting the data frame using sort().

However, in table form, the values of the table factors are ordered by their position in the table. Thus in the *JobSat* data, both income and satisfaction represent ordered factors, and the *positions* of the values in the rows and columns reflect their ordered nature, but only implicitly.

Yet, for analysis or graphing, there are occasions when you need *numeric* values for the levels of ordered factors in a table, e.g., to treat a factor as a quantitative variable. In such cases, you can simply re-assign the dimnames attribute of the table variables. For example, here, we assign numeric values to income as the middle of their ranges, and treat satisfaction as equally spaced with integer scores.

```
> dimnames(JobSat)$income <- c(7.5, 20, 32.5, 60)
> dimnames(JobSat)$satisfaction <- 1:4
```

A related case is when you want to preserve the character labels of table dimensions, but also allow them to be sorted in some particular order. A simple way to do this is to prefix each label with an integer index using paste().

```
> dimnames(JobSat)$income <-
+     paste(1:4, dimnames(JobSat)$income, sep = ":")
> dimnames(JobSat)$satisfaction <-
+     paste(1:4, dimnames(JobSat)$satisfaction, sep = ":")
```

A different situation arises with tables where you want to *permute* the levels of one or more variables to arrange them in a more convenient order without changing their labels. For example, in the *HairEyeColor* table, hair color and eye color are ordered arbitrarily.

For visualizing the data using mosaic plots and other methods described later, it turns out to be more useful to assure that both hair color and eye color are ordered from dark to light. Hair colors are actually ordered this way already: "Black", "Brown", "Red", "Blond". But eye colors are ordered as "Brown", "Blue", "Hazel", "Green". It is easiest to re-order the eye colors by indexing the columns (dimension 2) in this array to a new order, "Brown", "Hazel", "Green", "Blue", giving the indices of the old levels in the new order (here: 1,3,4,2). Again str() is your friend, showing the structure of the result to check that the result is what you want.

```
> data("HairEyeColor", package = "datasets")
> HEC <- HairEyeColor[, c(1, 3, 4, 2), ]
> str(HEC)

 num [1:4, 1:4, 1:2] 32 53 10 3 10 25 7 5 3 15 ...
 - attr(*, "dimnames")=List of 3
  ..$ Hair: chr [1:4] "Black" "Brown" "Red" "Blond"
  ..$ Eye : chr [1:4] "Brown" "Hazel" "Green" "Blue"
  ..$ Sex : chr [1:2] "Male" "Female"
```

Finally, there are situations where, particularly for display purposes, you want to re-order the *dimensions* of an *n*-way table, and/or change the labels for the variables or levels. This is easy when the data are in table form: aperm() permutes the dimensions, and assigning to names and dimnames changes variable names and level labels, respectively.

```
> str(UCBAdmissions)

 table [1:2, 1:2, 1:6] 512 313 89 19 353 207 17 8 120 205 ...
 - attr(*, "dimnames")=List of 3
  ..$ Admit : chr [1:2] "Admitted" "Rejected"
  ..$ Gender: chr [1:2] "Male" "Female"
  ..$ Dept  : chr [1:6] "A" "B" "C" "D" ...

> # vary along the 2nd, 1st, and 3rd dimension in UCBAdmissions
> UCB <- aperm(UCBAdmissions, c(2, 1, 3))
> dimnames(UCB)$Admit <- c("Yes", "No")
> names(dimnames(UCB)) <- c("Sex", "Admitted", "Department")
> str(UCB)

 table [1:2, 1:2, 1:6] 512 89 313 19 353 17 207 8 120 202 ...
 - attr(*, "dimnames")=List of 3
  ..$ Sex       : chr [1:2] "Male" "Female"
  ..$ Admitted  : chr [1:2] "Yes" "No"
  ..$ Department: chr [1:6] "A" "B" "C" "D" ...
```

2.4 Generating tables with table() and xtabs()

With data in case form or frequency form, you can generate frequency tables from factor variables in data frames using the table() function; for tables of proportions, use the prop.table() function, and for marginal frequencies (summing over some variables) use margin.table(). The examples below use the same case-form data frame mydata used earlier (Section 2.1.4).

```
> set.seed(12345)    # reproducibility
> n <- 100
> A <- factor(sample(c("a1", "a2"), n, replace = TRUE))
> B <- factor(sample(c("b1", "b2"), n, replace = TRUE))
> sex <- factor(sample(c("M", "F"), n, replace = TRUE))
> age <- round(rnorm(n, mean = 30, sd = 5))
> mydata <- data.frame(A, B, sex, age)
```

2.4.1 table()

table(...) takes a list of variables interpreted as factors, or a data frame whose columns are so interpreted. It does not take a data= argument, so either supply the names of columns in the data frame (possibly using with() for convenience), or select the variables using column indexes:

```
> # 2-Way Frequency Table
> table(mydata$A, mydata$B)              # A will be rows, B will be columns


      b1 b2
   a1 18 30
   a2 22 30

> ## same: with(mydata, table(A, B))
> (mytab <- table(mydata[,1:2]))         # same


      B
A     b1 b2
   a1 18 30
   a2 22 30
```

We can use `margin.table(X, margin)` to sum a table X for the indices in `margin`, i.e., over the dimensions not included in `margin`. A related function is `addmargins(X, margin, FUN = sum)`, which extends the dimensions of a table or array with the marginal values calculated by FUN.

```
> margin.table(mytab)         # sum over A & B

[1] 100

> margin.table(mytab, 1)      # A frequencies (summed over B)

A
a1 a2
48 52

> margin.table(mytab, 2)      # B frequencies (summed over A)

B
b1 b2
40 60

> addmargins(mytab)           # show all marginal totals

      B
A      b1  b2 Sum
   a1  18  30  48
   a2  22  30  52
   Sum 40  60 100
```

The function `prop.table()` expresses the table entries as a fraction of a given marginal table.

```
> prop.table(mytab)           # cell proportions

      B
A       b1   b2
   a1 0.18 0.30
   a2 0.22 0.30

> prop.table(mytab, 1)        # row proportions

      B
A          b1      b2
   a1 0.37500 0.62500
   a2 0.42308 0.57692
```

```
> prop.table(mytab, 2)      # column proportions

    B
A       b1    b2
  a1  0.45  0.50
  a2  0.55  0.50
```

`table()` can also generate multidimensional tables based on 3 or more categorical variables. In this case, use the `ftable()` or `structable()` function to print the results more attractively as a "flat" (2-way) table.

```
> # 3-Way Frequency Table
> mytab <- table(mydata[,c("A", "B", "sex")])
> ftable(mytab)

       sex  F   M
A  B
a1 b1        9   9
   b2       15  15
a2 b1       12  10
   b2       19  11
```

`table()` ignores missing values by default, but has optional arguments `useNA` and `exclude` that can be used to control this. See `help(table)` for the details.

2.4.2 `xtabs()`

The `xtabs()` function allows you to create cross tabulations of data using formula style input. This typically works with case-form or frequency-form data supplied in a data frame or a matrix. The result is a contingency table in array format, whose dimensions are determined by the terms on the right side of the formula. As shown below, the `summary` method for tables produces a simple χ^2 test of independence of all factors, and indicates the number of cases and dimensions.

```
> # 3-Way Frequency Table
> mytable <- xtabs(~ A + B + sex, data = mydata)
> ftable(mytable)      # print table

       sex  F   M
A  B
a1 b1        9   9
   b2       15  15
a2 b1       12  10
   b2       19  11

> summary(mytable)      # chi-squared test of independence

Call: xtabs(formula = ~A + B + sex, data = mydata)
Number of cases in table: 100
Number of factors: 3
Test for independence of all factors:
Chisq = 1.54, df = 4, p-value = 0.82
```

When the data have already been tabulated in frequency form, include the frequency variable (usually `count` or `Freq`) on the left side of the formula, as shown in the example below for the GSS data.

```
> (GSStab <- xtabs(count ~ sex + party, data = GSS))
```

```
          party
sex       dem indep rep
  female 279      73 225
  male   165      47 191
```

```
> summary(GSStab)
```

```
Call: xtabs(formula = count ~ sex + party, data = GSS)
Number of cases in table: 980
Number of factors: 2
Test for independence of all factors:
Chisq = 7, df = 2, p-value = 0.03
```

For **"table"** objects, the `plot` method produces basic mosaic plots using the `mosaicplot()` function from the **graphics** package.

2.5 Printing tables with `structable()` and `ftable()`

2.5.1 Text output

For 3-way and larger tables, the functions `ftable()` (in the **stats** package) and `structable()` (in **vcd**) provide a convenient and flexible tabular display in a "flat" (2-way) format.

With `ftable(X, row.vars=, col.vars=)`, variables assigned to the rows and/or columns of the result can be specified as the integer numbers or character names of the variables in the array `X`. By default, the last variable is used for the columns. The formula method, in the form `ftable(colvars ~ rowvars, data)` allows a formula where the left- and right-hand side of formula specify the column and row variables, respectively.

```
>   ftable(UCB)                          # default

                 Department   A   B   C   D   E   F
Sex    Admitted
Male   Yes                   512 353 120 138  53  22
       No                    313 207 205 279 138 351
Female Yes                    89  17 202 131  94  24
       No                     19   8 391 244 299 317

> #ftable(UCB, row.vars = 1:2)          # same result
>   ftable(Admitted + Sex ~ Department, data = UCB)    # formula method

                Admitted  Yes           No
                Sex      Male Female Male Female
Department
A                         512     89  313     19
B                         353     17  207      8
C                         120    202  205    391
D                         138    131  279    244
E                          53     94  138    299
F                          22     24  351    317
```

The `structable()` function is similar, but more general, and uses recursive splits in the vertical or horizontal directions (similar to the construction of mosaic displays). It works with both data frames and table objects.

```
> library(vcd)
> structable(HairEyeColor)              # show the table: default

               Eye Brown Blue Hazel Green
```

```
Hair  Sex
Black Male          32    11    10     3
      Female        36     9     5     2
Brown Male          53    50    25    15
      Female        66    34    29    14
Red   Male          10    10     7     7
      Female        16     7     7     7
Blond Male           3    30     5     8
      Female         4    64     5     8

> structable(Hair + Sex ~ Eye, HairEyeColor) # specify col ~ row variables

        Hair Black           Brown           Red           Blond
        Sex   Male Female  Male Female  Male Female  Male Female
Eye
Brown          32     36    53     66    10     16     3      4
Blue           11      9    50     34    10      7    30     64
Hazel          10      5    25     29     7      7     5      5
Green           3      2    15     14     7      7     8      8
```

It also returns an object of class `"structable"` for which there are a variety of special methods. For example, the transpose function `t()` interchanges rows and columns, so that a call like `t(structable(HairEyeColor))` produces the second result shown just above. There are also plot methods: for example, `plot()` produces mosaic plots from the vcd package.

2.6 Subsetting data

Often, the analysis of some data set is focused on a subset only. For example, the *HairEyeColor* data set introduced above tabulates frequencies of hair and eye colors for male and female students—the analysis could concentrate on one group only, or compare both groups in a stratified analysis. This section deals with extracting subsets of data in tables, structables, or data frames.

2.6.1 Subsetting tables

If data are available in tabular form created with `table()` or `xtabs()`, resulting in `table` objects, subsetting is done via indexing, either with integers or character strings corresponding to the factor levels. The following code extracts the female data from the `HairEyeColor` data set:

```
> HairEyeColor[,,"Female"]

       Eye
Hair    Brown Blue Hazel Green
  Black    36    9     5     2
  Brown    66   34    29    14
  Red      16    7     7     7
  Blond     4   64     5     8

> ##same using index: HairEyeColor[,,2]
```

Empty indices stand for taking all data of the corresponding dimension. The third one (Sex) is fixed at the second ("Female") level. Note that in this case, the dimensionality is reduced to a two-way table, since dimensions with only one level are dropped by default. Functions like `apply()` can iterate through all levels of one or several dimensions and apply a function to each subset. The following calculates the total amount of male and female students:

```
> apply(HairEyeColor, 3, sum)

  Male Female
   279    313
```

It is of course possible to select more than one level:

```
> HairEyeColor[c("Black", "Brown"), c("Hazel", "Green"),]

, , Sex = Male

        Eye
Hair     Hazel Green
  Black     10     3
  Brown     25    15

, , Sex = Female

        Eye
Hair     Hazel Green
  Black      5     2
  Brown     29    14
```

2.6.2 Subsetting structables

Structables work in a similar way, but take into account the hierarchical structure imposed by the "flattened" format, and also distinguish explicitly between subsetting levels and subsetting tables. In the following example, compare the different effects of applying the [and [[operators to the structable:

```
> hec <- structable(Eye ~ Sex + Hair, data = HairEyeColor)
> hec

               Eye Brown Blue Hazel Green
Sex    Hair
Male   Black        32    11    10     3
       Brown        53    50    25    15
       Red          10    10     7     7
       Blond         3    30     5     8
Female Black        36     9     5     2
       Brown        66    34    29    14
       Red          16     7     7     7
       Blond         4    64     5     8

> hec["Male",]

              Eye Brown Blue Hazel Green
Sex   Hair
Male  Black        32    11    10     3
      Brown        53    50    25    15
      Red          10    10     7     7
      Blond         3    30     5     8

> hec[["Male",]]

        Eye Brown Blue Hazel Green
Hair
Black        32    11    10     3
Brown        53    50    25    15
Red          10    10     7     7
Blond         3    30     5     8
```

The first form keeps the dimensionality, whereas the second conditions on the "Male" level and returns the corresponding subtable. The following does this twice, once for Sex, and once for Hair (restricted to the Male level):

```
> hec[[c("Male", "Brown"),]]

Eye Brown Blue Hazel Green

        53   50    25    15
```

2.6.3 Subsetting data frames

Data available in data frames (frequency or case form) can also be subsetted, either by using indexes on the rows and/or columns, or, more conveniently, by applying the subset() function. The following statement will extract the Treatment and Improved variables for all female patients older than 68:

```
> rows <- Arthritis$Sex == "Female" & Arthritis$Age > 68
> cols <- c("Treatment", "Improved")
> Arthritis[rows, cols]

   Treatment Improved
39   Treated     None
40   Treated     Some
41   Treated     Some
84   Placebo   Marked
```

Note the use of the single & for the logical expression selecting the rows. The same result can be achieved more conveniently using the subset() function, first taking the data set, followed by an expression for selecting the rows (evaluated in the context of the data frame), and then an expression for selecting the columns:

```
> subset(Arthritis, Sex == "Female" & Age > 68,
+        select = c(Treatment, Improved))

   Treatment Improved
39   Treated     None
40   Treated     Some
41   Treated     Some
84   Placebo   Marked
```

Note the non-standard evaluation of c(Treatment, Improved): the meaning of c() is not "combine the two columns into a single vector," but "select both from the data frame." Likewise, columns can be removed using − on column names, which is not possible using standard indexing in matrices or data frames:

```
> subset(Arthritis, Sex == "Female" & Age > 68,
+        select = -c(Age, ID))

   Treatment    Sex Improved
39   Treated Female     None
40   Treated Female     Some
41   Treated Female     Some
84   Placebo Female   Marked
```

2.7 Collapsing tables

2.7.1 Collapsing over table factors: `aggregate()`, `margin.table()`, and `apply()`

It sometimes happens that we have a data set with more variables or factors than we want to analyze, or else, having done some initial analyses, we decide that certain factors are not important, and so should be excluded from graphic displays by collapsing (summing) over them. For example, mosaic plots and fourfold displays are often simpler to construct from versions of the data collapsed over the factors that are not shown in the plots.

The appropriate tools to use again depend on the form in which the data are represented—a case-form data frame, a frequency-form data frame (`aggregate()`), or a table-form array or table object (`margin.table()` or `apply()`).

When the data are in frequency form, and we want to produce another frequency data frame, `aggregate()` is a handy tool, using the argument `FUN = sum` to sum the frequency variable over the factors *not* mentioned in the formula.

EXAMPLE 2.6: Dayton survey

The data frame *DaytonSurvey* in the **vcdExtra** package represents a 2^5 table giving the frequencies of reported use ("ever used?") of alcohol, cigarettes, and marijuana in a sample of 2276 high school seniors, also classified by sex and race.

```
> data("DaytonSurvey", package = "vcdExtra")
> str(DaytonSurvey)

'data.frame':  32 obs. of  6 variables:
 $ cigarette: Factor w/ 2 levels "Yes","No": 1 2 1 2 1 2 1 2 1 2 ...
 $ alcohol  : Factor w/ 2 levels "Yes","No": 1 1 2 2 1 1 2 2 1 1 ...
 $ marijuana: Factor w/ 2 levels "Yes","No": 1 1 1 1 2 2 2 2 1 1 ...
 $ sex      : Factor w/ 2 levels "female","male": 1 1 1 1 1 1 1 1 2 2 ...
 $ race     : Factor w/ 2 levels "white","other": 1 1 1 1 1 1 1 1 1 1 ...
 $ Freq     : num  405 13 1 1 268 218 17 117 453 28 ...

> head(DaytonSurvey)

  cigarette alcohol marijuana    sex  race Freq
1       Yes     Yes       Yes female white  405
2        No     Yes       Yes female white   13
3       Yes      No       Yes female white    1
4        No      No       Yes female white    1
5       Yes     Yes        No female white  268
6        No     Yes        No female white  218
```

To focus on the associations among the substances, we want to collapse over sex and race. The right-hand side of the formula used in the call to `aggregate()` gives the factors to be retained in the new frequency data frame, `Dayton_ACM_df`. The left-hand side is the frequency variable (`Freq`), and we aggregate using the `FUN = sum`.

```
> # data in frequency form: collapse over sex and race
> Dayton_ACM_df <- aggregate(Freq ~ cigarette + alcohol + marijuana,
+                            data = DaytonSurvey, FUN = sum)
> Dayton_ACM_df

  cigarette alcohol marijuana Freq
1       Yes     Yes       Yes  911
2        No     Yes       Yes   44
3       Yes      No       Yes    3
```

4	No	No	Yes	2
5	Yes	Yes	No	538
6	No	Yes	No	456
7	Yes	No	No	43
8	No	No	No	279

△

When the data are in table form, and we want to produce another table, `apply()` with `FUN = sum` can be used in a similar way to sum the table over dimensions not mentioned in the `MARGIN` argument. `margin.table()` is just a wrapper for `apply()` using the `sum()` function.

EXAMPLE 2.7: Dayton survey

To illustrate, we first convert the *DaytonSurvey* to a 5-way table using `xtabs()`, giving `Dayton_tab`.

```
> # convert to table form
> Dayton_tab <- xtabs(Freq ~ cigarette + alcohol + marijuana + sex + race,
+                     data = DaytonSurvey)
> structable(cigarette + alcohol + marijuana ~ sex + race,
+            data = Dayton_tab)
```

		cigarette	Yes				No			
		alcohol	Yes		No		Yes		No	
		marijuana	Yes	No	Yes	No	Yes	No	Yes	No
sex	race									
female	white		405	268	1	17	13	218	1	117
	other		23	23	0	1	2	19	0	12
male	white		453	228	1	17	28	201	1	133
	other		30	19	1	8	1	18	0	17

Then, use `apply()` on `Dayton_tab` to give the 3-way table `Dayton_ACM_tab` summed over sex and race. The elements in this new table are the column sums for `Dayton.tab` shown by `structable()` just above.

```
> # collapse over sex and race
> Dayton_ACM_tab <- apply(Dayton_tab, MARGIN = 1:3, FUN = sum)
> Dayton_ACM_tab <- margin.table(Dayton_tab, 1:3)    # same result
> structable(cigarette + alcohol ~ marijuana, data = Dayton_ACM_tab)
```

	cigarette	Yes		No	
	alcohol	Yes	No	Yes	No
marijuana					
Yes		911	3	44	2
No		538	43	456	279

△

(Note that `structable()` would do the collapsing job for us anyway.)

Many of these operations can be performed using the `**ply()` functions in the **plyr** (Wickham, 2014a) package. For example, with the data in a frequency form data frame, use `ddply()` to collapse over unmentioned factors, and `summarise()` as the function to be applied to each piece.

```
> library(plyr)
> Dayton_ACM_df <- ddply(DaytonSurvey, .(cigarette, alcohol, marijuana),
+                        summarise, Freq = sum(Freq))
```

2.7.2 Collapsing table levels: `collapse.table()`

A related problem arises when we have a table or array and for some purpose we want to reduce the number of levels of some factors by summing subsets of the frequencies. For example, we may have initially coded Age in 10-year intervals, and decide that, either for analysis or display purposes, we want to reduce Age to 20-year intervals. The `collapse.table()` function in **vcdExtra** was designed for this purpose.

EXAMPLE 2.8: Collapsing categories

Create a 3-way table, and collapse Age from 10-year to 20-year intervals and Education from three levels to two. To illustrate, we first generate a $2 \times 6 \times 3$ table of random counts from a Poisson distribution with mean of 100, with factors `sex`, `age`, and `education`.

```
> # create some sample data in frequency form
> set.seed(12345)    # reproducibility
> sex <- c("Male", "Female")
> age <- c("10-19", "20-29",  "30-39", "40-49", "50-59", "60-69")
> education <- c("low", "med", "high")
> dat <- expand.grid(sex = sex, age = age, education = education)
> counts <- rpois(36, 100)    # random Poisson cell frequencies
> dat <- cbind(dat, counts)
> # make it into a 3-way table
> tab1 <- xtabs(counts ~ sex + age + education, data = dat)
> structable(tab1)
```

```
                age 10-19 20-29 30-39 40-49 50-59 60-69
sex     education
Male    low           105    98   123    97    95   105
        med            74   113   114    82    95    85
        high          121   116   104   103    89   100
Female  low           107    95   105   116   103    92
        med            96    88    93   118    99   108
        high          120   102    96   103   127    84
```

Now collapse `age` to 20-year intervals, and `education` to 2 levels. In the arguments to `collapse.table()`, levels of `age` and `education` given the same label are summed in the resulting smaller table.

```
> # collapse age to 3 levels, education to 2 levels
> tab2 <- collapse.table(tab1,
+          age = c("10-29", "10-29",  "30-49", "30-49", "50-69", "50-69"),
+          education = c("<high", "<high", "high"))
> structable(tab2)
```

```
                age 10-29 30-49 50-69
sex     education
Male    <high         390   416   380
        high          237   207   189
Female  <high         386   432   402
        high          222   199   211
```

$\triangle$

2.8 Converting among frequency tables and data frames

As we've seen, a given contingency table can be represented equivalently in case form, frequency form, and table form. However, some R functions were designed for one particular representation. Table 2.1 gives an overview of some handy tools (with sketched usage) for converting from one form to another, discussed below.

Table 2.1: Tools for converting among different forms for categorical data

From this	To this		
	Case form	Frequency form	Table form
Case form	—	`Z <- xtabs(~ A+B)` `as.data.frame(Z)`	`table(A, B)`
Frequency form	`expand.dft(X)`	—	`xtabs(count ~ A+B)`
Table form	`expand.dft(X)`	`as.data.frame(X)`	—

2.8.1   Table form to frequency form

A contingency table in table form (an object of class "table") can be converted to a data frame in frequency form with `as.data.frame()`.[6] The resulting data frame contains columns representing the classifying factors and the table entries (as a column named by the `responseName` argument, defaulting to `Freq`). The function `as.data.frame()` is the inverse of `xtabs()`, which converts a data frame to a table.

EXAMPLE 2.9: General social survey

Convert the `GSStab` object in table form to a data.frame in frequency form. By default, the frequency variable is named `Freq`, and the variables `sex` and `party` are made factors.

```
> as.data.frame(GSStab)

     sex party Freq
1 female    dem  279
2   male    dem  165
3 female  indep   73
4   male  indep   47
5 female    rep  225
6   male    rep  191
```

△

In addition, there are situations where numeric table variables are represented as factors, but you need to convert them to numerics for calculation purposes.

EXAMPLE 2.10: Death by horse kick

For example, we might want to calculate the weighted mean of `nDeaths` in the *HorseKicks* data. Using `as.data.frame()` won't work here, because the variable `nDeaths` becomes a factor.

```
> str(as.data.frame(HorseKicks))

'data.frame': 5 obs. of  2 variables:
 $ nDeaths: Factor w/ 5 levels "0","1","2","3",..: 1 2 3 4 5
 $ Freq   : int  109 65 22 3 1
```

One solution is to use `data.frame()` directly and `as.numeric()` to coerce the table names to numbers.

---

[6]Because R is object-oriented, this is actually a shorthand for the function `as.data.frame.table()`, which is automatically selected for objects of class "table".

```
> horse.df <- data.frame(nDeaths = as.numeric(names(HorseKicks)),
+                          Freq = as.vector(HorseKicks))
> str(horse.df)

'data.frame': 5 obs. of  2 variables:
 $ nDeaths: num  0 1 2 3 4
 $ Freq   : int  109 65 22 3 1

> horse.df

  nDeaths Freq
1       0  109
2       1   65
3       2   22
4       3    3
5       4    1
```

Then, `weighted.mean()` works as we would like:

```
> weighted.mean(horse.df$nDeaths, weights=horse.df$Freq)

[1] 2
```

△

2.8.2   Case form to table form

Going the other way, we use `table()` to convert from case form to table form.

EXAMPLE 2.11: Arthritis treatment
   Convert the *Arthritis* data in case form to a 3-way table of `Treatment` × `Sex` × `Improved`.
We select the desired columns with their names, but could also use column numbers, e.g.,
`table(Arthritis[,c(2,3,5)])`.

```
> Art.tab <- table(Arthritis[,c("Treatment", "Sex", "Improved")])
> str(Art.tab)

'table' int [1:2, 1:2, 1:3] 19 6 10 7 7 5 0 2 6 16 ...
 - attr(*, "dimnames")=List of 3
  ..$ Treatment: chr [1:2] "Placebo" "Treated"
  ..$ Sex      : chr [1:2] "Female" "Male"
  ..$ Improved : chr [1:3] "None" "Some" "Marked"

> ftable(Art.tab)

                 Improved None Some Marked
Treatment Sex
Placebo   Female               19    7      6
          Male                 10    0      1
Treated   Female                6    5     16
          Male                  7    2      5
```

△

2.8.3   Table form to case form

There may also be times that you will need an equivalent case form data frame with factors representing the table variables rather than the frequency table. For example, the `mca()` function in

package MASS (for multiple correspondence analysis) only operates on data in this format. The
function `expand.dft()`[7] in vcdExtra does this, converting a table into a case form.

EXAMPLE 2.12: Arthritis treatment
    Convert the *Arthritis* data in table form (`Art.tab`) back to a `data.frame` in case form,
with factors `Treatment`, `Sex`, and `Improved`.

```
> library(vcdExtra)
> Art.df <- expand.dft(Art.tab)
> str(Art.df)

'data.frame':  84 obs. of  3 variables:
 $ Treatment: Factor w/ 2 levels "Placebo","Treated": 1 1 1 1 1 1 1 1 1 1 ...
 $ Sex      : Factor w/ 2 levels "Female","Male": 1 1 1 1 1 1 1 1 1 1 ...
 $ Improved : Factor w/ 3 levels "Marked","None",..: 2 2 2 2 2 2 2 2 2 2 ...
```

△

2.8.4 Publishing tables to LaTeX or HTML

OK, you've read your data into R, done some analysis, and now want to include some tables in a
LaTeX document or in a web page in HTML format. Formatting tables for these purposes is often
tedious and error-prone.

    There are a great many packages in R that provide for nicely formatted, publishable tables for
a wide variety of purposes; indeed, most of the tables in this book are generated using these tools.
See Leifeld (2013) for a description of the texreg (Leifeld, 2014) package and a comparison with
some of the other packages.

    Here, we simply illustrate the xtable (Dahl, 2014) package, which, along with capabilities for
statistical model summaries, time-series data, and so forth, has a `xtable.table` method for one-
way and two-way table objects.

    The *HorseKicks* data is a small one-way frequency table described in Example 3.4 and con-
tains the frequencies of 0, 1, 2, 3, 4 deaths per corps-year by horse-kick among soldiers in 20 corps
in the Prussian army.

```
> data("HorseKicks", package = "vcd")
> HorseKicks

nDeaths
  0   1   2   3   4
109  65  22   3   1
```

    By default, `xtable()` formats this in LaTeX as a vertical table, and prints the LaTeX markup to
the R console. This output is shown below.

```
> library(xtable)
> xtable(HorseKicks)

% latex table generated in R 3.2.1 by xtable 1.8-0 package
% Thu Nov 12 13:05:37 2015
\begin{table}[ht]
\centering
\begin{tabular}{rr}
  \hline
 & nDeaths \\
  \hline
0 & 109 \\
  1 &  65 \\
```

---

[7]The original code for this function was provided by Marc Schwarz on the R-Help mailing list.

```
  2 &   22 \\
  3 &    3 \\
  4 &    1 \\
    \hline
\end{tabular}
\end{table}
```

When this is rendered in a LaTeX document, the result of xtable() appears as shown in the table below.

```
> xtable(HorseKicks)
```

	nDeaths
0	109
1	65
2	22
3	3
4	1

The table above isn't quite right, because the column label "nDeaths" belongs to the first column, and the second column should be labeled "Freq." To correct that, we convert the *HorseKicks* table to a data frame (see Section 2.8 for details), add the appropriate colnames, and use the print.xtable method to supply some other options.

```
> tab <- as.data.frame(HorseKicks)
> colnames(tab) <- c("nDeaths", "Freq")
> print(xtable(tab), include.rownames = FALSE,
+        include.colnames = TRUE)
```

nDeaths	Freq
0	109
1	65
2	22
3	3
4	1

There are many more options to control the LaTeX details and polish the appearance of the table; see help(xtable) and vignette("xtableGallery", package = "xtable").

Finally, in Chapter 3, we display a number of similar one-way frequency tables in a transposed form to save display space. Table 3.3 is the finished version we show there. The code below uses the following techniques, giving the version shown in Table 2.2: (a) addmargins() is used to show the sum of all the frequency values; (b) t() transposes the data frame to have 2 rows; (c) rownames() assigns the labels we want for the rows; (d) using the caption argument provides a table caption, and a numbered table in LaTeX; (e) column alignment ("r" or "l") for the table columns is computed as a character string used for the align argument.

```
> horsetab <- t(as.data.frame(addmargins(HorseKicks)))
> rownames(horsetab) <- c( "Number of deaths", "Frequency" )
> horsetab <- xtable(horsetab, digits = 0, label="tab:xtable5",
+        caption = "von Bortkiewicz's data on deaths by horse kicks",
```

```
+           align = paste0 ("l|", paste(rep("r", ncol(horsetab)),
+                                                 collapse = ""))
+           )
> print(horsetab, include.colnames=FALSE, caption.placement="top")
```

Table 2.2: von Bortkiewicz's data on deaths by horse kicks

Number of deaths	0	1	2	3	4	Sum
Frequency	109	65	22	3	1	200

For use in a web page, blog, or Word document, you can use `type="HTML"` in the call to `print()` for "xtable" objects.

2.9   A complex example: TV viewing data⋆

If you have followed so far, congratulations! You are ready for a more complicated example that puts together a variety of the skills developed in this chapter: (a) reading raw data, (b) creating tables, (c) assigning level names to factors and (d) collapsing levels or variables for use in analysis.

For an illustration of these steps, we use the dataset `tv.dat`, supplied with the initial implementation of mosaic displays in R by Jay Emerson. In turn, they were derived from an early, compelling example of mosaic displays (Hartigan and Kleiner, 1984) that illustrated the method with data on a large sample of TV viewers whose behavior had been recorded for the Neilsen ratings. This data set contains sample television audience data from Neilsen Media Research for the week starting November 6, 1995.

The data file, `tv.dat`, is stored in frequency form as a file with 825 rows and 5 columns. There is no header line in the file, so when we use `read.table()` below, the variables will be named $V1 - V5$. This data represents a 4-way table of size $5 \times 11 \times 5 \times 3 = 825$ where the table variables are $V1 - V4$, and the cell frequency is read as $V5$.

The table variables are:
    $V1$— values 1:5 correspond to the days Monday–Friday;
    $V2$— values 1:11 correspond to the quarter-hour times 8:00 pm through 10:30 pm;
    $V3$— values 1:5 correspond to ABC, CBS, NBC, Fox, and non-network choices;
    $V4$— values 1:3 correspond to transition states: turn the television Off, Switch channels, or Persist in viewing the current channel.

2.9.1   Creating data frames and arrays

The file `tv.dat` is stored in the `doc/extdata` directory of **vcdExtra**; it can be read as follows:

```
> tv_data <- read.table(system.file("doc", "extdata", "tv.dat",
+                                         package = "vcdExtra"))
> str(tv_data)

'data.frame': 825 obs. of  5 variables:
 $ V1: int  1 2 3 4 5 1 2 3 4 5 ...
 $ V2: int  1 1 1 1 1 2 2 2 2 2 ...
 $ V3: int  1 1 1 1 1 1 1 1 1 1 ...
 $ V4: int  1 1 1 1 1 1 1 1 1 1 ...
 $ V5: int  6 18 6 2 11 6 29 25 17 29 ...
```

```
> head(tv_data, 5)

  V1 V2 V3 V4 V5
1  1  1  1  1  6
2  2  1  1  1 18
3  3  1  1  1  6
4  4  1  1  1  2
5  5  1  1  1 11
```

To read such data from a local file, just use `read.table()` in this form:

```
> tv_data <- read.table("C:/R/data/tv.dat")
```

or, to select the path using the file-chooser tool,

```
> tv_data <- read.table(file.choose())
```

We could use this data in frequency form for analysis by renaming the variables, and converting the integer-coded factors V1 – V4 to R factors. The lines below use the function `within()` to avoid having to use `TV.dat$Day <- factor(TV.dat$Day)` etc., and only supply labels for the first variable.

```
> TV_df <- tv_data
> colnames(TV_df) <- c("Day", "Time", "Network", "State", "Freq")
> TV_df <- within(TV_df, {
+           Day <- factor(Day,
+                         labels = c("Mon", "Tue", "Wed", "Thu", "Fri"))
+           Time <- factor(Time)
+           Network <- factor(Network)
+           State <- factor(State)
+           })
```

Alternatively, we could just reshape the frequency column (V5 or `tv_data[,5]`) into a 4-way array. In the lines below, we rely on the facts that (a) the table is complete—there are no missing cells, so `nrow(tv_data)` = 825; (b) the observations are ordered so that V1 varies most rapidly and V4 most slowly. From this, we can just extract the frequency column and reshape it into an array using the `dim` argument. The level names are assigned to `dimnames(TV)` and the variable names to `names(dimnames(TV))`.

```
> TV <- array(tv_data[,5], dim = c(5, 11, 5, 3))
> dimnames(TV) <-
+       list(c("Mon", "Tue", "Wed", "Thu", "Fri"),
+            c("8:00", "8:15", "8:30", "8:45", "9:00", "9:15",
+                "9:30", "9:45", "10:00", "10:15", "10:30"),
+            c("ABC", "CBS", "NBC", "Fox", "Other"),
+            c("Off", "Switch", "Persist"))
> names(dimnames(TV)) <- c("Day", "Time", "Network", "State")
```

More generally (even if there are missing cells), we can use `xtabs()` to do the cross-tabulation, using V5 as the frequency variable. Here's how to do this same operation with `xtabs()`:

```
> TV <- xtabs(V5 ~ ., data = tv_data)
> dimnames(TV) <-
+       list(Day = c("Mon", "Tue", "Wed", "Thu", "Fri"),
+            Time = c("8:00", "8:15", "8:30", "8:45", "9:00", "9:15",
+                     "9:30", "9:45", "10:00", "10:15", "10:30"),
+            Network = c("ABC", "CBS", "NBC", "Fox", "Other"),
+            State = c("Off", "Switch", "Persist"))
```

Note that in the lines above, the variable names are assigned directly as the names of the elements in the `dimnames` list.

2.9.2 Subsetting and collapsing

For many purposes, the 4-way table TV is too large and awkward to work with. Among the networks, Fox and Other occur infrequently, so we will remove them. We can also cut it down to a 3-way table by considering only viewers who persist with the current station.[8]

```
> TV <- TV[,,1:3,]        # keep only ABC, CBS, NBC
> TV <- TV[,,,3]          # keep only Persist -- now a 3 way table
> structable(TV)
```

		Time	8:00	8:15	8:30	8:45	9:00	9:15	9:30	9:45	10:00	10:15	10:30
Day	Network												
Mon	ABC		146	151	156	83	325	350	386	340	352	280	278
	CBS		337	293	304	233	311	251	241	164	252	265	272
	NBC		263	219	236	140	226	235	239	246	279	263	283
Tue	ABC		244	181	231	205	385	283	345	192	329	351	364
	CBS		173	180	184	109	218	235	256	250	274	263	261
	NBC		315	254	280	241	370	214	195	111	188	190	210
Wed	ABC		233	161	194	156	339	264	279	140	237	228	203
	CBS		158	126	207	59	98	103	122	86	109	105	110
	NBC		134	146	166	66	194	230	264	143	274	289	306
Thu	ABC		174	183	197	181	187	198	211	86	110	122	117
	CBS		196	185	195	104	106	116	116	47	102	84	84
	NBC		515	463	472	477	590	473	446	349	649	705	747
Fri	ABC		294	281	305	239	278	246	245	138	246	232	233
	CBS		130	144	154	81	129	153	136	126	138	136	152
	NBC		195	220	248	160	172	164	169	85	183	198	204

Finally, for some purposes, we might also want to collapse the 11 Time's into a smaller number. Here, we use collapse.table() (see Section 2.7.2), which was designed for this purpose.

```
> TV2 <- collapse.table(TV,
+                    Time = c(rep("8:00-8:59", 4),
+                             rep("9:00-9:59", 4),
+                             rep("10:00-10:44", 3)))
> structable(Day ~ Time + Network, TV2)
```

		Day	Mon	Tue	Wed	Thu	Fri
Time	Network						
8:00-8:59	ABC		536	861	744	735	1119
	CBS		1167	646	550	680	509
	NBC		858	1090	512	1927	823
9:00-9:59	ABC		1401	1205	1022	682	907
	CBS		967	959	409	385	544
	NBC		946	890	831	1858	590
10:00-10:44	ABC		910	1044	668	349	711
	CBS		789	798	324	270	426
	NBC		825	588	869	2101	585

Congratulations! If you followed the operations described above, you are ready for the material described in the rest of the book. If not, try working through some of exercises below.

2.10 Lab exercises

Exercise 2.1 The packages vcd and vcdExtra contain many data sets with some examples of analysis and graphical display. The goal of this exercise is to familiarize yourself with these resources.

---

[8]This relies on the fact that indexing an array drops dimensions of length 1 by default, using the argument drop = TRUE; the result is coerced to the lowest possible dimension.

You can get a brief summary of these using the function datasets() from vcdExtra. Use the following to get a list of these with some characteristics and titles.

```
> ds <- datasets(package = c("vcd", "vcdExtra"))
> str(ds, vec.len = 2)

'data.frame':   74 obs. of   5 variables:
 $ Package: chr   "vcd" "vcd" ...
 $ Item    : chr   "Arthritis" "Baseball" ...
 $ class   : chr   "data.frame" "data.frame" ...
 $ dim     : chr   "84x5" "322x25" ...
 $ Title   : chr   "Arthritis Treatment Data" "Baseball Data" ...
```

(a) How many data sets are there altogether? How many are there in each package?
(b) Make a tabular display of the frequencies by Package and class.
(c) Choose one or two data sets from this list, and examine their help files (e.g., help(Arthritis) or ?Arthritis). You can use, e.g., example(Arthritis) to run the R code for a given example.

Exercise 2.2 For each of the following data sets in the vcdExtra package, identify which are response variable(s) and which are explanatory. For factor variables, which are unordered (nominal) and which should be treated as ordered? Write a sentence or two describing substantive questions of interest for analysis of the data. (*Hint*: use data(foo, package="vcdExtra") to load, and str(foo), help(foo) to examine data set foo.)

(a) Abortion opinion data: *Abortion*
(b) Caesarian Births: *Caesar*
(c) Dayton Survey: *DaytonSurvey*
(d) Minnesota High School Graduates: *Hoyt*

Exercise 2.3 The data set *UCBAdmissions* is a 3-way table of frequencies classified by Admit, Gender, and Dept.

(a) Find the total number of cases contained in this table.
(b) For each department, find the total number of applicants.
(c) For each department, find the overall proportion of applicants who were admitted.
(d) Construct a tabular display of department (rows) and gender (columns), showing the proportion of applicants in each cell who were admitted relative to the total applicants in that cell.

Exercise 2.4 The data set *DanishWelfare* in vcd gives a 4-way, $3 \times 4 \times 3 \times 5$ table as a data frame in frequency form, containing the variable Freq and four factors, Alcohol, Income, Status, and Urban. The variable Alcohol can be considered as the response variable, and the others as possible predictors.

(a) Find the total number of cases represented in this table.
(b) In this form, the variables Alcohol and Income should arguably be considered *ordered* factors. Change them to make them ordered.
(c) Convert this data frame to table form, DanishWelfare.tab, a 4-way array containing the frequencies with appropriate variable names and level names.
(d) The variable Urban has 5 categories. Find the total frequencies in each of these. How would you collapse the table to have only two categories, City, Non-city?
(e) Use structable() or ftable() to produce a pleasing flattened display of the frequencies in the 4-way table. Choose the variables used as row and column variables to make it easier to compare levels of Alcohol across the other factors.

Exercise 2.5 The data set *UKSoccer* in vcd gives the distributions of number of goals scored by the 20 teams in the 1995/96 season of the Premier League of the UK Football Association.

```
> data("UKSoccer", package = "vcd")
> ftable(UKSoccer)

        Away  0  1  2   3  4
Home
0            27 29 10   8  2
1            59 53 14  12  4
2            28 32 14  12  4
3            19 14  7   4  1
4             7  8 10   2  0
```

This two-way table classifies all $20 \times 19 = 380$ games by the joint outcome (Home, Away), the number of goals scored by the Home and Away teams. The value 4 in this table actually represents 4 or more goals.

(a) Verify that the total number of games represented in this table is 380.
(b) Find the marginal total of the number of goals scored by each of the home and away teams.
(c) Express each of the marginal totals as proportions.
(d) Comment on the distribution of the numbers of home-team and away-team goals. Is there any evidence that home teams score more goals on average?

Exercise 2.6 The one-way frequency table *Saxony* in vcd records the frequencies of families with 0, 1, 2, ... 12 male children, among 6115 families with 12 children. This data set is used extensively in Chapter 3.

```
> data("Saxony", package = "vcd")
> Saxony

nMales
    0    1    2    3    4    5    6    7    8    9   10   11   12
    3   24  104  286  670 1033 1343 1112  829  478  181   45    7
```

Another data set, *Geissler*, in the vcdExtra package, gives the complete tabulation of all combinations of boys and girls in families with a given total number of children (size). The task here is to create an equivalent table, Saxony12 from the *Geissler* data.

```
> data("Geissler", package = "vcdExtra")
> str(Geissler)

'data.frame': 90 obs. of  4 variables:
 $ boys : int  0 0 0 0 0 0 0 0 0 0 ...
 $ girls: num  1 2 3 4 5 6 7 8 9 10 ...
 $ size : num  1 2 3 4 5 6 7 8 9 10 ...
 $ Freq : int  108719 42860 17395 7004 2839 1096 436 161 66 30 ...
```

(a) Use subset() to create a data frame, sax12 containing the *Geissler* observations in families with size==12.
(b) Select the columns for boys and Freq.
(c) Use xtabs() with a formula, Freq ~ boys, to create the one-way table.
(d) Do the same steps again to create a one-way table, Saxony11, containing similar frequencies for families of size==11.

Exercise 2.7 * *Interactive coding of table factors*: Some statistical and graphical methods for contingency tables are implemented only for two-way tables, but can be extended to 3+-way tables by recoding the factors to interactive combinations along the rows and/or columns, in a way similar to what ftable() and structable() do for printed displays.

For the *UCBAdmissions* data, produce a two-way table object, UCB.tab2, that has the combinations of Admit and Gender as the rows, and Dept as its columns, to look like the result below:

```
                   Dept
Admit:Gender        A    B    C    D    E    F
  Admitted:Female   89   17  202  131   94   24
  Admitted:Male    512  353  120  138   53   22
  Rejected:Female   19    8  391  244  299  317
  Rejected:Male    313  207  205  279  138  351
```

(a) Try this the long way: convert *UCBAdmissions* to a data frame (as.data.frame()), manipulate the factors (e.g., interaction()), then convert back to a table (as.data.frame()).

(b) Try this the short way: both ftable() and structable() have as.matrix() methods that convert their result to a matrix.

Exercise 2.8 The data set *VisualAcuity* in **vcd** gives a $4 \times 4 \times 2$ table as a frequency data frame.

```
> data("VisualAcuity", package = "vcd")
> str(VisualAcuity)

'data.frame': 32 obs. of  4 variables:
 $ Freq  : num  1520 234 117 36 266 ...
 $ right : Factor w/ 4 levels "1","2","3","4": 1 2 3 4 1 2 3 4 1 2 ...
 $ left  : Factor w/ 4 levels "1","2","3","4": 1 1 1 1 2 2 2 2 3 3 ...
 $ gender: Factor w/ 2 levels "male","female": 2 2 2 2 2 2 2 2 2 2 ...
```

(a) From this, use xtabs() to create two 4×4 frequency tables, one for each gender.

(b) Use structable() to create a nicely organized tabular display.

(c) Use xtable() to create a LaTeX or HTML table.

3

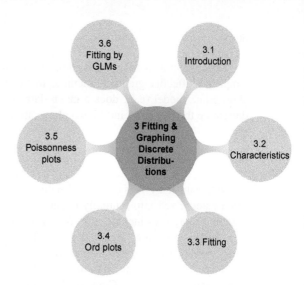

Fitting and Graphing Discrete Distributions

Discrete data often follow various theoretical probability models. Graphic displays are used to visualize goodness of fit, to diagnose an appropriate model, and determine the impact of individual observations on estimated parameters.

Not everything that counts can be counted, and not everything that can be counted counts.

Albert Einstein

Discrete frequency distributions often involve counts of occurrences of events, such as accident fatalities, incidents of terrorism or suicide, words in passages of text, or blood cells with some characteristic. Often interest is focused on how closely such data follow a particular probability distribution, such as the binomial, Poisson, or geometric distribution, which provide the basis for generating mechanisms that might give rise to the data. Understanding and visualizing such distributions in the simplest case of an unstructured sample provides a building block for generalized linear models (Chapter 11) where they serve as one component. They also provide the basis for a variety of recent extensions of regression models for count data (Chapter 11), some of which account for the excess counts of zeros (zero-inflated models) caused by left- or right-truncation often encountered in statistical practice.

This chapter describes the well-known discrete frequency distributions: the binomial, Poisson, negative binomial, geometric, and logarithmic series distributions in the simplest case of an unstructured sample. The chapter begins with simple graphical displays (line graphs and bar charts) to view the distributions of empirical data and theoretical frequencies from a specified discrete distribution.

The chapter then describes methods for fitting data to a distribution of a given form, and presents simple but effective graphical methods that can be used to visualize goodness of fit, to diagnose an appropriate model (e.g., does a given data set follow the Poisson or negative binomial?) and to determine the impact of individual observations on estimated parameters.

3.1   Introduction to discrete distributions

Discrete data analysis is concerned with the study of the tabulation of one or more types of events, often categorized into mutually exclusive and exhaustive categories. ***Binary Events*** having two outcome categories include the toss of a coin (head/tails), sex of a child (male/female), survival of a patient following surgery (lived/died), and so forth. ***Polytomous Events*** have more outcome categories, which may be *ordered* (rating of impairment: low/medium/high, by a physician) and possibly numerically valued (number of dots (pips), 1–6 on the toss of a die) or *unordered* (political party supported: Liberal, Conservative, Greens, Socialist).

In this chapter, we focus largely on one-way frequency tables for a single numerically valued variable. Probability models for such data provide the opportunity to describe or explain the *structure* in such data, in that they entail some data-generating mechanism and provide the basis for testing scientific hypotheses and predicting future results. If a given probability model does not fit the data, this can often be a further opportunity to extend understanding of the data, or the underlying substantive theory, or both.

The remainder of this section gives a few substantive examples of situations where the well-known discrete frequency distributions (binomial, Poisson, negative binomial, geometric, and log-arithmic series) might reasonably apply, at least approximately. The mathematical characteristics and properties of these theoretical distributions are postponed to Section 3.2.

In many cases, the data at hand pertain to two types of variables in a one-way frequency table. There is a basic outcome variable, k, taking integer values, $k = 0, 1, \ldots$, and called a ***count***. For each value of k, we also have a ***frequency***, n_k that the count k was observed in some sample. For example, in the study of children in families, the count variable k could be the total number of children or the number of male children; the frequency variable, n_k, would then give the number of families with that basic count k.

3.1.1   Binomial data

Binomial-type data arise as the discrete distribution of the number of "success" events in n independent binary trials, each of which yields a success (yes/no, head/tail, lives/dies, male/female) with a constant probability p.

Sometimes, as in Example 3.1 below, the available data record only the number of successes in n trials, with separate such observations recorded over time or space. More commonly, as in Example 3.2 and Example 3.3, we have available data on the frequency n_k of $k = 0, 1, 2, \ldots n$ successes in the n trials.

EXAMPLE 3.1: Arbuthnot data

Sex ratios, such as births of male to female children, have long been of interest in population studies and demography. Indeed, in 1710, John Arbuthnot (Arbuthnot, 1710) used data on the ratios of male to female christenings in London from 1629–1710 to carry out the first known significance test. The data for these 82 years showed that in *every* year there were more boys than girls. He calculated that, under the assumption that male and female births were equally likely, the probability of 82 years of more males than females was vanishingly small, ($\Pr \approx 4.14 \times 10^{-25}$). He used this to argue that a nearly constant birth ratio > 1 (or $\Pr(\text{Male}) > 0.5$) could be interpreted to show the guiding hand of a divine being.

Arbuthnot's data, along with some other related variables, are available in `Arbuthnot` in the

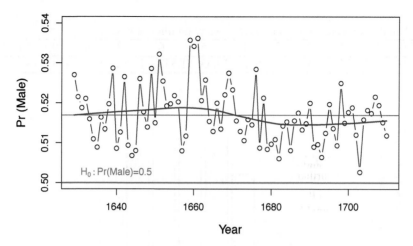

Figure 3.1: Arbuthnot's data on male/female sex ratios in London, 1629–1710, together with a (loess) smoothed curve (blue) over time and the mean Pr(Male).

HistData (Friendly, 2014a) package. For now, we simply display a plot of the probability of a male birth over time. The plot in Figure 3.1 shows the proportion of males over years, with horizontal lines at $Pr(Male) = 0.5$ and the mean, $Pr(Male) = 0.517$. Also shown is a (loess) smoothed curve, which suggests that any deviation from a constant sex ratio is relatively small, but also showed some systematic trend over time.

```
> data("Arbuthnot", package = "HistData")
> with(Arbuthnot, {
+    prob = Males / (Males + Females)
+    plot(x = Year, y = prob, type = "b",
+         ylim = c(0.5, 0.54), ylab = "Pr (Male)")
+    abline(h = 0.5, col = "red", lwd = 2)
+    abline(h = mean(prob), col = "blue")
+    lines(loess.smooth(Year, prob), col = "blue", lwd = 2)
+    text(x = 1640, y = 0.5, expression(H[0]: "Pr(Male)=0.5"),
+         pos = 3, col = "red")
+    })
```

Exercise 3.1 invites you to consider some other plots for this data. We return to this data in a later chapter where we ask whether the variation around the mean can be explained by any other considerations, or should just be considered random variation (see Exercise 7.1). △

EXAMPLE 3.2: Families in Saxony

A related example of sex ratio data that ought to follow a binomial distribution comes from a classic study by A. Geissler (1889). Geissler listed the data on the distributions of boys and girls in families in Saxony for the period 1876–1885. In total, over four million births were recorded, and the sex distribution in the family was available because the parents had to state the sex of all their children on the birth certificate.

The complete data, classified by number of boys and number of girls (each 0–12) appear in Edwards (1958, Table 1).[1] Lindsey (1995, Table 6.2) selected only the 6115 families with 12 children, and listed the frequencies by number of males. The data are shown in table form in Table 3.1 in the

---

[1] Edwards (1958) notes that over these 10 years, many parents will have had several children, and their family composition is therefore recorded more than once. However, in families with a given number of children, each family can appear only once.

Table 3.1: Number of male children in 6115 Saxony families of size 12

Males (k)	0	1	2	3	4	5	6	7	8	9	10	11	12
Families (n_k)	3	24	104	286	670	1,033	1,343	1,112	829	478	181	45	7

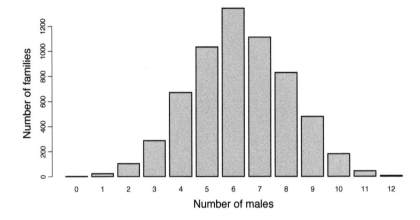

Figure 3.2: Number of males in Saxony families of size 12.

standard form of a complete discrete distribution. The basic outcome variable, $k = 0, 1, \ldots, 12$, is the number of male children in a family, and the frequency variable, n_k is the number of families with that number of boys.

Figure 3.2 shows a bar plot of the frequencies in Table 3.1. It can be seen that the distribution is quite symmetric. The questions of interest here are: (a) how close does the data follow a binomial distribution, with a constant $\Pr(\text{Male}) = p$? (b) is there evidence to reject the hypothesis that $p = 0.5$?

```
> data("Saxony", package = "vcd")
> barplot(Saxony, xlab = "Number of males", ylab = "Number of families",
+         col = "lightblue", cex.lab = 1.5)
```

$\triangle$

EXAMPLE 3.3: Weldon's dice

Common examples of binomial distributions involve tossing coins or dice, where some event outcome is considered a "success" and the number of successes (k) are tabulated in a long series of trials to give the frequency (n_k) of each basic count, k.

Perhaps the most industrious dice-tosser of all time, W. F. Raphael Weldon, an English evolutionary biologist and joint founding editor of *Biometrika* (with Francis Galton and Karl Pearson), tallied the results of throwing 12 dice 26,306 times. For his purposes, he considered the outcome of 5 or 6 pips showing on each die to be a success, and all other outcomes as failures.

Weldon reported his results in a letter to Francis Galton dated February 2, 1894, in order "to judge whether the differences between a series of group frequencies and a theoretical law ... were more than might be attributed to the chance fluctuations of random sampling" (Kemp and Kemp, 1991). In his seminal paper, Pearson (1900) used Weldon's data to illustrate the χ^2 goodness-of-fit test, as did Kendall and Stuart (1963, Table 5.1, p. 121).

These data are shown here as Table 3.2, in terms of the number of occurrences of a 5 or 6 in the throw of 12 dice. If the dice were all identical and perfectly fair (balanced), one would expect that $p = \Pr\{5 \text{ or } 6\} = \frac{1}{3}$ and the distribution of the number of 5 or 6 would be binomial.

Table 3.2: Frequencies of 5s or 6s in throws of 12 dice

# 5s or 6s (k)	0	1	2	3	4	5	6	7	8	9	10+
Frequency (n_k)	185	1,149	3,265	5,475	6,114	5,194	3,067	1,331	403	105	18

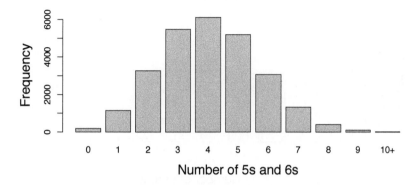

Figure 3.3: Weldon's dice data, frequency distribution of 5s and 6s in throws of 12 dice.

A peculiar feature of these data as presented by Kendall and Stuart (not uncommon in discrete distributions) is that the frequencies of 10–12 successes are lumped together.[2] This grouping must be taken into account in fitting the distribution. This dataset is available as *WeldonDice* in the vcd package. The distribution is plotted in Figure 3.3.

```
> data("WeldonDice", package = "vcd")
> dimnames(WeldonDice)$n56[11] <- "10+"
> barplot(WeldonDice, xlab = "Number of 5s and 6s", ylab = "Frequency",
+         col = "lightblue", cex.lab = 1.5)
```

△

3.1.2 Poisson data

Data of Poisson type arise when we observe the counts of events k within a fixed interval of time or space (length, area, volume) and tabulate their frequencies, n_k. For example, we may observe the number of radioactive particles emitted by a source per second or number of births per hour, or the number of tiger or whale sightings within some geographical regions.

In contrast to binomial data, where the counts are bounded below and above, in Poisson data the counts k are bounded below at 0, but can take integer values with no fixed upper limit. One defining characteristic for the Poisson distribution is for rare events, which occur independently with a small and constant probability, p, in small intervals, and we count the number of such occurrences.

Several examples of data of this general type are given below.

EXAMPLE 3.4: Death by horse kick

One of the oldest and best known examples of a Poisson distribution is the data from von Bortkiewicz (1898) on deaths of soldiers in the Prussian army from kicks by horses and mules,

---

[2]The unlumped entries are, for (number of 5s or 6s: frequency) — (10: 14); (11: 4), (12:0), given by Labby (2009). In this remarkable paper, Labby describes a mechanical device he constructed to repeat Weldon's experiment physically and automate the counting of outcomes. He created electronics to roll 12 dice in a physical box, and hooked that up to a webcam to capture an image of each toss and used image processing software to record the counts.

Table 3.3: von Bortkiewicz's data on deaths by horse kicks

Number of deaths (k)	0	1	2	3	4	Sum
Frequency (n_k)	109	65	22	3	1	200

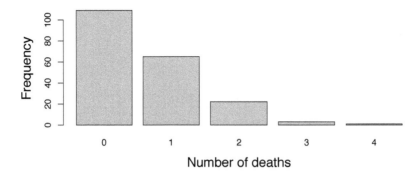

Figure 3.4: HorseKicks data, distribution of the number of deaths in 200 corps-years.

shown in Table 3.3. Ladislaus von Bortkiewicz, an economist and statistician, tabulated the number of soldiers in each of 14 army corps in the 20 years from 1875–1894 who died after being kicked by a horse (Andrews and Herzberg, 1985, p. 18). Table 3.3 shows the data used by Fisher (1925) for 10 of these army corps, summed over 20 years, giving 200 'corps-year' observations. In 109 corps-years, no deaths occurred; 65 corps-years had one death, etc.

The data set is available as *HorseKicks* in the vcd package. The distribution is plotted in Figure 3.4.

```
> data("HorseKicks", package = "vcd")
> barplot(HorseKicks, xlab = "Number of deaths", ylab = "Frequency",
+         col = "lightblue", cex.lab = 1.5)
```

△

EXAMPLE 3.5: Federalist Papers

In 1787–1788, Alexander Hamilton, John Jay, and James Madison wrote a series of newspaper essays to persuade the voters of New York State to ratify the U.S. Constitution. The essays were titled *The Federalist Papers* and all were signed with the pseudonym "Publius." Of the 77 papers published, the author(s) of 65 are known, but *both* Hamilton and Madison later claimed sole authorship of the remaining 12. Mosteller and Wallace (1963, 1984) investigated the use of statistical methods to identify authors of disputed works based on the frequency distributions of certain key function words, and concluded that Madison had indeed authored the 12 disputed papers.[3]

Table 3.4 shows the distribution of the occurrence of one of these "marker" words, the word *may* in 262 blocks of text (each about 200 words long) from issues of the *Federalist Papers* and other essays known to be written by James Madison. Read the table as follows: in 156 blocks, the word *may* did not occur; it occurred once in 63 blocks, etc. The distribution is plotted in Figure 3.5.

---

[3]It should be noted that this is a landmark work in the development of statistical methods and their application to the analysis of texts and cases of disputed authorship. In addition to *may*, they considered many such marker words, such as *any*, *by*, *from*, *upon*, and so forth. Among these, the word *upon* was the best discriminator between the works known by Hamilton (3 per 1000 words) and Madison (0.16 per 1000 words). In this work, they pioneered the use of Bayesian discriminant analysis, and the use of cross-validation to assess the stability of estimates and their conclusions.

Table 3.4: Number of occurrences of the word *may* in texts written by James Madison

Occurrences of *may* (k)	0	1	2	3	4	5	6	Sum
Blocks of text (n_k)	156	63	29	8	4	1	1	262

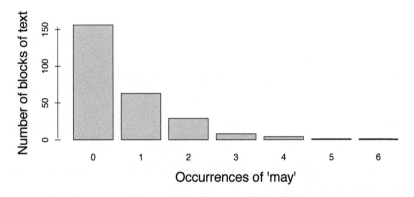

Figure 3.5: Federalist Papers data, distribution of the uses of the word *may*.

```
> data("Federalist", package = "vcd")
> barplot(Federalist,
+         xlab = "Occurrences of 'may'",
+         ylab = "Number of blocks of text",
+         col = "lightgreen", cex.lab = 1.5)
```

△

EXAMPLE 3.6: London cycling deaths

Aberdein and Spiegelhalter (2013) observed that from November 5–13, 2013, six people were killed while cycling in London. How unusual is this number of deaths in less than a two-week period? Was this a freak occurrence, or should Londoners petition for cycling lanes and greater road safety? To answer these questions, they obtained data from the UK Department of Transport *Road Safety Data* from 2005–2012 and selected all accident fatalities of cyclists within the city of London.

It seems reasonable to assume that, in any short period of time, deaths of people riding bicycles are independent events. If, in addition, the probability of such events is constant over this time span, the Poisson distribution should describe the distribution of $0, 1, 2, 3, \ldots$ deaths. Then, an answer to the main question can be given in terms of the probability of six (or more) deaths in a comparable period of time.

Their data, comprising 208 counts of deaths in the fortnightly periods from January 2005 to December 2012, are contained in the data set `CyclingDeaths` in **vcdExtra**. To work with the distribution, we first convert this to a one-way table.

```
> data("CyclingDeaths", package = "vcdExtra")
> CyclingDeaths.tab <- table(CyclingDeaths$deaths)
> CyclingDeaths.tab


  0   1   2   3
114  75  14   5
```

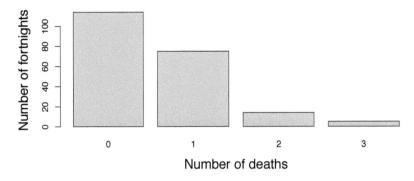

Figure 3.6: Frequencies of number of cyclist deaths in two-week periods in London, 2005–2012.

The maximum number of deaths was 3, which occurred in only 5 two-week periods. The distribution is plotted in Figure 3.6.

```
> barplot(CyclingDeaths.tab,
+         xlab = "Number of deaths", ylab = "Number of fortnights",
+         col = "pink", cex.lab = 1.5)
```

We return to this data in Example 3.10 and answer the question of how unusual six or more deaths would be in a Poisson distribution.

△

3.1.3 Type-token distributions

There are a variety of other types of discrete data distributions. One important class is ***type-token*** distributions, where the basic count k is the number of distinct types of some observed event, $k = 1, 2, \ldots$ and the frequency, n_k, is the number of different instances observed. For example, distinct words in a book, words that subjects list as members of the semantic category "fruit," musical notes that appear in a score, and species of animals caught in traps can be considered as types, and the occurrences of of those type comprise tokens.

This class differs from the Poisson type considered above in that the frequency for value $k = 0$ is *unobserved*. Thus, questions like (a) How many words did Shakespeare know? (b) How many words in the English language are members of the "fruit" category? (c) How many wolves remain in Canada's Northwest territories? depend on the unobserved count for $k = 0$. They cannot easily be answered without appeal to additional information or statistical theory.

EXAMPLE 3.7: Butterfly species in Malaya

In studies of the diversity of animal species, individuals are collected and classified by species. The distribution of the number of species (types) where $k = 1, 2, \ldots$ individuals (tokens) were collected forms a kind of type-token distribution. An early example of this kind of distribution was presented by Fisher et al. (1943). Table 3.5 lists the number of individuals of each of 501 species of butterfly collected in Malaya. There were thus 118 species for which just a single instance was found, 74 species for which two individuals were found, down to 3 species for which 24 individuals were collected. Fisher et al. note, however, that the distribution was truncated at $k = 24$. Type-token distributions are often J-shaped, with a long upper tail, as we see in Figure 3.7.

Table 3.5: Number of butterfly species n_k for which k individuals were collected

Individuals (k)	1	2	3	4	5	6	7	8	9	10	11	12	
Species (n_k)	118	74	44	24	29	22	20	19	20	15	12	14	
Individuals (k)	13	14	15	16	17	18	19	20	21	22	23	24	Sum
Species (n_k)	6	12	6	9	9	6	10	10	11	5	3	3	501

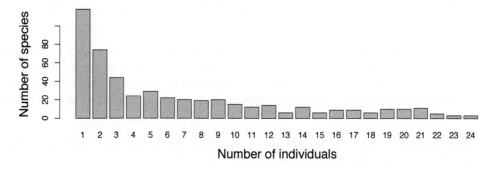

Figure 3.7: Butterfly species in Malaya.

```
> data("Butterfly", package = "vcd")
> barplot(Butterfly, xlab = "Number of individuals",
+         ylab = "Number of species", cex.lab = 1.5)
```

$\triangle$

3.2   Characteristics of discrete distributions

This section briefly reviews the characteristics of some of the important discrete distributions encountered in practice and illustrates their use with R. An overview of these distributions is shown in Table 3.6. For more detailed information on these and other discrete distributions, Johnson et al. (1992) and Wimmer and Altmann (1999) present the most comprehensive treatments; Zelterman (1999, Chapter 2) gives a compact summary.

For each distribution, we describe properties and generating mechanisms, and show how its

Table 3.6: Discrete probability distributions

Discrete distribution	Probability function, $p(k)$	Parameters
Binomial	$\binom{n}{k}p^k(1-p)^{n-k}$	$p = \Pr\,(\text{success})$; $n = \#\,\text{trials}$
Poisson	$e^{-\lambda}\lambda^k/k!$	$\lambda = \text{mean}$
Negative binomial	$\binom{n+k-1}{k}p^n(1-p)^k$	p; $n = \#\,\textit{successful}\,\text{trials}$
Geometric	$p(1-p)^k$	p
Logarithmic series	$\theta^k/[-k\log(1-\theta)]$	θ

Table 3.7: R functions for discrete probability distributions

Discrete distribution	Density (pmf) function	Cumulative (CDF)	Quantile CDF^{-1}	Random # generator
Binomial	dbinom()	pbinom()	qbinom()	rbinom()
Poisson	dpois()	ppois()	qpois()	rpois()
Negative binomial	dnbinom()	pnbinom()	qnbinom()	rnbinom()
Geometric	dgeom()	pgeom()	qgeom()	rgeom()
Logarithmic series	dlogseries()	plogseries()	qlogseries()	rlogseries()

parameters can be estimated and how to plot the frequency distribution. R has a wealth of functions for a wide variety of distributions. For ease of reference, their names and types for the distributions covered here are shown in Table 3.7. The naming scheme is simple and easy to remember: for each distribution, there are functions, with a prefix letter, d, p, q, r, followed by the name for that class of distribution:[4]

d a density function,[5] $\Pr\{X = k\} \equiv p(k)$ for the probability that the variable X takes the value k.
p a cumulative probability/density function, or CDF, $F(k) = \sum_{X \le k} p(k)$.
q a quantile function, the inverse of the CDF, $k = F^{-1}(p)$. The quantile is defined as the smallest value x such that $F(k) \ge p$.
r a random number generating function for that distribution.

In the R console, help(Distributions) gives an overview listing of the distribution functions available in the stats package.

3.2.1 The binomial distribution

The binomial distribution, $\text{Bin}(n, p)$, arises as the distribution of the number k of events of interest that occur in n independent trials when the probability of the event on any one trial is the constant value $p = \Pr(\text{event})$. For example, if 15% of the population has red hair, the number of redheads in randomly sampled groups of $n = 10$ might follow a binomial distribution, $\text{Bin}(10, 0.15)$; in Weldon's dice data (Example 3.3), the probability of a 5 or 6 should be $\frac{1}{3}$ on any one trial, and the number of 5s or 6s in tosses of 12 dice would follow $\text{Bin}(12, \frac{1}{3})$.

Over n independent trials, the number of events k may range from 0 to n; if X is a random variable with a binomial distribution, the probability that $X = k$ is given by

$$\text{Bin}(n, p) : \Pr\{X = k\} \equiv p(k) = \binom{n}{k} p^k (1 - p)^{n-k} \qquad k = 0, 1, \dots, n, \qquad (3.1)$$

where $\binom{n}{k} = n!/k!(n - k)!$ is the number of ways of choosing k out of n.

The first three (central) moments of the binomial distribution are as follows (letting $q = 1 - p$),

$$\begin{aligned} \text{Mean}(X) &= np \\ \text{Var}(X) &= npq \\ \text{Skew}(X) &= npq(q - p) . \end{aligned}$$

---

[4]The CRAN Task View on Probability Distributions, http://cran.r-project.org/web/views/Distributions.html, provides a general overview and lists a wide variety of contributed packages for specialized distributions, discrete and continuous.

[5]For discrete random variables this is usually called the probability mass function (pmf).

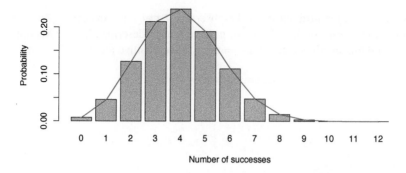

Figure 3.8: Binomial distribution for $k = 0, \ldots, 12$ successes in 12 trials and p=1/3.

It is easy to verify that the binomial distribution has its maximum variance when $p = \frac{1}{2}$. It is symmetric (Skew(x)=0) when $p = \frac{1}{2}$, and negatively (positively) skewed when $p < \frac{1}{2}$ ($p > \frac{1}{2}$).

If we are given data in the form of a discrete (binomial) distribution (and n is known), then the maximum likelihood estimator[6] of p can be obtained as the weighted mean of the values k with weights n_k,

$$\hat{p} = \frac{\bar{x}}{n} = \frac{(\sum_k k \times n_k)/\sum_k n_k}{n} ,$$

and has sampling variance $\mathcal{V}(\hat{p}) = pq/n$.

3.2.1.1 Calculation and visualization

As indicated in Table 3.7 (but without listing the parameters of these functions), binomial probabilities can be calculated with `dbinom(x, n, p)`, where x is a vector of the number of successes in n trials and p is the probability of success on any one trial. Cumulative probabilities, summed to a vector of quantiles, Q, can be calculated with `pbinom(Q, n, p)`, and the quantiles (the smallest value x such that $F(x) \geq P$) with `qbinom(P, n, p)`. To generate N random observations from a binomial distribution with n trials and success probability p use `rbinom(N, n, p)`[7].

For example, to find and plot the binomial probabilities corresponding to Weldon's tosses of 12 dice, with $k = 0, \ldots, 12$ and $p = \frac{1}{3}$, we could do the following (giving Figure 3.8):

```
> k <- 0 : 12
> Pk <- dbinom(k, 12, 1/3)
> b <- barplot(Pk, names.arg = k,
+             xlab = "Number of successes", ylab = "Probability")
> lines(x = b, y = Pk, col = "red")
```

We illustrate other styles for plotting in Section 3.2.2, Example 3.11 below.

EXAMPLE 3.8: Weldon's dice

Going a bit further, we can compare Weldon's data with the theoretical binomial distribution as shown below. Because the `WeldonDice` data collapsed the frequencies for 10–12 successes as 10+, we do the same with the binomial probabilities. The expected frequencies (`Exp`), if Weldon's dice tosses obeyed the binomial distribution, are calculated as $N \times p(k)$ for $N = 26,306$ tosses.

---

[6]For the purpose of this book, we assume at least a basic familiarity with the idea of maximum likelihood estimation. A useful brief introduction to this topic for binomial data is Fox (2015, Section D.6), available online.

[7]Note that the actual R function arguments differ from the ones used here.

In addition, we compute the differences of the observed (Freq) and expected (Exp) frequencies as column Diff, to be used for the χ^2 test for goodness of fit described later in Section 3.3, but a glance reveals that these are all negative for $k = 0, \ldots 4$ and positive thereafter.

```
> Weldon_df <- as.data.frame(WeldonDice) # convert to data frame
>
> k <- 0 : 12                            # same as seq(0, 12)
> Pk <- dbinom(k, 12, 1/3)               # binomial probabilities
> Pk <- c(Pk[1:10], sum(Pk[11:13]))      # sum values for 10+
> Exp <- round(26306 * Pk, 5)            # expected frequencies
> Diff <- Weldon_df$Freq - Exp           # raw residuals
> Chisq <- Diff^2 / Exp
> data.frame(Weldon_df, Prob = round(Pk, 5), Exp, Diff, Chisq)

      n56 Freq    Prob       Exp      Diff   Chisq
1       0  185 0.00771  202.749  -17.7495 1.55386
2       1 1149 0.04624 1216.497  -67.4968 3.74503
3       2 3265 0.12717 3345.366  -80.3661 1.93064
4       3 5475 0.21195 5575.610 -100.6102 1.81548
5       4 6114 0.23845 6272.561 -158.5614 4.00821
6       5 5194 0.19076 5018.049  175.9509 6.16947
7       6 3067 0.11127 2927.195  139.8047 6.67716
8       7 1331 0.04769 1254.512   76.4877 4.66346
9       8  403 0.01490  392.035   10.9649 0.30668
10      9  105 0.00331   87.119   17.8811 3.67008
11   10+   18 0.00054   14.305    3.6947 0.95424
```

$\triangle$

Finally, we can use programming features in R to calculate and plot probabilities for binomial distributions over a range of both k and p as follows, for the purposes of graphing the distributions as one or both varies. The following code uses outer() to create a 13×4 matrix Prob containing the result of dbinom() for all combinations of k = 0:12 and p = c(1/6, 1/3, 1/2, 2/3). These values are then supplied as arguments to dbinom().

```
> p <- c(1/6, 1/3, 1/2, 2/3)
> k <- 0 : 12
> Prob <- outer(k, p, function(k, p) dbinom(k, 12, p))
> str(Prob)

 num [1:13, 1:4] 0.1122 0.2692 0.2961 0.1974 0.0888 ...
```

In this form, each column of Prob can be most easily plotted against k using matplot(). The following code generates Figure 3.9.

```
> col <- palette()[2:5]
> matplot(k, Prob,
+         type = "o", pch = 15 : 17, col = col, lty = 1,
+         xlab = "Number of Successes", ylab = "Probability")
> legend("topright", legend = c("1/6","1/3","1/2","2/3"),
+        pch = 15 : 17, lty = 1, col = col, title = "Pr(Success)")
```

3.2.2 The Poisson distribution

The Poisson distribution gives the probability of an event occurring $k = 0, 1, 2, \ldots$ times over a large number of independent "trials," when the probability, p, that the event occurs on any one trial (in time or space) is small and constant. Hence, the Poisson distribution is usually applied to the study of rare events such as highway accidents at a particular location, deaths from horse kicks,

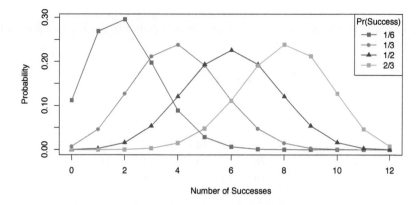

Figure 3.9: Binomial distributions for $k = 0, \ldots, 12$ successes in $n = 12$ trials, and four values of p.

or defects in a well-controlled manufacturing process. Other applications include: the number of customers contacting a call center per unit time; the number of insurance claims per unit region or unit time; number of particles emitted from a small radioactive sample.

For the Poisson distribution, the probability function is

$$\text{Pois}(\lambda) : \Pr\{X = k\} \equiv p(k) = \frac{e^{-\lambda} \lambda^k}{k!} \qquad k = 0, 1, \ldots \tag{3.2}$$

where the rate parameter, λ (> 0), turns out to be the mean of the distribution. The first three (central) moments of the Poisson distribution are:

$$
\begin{aligned}
\text{Mean}(X) &= \lambda \\
\text{Var}(X) &= \lambda \\
\text{Skew}(X) &= \lambda^{-1/2}
\end{aligned}
$$

So, the mean and variance of the Poisson distribution are always the same, which is sometimes used to identify a distribution as Poisson. For the binomial distribution, the mean (Np) is always greater than the variance (Npq); for other distributions (negative binomial and geometric) the mean is less than the variance. The Poisson distribution is always positively skewed, but skewness decreases as λ increases.

The maximum likelihood estimator of the parameter λ in Eqn. (3.2) is just the mean of the distribution,

$$\hat{\lambda} = \bar{x} = \frac{\sum_k k \, n_k}{\sum_k n_k} \, . \tag{3.3}$$

Hence, the expected frequencies can be estimated by substituting the sample mean into Eqn. (3.2) and multiplying by the total sample size N.

There are many useful properties of the Poisson distribution.[8] Among these are:

- Poisson variables have a nice reproductive property: if $X_1, X_2, \ldots X_m$ are independent Poisson variables with the same parameter λ, then their sum, $\sum X_i$ is a Poisson variate with parameter $m\lambda$; if the Poisson parameters differ, the sum is still Poisson with parameter $\sum \lambda_i$.
- For two or more independent Poisson variables, $X_1 \sim \text{Pois}(\lambda_1), X_2 \sim \text{Pois}(\lambda_2), \ldots$, with rate parameters $\lambda_1, \lambda_2 \ldots$, the distribution of any X_i, *conditional on their sum*, $\sum_j X_j = n$, is binomial, $\text{Bin}(n, p)$, where $p = \lambda_i / \sum_j \lambda_j$.

---

[8]See: http://en.wikipedia.org/wiki/Poisson_distribution.

Table 3.8: Goals scored by home and away teams in 380 games in the Premier Football League, 1995/96 season

Home Team Goals	Away Team Goals					Total
	0	1	2	3	4+	
0	27	29	10	8	2	76
1	59	53	14	12	4	142
2	28	32	14	12	4	90
3	19	14	7	4	1	45
4+	7	8	10	2	0	27
Total	140	136	55	38	11	380

- As λ increases, the Poisson distribution becomes increasingly symmetric, and approaches the normal distribution $N(\lambda, \lambda)$ with mean and variance λ as $\lambda \to \infty$. The approximation is quite good with $\lambda > 20$.
- If $X \sim \text{Pois}(\lambda)$, then $\sqrt{X}$ converges much faster to a normal distribution $N(\lambda, \frac{1}{4})$, with mean $\sqrt{\lambda}$ and constant variance $\frac{1}{4}$. Hence, the square root transformation is often recommended as a *variance stabilizing* transformation for count data when classical methods (ANOVA, regression) assuming normality are employed.

EXAMPLE 3.9: UK soccer scores

Table 3.8 gives the distributions of goals scored by the 20 teams in the 1995/96 season of the Premier League of the UK Football Association as presented originally by Lee (1997), and now available as the two-way table *UKSoccer* in the vcd package. Over a season each team plays each other team exactly once, so there are a total of $20 \times 19 = 380$ games. Because there may be an advantage for the home team, the goals scored have been classified as "home team" goals and "away team" goals in the table. Of interest for this example is whether the number of goals scored by home teams and away teams follow Poisson distributions, and how this relates to the distribution of the total number of goals scored.

If we assume that in any small interval of time there is a small, constant probability that the home team or the away team may score a goal, the distributions of the goals scored by home teams (the row totals in Table 3.8) may be modeled as $\text{Pois}(\lambda_H)$ and the distribution of the goals scored by away teams (the column totals) may be modeled as $\text{Pois}(\lambda_A)$.

If the number of goals scored by the home and away teams are independent[9], we would expect that the total number of goals scored in any game would be distributed as $\text{Pois}(\lambda_H + \lambda_A)$. These totals are shown in Table 3.9.

Table 3.9: Total goals scored in 380 games in the Premier Football League, 1995/95 season

Total goals	0	1	2	3	4	5	6	7
Number of games	27	88	91	73	49	31	18	3

As a preliminary check of the distributions for the home and away goals, we can determine if the

---

[9]This question is examined visually in Chapter 5 (Example 5.5) and Chapter 6 (Example 6.11), where we find that the answer is "basically, yes."

means and variances are reasonably close to each other. If so, then the total goals variable should also have a mean and variance equal to the sum of those statistics for the home and away goals.

In the R code below, we first convert the two-way frequency table *UKSoccer* to a data frame in frequency form. We use `within()` to convert Home and Away to numeric variables, and calculate `Total` as their sum.

```
> data("UKSoccer", package = "vcd")
>
> soccer.df <- as.data.frame(UKSoccer, stringsAsFactors = FALSE)
> soccer.df <- within(soccer.df, {
+    Home <- as.numeric(Home)        # make numeric
+    Away <- as.numeric(Away)        # make numeric
+    Total <- Home + Away            # total goals
+ })
> str(soccer.df)

'data.frame':   25 obs. of  4 variables:
 $ Home : num  0 1 2 3 4 0 1 2 3 4 ...
 $ Away : num  0 0 0 0 0 1 1 1 1 1 ...
 $ Freq : num  27 59 28 19 7 29 53 32 14 8 ...
 $ Total: num  0 1 2 3 4 1 2 3 4 5 ...
```

To calculate the mean and variance of these variables, first expand the data frame to 380 individual observations using `expand.dft()`. Then use `apply()` over the rows to calculate the mean and variance in each column.

```
> soccer.df <- expand.dft(soccer.df)     # expand to ungrouped form
> apply(soccer.df, 2, FUN = function(x)  c(mean = mean(x), var = var(x)))

        Home    Away   Total
mean  1.4868  1.0632  2.5500
var   1.3164  1.1728  2.6175
```

The means are all approximately equal to the corresponding variances. More to the point, the variance of the `Total` score is approximately equal to the sum of the individual variances. Note also there does appear to be an advantage for the home team, of nearly half a goal.

$\triangle$

EXAMPLE 3.10: London cycling deaths

A quick check of whether the number of deaths among London cyclists follows the Poisson distribution can be carried out by calculating the mean and variance. The ***index of dispersion***, the ratio of the variance to the mean, is commonly used to quantify whether a set of observed frequencies is more or less dispersed than a reference (Poisson) distribution.

```
> with(CyclingDeaths, c(mean = mean(deaths),
+                       var = var(deaths),
+                       ratio = mean(deaths) / var(deaths)))

   mean      var     ratio
0.56731  0.52685  1.07679
```

Thus, there was an average of about 0.57 deaths per fortnight, or a bit more than 1 per month, and no evidence for over- or underdispersion.

We can now answer the question of whether it was an extraordinary event to observe six deaths in a two-week period, by calculating the probability of more than 5 deaths using `ppois()`.

```
> mean.deaths <- mean(CyclingDeaths$deaths)
> ppois(5, mean.deaths, lower.tail = FALSE)

[1] 2.8543e-05
```

This probability is extremely small, so we conclude that the occurrence of six deaths was a singular event. The interpretation of this result might indicate an increased risk to cycling in London, and might prompt further study of road safety.   △

3.2.2.1 Calculation and visualization

For the Poisson distribution, you can generate probabilities using `dpois(x, lambda)` for the numbers of events in x with rate parameter `lambda`. As we did earlier for the binomial distribution, we can calculate these for a collection of values of `lambda` by using `expand.grid()` to create all combinations of the values of x we wish to plot.

EXAMPLE 3.11: Plotting styles for discrete distributions

In this example, we illustrate some additional styles for plotting discrete distributions, using both **lattice** `xyplot()` and the **ggplot2** package. The goal here is to visualize a collection of Poisson distributions for varying values of λ.

We first create the 63 combinations of x = 0:20 for three values of λ, `lambda = c(1, 4, 10)`, and use these columns as arguments to `dpois()`. Again, `lambda` is a numeric variable, but the plotting methods are easier if this variable is converted to a factor.

```
> KL <- expand.grid(k = 0 : 20, lambda = c(1, 4, 10))
> pois_df <- data.frame(KL, prob = dpois(KL$k, KL$lambda))
> pois_df$lambda = factor(pois_df$lambda)
> str(pois_df)

'data.frame': 63 obs. of  3 variables:
 $ k      : int  0 1 2 3 4 5 6 7 8 9 ...
 $ lambda: Factor w/ 3 levels "1","4","10": 1 1 1 1 1 1 1 1 1 1 ...
 $ prob   : num  0.3679 0.3679 0.1839 0.0613 0.0153 ...
```

Discrete distributions are often plotted as bar charts or in histogram-like form, as we did for the examples in Section 3.1, rather than the line-graph form used for the binomial distribution in Figure 3.9. With `xyplot()`, the plot style is controlled by the `type` argument, and the code below uses `type = c("h", "p")` to get *both* histogram-like lines to the origin and points. As well, the plot formula, `prob ~ k | lambda` instructs `xyplot()` to produce a multi-panel plot, conditioned on values of `lambda`. These lines produce Figure 3.10.

```
> library(lattice)
> xyplot(prob ~ k | lambda, data = pois_df,
+    type = c("h", "p"), pch = 16, lwd = 4, cex = 1.25, layout = c(3, 1),
+    xlab = list("Number of events (k)", cex = 1.25),
+    ylab = list("Probability", cex = 1.25))
```

The line-graph plot style of Figure 3.9 has the advantage that it is easier to compare the separate distributions in a single plot (using the `groups` argument) than across multiple panels (using a conditioning formula). It has the disadvantages that (a) a proper legend is difficult to construct with **lattice**, and (b) is difficult to read, because you have to visually coordinate the curves in the plot with the values shown in the legend. Figure 3.11 solves both problems using the **directlabels** package.

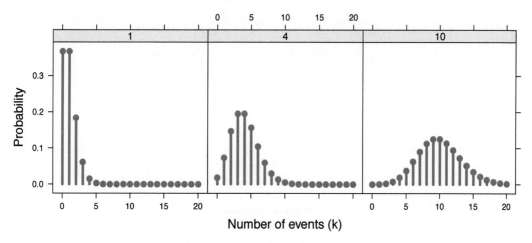

Figure 3.10: Poisson distributions for λ = 1, 4, 10, in a multi-panel display.

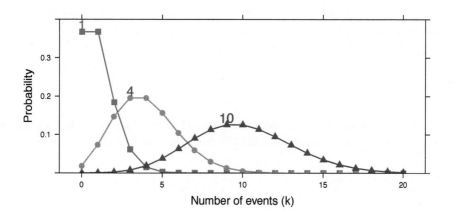

Figure 3.11: Poisson distributions for λ = 1, 4, 10, using direct labels.

```
> mycol <- palette()[2:4]
> plt <- xyplot(prob ~ k, data = pois_df, groups = lambda,
+    type = "b", pch = 15 : 17, lwd = 2, cex = 1.25, col = mycol,
+    xlab = list("Number of events (k)", cex = 1.25),
+    ylab = list("Probability",  cex = 1.25),
+    ylim = c(0, 0.4))
>
> library(directlabels)
> direct.label(plt, list("top.points", cex = 1.5, dl.trans(y = y + 0.1)))
```

Note that the plot constructed by xyplot() is saved as a ("trellis") object, plt. The function direct.label() massages this to add the labels directly to each curve. In the second argument above, "top.points" says to locate these at the maximum value on each curve.

Finally, we illustrate the use of **ggplot2** to produce a single-panel, multi-line plot of these distributions. The basic plot uses aes(x = k, y = prob, ...) to produce a plot of prob vs. k, assigning color and shape attributes to the values of lambda.

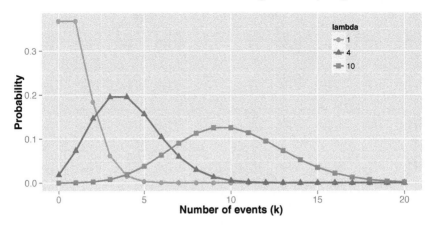

Figure 3.12: Poisson distributions for $\lambda = 1, 4, 10$, using ggplot2.

```
> library(ggplot2)
> gplt <- ggplot(pois_df,
+           aes(x = k, y = prob, colour = lambda, shape = lambda)) +
+     geom_line(size = 1) + geom_point(size = 3) +
+     xlab("Number of events (k)") +
+     ylab("Probability")
```

ggplot2 allows most details of the plot to be modified using theme(). Here we use this to move the legend inside the plot, and enlarge the axis labels and titles (producing Figure 3.12).

```
> gplt + theme(legend.position = c(0.8, 0.8)) +   # manually move legend
+           theme(axis.text = element_text(size = 12),
+                 axis.title = element_text(size = 14, face = "bold"))
```

△

3.2.3 The negative binomial distribution

The negative binomial distribution is a type of waiting-time distribution, but also arises in statistical applications as a generalization of the Poisson distribution, allowing for ***overdispersion*** (variance > mean). See Hilbe (2011) for a comprehensive treatment of negative binomial statistical models with many applications in R.

One form of the negative binomial distribution (also called the ***Pascal distribution***) arises when a series of independent Bernoulli trials is observed with constant probability p of some event, and we ask how many non-events (failures), k, it takes to observe n successful events. For example, in tossing one die repeatedly, we may consider the outcome "1" as a "success" (with $p = \frac{1}{6}$) and ask about the probability of observing $k = 0, 1, 2, \ldots$ failures before getting $n = 3$ 1s.

The probability function with parameters n (a positive integer, $0 < n < \infty$) and p ($0 < p < 1$) gives the probability that k non-events (failures) are observed before the n-th event (success), and can be written[10]

$$\text{NBin}(n, p) : \Pr\{X = k\} \equiv p(k) = \binom{n + k - 1}{k} p^n (1 - p)^k \qquad k = 0, 1, \ldots, \infty \qquad (3.4)$$

This formulation makes clear that a given sequence of events involves a total of $n + k$ trials, of which there are n successes, with probability p^n, and k are failures, with probability $(1 - p)^k$. The binomial coefficient $\binom{n+k-1}{k}$ gives the number of ways to choose the k successes from the remaining $n + k - 1$ trials preceding the last success.

The first three central moments of the negative binomial distribution are:

$$\begin{aligned} \text{Mean}(X) &= nq/p = \mu \\ \text{Var}(X) &= nq/p^2 \\ \text{Skew}(X) &= \frac{2 - p}{\sqrt{nq}}, \end{aligned}$$

where $q = 1 - p$. The variance of X is therefore greater than the mean, and the distribution is always positively skewed.

A more general form of the negative binomial distribution (the **Polya distribution**) allows n to take non-integer values and to be an unknown parameter. In this case, the combinatorial coefficient, $\binom{n+k-1}{k}$ in Eqn. (3.4), is calculated using the gamma function, $\Gamma(\bullet)$, a generalization of the factorial for non-integer values, defined so that $\Gamma(x + 1) = x!$ when x is an integer.

Then the probability function Eqn. (3.4) becomes

$$\Pr\{X = k\} \equiv p(k) = \frac{\Gamma(n + k)}{\Gamma(n)\Gamma(k + 1)} p^n (1 - p)^k \qquad k = 0, 1, \ldots, \infty. \qquad (3.5)$$

Greenwood and Yule (1920) developed the negative binomial distribution as a model for accident proneness or susceptibility of individuals to repeated attacks of disease. They assumed that for any individual, i, the number of accidents or disease occurrences has a Poisson distribution with parameter λ_i. If individuals vary in proneness, so that the λ_i have a gamma distribution, the resulting distribution is the negative binomial.

In this form, the negative binomial distribution is frequently used as an alternative to the Poisson distribution when the assumptions of the Poisson (constant probability and independence) are not satisfied, or when the variance of the distribution is greater than the mean (overdispersion). This gives rise to an alternative parameterization in terms of the mean (μ) of the distribution and its relation to the variance. From the relation of the mean and variance to the parameters n, p given above,

$$\text{Mean}(X) = \mu = \frac{n(1 - p)}{p} \quad \Longrightarrow \quad p = \frac{n}{n + \mu}, \qquad (3.6)$$

$$\text{Var}(X) = \frac{n(1 - p)}{p^2} \quad \Longrightarrow \quad \text{Var}(X) = \mu + \frac{\mu^2}{n}. \qquad (3.7)$$

This formulation allows the variance of the distribution to exceed the mean, and in these terms, the "size" parameter n is called the **dispersion parameter**.[11] Increasing this parameter corresponds to less heterogeneity, variance closer to the mean, and therefore greater applicability of the Poisson distribution.

---

[10]There are a variety of other parameterizations of the negative binomial distribution, but all of these can be converted to the form shown here, which is relatively standard, and consistent with R. They differ in whether the parameter n relates to the number of successes or the total number of trials, and whether the stopping criterion is defined in terms of failures or successes. See: http://en.wikipedia.org/wiki/Negative_binomial_distribution for details on these variations.

[11]Other terms are "shape parameter," with reference to the mixing distribution of Poissons with varying λ, "heterogeneity parameter," or "aggregation parameter."

3.2.3.1   Calculation and visualization

In R, the density (pmf), distribution (CDF), quantile, and random number functions for the negative binomial distribution are a bit special, in that the parameterization can be specified using either (n, p) or (n, μ) forms, where $\mu = n(1 - p)/p$. In our notation, probabilities can be calculated using dnbinom() using the call dbinom(k, n, p) or the call dbinom(k, n, mu=), as illustrated below:

```
> k <- 2
> n <- 2 : 4
> p <- 0.2
> dnbinom(k, n, p)

[1] 0.07680 0.03072 0.01024

> (mu <- n * (1 - p) / p)

[1]  8 12 16

> dnbinom(k, n, mu = mu)

[1] 0.07680 0.03072 0.01024
```

Thus, for the distribution with k = 2 failures and n = 2:4 successes with probability p = 0.2, the values n = 2:4 correspond to means $\mu = 8, 12, 16$ as shown above.

As before, we can calculate these probabilities for a range of the combinations of arguments using expand.grid(). In the example below, we allow three values for each of n and p and calculate all probabilities for all values of k from 0 to 20. The result, nbin_df, is like a 3-way, $21 \times 3 \times 3$ array of prob values, but in data frame format.

```
> XN <- expand.grid(k = 0 : 20, n = c(2, 4, 6), p = c(0.2, 0.3, 0.4))
> nbin_df <- data.frame(XN, prob = dnbinom(XN$k, XN$n, XN$p))
> nbin_df$n <- factor(nbin_df$n)
> nbin_df$p <- factor(nbin_df$p)
> str(nbin_df)

'data.frame': 189 obs. of  4 variables:
 $ k   : int  0 1 2 3 4 5 6 7 8 9 ...
 $ n   : Factor w/ 3 levels "2","4","6": 1 1 1 1 1 1 1 1 1 1 ...
 $ p   : Factor w/ 3 levels "0.2","0.3","0.4": 1 1 1 1 1 1 1 1 1 1 ...
 $ prob: num  0.04 0.064 0.0768 0.0819 0.0819 ...
```

With 9 combinations of the parameters, it is most convenient to plot these in separate panels, in a 3×3 display, as in Figure 3.13. The formula prob ~ k | n + p in the call to xyplot() constructs plots of prob vs. k conditioned on the combinations of n and p.

```
> xyplot(prob ~ k | n + p, data = nbin_df,
+     xlab = list("Number of failures (k)", cex = 1.25),
+     ylab = list("Probability",   cex = 1.25),
+     type = c("h", "p"), pch = 16, lwd = 2,
+     strip = strip.custom(strip.names = TRUE)
+     )
```

It can be readily seen that the mean increases from left to right with n, and increases from top to bottom with decreasing p. For these distributions, we can also calculate the theory-implied means, μ, across the entire distributions, $k = 0, 1, \ldots \infty$, as shown below.

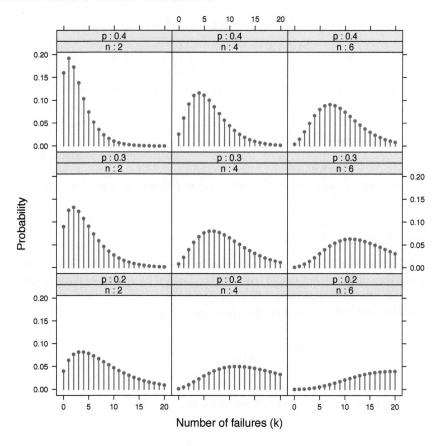

Figure 3.13: Negative binomial distributions for $n = 2, 4, 6$ and $p = 0.2, 0.3, 0.4$, using xyplot().

```
> n <- c(2, 4, 6)
> p <- c(0.2, 0.3, 0.4)
> NP <- outer(n, p, function(n, p) n * (1 - p) / p)
> dimnames(NP) <- list(n = n, p = p)
> NP

    p
n    0.2      0.3 0.4
  2    8   4.6667   3
  4   16   9.3333   6
  6   24  14.0000   9
```

3.2.4 The geometric distribution

The special case of the negative binomial distribution when $n = 1$ is a geometric distribution. We observe a series of independent trials and count the number of non-events (failures) preceding the first successful event. The probability that there will be k failures before the first success is given by

$$\text{Geom}(p) : \Pr\{X = k\} \equiv p(k) = p(1 - p)^k \qquad k = 0, 1, \dots . \tag{3.8}$$

For this distribution the central moments are:

$$\text{Mean}(X) \;=\; 1/p$$

$$\text{Var}(X) = (1-p)/p^2$$
$$\text{Skew}(X) = (2-p)/\sqrt{1-p}$$

Note that estimation of the parameter p for the geometric distribution can be handled as the special case of the negative binomial by fixing $n = 1$, so no special software is needed. Going the other way, if $X_1, X_2, \ldots X_n$ are independent geometrically distributed as $\text{Geom}(p)$, then their sum, $Y = \sum_j^n X_j$ is distributed as $\text{NBin}(p, n)$.

In R, the standard set of functions for the geometric distribution are available as `dgeom(x, prob)`, `pgeom(q, prob)`, `qgeom(p, prob)`, and `rgeom(n, prob)`, where `prob` represents p here. Visualization of the geometric distribution follows the pattern used earlier for other discrete distributions.

3.2.5 The logarithmic series distribution

The logarithmic series distribution is a long-tailed distribution introduced by Fisher et al. (1943) in connection with data on the abundance of individuals classified by species of the type shown for the distribution of butterfly species in Table 3.5.

The probability distribution function with parameter p is given by

$$\text{LogSer}(p) : \Pr\{X = k\} \equiv p(k) = \frac{p^k}{-(k \log(1-p))} = \alpha \, p^k / k \qquad k = 1, 2, \ldots, \infty, \qquad (3.9)$$

where $\alpha = -1/\log(1-p)$ and $0 < p < 1$.

For this distribution, the first two central moments are:

$$\text{Mean}(X) = \alpha \left(\frac{p}{1-p} \right),$$
$$\text{Var}(X) = -p \frac{p + \log(1-p)}{(1-p)^2 \log^2(1-p)}.$$

Fisher derived the logarithmic series distribution by assuming that for a given species the number of individuals trapped has a Poisson distribution with parameter $\lambda = \gamma t$, where γ is a parameter of the species (susceptibility to entrapment) and t is a parameter of the trap. If different species vary so that the parameter γ has a gamma distribution, then the number of representatives of each species trapped will have a negative binomial distribution. However, the observed distribution is necessarily truncated on the left, because one cannot observe the number of species never caught (where $k = 0$). The logarithmic series distribution thus arises as a limiting form of the zero-truncated negative binomial.

Maximum likelihood estimation of the parameter p in the log-series distribution is described by Böhning (1983), extending a simpler Newton's method approximation by Birch (1963a). The vcdExtra package contains the set of R functions, `dlogseries(x, prob)`, `plogseries(q, prob)`, `qlogseries(p, prob)`, and `rlogseries(n, prob)`, where `prob` represents p here.

3.2.6 Power series family

We mentioned earlier that the Poisson distribution was unique among all discrete (one-parameter) distributions, in that it is the only one whose mean and variance are equal (Kosambi, 1949). The relation between mean and variance of discrete distributions also provides the basis for integrating them into a general family. All of the discrete distributions described in this section are in fact

Table 3.10: The Power Series family of discrete distributions

Discrete Distributiion	Probability function, $p(k)$	Series parameter, θ	Series function, $f(\theta)$	Series coefficient, $a(k)$
Poisson	$e^{-\lambda}\lambda^k/k!$	$\theta = \lambda$	e^θ	$1/k!$
Binomial	$\binom{n}{k}p^k(1-p)^{n-k}$	$\theta = p/(1-p)$	$(1+\theta)^n$	$\binom{n}{k}$
Negative binomial	$\binom{n+k-1}{k}p^n(1-p)^k$	$\theta = (1-p)$	$(1-\theta)^{-k}$	$\binom{n+k-1}{k}$
Geometric	$p(1-p)^k$	$\theta = (1-p)$	$(1-\theta)^{-k}$	1
Logarithmic series	$\theta^k/[-k\log(1-\theta)]$	$\theta = \theta$	$-\log(1-\theta)$	$1/k$

special cases of a family of discrete distributions called the power series distributions by Noack (1950) and defined by

$$p(k) = a(k)\theta^k/f(\theta) \qquad k = 0, 1, \ldots,$$

with parameter $\theta > 0$, where $a(k)$ is a coefficient function depending only on k and $f(\theta) = \sum_k a(k)\theta^k$ is called the series function. The definitions of these functions are shown in Table 3.10. These relations among the discrete distribution provide the basis for graphical techniques for diagnosing the form of discrete data described later in this chapter (Section 3.5.4).

3.3 Fitting discrete distributions

In applications to discrete data such as the examples in Section 3.1, interest is often focused on how closely such data follow a particular distribution, such as the Poisson, binomial, or geometric distribution. A close fit provides for interpretation in terms of the underlying mechanism for the distribution; conversely, a bad fit can suggest the possibility for improvement by relaxing one or more of the assumptions. We examine more detailed and nuanced methods for diagnosing and testing discrete distributions in Section 3.4 and Section 3.5 below.

Fitting a discrete distribution involves three basic steps:

1. Estimating the parameter(s) of the distribution from the data; for example, p for the binomial, λ for the Poisson, n and p for the negative binomial. Typically, this is carried out by maximum likelihood methods, or a simpler method of moments, which equates sample moments (mean, variance, skewness) to those of the theoretical distribution, and solves for the parameter estimates. These methods are illustrated in Section 3.3.1.
2. From this, we can calculate the fitted probabilities, $\hat{p}_k$ that apply for the given distribution, or equivalently, the model expected frequencies, $N\hat{p}_k$, where N is the total sample size.
3. Finally, we can calculate goodness-of-fit tests measuring the departure between the observed and fitted frequencies.

Often goodness-of-fit is examined with a classical (Pearson) ***goodness-of-fit*** (GOF) chi-squared test,

$$X^2 = \sum_{k=1}^{K} \frac{(n_k - N\hat{p}_k)^2}{N\hat{p}_k} \sim \chi^2_{(K-s-1)}, \tag{3.10}$$

where there are K frequency classes, s parameters have been estimated from the data, and $\hat{p}_k$ is the estimated probability of each basic count, under the null hypothesis that the data follows the chosen distribution.

An alternative test statistic is the likelihood-ratio G^2 statistic,

$$G^2 = \sum_{k=1}^{K} n_k \log(n_k/N\hat{p}_k), \qquad (3.11)$$

when the $\hat{p}_k$ are estimated by maximum likelihood, which also has an asymptotic $\chi^2_{(K-s-1)}$ distribution. "Asymptotic" means that these are *large sample tests*, meaning that the test statistic follows the χ^2 distribution increasingly well as $N \to \infty$. A common rule of thumb is that all *expected* frequencies should exceed one and that fewer than 20% should be less than 5.

EXAMPLE 3.12: Death by horse kick

We illustrate the basic ideas of goodness-of fit tests with the `HorseKicks` data, where we expect a Poisson distribution with parameter λ = mean number of deaths. As shown in Eqn. (3.3), this is calculated as the frequency (n_k) weighted mean of the k values, here, number of deaths.

In R, such one-way frequency distributions should be converted to data frames with numeric variables. The calculation below uses `weighted.mean()` with the frequencies as weights, and finds $\lambda = 0.61$ as the mean number of deaths per corps-year.

```
> # goodness-of-fit test
> tab <- as.data.frame(HorseKicks, stringsAsFactors = FALSE)
> colnames(tab) <- c("nDeaths", "Freq")
> str(tab)

'data.frame': 5 obs. of 2 variables:
 $ nDeaths: chr  "0" "1" "2" "3" ...
 $ Freq   : int  109 65 22 3 1

> (lambda <- weighted.mean(as.numeric(tab$nDeaths), w = tab$Freq))

[1] 0.61
```

From this, we can calculate the probabilities (`phat`) of k = `0:4` deaths, and hence the expected (`exp`) frequencies in a Poisson distribution.

```
> phat <- dpois(0 : 4, lambda = lambda)
> exp <- sum(tab$Freq) * phat
> chisq <- (tab$Freq - exp)^2 / exp
>
> GOF <- data.frame(tab, phat, exp, chisq)
> GOF

  nDeaths Freq      phat       exp      chisq
1       0  109 0.5433509 108.67017 0.0010011
2       1   65 0.3314440  66.28881 0.0250573
3       2   22 0.1010904  20.21809 0.1570484
4       3    3 0.0205551   4.11101 0.3002534
5       4    1 0.0031346   0.62693 0.2220057
```

Finally, the Pearson χ^2 is just the sum of the `chisq` values and `pchisq()` is used to calculate the p-value of this test statistic—the probability of obtaining this χ^2 or a more extreme value if our assumption on the underlying distribution is true.

```
> sum(chisq)  # chi-square value

[1] 0.70537

> pchisq(sum(chisq), df = nrow(tab) - 2, lower.tail = FALSE)

[1] 0.87194
```

The result, $\chi_3^2 = 0.70537$, shows an extremely good fit of these data to the Poisson distribution, perhaps exceptionally so.[12]

$\triangle$

3.3.1 R tools for discrete distributions

In R, the function `fitdistr()` in the MASS package is a basic work horse for fitting a variety of distributions by maximum likelihood and other methods, giving parameter estimates and standard errors. Among discrete distributions, the binomial, Poisson and geometric distributions have closed-form maximum likelihood estimates; the negative binomial distribution, parameterized by (n, μ), is estimated iteratively by direct optimization.

These basic calculations are extended and enhanced for one-way discrete distributions in the vcd function `goodfit()`, which computes the fitted values of a discrete distribution (either Poisson, binomial, or negative binomial) to the count data. If the parameters are not specified they are estimated either by maximum likelihood (ML) or Minimum Chi-squared. `print()` and `summary()` methods for the "goodfit" objects give, respectively, a table of observed and fitted frequencies, and the Pearson and/or likelihood ratio goodness-of-fit statistics. Plotting methods for visualizing the discrepancies between observed and fitted frequencies are described and illustrated in Section 3.3.2.

EXAMPLE 3.13: Families in Saxony

This example uses `goodfit()` to fit the binomial to the distribution of the number of male children in families of size 12 in Saxony. Note that for the binomial, both n and p are considered as parameters, and by default n is taken as the maximum count.

```
> data("Saxony", package = "vcd")
> Sax_fit <- goodfit(Saxony, type = "binomial")
> unlist(Sax_fit$par) # estimated parameters

    prob      size
 0.51922 12.00000
```

So, we estimate the probability of a male in these families to be $p = 0.519$, a value that is quite close to the value found in Arbuthnot's data ($p = 0.517$).

It is useful to know that `goodfit()` returns a list structure of named components that are used by method functions for class "goodfit" objects. The `print.goodfit()` method prints the table of observed and fitted frequencies. `summary.goodfit()` calculates and prints the likelihood ratio χ^2 GOF test when the ML estimation method is used.

```
> names(Sax_fit)        # components of "goodfit" objects

[1] "observed" "count"    "fitted"   "type"     "method"
[6] "df"       "par"

> Sax_fit               # print method


Observed and fitted values for binomial distribution
with parameters estimated by `ML'

 count observed     fitted pearson residual
     0        3    0.93284          2.14028
```

---

[12] An exceptionally good fit occurs when the p-value for the test χ^2 statistic is so high, as to suggest that something unreasonable under random sampling might have occurred. The classic example of this is the controversy over Gregor Mendel's experiments of cross-breeding garden peas with various observed (phenotype) characteristics, where R. A. Fisher (1936a) suggested that observed frequencies of combinations like (smooth/wrinkled), (green/yellow) in a second generation were uncomfortably too close to the 3 : 1 ratio predicted by genetic theory.

```
 1       24    12.08884              3.42580
 2      104    71.80317              3.79963
 3      286   258.47513              1.71205
 4      670   628.05501              1.67371
 5     1033  1085.21070             -1.58490
 6     1343  1367.27936             -0.65661
 7     1112  1265.63031             -4.31841
 8      829   854.24665             -0.86380
 9      478   410.01256              3.35761
10      181   132.83570              4.17896
11       45    26.08246              3.70417
12        7     2.34727              3.03687

> summary(Sax_fit)    # summary method


Goodness-of-fit test for binomial distribution

                     X^2 df   P(> X^2)
Likelihood Ratio 97.007 11  6.9782e-16
```

Note that the GOF test gives a highly significant p-value (pratically zero), indicating significant lack of fit to the binomial distribution.[13] Some further analysis of this result is explored in examples below.   △

EXAMPLE 3.14: Weldon's dice

Weldon's dice data, explored in Example 3.3, are also expected to follow a binomial distribution, here with $p = \frac{1}{3}$. However, as given in the data set WeldonDice, the frequencies for counts 10–12 were grouped as "10+." In this case, it is necessary to supply the correct value of $n = 12$ as the value of the size parameter in the call to goodfit().

```
> data("WeldonDice", package = "vcd")
> dice_fit <- goodfit(WeldonDice, type = "binomial",
+                     par = list(size = 12))
> unlist(dice_fit$par)

   prob      size
0.33769 12.00000
```

The probability of a success (a 5 or 6) is estimated as $\hat{p} = 0.3377$, not far from the theoretical value, $p = 1/3$.

```
> print(dice_fit, digits = 0)


Observed and fitted values for binomial distribution
with parameters estimated by `ML'

 count observed fitted pearson residual
    0      185    187               -0
    1     1149   1147                0
    2     3265   3216                1
    3     5475   5465                0
    4     6114   6269               -2
    5     5194   5114                1
    6     3067   3042                0
```

---

[13] A handy rule-of-thumb is to think of the ratio of χ^2/df, because, under the null hypothesis of acceptable fit, $\mathcal{E}(\chi^2/df) = 1$, so ratios exceeding ≈ 2.5 are troubling. Here, the ratio is $97/11 = 8.8$, so the lack of fit is substantial.

```
        7      1331   1330                      0
        8       403    424                     -1
        9       105     96                      1
       10        18     15                      1
       11         0      1                     -1
       12         0      0                     -0
```

```
> summary(dice_fit)


 Goodness-of-fit test for binomial distribution

                    X^2 df P(> X^2)
 Likelihood Ratio 11.506  9   0.2426
```

Here, we find an acceptable fit for the binomial distribution.   △

EXAMPLE 3.15: Death by horse kick

This example reproduces the calculations done "manually" in Example 3.12 above. We fit the Poisson distribution to the *HorseKicks* data by specifying `type = "poisson"` (actually, that is the default for `goodfit()`).

```
> data("HorseKicks", package = "vcd")
> HK_fit <- goodfit(HorseKicks, type = "poisson")
> HK_fit$par

$lambda
[1] 0.61

> HK_fit


 Observed and fitted values for poisson distribution
 with parameters estimated by `ML'

 count observed     fitted pearson residual
     0      109  108.67017          0.03164
     1       65   66.28881         -0.15830
     2       22   20.21809          0.39629
     3        3    4.11101         -0.54795
     4        1    0.62693          0.34142
```

The `summary` method uses the LR test by default, so the `X^2` value reported below differs slightly from the Pearson χ^2 value shown earlier.

```
> summary(HK_fit)


 Goodness-of-fit test for poisson distribution

                     X^2 df P(> X^2)
 Likelihood Ratio 0.86822  3  0.83309
```

△

EXAMPLE 3.16: Federalist Papers

In Example 3.5 we examined the distribution of the marker word "may" in blocks of text in the *Federalist Papers* written by James Madison. A naive hypothesis is that these occurrences might follow a Poisson distribution, that is, as independent occurrences with constant probability across the 262 blocks of text. Using the same methods as above, we fit these data to the Poisson distribution:

```
> data("Federalist", package = "vcd")
> Fed_fit0 <- goodfit(Federalist, type = "poisson")
> unlist(Fed_fit0$par)

 lambda
0.65649

> Fed_fit0


Observed and fitted values for poisson distribution
with parameters estimated by `ML'

 count observed       fitted pearson residual
     0      156 135.891389         1.724988
     1       63  89.211141        -2.775086
     2       29  29.283046        -0.052306
     3        8   6.407995         0.628903
     4        4   1.051694         2.874934
     5        1   0.138085         2.319483
     6        1   0.015109         7.620613
```

The GOF test below shows a substantial lack of fit, rejecting the assumptions of the Poisson model.

```
> summary(Fed_fit0)


Goodness-of-fit test for poisson distribution

                   X^2 df   P(> X^2)
Likelihood Ratio 25.243  5 0.00012505
```

Mosteller and Wallace (1963) determined that the negative binomial distribution provided a better fit to these data than the Poisson. We can verify this as follows:

```
> Fed_fit1 <- goodfit(Federalist, type = "nbinomial")
> unlist(Fed_fit1$par)

  size    prob
1.18633 0.64376

> summary(Fed_fit1)


Goodness-of-fit test for nbinomial distribution

                  X^2 df P(> X^2)
Likelihood Ratio 1.964  4  0.74238
```

Recall that the Poisson distribution assumes that the probability of a word like *may* appearing in a block of text is small and constant, and that for the Poisson, $\mathcal{E}(x) = \mathcal{V}(x) = \lambda$. One interpretation of the better fit of the negative binomial is that the use of a given word occurs with Poisson frequencies, but Madison varied its rate λ_i from one block of text to another according to a gamma distribution, allowing the variance to be greater than the mean.   △

3.3.2   Plots of observed and fitted frequencies

In the examples of the last section, we saw cases where the GOF tests showed close agreement between the observed and model-fitted frequencies, and cases where they diverged significantly, to cause rejection of a hypothesis that the data followed the specified distribution.

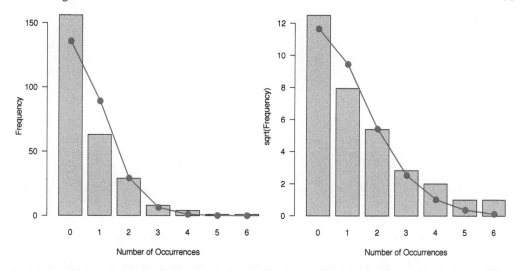

Figure 3.14: Plots for the Federalist Papers data, fitting the Poisson model. Each panel shows the observed frequencies as bars and the fitted frequencies as a smooth curve. Left: raw frequencies; right: plotted on a square-root scale to emphasize smaller frequencies.

What is missing from such numerical summaries is any appreciation of the *details* of this statistical comparison. Plots of the observed and fitted frequencies can help to show both the shape of the theoretical distribution we have fitted and the pattern of any deviations between our data and theory.

In this section we illustrate some simple plotting tools for these purposes, using the `plot.goodfit()` method for **"goodfit"** objects. [14] The left panel of Figure 3.14 shows the fit of the Poisson distribution to the Federalist Papers data, using one common form of plot that is sometimes used for this purpose. In this plot, observed frequencies are shown by bars and fitted frequencies are shown by points, connected by a smooth (spline) curve.

Such a plot, however, is dominated by the largest frequencies, making it hard to assess the deviations among the smaller frequencies. To make the smaller frequencies more visible, Tukey (1977) suggests plotting the frequencies on a square-root scale, which he calls a **rootogram**. This plot is shown in the right panel of Figure 3.14.

```
> plot(Fed_fit0, scale = "raw", type = "standing")
> plot(Fed_fit0, type = "standing")
```

Additional improvements over the standard plot on the scale of raw frequencies are shown in Figure 3.15, both of which use the square root scale. The left panel moves the rootogram bars so their tops are at the expected frequencies (giving a **hanging rootogram**). This has the advantage that we can more easily judge the pattern of departures against the horizontal reference line at 0, than against the curve.

```
> plot(Fed_fit0, type = "hanging", shade = TRUE)
> plot(Fed_fit0, type = "deviation", shade = TRUE)
```

A final variation is to emphasize the differences between the observed and fitted frequencies by drawing the bars to show the gaps between the 0 line and the (observed−expected) difference (Figure 3.15, right).

---

[14]Quantile–quantile (QQ) plots are a common alternative for the goal of comparing observed and expected values under some distribution. These plots are useful for unstructured samples, but less so when we also want to see the shape of a distribution, as is the case here.

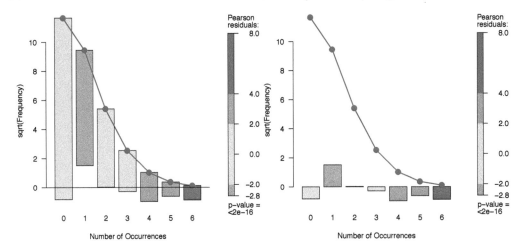

Figure 3.15: Plots for the Federalist Papers data, fitting the Poisson model. Left: hanging rootogram; Right: deviation rootogram. Color reflects the sign and magnitude of the contributions to lack of fit.

All of these plots are actually produced by the `rootogram()` function in **vcd**, the `plot()` method for `"goodfit"` objects. The default is `type = "hanging"`, and there are many options to control the plot details. For example, the plots in Figure 3.15 use `shade=TRUE` to color the bars according to the contribution to the Pearson chi-square.[15]

The plots in Figure 3.14 and Figure 3.15 used the ill-fitting Poisson model on purpose to highlight how these plots show the departure between the observed and fitted frequencies. Figure 3.16 compares this with the negative binomial model, `Fed_fit1`, which we saw has a much better, and acceptable fit.

```
> plot(Fed_fit0, main = "Poisson", shade = TRUE, legend = FALSE)
> plot(Fed_fit1, main = "Negative binomial", shade = TRUE, legend = FALSE)
```

Comparing the two plots in Figure 3.16, we can see that the Poisson model overestimates the frequency of counts $k = 1$ and underestimates the larger counts for $k = 4$–6 occurrences. The surprising feature here is that the greatest contribution to lack of fit for the Poisson model is the frequency for $k = 6$. The deviations for the negative binomial are small and unsystematic.

Finally, Figure 3.17 shows hanging rootograms for two atrociously bad models for the data on butterfly species in Malaya considered in Example 3.7. As we will see in Section 3.4, this long-tailed distribution is better approximated by the logarithmic series distribution, but this distribution is presently not handled by `goodfit()`.

```
> data("Butterfly", package = "vcd")
> But_fit1 <- goodfit(Butterfly, type = "poisson")
> But_fit2 <- goodfit(Butterfly, type = "nbinomial")
> plot(But_fit1, main = "Poisson", shade = TRUE, legend = FALSE)
> plot(But_fit2, main = "Negative binomial", shade = TRUE, legend = FALSE)
```

---

[15]The bipolar color scheme uses blue for positive standard Pearson residuals, $(n_k - N\hat{p}_k)/\sqrt{N\hat{p}_k}$, and red for negative residuals, with shading intensity proportional to the categorized absolute values shown in the legend.

3.4 Diagnosing discrete distributions: Ord plots

Ideally, the general form chosen for a discrete distribution should be dictated by substantive knowledge of a plausible mechanism for generating the data. When such knowledge is lacking, however, we may not know which distribution is most appropriate for some particular set of data. In these cases, the question is often turned around, so that we seek a distribution that fits well, and then try to understand the mechanism in terms of aspects of the underlying probability theory (independent trials, rare events, waiting-time to an occurrence, and so forth).

Although it is possible to fit each of several possibilities, the summary goodness-of-fit statistics can easily be influenced by one or two disparate cells, or additional (ignored or unknown) factors. One simple alternative is a plot suggested by Ord (1967), which may be used to diagnose the form of the discrete distribution.

Ord showed that a linear relationship of the form:

$$\frac{k\,p(k)}{p(k-1)} \equiv \frac{k\,n_k}{n_{k-1}} = a + b\,k \tag{3.12}$$

holds for each of the Poisson, binomial, negative binomial, and logarithmic series distributions, and these distributions are distinguished by the signs of the intercept, a, and slope, b, as shown in Table 3.11.

Table 3.11: Diagnostic slope and intercept for four discrete distributions. The ratios kn_k/n_{k-1} plotted against k should appear as a straight line, whose slope and intercept determine the particular distribution.

Slope (b)	Intercept (a)	Distribution (parameter)	Parameter estimate
0	+	Poisson (λ)	$\lambda = a$
−	+	Binomial (n, p)	$p = b/(b-1)$
+	+	Negative binomial (n, p)	$p = 1 - b$
+	−	Log. series (θ)	$\theta = b$
			$\theta = -a$

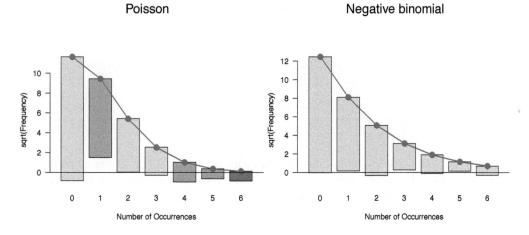

Figure 3.16: Hanging rootograms for the Federalist Papers data, comparing the Poisson and negative binomial models.

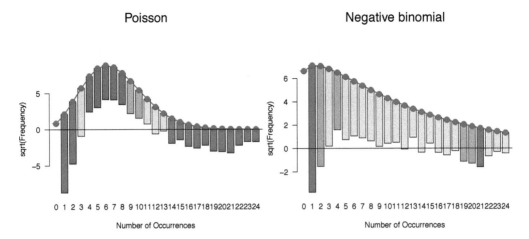

Figure 3.17: Hanging rootograms for the Butterfly data, comparing the Poisson and negative binomial models. The lack of fit for both is readily apparent.

The slope, b, in Eqn. (3.12) is zero for the Poisson, negative for the binomial, and positive for the negative binomial and logarithmic series distributions; the latter two are distinguished by their intercepts. In practical applications of this idea, the details are important: how to fit the line, and how to determine if the pattern of signs are sufficient to reasonably provide a diagnosis of the distribution type.

One difficulty in applying this technique is that the number of points (distinct values of k) in the Ord plot is often small, and the sampling variances of $k\, n_k/n_{k-1}$ can vary enormously. A little reflection indicates that points where n_k is small should be given less weight in determining the slope of the line (and hence determining the form of the distribution). In applications it has been found that using a weighted least squares fit of $k\, n_k/n_{k-1}$ on k, using weights of $w_k = \sqrt{n_k - 1}$ produces reasonably good automatic diagnosis of the form of a probability distribution. Moreover, to judge whether a coefficient is positive or negative, a small tolerance is used; if none of the distributions can be classified, no parameters are estimated. Caution is advised in accepting the conclusion, because it is based on these simple heuristics.

In the vcd package this method is implemented in the `Ord_plot()` function. The essential ideas are illustrated using the `Butterfly` data below, which produces Figure 3.18. Note that the function returns (invisibly) the values of the intercept and slope in the weighted least squares (WLS) regression.

```
> ord <- Ord_plot(Butterfly,
+                 main = "Butterfly species collected in Malaya",
+                 gp = gpar(cex = 1), pch = 16)
> ord

Intercept      Slope
 -0.70896    1.06082
```

In this plot, the black line shows the usual ordinary least squares (OLS) regression fit of frequency, n_k, on number of occurrences, k; the red line shows the weighted least squares fit, using weights of $\sqrt{n_k - 1}$. In this case, the two lines are fairly close together, in regards to their intercepts and slopes. The positive slope and negative intercept diagnoses this as a log-series distribution.

In other cases, the number of distinct points (values of k) is small, and the sampling variances of the ratios $k\, n_k/n_{k-1}$ can vary enormously. The following examples illustrate some other distributions and some of the details of the heuristics.

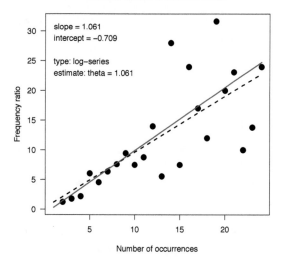

Figure 3.18: Ord plot for the Butterfly data. The slope and intercept in the plot correctly diagnoses the log-series distribution.

3.4.0.1 Ord plot examples

EXAMPLE 3.17: Death by horse kick

The results below show the calculations for the horse kicks data, with the frequency ratio $k\, n_k/n_{k-1}$ labeled y.

```
> data("HorseKicks", package = "vcd")
> nk <- as.vector(HorseKicks)
> k <- as.numeric(names(HorseKicks))
> nk1 <- c(NA, nk[-length(nk)])
> y <- k * nk / nk1
> weight <- sqrt(pmax(nk, 1) - 1)
> (ord_df <- data.frame(k, nk, nk1, y, weight))

  k  nk nk1       y  weight
1 0 109  NA      NA 10.3923
2 1  65 109 0.59633  8.0000
3 2  22  65 0.67692  4.5826
4 3   3  22 0.40909  1.4142
5 4   1   3 1.33333  0.0000

> coef(lm(y ~ k, weights = weight, data = ord_df))

(Intercept)           k
   0.656016   -0.034141
```

The weighted least squares line, with weights w_k, has a slope (-0.03) close to zero, indicating the Poisson distribution.[16] The estimate $\lambda = a = .656$ compares favorably with the maximum likelihood estimate (MLE), $\lambda = 0.610$ and the value from the Poissonness plot, shown in the following section. The call to Ord_plot() below produces Figure 3.19.

```
> Ord_plot(HorseKicks,
+          main = "Death by horse kicks", gp = gpar(cex = 1), pch = 16)
```

---

[16]The heuristic adopted in Ord_plot() uses a tolerance of 0.1 to decide if a coefficient is negative, zero, or positive.

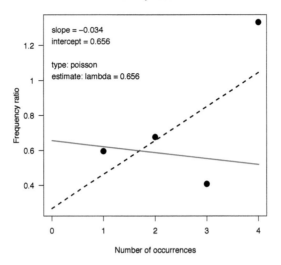

Figure 3.19: Ord plot for the HorseKicks data. The plot correctly diagnoses the Poisson distribution.

$\triangle$

EXAMPLE 3.18: Federalist Papers

Figure 3.20 (left) shows the Ord plot for the *Federalist* data. The slope is positive, so either the negative binomial or log series are possible, according to Table 3.11. The intercept is essentially zero, which is ambiguous. However, the logarithmic series requires $b \approx -a$, so the negative binomial is a better choice. Mosteller and Wallace (1963, 1984) did in fact find a reasonably good fit to this distribution. Note that there is one apparent outlier, at $k = 6$, whose effect on the OLS line is to increase the slope and decrease the intercept. $\triangle$

```
> Ord_plot(Federalist, main = "Instances of 'may' in Federalist Papers",
+            gp = gpar(cex = 1), pch = 16)
```

EXAMPLE 3.19: Women in queues

Jinkinson and Slater (1981) and Hoaglin and Tukey (1985) give the frequency distribution of the number of females observed in 100 queues of length 10 in a London Underground station, recorded in the data set *WomenQueue* in **vcd**.

```
> data("WomenQueue", package = "vcd")
> WomenQueue

nWomen
 0  1  2  3  4  5  6  7  8  9 10
 1  3  4 23 25 19 18  5  1  1  0
```

If it is assumed that people line up independently, and that men and women are equally likely to be found in a queue (not necessarily reasonable assumptions), then the number of women out of 10 would have a (symmetric) binomial distribution with parameters $n = 10$ and $p = \frac{1}{2}$. However, there is no real reason to expect that males and females are equally likely to be found in queues in the London underground, so we may be interested in estimating p from the data and determining if a binomial distribution fits.

```
> Ord_plot(WomenQueue, main = "Women in queues of length 10",
+          gp = gpar(cex = 1), pch = 16)
```

Figure 3.20 (right) shows the Ord plot for these data. The negative slope and positive intercept clearly diagnose this distribution as binomial. The rough estimate of $\hat{p} = b/(1-b) = 0.53$ indicates that women are slightly more prevalent than men in these data for the London underground. △

3.4.0.2 Limitations of Ord plots

Using a single simple diagnostic plot to determine one of four common discrete distributions is advantageous, but your enthusiasm should be dampened by several weaknesses:

- The Ord plot lacks resistance, because a single discrepant frequency affects the points n_k/n_{k-1} for both k and $k + 1$.
- The sampling variance of $k\, n_k/n_{k-1}$ fluctuates widely (Hoaglin and Tukey, 1985, Jinkinson and Slater, 1981). The use of weights w_k helps, but is purely a heuristic device. The `Ord_plot()` function explicitly shows both the OLS line and the WLS line, which provides some indication of the effect of the points on the estimation of slope and intercept.

3.5 Poissonness plots and generalized distribution plots

The **Poissonness plot** (Hoaglin, 1980) is a robust plot to sensitively determine how well a one-way table of frequencies follows a Poisson distribution. It plots a quantity called a against k, designed so that the result will be points along a straight line when the data follow a Poisson distribution. When the data deviate from a Poisson, the points will be curved. Hoaglin and Tukey (1985) developed similar plots for other discrete distributions, including the binomial, negative binomial, and logarithmic series distributions. We first describe the features and construction of these plots for the Poisson distribution; then (Section 3.5.4) the extension to other distributions.

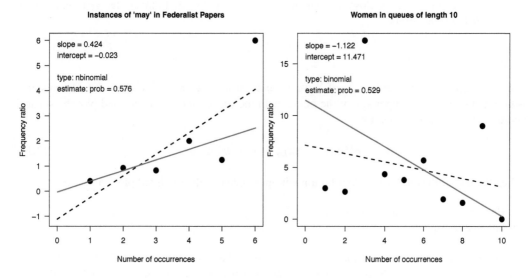

Figure 3.20: Ord plots for the Federalist (left) and WomenQueue (right) data sets.

3.5.1 Features of the Poissonness plot

The Poissonness plot has the following desirable features:

- **Resistance**: a single discrepant value of n_k affects only the point at value k. (In the Ord plot it affects each of its neighbors.)

- **Comparison standard**: An approximate confidence interval can be found for each point, indicating its inherent variability and helping to judge whether each point is discrepant.

- **Influence**: Extensions of the method result in plots that show the effect of each point on the estimate of the main parameter of the distribution (λ in the Poisson).

3.5.2 Plot construction

Assume, for some fixed λ, each observed frequency, n_k equals the expected frequency, $m_k = N p_k$. Then, setting $n_k = N p_k = N e^{-\lambda} \lambda^k / k!$, and taking logs of both sides gives

$$\log(n_k) = \log N - \lambda + k \log \lambda - \log k! \, .$$

This can be rearranged to a linear equation in k,

$$\phi\left(n_k\right) \equiv \log\left(\frac{k! \, n_k}{N}\right) = -\lambda + (\log \lambda)\, k \, . \tag{3.13}$$

The left side of Eqn. (3.13) is called the **count metameter**, and denoted $\phi\left(n_k\right)$. Hence, plotting $\phi(n_k)$ against k should give a straight line of the form $\phi(n_k) = a + bk$ with

- slope = $\log \lambda$
- intercept = $-\lambda$

when the observed frequencies follow a Poisson distribution. If the points in this plot are close enough to a straight line, then an estimate of λ may be obtained from the slope b of the line, and $\hat{\lambda} = e^b$ should be reasonably close in value to the MLE of λ, $\hat{\lambda} = \bar{x}$. In this case, we might as well use the MLE as our estimate.

3.5.2.1 Leveled plot

If we have a preliminary estimate λ_0 of λ, we can use this to give a new plot where the reference line is horizontal, making comparison of the points with the line easier. In this leveled plot the vertical coordinate $\phi(n_k)$ is modified to

$$\phi'\left(n_k\right) = \phi(n_k) + \lambda_0 - k \log \lambda_0 \, . \tag{3.14}$$

When the data follow a Poisson distribution with parameter λ, the modified plot will have

- slope = $\log \lambda - \log \lambda_0 = \log(\lambda/\lambda_0)$
- intercept = $\lambda_0 - \lambda$

In the ideal case, where our estimate of λ_0 is close to the true λ, the line will be approximately horizontal at $\phi(n_k)' = 0$. The modified plot is particularly useful in conjunction with the confidence intervals for individual points described below.

3.5.2.2 Confidence intervals

The goal of the Poissonness plot is to determine whether the points are "sufficiently linear" to conclude that the Poisson distribution is adequate for the data. Confidence intervals for the points can help you decide, and also show the relative precision of the points in these plots.

For example, when one or two points deviate from an otherwise nearly linear relation, it is helpful to determine whether the discrepancy is consistent with chance variation. As well, we must recognize that classes with small frequencies n_k are less precise than classes with large frequencies.

Hoaglin and Tukey (1985) develop approximate confidence intervals for $\log(m_k)$ for each point in the Poissonness plot. These are calculated as

$$\phi\left(n_k^*\right) \pm h_k \tag{3.15}$$

where the count metameter function is calculated using a modified frequency n_k^*, defined as

$$n_k^* = \begin{cases} n_k - .8n_k - .67 & n \geq 2 \\ 1/e & n = 1 \\ \text{undefined} & n = 0 \end{cases}$$

and h_k is the half-width of the 95% confidence interval,

$$h_k = 1.96 \frac{\sqrt{1 - \widehat{p}_k}}{[n_k - (.25\widehat{p}_k + .47)\sqrt{n_k}]^{1/2}}$$

and $\widehat{p}_k = n_k/N$. A more modern approach could use a bootstrap estimate.

3.5.3 The `distplot()` function

Poissonness plots (and versions for other distributions) are produced by the function `distplot()` in **vcd**. As with `Ord_plot()`, the first argument is either a vector of counts, a one-way table of frequencies of counts, or a data frame or matrix with frequencies in the first column and the corresponding counts in the second column. Nearly all of the examples in this chapter use one-way tables of counts.

The `type` argument specifies the type of distribution. For `type = "poisson"`, specifying a value for `lambda = `λ_0 gives the leveled version of the plot.

EXAMPLE 3.20: Death by horse kick

The calculations for the Poissonness plot, including confidence intervals, are shown below for the *HorseKicks* data. The call to `distplot()` produces the plot in the left panel of Figure 3.21.

```
> data("HorseKicks", package = "vcd")
> dp <- distplot(HorseKicks, type = "poisson",
+     xlab = "Number of deaths", main = "Poissonness plot: HorseKicks data")
> print(dp, digits = 4)

  Counts Freq Metameter CI.center CI.width CI.lower CI.upper
1      0  109    -0.607   -0.6131   0.1305  -0.7436  -0.4827
2      1   65    -1.124   -1.1343   0.2069  -1.3412  -0.9274
3      2   22    -1.514   -1.5451   0.4169  -1.9620  -1.1281
4      3    3    -2.408   -2.6607   1.3176  -3.9783  -1.3431
5      4    1    -2.120   -3.1203   2.6887  -5.8089  -0.4316
```

In this plot, the open circles show the calculated observed values of the count Metameter $= \phi\left(n_k\right)$. The smaller filled points show the centers of the confidence intervals, CI.center $= \phi\left(n_k^*\right)$ (Eqn. (3.15)), and the dashed lines show the extent of the confidence intervals.

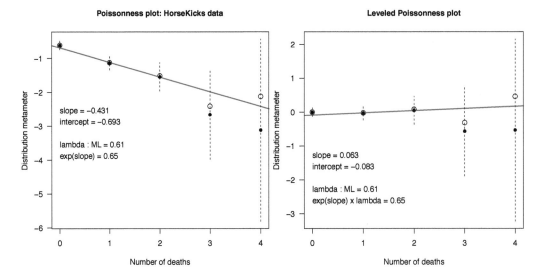

Figure 3.21: Poissonness plots for the HorseKick data. Left: standard plot; right: leveled plot.

The fitted least squares line has a slope of -0.431, which would indicate $\lambda = e^{-0.431} = 0.65$. This compares well with the MLE, $\lambda = \bar{x} = 0.61$.

Using `lambda = 0.61` as below gives the leveled version shown in the right panel of Figure 3.21.

```
> # leveled version, specifying lambda
> distplot(HorseKicks, type = "poisson", lambda = 0.61,
+    xlab = "Number of deaths", main = "Leveled Poissonness plot")
```

In both plots the fitted line is within the confidence intervals, indicating the adequacy of the Poisson model for these data. The widths of the intervals for $k > 2$ are graphic reminders that these observations have decreasingly low precision where the counts n_k are small.

$\triangle$

3.5.4   Plots for other distributions

As described in Section 3.2.6, the binomial, Poisson, negative binomial, geometric, and logseries distributions are all members of the general power series family of discrete distributions. For this family, Hoaglin and Tukey (1985) developed similar plots of a count metameter against k, which appear as a straight line when a data distribution follows a given family member.

The distributions which can be analyzed in this way are shown in Table 3.12, with the interpretation given to the slope and intercept in each case.

For example, for the Binomial distribution, a "binomialness" plot is constructed by plotting $\log n_k^* / N \binom{n}{k}$ against k. If the points in this plot approximate a straight line, the slope is interpreted as $\log(p/(1-p))$, so the binomial parameter p may be estimated as $p = e^b/(1 + e^b)$.

Unlike the Ord plot, a different plot is required for each distribution, because the count metameter, $\phi(n_k)$, differs from distribution to distribution. Moreover, systematic deviation from a linear relationship does not indicate which distribution provides a better fit. However, the attention to robustness, and the availability of confidence intervals and influence diagnostics, make this a highly useful tool for visualizing discrete distributions.

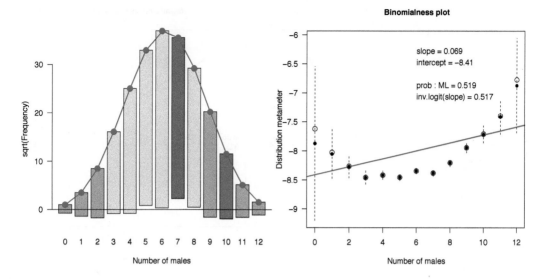

Figure 3.22: Diagnostic plots for males in Saxony families. Left: `goodfit()` plot; right: `distplot()` plot. Both plots show heavier tails than in a binomial distribution.

EXAMPLE 3.21: Families in Saxony

Our analysis in Example 3.2 and Example 3.13 of the *Saxony* data showed that the distribution of male children had slightly heavier tails than the binomial, meaning the observed distribution is overdispersed. We can see this in the `goodfit()` plot shown in Figure 3.22 (left), and even more clearly in the distribution diagnostic plot produced by `distplot()` in the right panel of Figure 3.22. For a binomial distribution, we call this distribution plot a "binomialness plot."

```
> plot(goodfit(Saxony, type = "binomial", par = list(size=12)),
+       shade=TRUE, legend=FALSE,
+       xlab = "Number of males")
> distplot(Saxony, type = "binomial", size = 12,
+    xlab = "Number of males")
```

Table 3.12: Plot parameters for five discrete distributions. In each case the count metameter, $\phi(n_k^*)$ is plotted against k, yielding a straight line when the data follow the given distribution.

Distribution	Probability function, $p(k)$	Count) metameter, $\phi(n_k^*)$	Theoretical slope (b)	Theoretical intercept (a)
Poisson	$e^{-\lambda}\lambda^k/k!$	$\log(k!n_k^*/N)$	$\log(\lambda)$	$-\lambda$
Binomial	$\binom{n}{k}p^k(1-p)^{n-k}$	$\log\left(n_k^*/N\binom{n}{k}\right)$	$\log\left(\frac{p}{1-p}\right)$	$n\log(1-p)$
Negative bino-mial	$\binom{n+k-1}{k}p^n(1-p)^k$	$\log\left(n_k^*/N\binom{n+k-1}{k}\right)$	$\log(1-p)$	$n\log(p)$
Geometric	$p(1-p)^k$	$\log(n_k^*/N)$	$\log(1-p)$	$\log(p)$
Log series	$\theta^k/[-k\log(1-\theta)]$	$\log(kn_k^*/N)$	$\log(\theta)$	$-\log(-\log(1-\theta))$

Source: adapted from Hoaglin and Tukey (1985), Table 9-15.

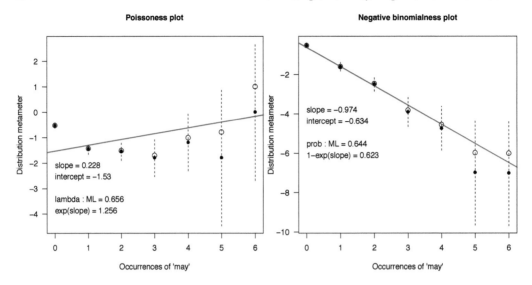

Figure 3.23: Diagnostic plots for the Federalist Papers data. Left: Poissonness plot; right: negative binomialness plot.

The weight of evidence is thus that, as simple as the binomial might be, it is inadequate to fully explain the distribution of sex ratios in this large sample of families of 12 children. To understand this data better, it is necessary to question the assumptions of the binomial (births of males are independent Bernoulli trials with constant probability p) as a model for this birth distribution and/or find a more adequate model.[17]   △

EXAMPLE 3.22: Federalist Papers
In Example 3.16 we carried out GOF tests for the Poisson and negative binomial models with the Federalist Papers data; Figure 3.16 showed the corresponding rootogram plots. Figure 3.23 compares these two using the diagnostic plots of this section. Again the Poisson shows systematic departure from the linear relation required in the Poissonness plot, while the negative binomial model provides an acceptable fit to these data.

```
> distplot(Federalist, type = "poisson", xlab = "Occurrences of 'may'")
> distplot(Federalist, type = "nbinomial", xlab = "Occurrences of 'may'")
```

                                                                                              △

3.6 Fitting discrete distributions as generalized linear models*

In Section 3.2.6, we described how the common discrete distributions are all members of the general power series family. This provides the basis for the generalized distribution plots described in Section 3.5.4. Another general family of distributions—the **exponential family**—includes most of the common continuous distributions: the normal, gamma, exponential, and others, and is the basis of the class of generalized linear models (GLMs) fit by `glm()`.

---

[17]On these questions, Edwards (1958) reviews numerous other studies of these Geissler's data, and fits a so-called β-**binomial** model proposed by Skellam (1948), where p varies among families according to a β distribution. He concludes that there is evidence that p varies between families of the same size. One suggested explanation is that family decisions to have a further child is influenced by the balance of boys and girls among their earlier children.

Table 3.13: Poisson loglinear representations for some discrete distributions

Distribution	Sufficient statistics	Offset
Geometric	k	
Poisson	k	$-\log(k!)$
Binomial	k	$\log\binom{n}{k}$
Double binomial	$k, k\log(k) + (n-k)\log(n-k)$	$\log\binom{n}{k}$

A clever approach by Lindsey and Mersch (1992), Lindsey (1995, Section 6.1) shows how various discrete (and continuous) distributions can be fit to frequency data using generalized linear models for log frequency (which are equivalent to Poisson loglinear models). The uniform, geometric, binomial, and the Poisson distributions may all be fit easily in this way, but the idea extends to some other distributions, such as the ***double binomial distribution***, that allows a separate parameter for overdispersion relative to the binomial. A clear advantage is that this method gives estimated standard errors for the distribution parameters as well as estimated confidence intervals for fitted probabilities.

The essential idea is that, for frequency data, any distribution in the exponential family may be represented by a linear model for the logarithm of the cell frequency, with a Poisson distribution for errors, otherwise known as a "Poisson loglinear regression model." These have the form

$$\log(N\pi_k) = \text{offset} + \beta_0 + \boldsymbol{\beta}^{\mathsf{T}}\boldsymbol{S}(k),$$

where N is the total frequency, π_k is the modeled probability of count k, $\boldsymbol{S}(k)$ is a vector of zero or more sufficient statistics for the canonical parameters of the exponential family distribution, and the offset term is a value that does not depend on the parameters.

Table 3.13 shows the sufficient statistics and offsets for several discrete distributions. See Lindsey and Mersch (1992) for further details, and definitions for the double-binomial distribution,[18] and Lindsey (1995, pp. 130–133) for his analysis of the *Saxony* data using this distribution. Lindsey and Altham (1998) provide an analysis of the complete Geissler data (provided in the data set *Geissler* in vcdExtra) using several different models to handle overdispersion.

EXAMPLE 3.23: Families in Saxony

The binomial distribution and the double binomial can both be fit to frequency data as a Poisson regression via glm() using $\log\binom{n}{k}$ as an offset. First, we convert *Saxony* into a numeric data frame for use with glm().

```
> data("Saxony", package = "vcd")
> Males <- as.numeric(names(Saxony))
> Families <- as.vector(Saxony)
> Sax.df <- data.frame(Males, Families)
```

To calculate the offset for glm() in R, note that choose(12,0:12) returns the binomial coefficients, and lchoose(12,0:12) returns their logs.

---

[18]In R, the double binomial distribution is implemented in the rmutil package, providing the standard complement of density function (ddoublebinom()), CDF (pdoublebinom()), quantiles (qdoublebinom()), and random generation (rdoublebinom()). This package is not on CRAN, but is available at http://www.commanster.eu/rcode.html.

```
> # fit binomial (12, p) as a glm
> Sax.bin <- glm(Families ~ Males, offset = lchoose(12, 0:12),
+                  family = poisson, data = Sax.df)
>
> # brief model summaries
> LRstats(Sax.bin)

Likelihood summary table:
          AIC BIC LR Chisq Df Pr(>Chisq)
Sax.bin 191 192       97 11     7e-16 ***
---
Signif. codes:  0 '***' 0.001 '**' 0.01 '*' 0.05 '.' 0.1 ' ' 1

> coef(Sax.bin)

(Intercept)        Males
 -0.069522     0.076898
```

As we have seen, this model fits badly. The parameter estimate for Males, $\beta_1 = 0.0769$ is actually estimating the logit of p, $\log p/(1-p)$, so the inverse transformation gives $\hat{p} = \frac{\exp(\beta_1)}{1+\exp(\beta_1)} = 0.5192$, as we had before.

The double binomial model can be fitted as follows. The term YlogitY calculates $k \log(k) + (n-k) \log(n-k)$, the second sufficient statistic for the double binomial (see Table 3.13) fitted via glm().

```
> # double binomial, (12, p, psi)
> Sax.df$YlogitY <-
+   Males        * log(ifelse(Males == 0, 1, Males)) +
+        (12-Males) * log(ifelse(12-Males == 0, 1, 12-Males))
>
> Sax.dbin <- glm(Families ~ Males + YlogitY, offset = lchoose(12,0:12),
+         family = poisson, data = Sax.df)
> coef(Sax.dbin)

(Intercept)        Males      YlogitY
 -3.096918     0.065977     0.140205

> LRstats(Sax.bin, Sax.dbin)

Likelihood summary table:
           AIC BIC LR Chisq Df Pr(>Chisq)
Sax.bin  191 192     97.0 11      7e-16 ***
Sax.dbin 109 111     13.1 10       0.22
---
Signif. codes:  0 '***' 0.001 '**' 0.01 '*' 0.05 '.' 0.1 ' ' 1
```

From the above, we can see that the double binomial model Sax.dbin with one more parameter is significantly better than the simple binomial and represents an adequate fit to the data. The table below displays the fitted values and standardized residuals for both models.

```
> results <- data.frame(Sax.df,
+          fit.bin = fitted(Sax.bin), res.bin = rstandard(Sax.bin),
+          fit.dbin = fitted(Sax.dbin), res.dbin = rstandard(Sax.dbin))
> print(results, digits = 2)
```

	Males	Families	YlogitY	fit.bin	res.bin	fit.dbin	res.dbin
1	0	3	30	0.93	1.70	3.0	0.026
2	1	24	26	12.09	3.05	23.4	0.136
3	2	104	24	71.80	3.71	104.3	-0.036
4	3	286	23	258.48	1.87	307.8	-1.492
5	4	670	22	628.06	1.94	652.9	0.778

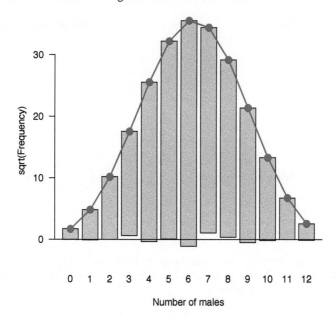

Figure 3.24: Rootogram for the double binomial model for the Saxony data. This now fits well in the tails of the distribution.

6	5	1033	22	1085.21	-1.87	1038.5	-0.202
7	6	1343	22	1367.28	-0.75	1264.2	2.635
8	7	1112	22	1265.63	-5.09	1185.0	-2.550
9	8	829	22	854.25	-1.03	850.1	-0.846
10	9	478	23	410.01	3.75	457.2	1.144
11	10	181	24	132.84	4.23	176.8	0.371
12	11	45	26	26.08	3.42	45.2	-0.039
13	12	7	30	2.35	2.45	6.5	0.192

Finally, Figure 3.24 shows the rootogram for the double binomial, which can be compared with the binomial model shown in Figure 3.22. We can see that the fit is now quite good, particularly in the tails. The positive coefficient for the term `YlogitY` gives additional weight in the tails.

```
> with(results, vcd::rootogram(Families, fit.dbin, Males,
+                      xlab = "Number of males"))
```

△

3.6.1 Covariates, overdispersion, and excess zeros

All of the examples in this chapter are somewhat special, in that in each case the data consist only of a one-way frequency distribution of a basic count variable. In more general and realistic settings, there may also be one or more explanatory variables or ***covariate***s that influence the frequency distributions of the counts. For example, in the `Saxony` data, the number of boys in families of size 12 was aggregated over the years 1876–1885, and it is possible that any deviation from a binomial distribution could be due to variation over time or unmeasured predictors (e.g., rural vs. urban, age of parents).

This is where the generalized linear model approach introduced here (treated in detail in Chapter 11), begins to shine—because it allows such covariates to be taken into account, and then questions regarding the *form* of the distribution pertain only to the variation of the frequencies not fitted

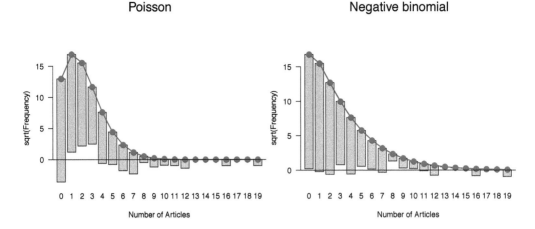

Figure 3.25: Hanging rootograms for publications by PhD candidates, comparing the Poisson and negative binomial models. The Poisson model clearly does not fit. The the negative binomial is better, but still has significant lack of fit.

by the model. The next example illustrates what can go wrong when important predictors are omitted from the analysis.

EXAMPLE 3.24: Publications of PhD candidates

Long (1990, 1997) gave data on the number of publications by 915 doctoral candidates in biochemistry in the last three years of their PhD studies, contained in the data set *PhdPubs* in **vcdExtra**. The data set also includes information on gender, marital status, number of young children, prestige of the doctoral department, and number of publications by the student's mentor. The frequency distribution of number of publications by these students is shown below.

```
> data("PhdPubs", package = "vcdExtra")
> table(PhdPubs$articles)
```

0	1	2	3	4	5	6	7	8	9	10	11	12	16	19
275	246	178	84	67	27	17	12	1	2	1	1	2	1	1

The naive approach, ignoring the potential predictors, is just to try fitting various probability models to this one-way distribution. Rootograms for the simpler Poisson distribution and the negative binomial that allows for overdispersion are shown in Figure 3.25.

```
> library(vcd)
> plot(goodfit(PhdPubs$articles), xlab = "Number of Articles",
+       main = "Poisson")
> plot(goodfit(PhdPubs$articles, type = "nbinomial"),
+       xlab = "Number of Articles", main = "Negative binomial")
```

From these plots it is clear that the Poisson distribution doesn't fit well at all, because there is a large excess of zero counts—candidates with no publications, and most of the counts of four or more publications are larger than the Poisson model predicts. The fit of the negative binomial model in the right panel of Figure 3.25 looks much better, except that for eight or more publications, there is a systematic tendency of overfitting for 8–10 and underfittting for the observed counts of 12 or more. This lack of fit is confirmed by the formal test.

```
> summary(goodfit(PhdPubs$articles, type = "nbinomial"))


Goodness-of-fit test for nbinomial distribution

                      X^2 df   P(> X^2)
Likelihood Ratio 31.098 12 0.0019033
```

The difficulty with this simple analysis is not only that it ignores the possible predictors of publishing by these PhD candidates, but also, by doing so, it prevents a better, more nuanced explanation of the phenomenon under study. This example is re-visited in Chapter 11, Example 11.1, where we consider generalized linear models taking potential predictors into account, as well as extended *zero-inflated* models allowing special consideration of zero counts. $\triangle$

3.7 Chapter summary

- Discrete distributions typically involve basic *counts* of occurrences of some event occurring with varying *frequency*. The ideas and methods for one-way tables described in this chapter are building blocks for the analysis of more complex data.

- The most commonly used discrete distributions include the binomial, Poisson, negative binomial, geometric, and logarithmic series distributions. Happily, these are all members of a family called the power series distributions. Methods of fitting an observed data set to any of these distributions are described, and implemented in the `goodfit()` function.

- After fitting an observed distribution it is useful to plot the observed and fitted frequencies. Several ways of making these plots are described, and implemented in the `rootogram()` function.

- A heuristic graphical method for identifying which discrete distribution is most appropriate for a given set of data involves plotting ratios kn_k/n_{k-1} against k. These plots are constructed by the function `Ord_plot()`.

- A more robust plot for a Poisson distribution involves plotting a count metameter, $\phi(n_k)$ against k, which gives a straight line (whose slope estimates the Poisson parameter) if the data follow a Poisson distribution. This plot provides robust confidence intervals for individual points and provides a means to assess the influence of individual points on the Poisson parameter. These plots are provided by the function `distplot()`.

- The ideas behind the Poissonness plot can be applied to the other discrete distributions.

3.8 Lab exercises

Exercise 3.1 The `Arbuthnot` data in HistData (Example 3.1) also contains the variable `Ratio`, giving the ratio of male to female births.

(a) Make a plot of `Ratio` over `Year`, similar to Figure 3.1. What features stand out? Which plot do you prefer to display the tendency for more male births?
(b) Plot the total number of christenings, `Males + Females` or `Total` (in 000s) over time. What unusual features do you see?

Exercise 3.2 Use the graphical methods illustrated in Section 3.2 to plot a collection of geometric distributions for $p = 0.2, 0.4, 0.6, 0.8$, over a range of values of $k = 0, 1, \ldots 10$.

(a) With `xyplot()`, try the different plot formats using points connected with lines, as in Figure 3.9, or using points and lines down to the origin, as in the panels of Figure 3.10.

(b) Also with `xyplot()`, produce one version of a multi-line plot in a single panel that you think shows well how these distributions change with the probability p of success.

(c) Do the same in a multi-panel version, conditional on p.

Exercise 3.3 Use the data set `WomenQueue` to:

(a) Produce plots analogous to those shown in Section 3.1 (some sort of bar graph of frequencies).

(b) Check for goodness-of-fit to the binomial distribution using the `goodfit()` methods described in Section 3.3.2.

(c) Make a reasonable plot showing departure from the binomial distribution.

(d) Suggest some reasons why the number of women in queues of length 10 might depart from a binomial distribution, $\text{Bin}(n = 10, p = 1/2)$.

Exercise 3.4 Continue Example 3.13 on the distribution of male children in families in Saxony by fitting a binomial distribution, $\text{Bin}(n = 12, p = \frac{1}{2})$, specifying equal probability for boys and girls. [*Hint*: you need to specify both `size` and `prob` values for `goodfit()`.]

(a) Carry out the GOF test for this fixed binomial distribution. What is the ratio of χ^2/df? What do you conclude?

(b) Test the additional lack of fit for the model $\text{Bin}(n = 12, p = \frac{1}{2})$ compared to the model $\text{Bin}(n = 12, p = \hat{p})$ where $\hat{p}$ is estimated from the data.

(c) Use the `plot.gootfit()` method to visualize these two models.

Exercise 3.5 For the `Federalist` data, the examples in Section 3.3.1 and Section 3.3.2 showed the negative binomial to provide an acceptable fit. Compare this with the simpler special case of geometric distribution, corresponding to $n = 1$.

(a) Use `goodfit()` to fit the geometric distribution. [Hint: use `type="nbinomial"`, but specify `size=1` as a parameter.]

(b) Compare the negative binomial and the geometric models statistically, by a likelihood-ratio test of the difference between these two models.

(c) Compare the negative binomial and the geometric models visually by hanging rootograms or other methods.

Exercise 3.6 Mosteller and Wallace (1963, Table 2.4) give the frequencies, n_k, of counts $k = 0, 1, \ldots$ of other selected marker words in 247 blocks of text known to have been written by Alexander Hamilton. The data below show the occurrences of the word *upon*, that Hamilton used much more than did James Madison.

```
> count <- 0 : 5
> Freq <- c(129, 83, 20, 9, 5, 1)
```

(a) Read these data into R and construct a one-way table of frequencies of counts or a matrix or data frame with frequencies in the first column and the corresponding counts in the second column, suitable for use with `goodfit()`.

(b) Fit and plot the Poisson model for these frequencies.

(c) Fit and plot the negative binomial model for these frequencies.

(d) What do you conclude?

Exercise 3.7 The data frame `Geissler` in the vcdExtra package contains the complete data from Geissler's (1889) tabulation of family sex composition in Saxony. The table below gives the number of boys in families of size 11.

boys	0	1	2	3	4	5	6	7	8	9	10	11
Freq	8	72	275	837	1,540	2,161	2,310	1,801	1,077	492	93	24

(a) Read these data into R.

(b) Following Example 3.13, use `goodfit()` to fit the binomial model and plot the results. Is there an indication that the binomial does not fit these data?

(c) Diagnose the form of the distribution using the methods described in Section 3.4.

(d) Try fitting the negative binomial distribution, and use `distplot()` to diagnose whether the negative binomial is a reasonable fit.

Exercise 3.8 The data frame `Bundesliga` gives a similar data set to that for UK soccer scores (`UKSoccer`) examined in Example 3.9, but over a wide range of years. The following lines calculate a two-way table, `BL1995`, of home-team and away-team goals for the 306 games in the year 1995.

```
> data("Bundesliga", package = "vcd")
> BL1995 <- xtabs(~ HomeGoals + AwayGoals, data = Bundesliga,
+               subset = (Year == 1995))
> BL1995

         AwayGoals
HomeGoals  0  1  2  3  4  5  6
        0 26 16 13  5  0  1  0
        1 19 58 20  5  4  0  1
        2 27 23 20  5  1  1  1
        3 14 11 10  4  2  0  0
        4  3  5  3  0  0  0  0
        5  4  1  0  1  0  0  0
        6  1  0  0  1  0  0  0
```

(a) As in Example 3.9, find the one-way distributions of `HomeGoals`, `AwayGoals`, and `TotalGoals = HomeGoals + AwayGoals`.

(b) Use `goodfit()` to fit and plot the Poisson distribution to each of these. Does the Poisson seem to provide a reasonable fit?

(c) Use `distplot()` to assess fit of the Poisson distribution.

(d) What circumstances of scoring goals in soccer might cause these distributions to deviate from Poisson distributions?

Exercise 3.9 * Repeat the exercise above, this time using the data for all years in which there was the standard number (306) of games, that is for `Year>1965`, tabulated as shown below.

```
> BL <- xtabs(~ HomeGoals + AwayGoals, data = Bundesliga,
+               subset = (Year > 1965))
```

Exercise 3.10 Using the data `CyclingDeaths` introduced in Example 3.6 and the one-way frequency table `CyclingDeaths.tab = table(CyclingDeaths$deaths)`,

(a) Make a sensible plot of the number of deaths over time. For extra credit, add a smoothed curve (e.g., using `lines(lowess(...))`).

(b) Test the goodness of fit of the table `CyclingDeaths.tab` to a Poisson distribution statistically using `goodfit()`.

(c) Continue this analysis using a `rootogram()` and `distplot()`.

(d) Write a one-paragraph summary of the results of these analyses and your conclusions.

Exercise 3.11 * The one-way table, `Depends`, in vcdExtra and shown below gives the frequency distribution of the number of dependencies declared in 4, 983 R packages maintained on the CRAN distribution network on January 17, 2014. That is, there were 986 packages that had no dependencies, 1, 347 packages that depended on one other package, ... up to 2 packages that depended on 14 other packages.

Depends	0	1	2	3	4	5	6	7	8	9	10	11	12	13	14
# Pkgs	986	1,347	993	685	375	298	155	65	32	19	9	4	9	4	2

(a) Make a bar plot of this distribution.

(b) Use `Ord_plot()` to see if this method can diagnose the form of the distribution.

(c) Try to fit a reasonable distribution to describe dependencies among R packages.

Exercise 3.12 * How many years does it take to get into the baseball Hall of Fame? The Lahman (Friendly, 2014b) package provides a complete record of historical baseball statistics from 1871 to the present. One table, `HallOfFame`, records the history of players nominated to the Baseball Hall of Fame, and those eventually inducted. The table below, calculated in `help(HallOfFame, package="Lahman")`, records the distribution of the number of years taken (from first nomination) for the 109 players in the Hall of Fame to be inducted (1936–present). Note that `years==0` does not, and cannot, occur in this table, so the distribution is restricted to positive counts. Such distributions are called **zero-truncated distribution**s. Such distributions are like the ordinary ones, but with the probability of zero being zero. Thus the other probabilities are scaled up (i.e., divided by $1 - \Pr(Y = 0)$) so they sum to 1.

years	1	2	3	4	5	6	7	8	9	10	11	12	13	14	15
inducted	46	10	8	7	8	4	2	4	6	3	3	1	4	1	2

(a) For the Poisson distribution, show that the zero-truncated probability function can be expressed in the form

$$\Pr\{X = k \mid k > 0\}) = \frac{1}{1 - e^{-\lambda}} \times \frac{e^{-\lambda} \lambda^k}{k!} \qquad k = 1, 2, \ldots$$

(b) Show that the mean is $\lambda/(1 - \exp(-\lambda))$.

(c) Enter these data into R as a one-way table, and use `goodfit()` to fit the standard Poisson distribution, as if you hadn't encountered the problem of zero truncation.

Part II

Exploratory and Hypothesis-Testing Methods

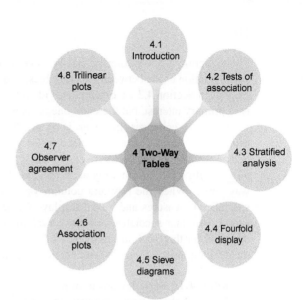

Two-Way Contingency Tables

The analysis of two-way frequency tables concerns the association between two variables. A variety of specialized graphical displays help us to visualize the pattern of association, using area of some region to represent the frequency in a cell. Some of these methods are focused on visualizing an odds ratio (for 2×2 tables), or the general pattern of association, or the agreement between row and column categories in square tables.

4.1   Introduction

> Tables are like cobwebs, like the sieve of Danaides; beautifully reticulated, orderly to look upon, but which will hold no conclusion. Tables are abstractions, and the object a most concrete one, so difficult to read the essence of.
>
> From *Chartism* by Thomas Carlyle (1840), Chapter II, Statistics

Most methods of statistical analysis are concerned with understanding relationships or dependence among variables. With categorical variables, these relationships are often studied from data that has been summarized by a ***contingency table*** in table form or frequency form, giving the frequencies of observations cross-classified by two or more such variables. As Thomas Carlyle said, it is often difficult to appreciate the message conveyed in numerical tables.

This chapter is concerned with simple graphical methods for understanding the association between two categorical variables. Some examples are also presented that involve a third, ***stratifying variable***, where we wish to determine if the relationship between two primary variables is the same or different for all levels of the stratifying variable. More general methods for fitting models and displaying associations for three-way and larger tables are described in Chapter 5.

In Section 4.2, we describe briefly some numerical and statistical methods for testing whether an association exists between two variables, and measures for quantifying the strength of this association. In Section 4.3 we extend these ideas to situations where the relation between two variables is of primary interest, but there are one or more background variables to be controlled.

The main emphasis, however, is on graphical methods that help to describe the *pattern* of an association between variables. Section 4.4 presents the fourfold display, designed to portray the odds ratio in 2×2 tables or a set of k such tables. *Sieve diagrams* (Section 4.5) and *association plots* (Section 4.6) are more general methods for depicting the pattern of associations in any two-way table. When the row and column variables represent the classifications of different raters, specialized measures and visual displays for *inter-rater agreement* (Section 4.7) are particularly useful. Another specialized display, a *trilinear plot* or *ternary plot*, described in Section 4.8, is designed for three-column frequency tables or compositional data. In order to make clear some of the distinctions that occur in contingency table analysis, we begin with several examples.

EXAMPLE 4.1: Berkeley admissions

Table 4.1 shows aggregate data on applicants to graduate school at Berkeley for the six largest departments in 1973 classified by admission and gender (Bickel et al., 1975). See *UCBAdmissions* (in package **datasets**) for the complete data set. For such data we might wish to study whether there is an association between admission and gender. Are male (or female) applicants more likely to be admitted? The presence of an association might be considered as evidence of sex bias in admission practices.

Table 4.1 is an example of the simplest kind of contingency table, a 2×2 classification of individuals according to two dichotomous (binary) variables. For such a table, the question of whether there is an association between admission and gender is equivalent to asking if the proportions of males and females who are admitted to graduate school are different, or whether the difference in proportions admitted is not zero.   △

Table 4.1: Admissions to Berkeley graduate programs

Gender	Admitted	Rejected	Total	% Admit
Males	1198	1493	2691	44.52
Females	557	1278	1835	30.35
Total	1755	2771	4526	38.78

Although the methods for quantifying association in larger tables can be used for 2×2 tables, there are specialized measures (described in Section 4.2) and graphical methods for these simpler tables.

As we mentioned in Section 1.2.4 it is often useful to make a distinction between *response*, or outcome variables, on the one hand, and possible *explanatory* or predictor variables on the other. In Table 4.1, it is natural to consider admission as the outcome, and gender as the explanatory variable. In other tables, no variable may be clearly identified as *the* outcome, or there may be several response variables, giving a multivariate problem.

EXAMPLE 4.2: Hair color and eye color

Table 4.2 shows data collected by Snee (1974) on the relation between hair color and eye color among 592 students in a statistics course (a two-way margin of *HairEyeColor*).

Neither hair color nor eye color is considered a response in relation to the other; our interest concerns whether an association exists between them. Hair color and eye color have both been classified into four categories. Although the categories used are among the most common, they are

Table 4.2: Hair-color eye-color data

Eye Color	Hair Color				Total
	Black	Brown	Red	Blond	
Brown	68	119	26	7	220
Blue	20	84	17	94	215
Hazel	15	54	14	10	93
Green	5	29	14	16	64
Total	108	286	71	127	592

not the only categories possible.[1]   A common, albeit deficient, representation of such a table is a *grouped barchart*, as shown in the left of Figure 4.1:

```
> hec <- margin.table(HairEyeColor, 2:1)
> barplot(hec, beside = TRUE, legend = TRUE)
```

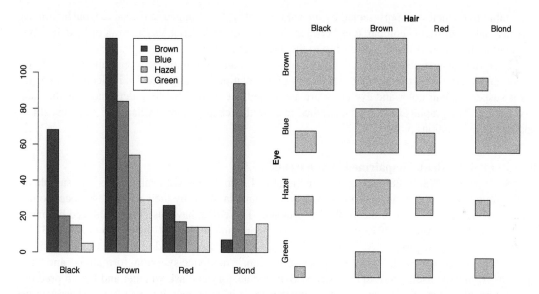

Figure 4.1: Two basic displays for the Hair-color Eye-color data. Left: grouped barchart; right: tile plot.

For each hair color, a group of bars represent the corresponding eye colors, the heights being proportional to the absolute frequencies. Bar graphs do not extend well to more than one dimension since

- the graphical representation does not match the tabular data structure, complicating comparisons with the raw data;

- it is harder to compare bars accross groups than within groups;

- by construction, the grouping suggests a conditional or causal relationship of the variables (here: "what is the eye color *given* the hair color?," "how does eye color influence hair color?"), even though such an interpretation may be inappropriate (as in this example);

---

[1]If students had been asked to write down their hair and eye colors, it is likely that many more than four categories of each would appear in a sample of nearly 600.

Table 4.3: Mental impairment and parents' SES

| | \multicolumn{4}{c}{Mental impairment} | | | |
SES	Well	Mild	Moderate	Impaired
1	64	94	58	46
2	57	94	54	40
3	57	105	65	60
4	72	141	77	94
5	36	97	54	78
6	21	71	54	71

- labeling may become increasingly complex.

A somewhat better approach is a ***tile plot*** (using `tile()` in **vcd**), as shown next to the bar plot in Figure 4.1:

```
> tile(hec)
```

The table frequencies are represented by the area of rectangles arranged in the same tabular form as the raw data, facilitating comparisons between tiles accross both variables (by rows or by columns), by maintaining a one-to-one relationship to the underlying table[2].

Everyday observation suggests that there probably is an association between hair color and eye color, and we will describe tests and measures of associations for larger tables in Section 4.2.3. If, as is suspected, hair color and eye color are associated, we would like to understand *how* they are associated. The graphical methods described later in this chapter and in Chapter 5 help reveal the pattern of associations present.   △

EXAMPLE 4.3: Mental impairment and parents' SES

Srole et al. (1978, p. 289) gave the data in Table 4.3 on the mental health status of a sample of 1660 young New York residents in midtown Manhattan classified by their parents' socioeconomic status (SES); see *Mental* in the **vcdExtra** package. These data have also been analyzed by many authors, including Agresti (2013, Section 10.5.3), Goodman (1979), and Haberman (1979, p. 375).

There are six categories of SES (from 1 = "High" to 6 = "Low"), and mental health is classified in the four categories "well," "mild symptom formation," "moderate symptom formation," and "impaired." It may be useful here to consider SES as explanatory and ask whether and how it predicts mental health status as a response, that is, whether there is an association, and if so, investigate its nature.

```
> data("Mental", package = "vcdExtra")
> mental <- xtabs(Freq ~ ses + mental, data = Mental)
> spineplot(mental)
```

Figure 4.2 shows a ***spineplot*** of this data—basically a stacked barchart of the row percentages of mental impairment for each SES category, the width of each bar being proportional to the overall SES percentages.[3] From this graph, it is apparant that the "well" mental state decreases with social-economic status, while the "impaired" state increases. This pattern is more specific than overall association (as suspected for the hair-color eye-color data), and indeed, more powerful and focused tests are available when we treat these variables as *ordinal*, as we will see in Section 4.2.4.   △

---

[2]This kind of display is more generally known as a *fluctuation diagram* (Hofmann, 2000), flexibly implemented by function `fluctile()` in the package **extracat** (Pilhoefer, 2014).

[3]Thus, in the more technical terms introduced in 4.2.1, this spineplot shows the conditional distribution of impairment, given the categories of SES.

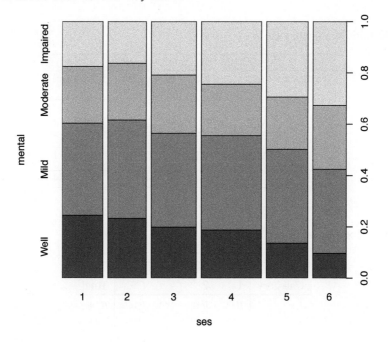

Figure 4.2: Spineplot of the Mental data.

EXAMPLE 4.4: Arthritis treatment

The data in Table 4.4 compares an active treatment for rheumatoid arthritis to a placebo (Koch and Edwards, 1988), used in examples in Chapter 2 (Example 2.2). The outcome reflects whether individuals showed no improvement, some improvement, or marked improvement. Here, the outcome variable is an ordinal one, and it is probably important to determine if the relation between treatment and outcome is the same for males and females. The data set is given in case form in `Arthritis` (in package **vcd**).

This is, of course, a three-way table, with factors `Treatment`, `Sex`, and `Improvement`. If the relation between treatment and outcome is the same for both genders, an analysis of the Treatment by Improvement table (collapsed over sex) could be carried out. Otherwise we could perform separate analyses for men and women, or treat the combinations of treatment and sex as four levels of a "population" variable, giving a 4×3 two-way table. These simplified approaches each ignore certain information available in an analysis of the full three-way table. △

4.2 Tests of association for two-way tables

4.2.1 Notation and terminology

To establish notation, let $N = \{n_{ij}\}$ be the observed frequency table of variables A and B with r rows and c columns, as shown in Table 4.5. In what follows, a subscript is replaced by a "+" when summed over the corresponding variable, so $n_{i+} = \sum_j n_{ij}$ gives the total frequency in row i, $n_{+j} = \sum_i n_{ij}$ gives the total frequency in column j, and $n_{++} = \sum_i \sum_j n_{ij}$ is the grand total; for convenience, n_{++} is also symbolized by n.

Table 4.4: Arthritis treatment data

Treatment	Sex	Improvement None	Some	Marked	Total
Active	Female	6	5	16	27
	Male	7	2	5	14
Placebo	Female	19	7	6	32
	Male	10	0	1	11
Total		42	14	28	84

Table 4.5: The $R \times C$ contingency table

Row Category	Column category 1	2	$\cdots$	C	Total
1	n_{11}	n_{12}	$\cdots$	n_{1C}	n_{1+}
2	n_{21}	n_{22}	$\cdots$	n_{2C}	n_{2+}
$\vdots$	$\vdots$	$\vdots$	$\cdots$	$\vdots$	$\vdots$
R	n_{R1}	n_{R2}	$\cdots$	n_{RC}	n_{R+}
Total	n_{+1}	n_{+2}	$\cdots$	n_{+C}	n_{++}

When each observation is randomly sampled from some population and classified on two categorical variables, A and B, we refer to the ***joint distribution*** of these variables, and let $\pi_{ij} = \Pr(A = i, B = j)$ denote the population probability that an observation is classified in row i, column j (or cell (ij)) in the table. Corresponding to these population joint probabilities, the cell proportions, $p_{ij} = n_{ij}/n$, give the sample joint distribution.

The row totals n_{i+} and column totals n_{+j} are called ***marginal frequencies*** for variables A and B, respectively. These describe the distribution of each variable *ignoring* the other. For the population probabilities, the ***marginal distributions*** are defined analogously as the row and column totals of the joint probabilities, $\pi_{i+} = \sum_j \pi_{ij}$, and $\pi_{+j} = \sum_i \pi_{ij}$. The sample marginal proportions are, correspondingly, $p_{i+} = \sum_j p_{ij} = n_{i+}/n$, and $p_{+j} = \sum_i p_{ij} = n_{+j}/n$.

When one variable (the column variable, B, for example) is a response variable, and the other (A) is an explanatory variable, it is most often useful to examine the distribution of the response B for *each* level of A separately. These define the ***conditional distributions*** of B, given the level of A, and are defined for the population as $\pi_{j \mid i} = \pi_{ij}/\pi_{i+}$.

These definitions are illustrated for the Berkeley data (Table 4.1) below, using the function `CrossTable()`.

```
> Berkeley <- margin.table(UCBAdmissions, 2:1)
> library(gmodels)
> CrossTable(Berkeley, prop.chisq = FALSE, prop.c = FALSE,
+            format = "SPSS")


  Cell Contents
|-----------------------|
|                 Count |
|           Row Percent |
|         Total Percent |
```

```
|-------------------------------|

Total Observations in Table:   4526

           | Admit
    Gender | Admitted  | Rejected  | Row Total |
-----------|-----------|-----------|-----------|
      Male |      1198 |      1493 |      2691 |
           |   44.519% |   55.481% |   59.456% |
           |   26.469% |   32.987% |           |
-----------|-----------|-----------|-----------|
    Female |       557 |      1278 |      1835 |
           |   30.354% |   69.646% |   40.544% |
           |   12.307% |   28.237% |           |
-----------|-----------|-----------|-----------|
Column Total |     1755 |      2771 |      4526 |
-----------|-----------|-----------|-----------|
```

The output shows the joint frequencies, n_{ij}, and joint sample percentages, $100 \times p_{ij}$, in the first row within each table cell. The second row in each cell ("Row percent") gives the conditional percentage of admission or rejection, $100 \times p_{j \mid i}$ for males and females separately. The row and column labelled "Total" give the marginal frequencies, n_{i+} and n_{+j}, and marginal percentages, p_{i+} and p_{+j}.

4.2.2   2 by 2 tables: Odds and odds ratios

The 2 by 2 contingency table of applicants to Berkeley graduate programs in Table 4.1 may be regarded as an example of a ***cross-sectional study***.   The total of $n = 4,526$ applicants in 1973 has been classified by both gender and admission status. Here, we would probably consider the total n to be fixed, and the cell frequencies n_{ij}, $i = 1, 2; j = 1, 2$ would then represent a single ***multinomial sample*** for the cross-classification by two binary variables, with probabilities cell p_{ij}, $i = 1, 2; j = 1, 2$ such that

$$p_{11} + p_{12} + p_{21} + p_{22} = 1 \,.$$

The basic null hypothesis of interest for a multinomial sample is that of *independence*. Are admission and gender independent of each other?

Alternatively, if we consider admission the response variable, and gender an explanatory variable, we would treat the numbers of male and female applicants as fixed and consider the cell frequencies to represent two independent ***binomial samples*** for a binary response. In this case, the null hypothesis is described as that of *homogeneity* of the response proportions across the levels of the explanatory variable.

Measures of association are used to quantify the strength of association between variables. Among the many measures of association for contingency tables, the ***odds ratio*** is particularly useful for 2×2 tables, and is a fundamental parameter in several graphical displays and models described later. Other measures of strength of association for 2×2 tables are described in Stokes et al. (2000, Chapter 2) and Agresti (1996, Section 2.2).

For a binary response, where the probability of a "success" is π, the ***odds*** of a success is defined as

$$\text{odds} = \frac{\pi}{1 - \pi} \,.$$

Hence, odds $= 1$ corresponds to $\pi = 0.5$, or success and failure equally likely. When success is more likely than failure $\pi > 0.5$, and the odds > 1; for instance, when $\pi = 0.75$, odds $= .75/.25 = 3$, so a success is three times as likely as a failure. When failure is more likely, $\pi < 0.5$, and the odds < 1; for instance, when $\pi = 0.25$, odds $= .25/.75 = \frac{1}{3}$.

The odds of success thus vary *multiplicatively* around 1. Taking logarithms gives an equivalent measure that varies *additively* around 0, called the **log odds** or **logit**:

$$\text{logit}(\pi) \equiv \log(\text{odds}) = \log\left(\frac{\pi}{1-\pi}\right). \tag{4.1}$$

The logit is symmetric about $\pi = 0.5$, in that $\text{logit}(\pi) = -\text{logit}(1-\pi)$. The following lines calculate the odds and log odds for a range of probabilities. As you will see in Chapter 7, the logit transformation of a probability is fundamental in logistic regression.

```
> p <- c(0.05, .1, .25, .50, .75, .9, .95)
> odds <- p / (1 - p)
> logodds <- log(odds)
> data.frame(p, odds, logodds)

     p       odds logodds
1 0.05  0.052632 -2.9444
2 0.10  0.111111 -2.1972
3 0.25  0.333333 -1.0986
4 0.50  1.000000  0.0000
5 0.75  3.000000  1.0986
6 0.90  9.000000  2.1972
7 0.95 19.000000  2.9444
```

A binary response for two groups gives a 2×2 table, with Group as the row variable, say. Let π_1 and π_2 be the success probabilities for Group 1 and Group 2. The **odds ratio**, θ, is just the ratio of the odds for the two groups:

$$\text{odds ratio} \equiv \theta = \frac{\text{odds}_1}{\text{odds}_2} = \frac{\pi_1/(1-\pi_1)}{\pi_2/(1-\pi_2)}.$$

Like the odds itself, the odds ratio is always non-negative, between 0 and ∞. When $\theta = 1$, the distributions of success and failure are the same for both groups (so $\pi_1 = \pi_2$); there is no association between row and column variables, or the response is independent of group. When $\theta > 1$, Group 1 has a greater success probability; when $\theta < 1$, Group 2 has a greater success probability.

Similarly, the odds ratio may be transformed to a log scale, to give a measure that is symmetric about 0. The **log odds ratio**, symbolized by ψ, is just the difference between the logits for Groups 1 and 2:

$$\text{log odds ratio} \equiv \psi = \log(\theta) = \log\left[\frac{\pi_1/(1-\pi_1)}{\pi_2/(1-\pi_2)}\right] = \text{logit}(\pi_1) - \text{logit}(\pi_2).$$

Independence corresponds to $\psi = 0$, and reversing the rows or columns of the table merely changes the sign of ψ.

For sample data, the **sample odds ratio** is the ratio of the sample odds for the two groups:

$$\hat{\theta} = \frac{p_1/(1-p_1)}{p_2/(1-p_2)} = \frac{n_{11}/n_{12}}{n_{21}/n_{22}} = \frac{n_{11}n_{22}}{n_{12}n_{21}}. \tag{4.2}$$

The sample estimate $\hat{\theta}$ in Eqn. (4.2) is the maximum likelihood estimator of the true θ. The sampling distribution of $\hat{\theta}$ is asymptotically normal as $n \to \infty$, but may be highly skewed in small to moderate samples.

Consequently, inference for the odds ratio is more conveniently carried out in terms of the log odds ratio, whose sampling distribution is more closely normal, with mean $\psi = \log(\theta)$, and asymptotic standard error (ASE)

$$\text{ASE}_{\log(\theta)} \equiv \hat{s}(\hat{\psi}) = \sqrt{\frac{1}{n_{11}} + \frac{1}{n_{12}} + \frac{1}{n_{21}} + \frac{1}{n_{22}}} = \sqrt{\sum_{i,j} n_{ij}^{-1}}. \tag{4.3}$$

A large-sample $100(1 - \alpha)\%$ confidence interval for $\log(\theta)$ may therefore be calculated as

$$\log(\theta) \pm z_{1-\alpha/2} \, \text{ASE}_{\log(\theta)} = \hat{\psi} \pm z_{1-\alpha/2} \, \hat{s}(\hat{\psi})$$

where $z_{1-\alpha/2}$ is the cumulative normal quantile with $1 - \alpha/2$ in the lower tail. Confidence intervals for θ itself are obtained by exponentiating the end points of the interval for $\psi = \log(\theta)$,[4]

$$\exp\left(\hat{\psi} \pm z_{1-\alpha/2}\hat{s}(\hat{\psi})\right) \ .$$

EXAMPLE 4.5: Berkeley admissions

As an illustratation, we apply these formulae to the UCB Admissions data, using the `loddsratio()` function in vcd, which by default calculates log-odds:

```
> data("UCBAdmissions")
> UCB <- margin.table(UCBAdmissions, 1:2)
> (LOR <- loddsratio(UCB))

log odds ratios for Admit and Gender

[1] 0.61035

> (OR <- loddsratio(UCB, log = FALSE))

 odds ratios for Admit and Gender

[1] 1.8411
```

The function returns an object for which the `summary()` method computes the ASE and carries out the significance test (for the log odds):

```
> summary(LOR)


z test of coefficients:

                              Estimate Std. Error z value
Admitted:Rejected/Male:Female   0.6104     0.0639    9.55
                              Pr(>|z|)
Admitted:Rejected/Male:Female   <2e-16 ***
---
Signif. codes:  0 '***' 0.001 '**' 0.01 '*' 0.05 '.' 0.1 ' ' 1
```

Clearly, the hypothesis of independence has to be rejected, suggesting the presence of gender bias. `confint()` computes confidence intervals for (log) odds ratios:

```
> confint(OR)

                              2.5 % 97.5 %
Admitted:Rejected/Male:Female 1.6244 2.0867

> confint(LOR)

                               2.5 %  97.5 %
Admitted:Rejected/Male:Female 0.48512 0.73558
```

---

[4]Note that $\hat{\theta}$ is 0 or ∞ if any $n_{ij} = 0$. Haldane (1955) and Gart and Zweifel (1967) showed that improved estimators of θ and $\psi = \log(\theta)$ are obtained by replacing each n_{ij} by $[n_{ij} + \frac{1}{2}]$ in Eqn. (4.2) and Eqn. (4.3). This adjustment is preferred in small samples, and required if any zero cells occur. In large samples, the effect of adding 0.5 to each cell becomes negligible.

Finally, we note that an exact test (based on the hypergeometric distribution) is provided by
`fisher.test()` (see the help page for the details):

```
> fisher.test(UCB)


Fisher's Exact Test for Count Data

data:  UCB
p-value <2e-16
alternative hypothesis: true odds ratio is not equal to 1
95 percent confidence interval:
 1.6214 2.0912
sample estimates:
odds ratio
    1.8409
```

In general, exact tests are to be prefered over asymptotic tests like the one described above. Note,
however, that the results are very similar in this example.   △

4.2.3   Larger tables: Overall analysis

For two-way tables, overall tests of association can be carried out using `assocstats()`. If the
data set has more than two factors (as in the `Arthritis` data), the other factors will be ignored
(and collapsed) if not included when the table is constructed. This simplified analysis may be
misleading if the excluded factors interact with the factors used in the analysis.

EXAMPLE 4.6: Arthritis treatment
   Since the main interest is in the relation between `Treatment` and `Improved`, an overall
analysis (that ignores `Sex`) can be carried out by creating a two-way table with `xtabs()` as shown
below.

```
> data("Arthritis", package = "vcd")
> Art <- xtabs(~ Treatment + Improved, data = Arthritis)
> Art

          Improved
Treatment None Some Marked
  Placebo   29    7      7
  Treated   13    7     21

> round(100 * prop.table(Art, margin = 1), 2)

          Improved
Treatment  None  Some Marked
  Placebo 67.44 16.28  16.28
  Treated 31.71 17.07  51.22
```

   The row proportions show a clear difference in the outcome for the two groups: For those given
the placebo, 67% reported no improvement; in the treated group, 51% reported marked improvement. χ^2 tests and measures of association are provided by `assocstats()` as shown below:

```
> assocstats(Art)

                  X^2 df  P(> X^2)
Likelihood Ratio 13.530  2 0.0011536
Pearson          13.055  2 0.0014626
```

```
Phi-Coefficient    : NA
Contingency Coeff.: 0.367
Cramer's V         : 0.394
```

△

The measures of association are normalized variants of the χ^2 statistic. Caution is needed for interpretation since the maximum values depend on the table dimensions.

4.2.4 Tests for ordinal variables

For $r \times c$ tables, more sensitive tests than the test for general association (independence) are available if either or both of the row and column variables are ordinal. Generalized **Cochran–Mantel–Haenszel tests** (Landis et al., 1978), which take the ordinal nature of a variable into account, are provided by the CMHtest() in **vcdExtra**. These tests are based on assigning numerical scores to the table categories; the default (table) scores treat the levels as equally spaced. They generally have higher power when the pattern of association is determined by the order of an ordinal variable.

EXAMPLE 4.7: Mental impairment and parents' SES

We illustrate these tests using the data on mental impairment and SES introduced in Example 4.3, where both variables can be considered ordinal.

```
> data("Mental", package = "vcdExtra")
> mental <- xtabs(Freq ~ ses + mental, data = Mental)
> assocstats(mental)     # standard chisq tests

                  X^2 df   P(> X^2)
Likelihood Ratio 47.418 15 3.1554e-05
Pearson          45.985 15 5.3458e-05

Phi-Coefficient    : NA
Contingency Coeff.: 0.164
Cramer's V         : 0.096

> CMHtest(mental)          # CMH tests

Cochran-Mantel-Haenszel Statistics for ses by mental

                 AltHypothesis Chisq Df      Prob
cor         Nonzero correlation  37.2  1  1.09e-09
rmeans  Row mean scores differ   40.3  5  1.30e-07
cmeans  Col mean scores differ   40.7  3  7.70e-09
general     General association   46.0 15  5.40e-05
```

In this data set, all four tests show a highly significant association. However, the cor test for nonzero correlation uses only one degree of freedom, whereas the test of general association requires 15 df.   △

The four tests differ in the types of departure from independence they are sensitive to:

General Association When the row and column variables are both nominal (unordered), the only alternative hypothesis of interest is that there is *some* association between the row and column variables. The CMH test statistic is similar to the (Pearson) Chi-Square and Likelihood Ratio Chi-Square in the result from assocstats(); all have $(r - 1)(c - 1)$ df.

Row Mean Scores Differ If the column variable is ordinal, assigning scores to the column variable produces a mean for each row. The association between row and column variables can be expressed as a test of whether these means differ over the rows of the table, with $r - 1$ df. This is analogous to the Kruskal-Wallis non-parametric test (ANOVA based on rank scores).

Column Mean Scores Differ Same as the above, assigning scores to the row variable.

Nonzero Correlation (Linear association) When *both* row and column variables are ordinal, we
could assign scores to both variables and compute the correlation (r), giving Spearman's rank
correlation coefficient. The CMH χ^2 is equal to $(N-1)r^2$, where N is the total sample size.
The test is most sensitive to a pattern where the row mean score changes linearly over the rows.

4.2.5 Sample CMH profiles

Two contrived examples may make the differences among these tests more apparent. Visualizations of the patterns of association reinforces the aspects to which the tests are most sensitive, and
introduces the sieve diagram described more fully in Section 4.5.

4.2.5.1 General association

The table below exhibits a general association between variables A and B, but no difference in row
means or linear association. The row means for category j are calculated by assigning integer scores,
$b_i = i$, to the column categories, and using the corresponding frequencies of row j as weights. The
column means are obtained analogously. Figure 4.3 (left) shows the pattern of association in this
table graphically, as a sieve diagram (described in Section 4.5).

	b1	b2	b3	b4	b5	Total	Mean
a1	0	15	25	15	0	55	3.0
a2	5	20	5	20	5	55	3.0
a3	20	5	5	5	20	55	3.0
Total	25	40	35	40	25	165	3.0
Mean	2.8	1.6	1.4	1.6	2.8	2.1	

This is reflected in the `CMHtest()` output shown below (`cmhdemo1` contains the data shown
above).

```
> CMHtest(cmhdemo1)

Cochran-Mantel-Haenszel Statistics

                AltHypothesis Chisq Df      Prob
cor          Nonzero correlation   0.0  1 1.00e+00
rmeans   Row mean scores differ   0.0  2 1.00e+00
cmeans   Col mean scores differ  72.2  4 7.78e-15
general    General association    91.8  8 2.01e-16
```

The chi-square values for non-zero correlation and different row mean scores are exactly zero
because the row means are all equal. Only the general association test shows that A and B are
associated.

4.2.5.2 Linear association

The table below contains a weak, non-significant general association, but significant row mean
differences and linear associations. The unstructured test of general association would therefore
lead to the conclusion that no association exists, while the tests taking ordinal factors into account
would conclude otherwise. Note that the largest frequencies shift towards lower levels of B as the
level of variable A increases. See Figure 4.3 (right) for a visual representation of this pattern.

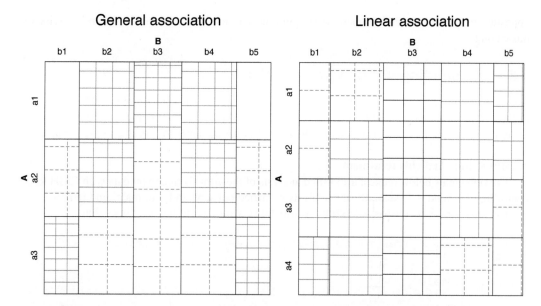

Figure 4.3: Sieve diagrams for two patterns of association: Left: general association; right: linear association.

	b1	b2	b3	b4	b5	Total	Mean
a1	2	5	8	8	8	31	3.48
a2	2	8	8	8	5	31	3.19
a3	5	8	8	8	2	31	2.81
a4	8	8	8	5	2	31	2.52
Total	17	29	32	29	17	124	3.00
Mean	3.1	2.7	2.5	2.3	1.9	2.5	

Note that the χ^2-values for the row-means and non-zero correlation tests from CMHtest() are very similar, but the correlation test is more highly significant since it is based on just one degree of freedom. In the following example, cmhdemo2 corresponds to the table above:

```
> CMHtest(cmhdemo2)

Cochran-Mantel-Haenszel Statistics

                  AltHypothesis Chisq Df     Prob
cor          Nonzero correlation  10.6  1 0.00111
rmeans   Row mean scores differ  10.7  3 0.01361
cmeans   Col mean scores differ  11.4  4 0.02241
general      General association  13.4 12 0.34064
```

The difference in sensitivity and power among these tests for categorical data is analogous to the difference between general ANOVA tests and tests for linear trend (contrasts) in experimental designs with quantitative factors: The more specific test has greater power, but is sensitive to a narrower range of departures from the null hypothesis. The more focused tests for ordinal factors are a better bet when we believe that the association depends on the ordered nature of the factor levels.

4.3 Stratified analysis

An overall analysis ignores other variables (like sex), by collapsing over them. In the *Arthritis* data, it is possible that the treatment is effective only for one gender, or even that the treatment has opposite effects for men and women. If so, pooling over the ignored variable(s) can be seriously misleading.

4.3.1 Computing strata-wise statistics

A *stratified analysis* controls for the effects of one or more background variables. This is similar to the use of a blocking variable in an ANOVA design. Tests for association can be obtained by applying a function (`assocstats()`, `CMHtest()`) over the levels of the stratifying variables.

EXAMPLE 4.8: Arthritis treatment
The statements below request a stratified analysis of the arthritis treatment data with CMH tests, controlling for gender. Essentially, the analysis is carried out separately for males and females.

The table `Art2` is constructed as a three-way table, with `Sex` as the last dimension.

```
> Art2 <- xtabs(~ Treatment + Improved + Sex, data = Arthritis)
> Art2

, , Sex = Female

          Improved
Treatment None Some Marked
  Placebo   19    7      6
  Treated    6    5     16

, , Sex = Male

          Improved
Treatment None Some Marked
  Placebo   10    0      1
  Treated    7    2      5
```

Both `assocstats()` and `CMHtest()` are designed for stratified tables, and use all dimensions after the first two as strata.

```
> assocstats(Art2)

$`Sex:Female`
                   X^2 df  P(> X^2)
Likelihood Ratio 11.731  2 0.0028362
Pearson          11.296  2 0.0035242

Phi-Coefficient    : NA
Contingency Coeff.: 0.401
Cramer's V         : 0.438

$`Sex:Male`
                  X^2 df P(> X^2)
Likelihood Ratio 5.8549  2 0.053532
Pearson          4.9067  2 0.086003

Phi-Coefficient    : NA
Contingency Coeff.: 0.405
Cramer's V         : 0.443
```

Note that even though the strength of association (Cramer's V) is similar in the two groups, the χ^2 tests show significance for females, but not for males. This is true even using the more powerful CMH tests below, treating `Treatment` as ordinal. The reason is that there were more than twice as many females as males in this sample.

```
> CMHtest(Art2)

$`Sex:Female`
Cochran-Mantel-Haenszel Statistics for Treatment by Improved
in stratum Sex:Female

                  AltHypothesis Chisq Df      Prob
cor          Nonzero correlation  10.9  1 0.000944
rmeans   Row mean scores differ   10.9  1 0.000944
cmeans   Col mean scores differ   11.1  2 0.003878
general      General association  11.1  2 0.003878


$`Sex:Male`
Cochran-Mantel-Haenszel Statistics for Treatment by Improved
in stratum Sex:Male

                  AltHypothesis Chisq Df    Prob
cor          Nonzero correlation  3.71  1 0.0540
rmeans   Row mean scores differ   3.71  1 0.0540
cmeans   Col mean scores differ   4.71  2 0.0949
general      General association  4.71  2 0.0949

> apply(Art2, 3, sum)

Female   Male
    59     25
```

$\triangle$

4.3.2 Assessing homogeneity of association

In a stratified analysis it is often crucial to know if the association between the primary table variables is the same over all strata. For $2 \times 2 \times k$ tables this question reduces to whether the odds ratio is the same in all k strata. The **vcd** package implements Woolf's test (Woolf, 1995) in `woolf_test()` for this purpose.

For larger n-way tables, this question is equivalent to testing whether the association between the primary variables, A and B, say, is the same for all levels of the stratifying variables, $C, D, \ldots$.

EXAMPLE 4.9: Berkeley admissions

Here we illustrate the use of Woolf's test for the *UCBAdmissions* data. The test is significant, indicating that the odds ratios cannot be considered equal across departments. We will see why when we visualize the data by department in the next section.

```
> woolf_test(UCBAdmissions)


Woolf-test on Homogeneity of Odds Ratios (no 3-Way
assoc.)

data:  UCBAdmissions
X-squared = 17.9, df = 5, p-value = 0.0031
```

$\triangle$

EXAMPLE 4.10: Arthritis treatment

For the arthritis data, homogeneity means the association between treatment and outcome (`improve`) is the same for both men and women. Again, we are using `woolf_test()` to test if this assumption holds.

```
> woolf_test(Art2)


Woolf-test on Homogeneity of Odds Ratios (no 3-Way
assoc.)

data:   Art2
X-squared = 0.318, df = 1, p-value = 0.57
```

Even though we found in the CMH analysis above that the association between `Treatment` and `Improved` was stronger for females than males, the analysis using `woolf_test()` is clearly non-significant, so we cannot reject homogeneity of association. △

Remark

As will be discussed later (Section 5.4) in the case of a 3-way table, the hypothesis of homogeneity of association among three variables A, B and C can be stated as the *loglinear model* of no three-way association, [AB][AC][BC]. This notation (described in Section 5.4.1 and Section 9.2) lists only the high-order association terms in a linear model for log frequency.

This hypothesis can be stated as the loglinear model,

$$[\text{SexTreatment}] \; [\text{SexImproved}] \; [\text{TreatmentImproved}]. \tag{4.4}$$

Such tests can be carried out most conveniently using `loglm()` in the **MASS** package. The model formula uses the standard **R** notation `()^2` to specify all terms of order 2.

```
> library(MASS)
> loglm(~ (Treatment + Improved + Sex)^2, data = Art2)

Call:
loglm(formula = ~(Treatment + Improved + Sex)^2, data = Art2)

Statistics:
                    X^2 df P(> X^2)
Likelihood Ratio 1.7037  2  0.42663
Pearson          1.1336  2  0.56735
```

Consistent with the Woolf test, the interaction terms are not significant.

4.4   Fourfold display for 2 x 2 tables

The *fourfold display* is a special case of a *radial diagram* (or "polar area chart") designed for the display of 2×2 (or $2 \times 2 \times k$) tables (Fienberg, 1975, Friendly, 1994a,c). In this display the frequency n_{ij} in each cell of a fourfold table is shown by a quarter circle, whose radius is proportional to $\sqrt{n_{ij}}$, so the area is proportional to the cell count. The fourfold display is similar to a pie chart in using segments of a circle to show frequencies. It differs from a pie chart in that it keeps the angles of the segments constant and varies the radius, whereas the pie chart varies the angles and keeps the radius constant.

The main purpose of this display is to depict the sample odds ratio, $\hat{\theta} = (n_{11}/n_{12}) \div (n_{21}/n_{22})$. An association between the variables ($\theta \neq 1$) is shown by the tendency of diagonally opposite cells in one direction to differ in size from those in the opposite direction, and the display uses color or shading to show this direction. Confidence rings for the observed θ allow a visual test of the hypothesis of independence, $H_0 : \theta = 1$. They have the property that (in a standardized display) the rings for adjacent quadrants overlap *iff* the observed counts are consistent with the null hypothesis.

EXAMPLE 4.11: Berkeley admissions

Figure 4.4 (left) shows the basic, unstandardized fourfold display for the Berkeley admissions data (Table 4.1). Here, the area of each quadrant is proportional to the cell frequency, shown numerically in each corner. The odds ratio is proportional to the product of the areas shaded dark, divided by the product of the areas shaded light. The sample odds ratio, Odds(Admit|Male) / Odds(Admit|Female) is 1.84 (see Example 4.9) indicating that males were nearly twice as likely to be admitted.

```
> fourfold(Berkeley, std = "ind.max")     # unstandardized
> fourfold(Berkeley, margin = 1)          # equating gender
```

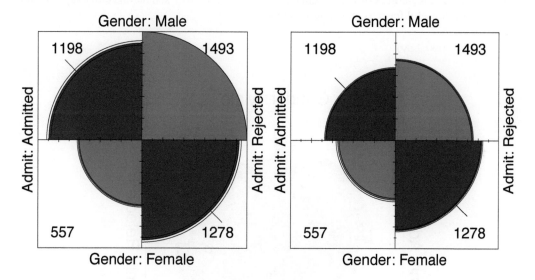

Figure 4.4: Fourfold displays for the Berkeley admission data. Left: unstandardized; right: equating the proportions of males and females.

However, it is difficult to make these visual comparisons because there are more men than women, and because the proportions admitted and rejected are unequal. In the unstandardized display the confidence bands have no interpretation as a test of $H_0 : \theta = 1$.

The data in a 2×2 table can be standardized to make these visual comparisons easier. Table 4.6 shows the Berkeley data with the addition of row percentages (which equate for the number of men and women applicants) indicating the proportion of each gender accepted and rejected. We see that 44.52% of males were admitted, while only 30.35% of females were admitted. Moreover, the row percentages have the same odds ratio as the raw data: $44.52 \times 69.65/30.35 \times 55.48 = 1.84$. Figure 4.4 (right) shows the fourfold display where the area of each quarter circle is proportional to these row percentages.

Table 4.6: Admissions to Berkeley graduate programs, frequencies and row percentages.

	Frequencies		Row Percents	
	Admitted	Rejected	Admitted	Rejected
Males	1198	1493	44.52	55.48
Females	557	1278	30.35	69.65

With this standardization, the confidence rings have the property that the confidence rings for each upper quadrant will overlap with those for the quadrant below it if the odds ratio does not differ from 1.0. (Details of the calculation of confidence rings are described in the next section.) No similar statement can be made about the corresponding left and right quadrants, however, because the overall rate of admission has not been standardized.

As a final step, we can standardize the data so that *both* table margins are equal, while preserving the odds ratio. Each quarter circle is then drawn to have an area proportional to this standardized cell frequency. This makes it easier to see the association between admission and sex without being influenced by the overall admission rate or the differential tendency of males and females to apply. With this standardization, the four quadrants will align (overlap) horizontally and vertically when the odds ratio is 1, regardless of the marginal frequencies. The fully standardized display, which is usually the most useful form, is shown in Figure 4.5.

```
> fourfold(Berkeley)    # standardize both margins
```

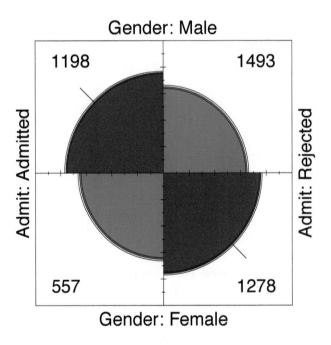

Figure 4.5: Fourfold display for Berkeley admission data with margins for gender and admission equated. The area of each quadrant shows the standardized frequency in each cell.

$\triangle$

These displays also use color (blue) and diagonal tick marks to show the direction of positive association. The visual interpretation (also conveyed by area) is that males are more likely to be accepted, females more likely to be rejected.

The quadrants in Figure 4.5 do not align and the 95% confidence rings around each quadrant do not overlap, indicating that the odds ratio differs significantly from 1—putative evidence of gender bias. The very narrow width of the confidence rings gives a visual indication of the precision of the data—if we stopped here, we might feel quite confident of this conclusion.

4.4.1 Confidence rings for odds ratio

Confidence rings for the fourfold display are computed from a confidence interval for θ, whose endpoints can each be mapped into a 2×2 table. Each such table is then drawn in the same way as the data.

The interval for θ is most easily found by considering the distribution of $\hat{\psi} = \log \hat{\theta}$, whose standard error may be estimated by Eqn. (4.3). Then an approximate $1 - \alpha$ confidence interval for ψ is given by

$$\hat{\psi} \pm \hat{s}(\hat{\psi}) \, z_{1-\alpha/2} = \{\hat{\psi}_l, \hat{\psi}_u\} \, ,$$

as described in Section 4.2.2. The corresponding limits for the odds ratio θ are $\{\exp(\hat{\psi}_l), \exp(\hat{\psi}_u)\}$. For the data shown in Figure 4.5, $\hat{\psi} = \log \hat{\theta} = .6104$, and $\hat{s}(\hat{\psi}) = 0.0639$, so the 95%, limits for θ are $\{1.624, 2.087\}$, as shown by the calculations below. The same result is returned by `confint()` for a "loddsratio" object.

```
> summary(loddsratio(Berkeley))


z test of coefficients:

                                Estimate Std. Error z value
Male:Female/Admitted:Rejected     0.6104     0.0639    9.55
                                Pr(>|z|)
Male:Female/Admitted:Rejected     <2e-16 ***
---
Signif. codes:  0 '***' 0.001 '**' 0.01 '*' 0.05 '.' 0.1 ' ' 1

> exp(.6103 + c(-1, 1) * qnorm(.975) * 0.06398)

[1] 1.6240 2.0869

> confint(loddsratio(Berkeley, log = FALSE))

                              2.5 % 97.5 %
Male:Female/Admitted:Rejected 1.6244 2.0867
```

Now consider how to find a 2×2 table whose frequencies correspond to the odds ratios at the limits of the confidence interval. A table standardized to equal row and column margins can be represented by the 2×2 matrix with entries

$$\begin{bmatrix} p & (1-p) \\ (1-p) & p \end{bmatrix} \, ,$$

whose odds ratio is $\theta = p^2/(1-p)^2$. Solving for p gives $p = \sqrt{\theta}/(1 + \sqrt{\theta})$. The corresponding frequencies can then be found by adjusting the standardized table to have the same row and column margins as the data. The results of these computations, which generate the confidence rings in Figure 4.5, are shown in Table 4.7.

4.4.2 Stratified analysis for $2 \times 2 \times k$ tables

In a $2 \times 2 \times k$ table, the last dimension often corresponds to "strata" or populations, and it is typically of interest to see if the association between the first two variables is homogeneous across strata. For such tables, simply make one fourfold panel for each stratum. The standardization of marginal frequencies is designed to allow easy visual comparison of the pattern of association when the marginal frequencies vary across two or more populations.

Table 4.7: Odds ratios and equivalent tables for 95% confidence rings for the Berkeley data.

	Odds Ratio	Standardized Table		Equivalent Frequencies	
Lower	1.624	0.560	0.440	1,167.1	587.9
limit		0.440	0.560	1,523.9	1,247.1
Data	1.841	0.576	0.424	1,198.0	557.0
		0.424	0.576	1,493.0	1,278.0
Upper	2.087	0.591	0.409	1,228.4	526.6
limit		0.409	0.591	1,462.6	1,308.4

4.4.2.1 Stratified displays

The admissions data shown in Figure 4.4 and Figure 4.5 were actually obtained from six departments—the six largest at Berkeley (Bickel et al., 1975). To determine the source of the apparent sex bias in favor of males, we make a new plot, Figure 4.6, stratified by department.

```
> # fourfold display
> UCB <- aperm(UCBAdmissions, c(2, 1, 3))
> fourfold(UCB, mfrow = c(2, 3))
```

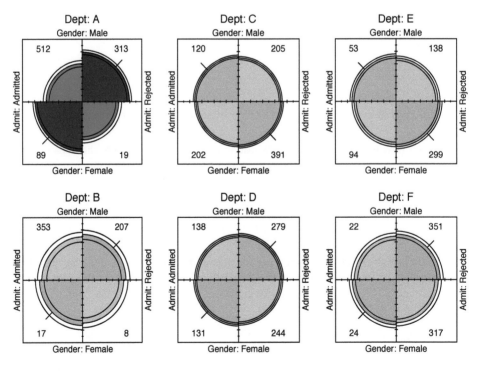

Figure 4.6: Fourfold displays for Berkeley admissions data, stratified by department. The more intense shading for Dept. A indicates a significant association.

Surprisingly, Figure 4.6 shows that, for five of the six departments, the odds of admission is approximately the same for both men and women applicants. Department A appears to differs from the others, with women approximately 2.86 ($= (313/19)/(512/89)$) times as likely to gain admission. This appearance is confirmed by the confidence rings, which in Figure 4.6 are joint[5] 95% intervals for θ_c, $c = 1, \ldots, k$.

This result, which contradicts the display for the aggregate data in Figure 4.4, is a nice example of **Simpson's paradox**,[6] and illustrates clearly why an overall analysis of a three- (or higher-) way table can be misleading.

The resolution of this contradiction can be found in the large differences in admission rates among departments. Men and women apply to different departments differentially, and in these data women happen to apply in larger numbers to departments that have a low acceptance rate. The aggregate results are misleading because they falsely assume men and women are equally likely to apply in each field.[7]

4.4.2.2 Visualization principles for complex data

An important principle in the display of large, complex data sets is **controlled comparison**—we want to make comparisons against a clear standard, with other things held constant. The fourfold display differs from a pie chart in that it holds the angles of the segments constant and varies the radius. An important consequence is that we can quite easily compare a series of fourfold displays for different strata, since corresponding cells of the table are always in the same position. As a result, an array of fourfold displays serve the goals of comparison and detection better than an array of pie charts.

Moreover, it allows the observed frequencies to be standardized by equating either the row or column totals, while preserving the design goal for this display—the odds ratio. In Figure 4.6, for example, the proportion of men and women, and the proportion of accepted applicants were equated visually in each department. This provides a clear standard that also greatly facilitates controlled comparison.

As mentioned in the introduction, another principle is **visual impact**—we want the important features of the display to be easily distinguished from the less important (Tukey, 1993). Figure 4.6 distinguishes the one department for which the odds ratio differs significantly from 1 by shading intensity, even though the same information can be found by inspection of the confidence rings.

EXAMPLE 4.12: Breathlessness and wheeze in coal miners

The various ways of standardizing a collection of 2×2 tables allows visualizing relations with different factors (row percentages, column percentages, strata totals) controlled. However, different kinds of graphs can speak more eloquently to other questions by focusing more directly on the odds ratio.

Agresti (2002, Table 9.8) cites data from Ashford and Sowden (1970) on the association between two pulmonary conditions, breathlessness and wheeze, in a large sample of coal miners. The miners are classified into age groups, and the question treated by Agresti is whether the association between these two symptoms is homogeneous over age. These data are available in the `CoalMiners` data in **vcd**, a $2 \times 2 \times 9$ frequency table. The first group, aged 20–24 has been omitted from these analyses.

---

[5]For multiple-strata plots, `fourfold()` by default adjusts the significance level for multiple testing, using Holm's (1979) method provided by `p.adjust()`.

[6]Simpson's paradox (Simpson, 1951) occurs in a three-way table, $[A, B, C]$, when the marginal association between two variables, A, B collapsing over C, differs in *direction* from the partial association A, $B|C = c_k$ at the separate levels of C. Strictly speaking, Simpson's paradox would require that for all departments separately the odds ratio $\theta_k < 1$ (which occurs for Departments A, B, D, and F in Figure 4.6) while in the aggregate data $\theta > 1$.

[7]This explanation ignores the possibility of structural bias against women, e.g., lack of resources allocated to departments that attract women applicants.

```
> data("CoalMiners", package = "vcd")
> CM <- CoalMiners[, , 2 : 9]
> structable(. ~ Age, data = CM)

        Breathlessness    B           NoB
        Wheeze           W   NoW      W   NoW
Age
25-29                    23    9    105  1654
30-34                    54   19    177  1863
35-39                   121   48    257  2357
40-44                   169   54    273  1778
45-49                   269   88    324  1712
50-54                   404  117    245  1324
55-59                   406  152    225   967
60-64                   372  106    132   526
```

The question of interest can be addressed by displaying the odds ratio in the 2×2 tables with the margins of breathlessness and wheeze equated (i.e., with the default `std='margins'` option), which gives the graph shown in Figure 4.7. Although the panels for all age groups show an overwhelmingly positive association between these two symptoms, one can also (by looking carefully) see that the strength of this association declines with increasing age.

```
> fourfold(CM, mfcol = c(2, 4))
```

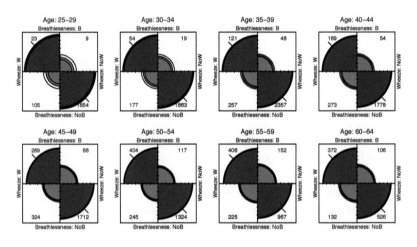

Figure 4.7: Fourfold display for CoalMiners data, both margins equated.

However, note that the pattern of change over age is somewhat subtle compared to the dominant positive association within each panel. When the goal is to display how the odds ratio varies with a quantitative factor such as age, it is often better to simply calculate and plot the odds ratio directly.

The `loddsratio()` function in **vcd** calculates odds ratios. By default, it returns the log odds. Use the option `log=FALSE` to get the odds ratios themselves. It is easy to see that the (log) odds ratios decline with age.

```
> loddsratio(CM)

log odds ratios for Breathlessness and Wheeze by Age

 25-29  30-34  35-39  40-44  45-49  50-54  55-59  60-64
3.6953 3.3983 3.1407 3.0147 2.7820 2.9264 2.4406 2.6380

> loddsratio(CM, log = FALSE)
```

```
odds ratios for Breathlessness and Wheeze by Age

 25-29   30-34   35-39   40-44   45-49   50-54   55-59   60-64
40.256  29.914  23.119  20.383  16.152  18.660  11.480  13.985
```

When the analysis goal is to understand how the odds ratio varies with a stratifying factor (which could be a quantitative variable), it is often better to plot the odds ratio directly.

The lines below use the `plot()` method for **"oddsratio"** objects. This produces a line graph of the log odds ratio against the stratum variable, together with confidence interval error bars. In addition, because age is a quantitative variable, we can calculate and display the fitted relation for a linear model relating `lodds` to `age`. Here, we try using a quadratic model (`poly(age, 2)`) mainly to see if the trend is nonlinear.

```
> lor_CM <- loddsratio(CM)
> plot(lor_CM, bars=FALSE, baseline=FALSE, whiskers=.2)
>
> lor_CM_df <- as.data.frame(lor_CM)
> age <- seq(25, 60, by = 5) + 2
> lmod <- lm(LOR ~ poly(age, 2), weights = 1 / ASE^2, data = lor_CM_df)
> grid.lines(seq_along(age), fitted(lmod),
+               gp = gpar(col = "red", lwd = 2), default.units = "native")
```

log odds ratios for Breathlessness and Wheeze by Age

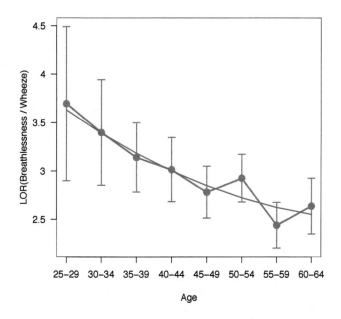

Figure 4.8: Log odds plot for the CoalMiners data. The smooth curve shows a quadratic fit to age.

In Figure 4.8, it appears that the decline in the log odds ratio levels off with increasing age. One virtue of fitting the model in this way is that we can test the additional contribution of the quadratic term, which turns out to be insignificant.

```
> summary(lmod)


Call:
lm(formula = LOR ~ poly(age, 2), data = lor_CM_df, weights = 1/ASE^2)
```

```
Weighted Residuals:
       1       2       3       4       5       6       7       8
  0.1617  0.0162 -0.2443  0.0627 -0.4971  1.6115 -1.5228  0.5851

Coefficients:
               Estimate Std. Error t value Pr(>|t|)
(Intercept)      2.9953     0.0783   38.28 2.3e-07 ***
poly(age, 2)1   -0.9977     0.2513   -3.97   0.011 *
poly(age, 2)2    0.1768     0.2171    0.81   0.452
---
Signif. codes:  0 '***' 0.001 '**' 0.01 '*' 0.05 '.' 0.1 ' ' 1

Residual standard error: 1.06 on 5 degrees of freedom
Multiple R-squared:  0.782, Adjusted R-squared:  0.694
F-statistic: 8.94 on 2 and 5 DF,  p-value: 0.0223
```

△

4.5   Sieve diagrams

> The wise ones fashioned speech with their thought, sifting it as grain is sifted through a sieve.

> Buddha

For two- (and higher-) way contingency tables, the design principles of perception, detection, and comparison (see Chapter 1) suggest that we should try to show the observed frequencies in relation to what we would expect those frequencies to be under a reasonable null model—for example, the hypothesis that the row and column variables are unassociated.

To this end, several schemes for representing contingency tables graphically are based on the fact that when the row and column variables are independent, the estimated expected frequencies, m_{ij}, are products of the row and column totals (divided by the grand total).

$$m_{ij} = \frac{n_{i+}n_{+j}}{n_{++}} .$$

Then, each cell can be represented by a rectangle whose area shows the observed cell frequency, n_{ij}, expected frequency, m_{ij}, or deviation (residual) from independence, $n_{ij} - m_{ij}$. Visual attributes (color, shading) of the rectangles can be used to highlight the pattern of association.

4.5.1   Two-way tables

For example, for any two-way table, the expected frequencies under independence can be represented by rectangles whose widths are proportional to the total frequency in each column, n_{+j}, and whose heights are proportional to the total frequency in each row, n_{i+}; the area of each rectangle is then proportional to m_{ij}. Figure 4.9 (left) shows the expected frequencies for the hair and eye color data (Table 4.2), calculated using `independence_table()` in vcd.

```
> haireye <- margin.table(HairEyeColor, 1:2)
> expected = independence_table(haireye)
> round(expected, 1)

       Eye
Hair     Brown  Blue Hazel Green
  Black   40.1  39.2  17.0  11.7
  Brown  106.3 103.9  44.9  30.9
  Red     26.4  25.8  11.2   7.7
  Blond   47.2  46.1  20.0  13.7
```

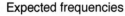

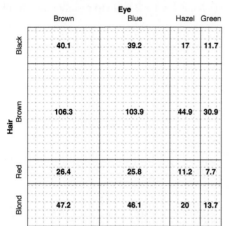
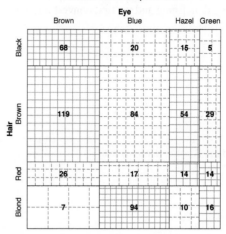

Figure 4.9: Sieve diagrams for the `HairEyeColor` data. Left: expected frequencies shown in cells as numbers and the number of boxes; right: observed frequencies shown in cells.

Figure 4.9 (left) simply represents the model—what the frequencies would be if hair color and eye color were independent—not the data. Note, however, that the rectangles are cross-ruled so that the number of boxes in each (counting up the fractional bits) equals the expected frequency with which the cell is labeled, and moreover, the rulings are equally spaced in all cells. Hence, cross-ruling the cells to show the observed frequency would give a data display that implicitly compares observed and expected frequencies as shown in Figure 4.9 (right).

Riedwyl and Schüpbach (1983, 1994) proposed a *sieve diagram* (later called a *parquet diagram*) based on this principle. In this display the area of each rectangle is always proportional to expected frequency but observed frequency is shown by the number of squares in each rectangle, as in Figure 4.9 (right).

Hence, the difference between observed and expected frequency appears as variations in the density of shading. Cells whose observed frequency n_{ij} exceeds the expected m_{ij} appear denser than average. The pattern of positive and negative deviations from independence can be more easily seen by using color, say, red for negative deviations, and blue for positive.[8]

EXAMPLE 4.13: Hair color and eye color

The sieve diagram for hair color and eye color shown in Figure 4.9 (right) can be interpreted as follows: The pattern of color and shading shows the high frequency of blue-eyed blonds and people with brown eyes and dark hair. People with hazel eyes are also more likely to have red or brown hair, and those with green eyes more likely to have red or blond hair, than would be observed under independence.   △

EXAMPLE 4.14: Visual acuity

In World War II, all workers in the UK Royal Ordnance factories were given test of visual acuity (unaided distance vision) of their left and right eyes on a 1 (high) to 4 (low) scale. The dataset `VisualAcuity` in vcd gives the results for 10,719 workers (3,242 men, 7,477 women) aged 30–39.

Figure 4.10 shows the sieve diagram for data from the larger sample of women (Kendall and

---

[8]Positive residuals are also shown by solid lines, negative residuals by broken lines, so that they may still be distinguished in monochrome versions.

Stuart (1961, Table 33.5) and Bishop et al. (1975, p. 284)). The `VisualAcuity` data is a frequency data frame and we first convert it to table form (VA), a $4 \times 4 \times 2$ table to re-label the variables and levels.

```
> # re-assign names/dimnames
> data("VisualAcuity", package = "vcd")
> VA <- xtabs(Freq ~ right + left + gender, data = VisualAcuity)
> dimnames(VA)[1:2] <- list(c("high", 2, 3, "low"))
> names(dimnames(VA))[1:2] <- paste(c("Right", "Left"), "eye grade")
> structable(aperm(VA))
```

```
                          Left eye grade high     2    3  low
gender Right eye grade
male   high                              821   112   85   35
       2                                 116   494  145   27
       3                                  72   151  583   87
       low                                43    34  106  331
female high                             1520   266  124   66
       2                                 234  1512  432   78
       3                                 117   362 1772  205
       low                               36    82  179  492
```

```
> sieve(VA[, , "female"], shade = TRUE)
```

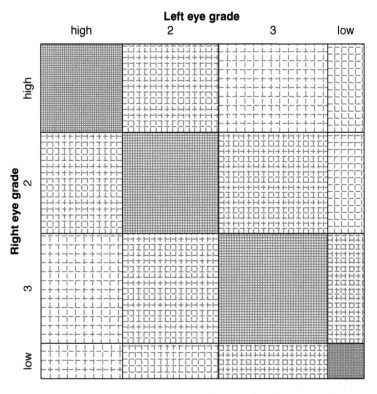

Figure 4.10: Vision classification for 7,477 women in Royal Ordnance factories. The high frequencies in the diagonal cells indicate the main association, but a subtler pattern also appears in the symmetric off-diagonal cells.

The diagonal cells show the obvious: people tend to have the same visual acuity in both eyes, and there is strong lack of independence. The off diagonal cells show a more subtle pattern that suggests symmetry—the cells below the diagonal are approximately equally dense as the corresponding cells above the diagonal. Moreover, the relatively consistent pattern on the diagonals $\pm 1, \pm 2, \ldots$ away from the main diagonals suggests that the association may be explained in terms of the *difference* in visual acuity between the two eyes.

These suggestions can be tested by fitting intermediate models between the null model of independence (which fits terribly) and the saturated model (which fits perfectly), as we shall see later in this book. A model of ***quasi-independence***, for example (see Example 10.5 in Chapter 9) ignores the diagonal cells and tests whether independence holds for the remainder of the table. The ***symmetry*** model for a square table allows association, but constrains the expected frequencies above and below the main diagonal to be equal. Such models provide a way of testing *specific* explanatory models that relate to substantive hypotheses and what we observe in our visualizations. These and other models for square tables are discussed further in Section 10.2. △

4.5.2 Larger tables: The strucplot framework

The implementation of sieve diagrams in vcd is far more general than illustrated in the examples above. For one thing, the `sieve` function has a formula method, which allows one to specify the variables in the display as a model formula. For example, for the `VisualAcuity` data, a plot of the (marginal) frequencies for left and right eye grades pooling over gender can be obtained with the call below (this plot is not shown).

```
> sieve(Freq ~ right + left, data = VisualAcuity, shade = TRUE)
```

More importantly, sieve diagrams are just one example of the ***strucplot framework***, a general system for visualizing n-way frequency tables in a hierarchical way. We describe this framework in more detail in Section 5.3 in the context of mosaic displays. For now, we just illustrate the extension of the formula method to provide for conditioning variables. In the call below, the formula `Freq ~ right + left | gender` means to produce a separate block in the plot for the levels of `gender`. The `set_varnames` argument relabels the variable names.

```
> sieve(Freq ~ right + left | gender, data = VisualAcuity,
+        shade = TRUE, set_varnames = c(right = "Right eye grade",
+                                       left = "Left eye grade"))
```

In Figure 4.11, the relative sizes of the blocks for the conditioning variable (`gender`) show the much larger number of women than men in this data. Within each block, color and density of the box rules shows the association of left and right acuity, and it appears that the pattern for men is similar to that observed for women.

An alternative way of visualizing stratified data is a ***coplot*** or ***conditioning plot***, which, for each stratum, shows an appropriate display for a subset of the data. Figure 4.12 visualizes separate sieve plots for men and women:

```
> cotabplot(VA, cond = "gender", panel = cotab_sieve, shade = TRUE)
```

The main difference to the extended sieve plots is that the distribution of the conditioning variable is not shown, which basically is a loss of information, but advantageous if the distribution of the conditioning variable(s) is highly skewed, since the partial displays of small strata will not be distorted.

The methods described in Section 4.3.2 can be used to test the hypothesis of homogeneity of association, and loglinear models described in Chapter 9 provide specific tests of hypotheses of ***symmetry***, ***quasi-independence***, and other models for structured associations.

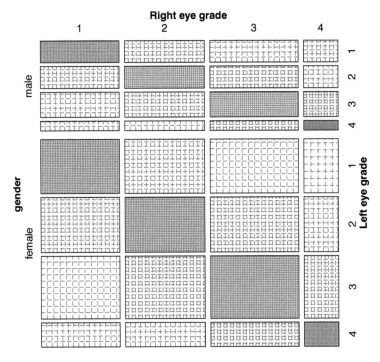

Figure 4.11: Sieve diagram for the three-way table of VisualAcuity, conditioned on gender.

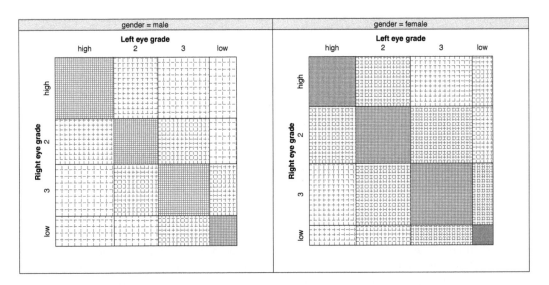

Figure 4.12: Conditional Sieve diagram for the three-way table of VisualAcuity, conditioned on gender.

EXAMPLE 4.15: Berkeley admissions

This example illustrates some additional flexibility of sieve plots with the strucplot framework, using the Berkeley admissions data. The left panel of Figure 4.13 shows the sieve diagrams for the relation between department and admission, conditioned by gender. It can easily be seen that (a) overall, there were more male applicants than female; (b) there is a moderately similar pattern of observed > expected (blue) for males and females.

```
> # conditioned on gender
> sieve(UCBAdmissions, shade = TRUE, condvar = 'Gender')
> # three-way table, Department first, with cell labels
> sieve(~ Dept + Admit + Gender, data = UCBAdmissions,
+       shade = TRUE, labeling = labeling_values,
+       gp_text = gpar(fontface = 2), abbreviate_labs = c(Gender = TRUE))
```

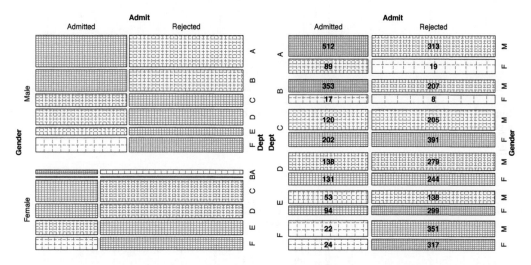

Figure 4.13: Sieve diagrams for the three-way table of the Berkeley admissions data. Left: Admit by Dept, conditioned on Gender; right: Dept re-ordered as the first splitting variable.

In the right panel of Figure 4.13, the three-way table was first permuted to make `Dept` the first splitting variable. Each 2 × 2 table of `Admit` by `Gender` then appears, giving a sieve diagram version of what we showed earlier in fourfold displays (Figure 4.6). The `labeling` argument is used here to write the cell frequency in each rectangle. `gp_text` renders them in bold font, and `abbreviate_labs` abbreviates the gender labels to avoid overplotting.

Alternatively, we can again use coplots to visualize conditioned sieve plots for this data. The following calls produce Figure 4.14 and Figure 4.15, with different conditioning and styles.

```
> cotabplot(UCBAdmissions, cond = "Gender", panel = cotab_sieve,
+           shade = TRUE)
```

```
> cotabplot(UCBAdmissions, cond = "Dept", panel = cotab_sieve,
+           shade = TRUE, labeling = labeling_values,
+           gp_text = gpar(fontface = "bold"))
```

Remark

Finally, for tables of more than two dimensions, there is a variety of different models for "independence" (discussed in Chapter 9 on log-linear models), and the strucplot framework allows these to

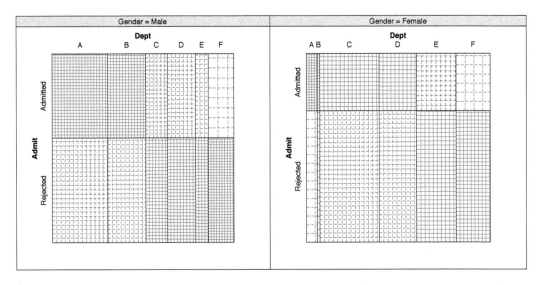

Figure 4.14: Conditional Sieve diagram for the three-way table of the Berkeley data, conditioned on gender.

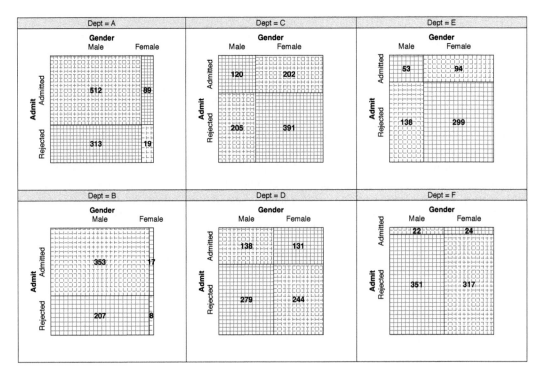

Figure 4.15: Conditional Sieve diagram for the three-way table of the Berkeley data, conditioned on department.

be specified with the `expected` argument, either as an array of numbers conforming to the `data` argument, or as a model formula for `loglm()`.

For example, a sieve diagram may be used to determine if the association between gender and department is the same across departments by fitting the model ~ `Admit * Gender + Dept`, which says that `Dept` is independent of the combinations of `Admit` and `Gender`. This is done as shown below, giving the plot in Figure 4.16.

```
> UCB2 <- aperm(UCBAdmissions, c(3, 2, 1))
> sieve(UCB2, shade = TRUE, expected = ~ Admit * Gender + Dept,
+       split_vertical = c(FALSE, TRUE, TRUE))
```

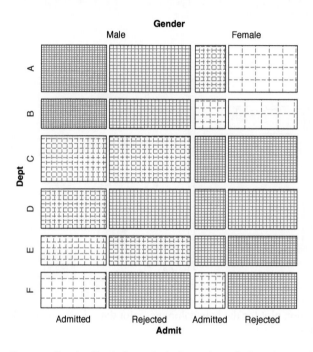

Figure 4.16: Sieve diagram for the Berkeley admissions data, fitting the model of joint independence, Admit * Gender + Dept.

In terms of the loglinear models discussed in Chapter 5, this is equivalent to fitting the model of *joint independence*, [Admit Gender][Dept]. Figure 4.16 shows the greater numbers of male applicants in departments A and B (whose overall rate of admission is high) and greater numbers of female applicants in the remaining departments (where the admission rate is low).

$\triangle$

4.6 Association plots

In the *sieve diagram* the foreground (rectangles) shows expected frequencies; deviations from independence are shown by color and density of shading. The *association plot* (Cohen, 1980, Friendly, 1991) puts deviations from independence in the foreground: the area of each box is made proportional to the (observed − expected) frequency.

For a two-way contingency table, the signed contribution to Pearson χ^2 for cell i, j is

$$r_{ij} = \frac{n_{ij} - m_{ij}}{\sqrt{m_{ij}}} = \text{Pearson residual}, \qquad \chi^2 = \sum_{i,j} r_{ij}^2. \qquad (4.5)$$

In the association plot, each cell is shown by a rectangle, having:

- (signed) height $\sim r_{ij}$,
- width = $\sqrt{m_{ij}}$,

so, the area of each cell is proportional to the raw residual, $n_{ij} - m_{ij}$. The rectangles for each row in the table are positioned relative to a baseline representing independence ($r_{ij} = 0$) shown by a dotted line. Cells with observed > expected frequency rise above the line (and are colored blue); cells that contain less than the expected frequency fall below it (and are shaded red).

```
> assoc(~ Hair + Eye, data = HairEyeColor, shade = TRUE)
> assoc(HairEyeColor, shade = TRUE)
```

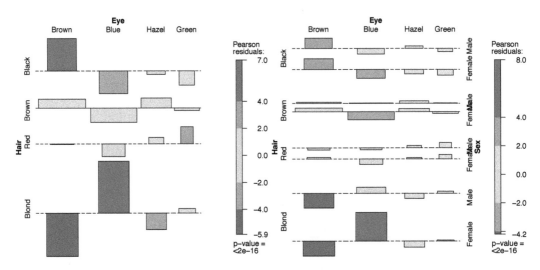

Figure 4.17: Association plot for the hair-color eye-color data. Left: marginal table, collapsed over gender; right: full table.

Figure 4.17 (left) shows the association plot for the data on hair color and eye color. In constructing this plot, each rectangle is shaded according to the value of the Pearson residual from Eqn. (4.5), using a simple scale shown in the legend, where residuals $|r_{ij}| > 2$ are shaded blue or red depending on their sign, and residuals $|r_{ij}| > 4$ are shaded with a more saturated color.

One virtue of the association plot is that it is quite simple to interpret in terms of the pattern of positive and negative r_{ij} values. Bertin (1981) uses similar graphics to display large complex contingency tables. Like the sieve diagram, however, patterns of association are most apparent when the rows and columns of the display are ordered in a sensible way.

We note here that the association plot also belongs to the strucplot framework and thus extends to higher-way tables. For example, the full *HairEyeColor* table is also classified by Sex. The plot for the three-way table is shown in Figure 4.17 (right). In this plot the third table variable (Sex here) is shown nested within the first two, allowing easy comparison of the profiles of hair and eye color for males and females.

4.7 Observer agreement

When the row and column variables represent different observers' rating the same subjects or objects, interest is focused on ***observer agreement*** rather than mere association. In this case, measures and tests of agreement provide a method of assessing the reliability of a subjective classification or assessment procedure.

For example, two (or more) clinical psychologists might classify patients on a scale with categories (a) normal, (b) mildly impaired, (c) severely impaired. Or, ethologists might classify the behavior of animals in categories of cooperation, dominance and so forth, or paleologists might classify pottery fragments according to categories of antiquity or cultural groups. As these examples suggest, the rating categories are often ordered, but not always.

For two raters, a contingency table can be formed by classifying all the subjects/objects rated according to the rating categories used by the two observers. In most cases, the same categories are used by both raters, so the contingency table is square, and the entries in the diagonal cells are the cases where the raters agree.

In this section we describe some measures of the strength of agreement and then a method for visualizing the pattern of agreement. But first, the following examples show some typical agreement data.

EXAMPLE 4.16: Sex is fun

The *SexualFun* table in **vcd** (Agresti (1990, Table 2.10), from Hout et al. (1987)) summarizes the responses of 91 married couples to a questionnaire item: "Sex is fun for me and my partner: (a) Never or occasionally, (b) Fairly often, (c) Very often, (d) Almost always."

```
> data("SexualFun", package = "vcd")
> SexualFun

                Wife
Husband        Never Fun Fairly Often Very Often Always fun
  Never Fun            7            7          2          3
  Fairly Often         2            8          3          7
  Very Often           1            5          4          9
  Always fun           2            8          9         14
```

In each row the diagonal entry is not always the largest, though it appears that the partners tend to agree more often when either responds "Almost always."   △

EXAMPLE 4.17: Diagnosis of MS patients

Landis and Koch (1977) gave data on the diagnostic classification of multiple sclerosis (MS) patients by two neurologists, one from Winnipeg and one from New Orleans. There were two samples of patients, 149 from Winnipeg and 69 from New Orleans, and each neurologist classified all patients into one of four diagnostic categories: (a) Certain MS, (b) Probable MS, (c) Possible MS, (d) Doubtful, unlikely, or definitely not MS.

These data are available in *MSPatients*, a $4 \times 4 \times 2$ table, as shown below. It is convenient to show the data in separate slices for the Winnipeg and New Orleans patients:

```
> MSPatients[, , "Winnipeg"]

                        Winnipeg Neurologist
New Orleans Neurologist Certain Probable Possible Doubtful
            Certain          38        5        0        1
            Probable         33       11        3        0
            Possible         10       14        5        6
            Doubtful          3        7        3       10

> MSPatients[, , "New Orleans"]

                        Winnipeg Neurologist
New Orleans Neurologist Certain Probable Possible Doubtful
```

```
             Certain       5         3          0          0
             Probable      3        11          4          0
             Possible      2        13          3          4
             Doubtful      1         2          4         14

> apply(MSPatients, 3, sum)        # show sample sizes

  Winnipeg New Orleans
       149          69
```

In this example, note that the distribution of degree of severity of MS may differ between the two patient samples. As well, for a given sample, the two neurologists may be more or less strict about the boundaries between the rating categories.

△

4.7.1   Measuring agreement

In assessing the strength of *agreement* we usually have a more stringent criterion than in measuring the strength of *association*, because observers ratings can be strongly associated without strong agreement. For example, one rater could use a more stringent criterion and thus consistently rate subjects one category lower (on an ordinal scale) than another rater.

More generally, measures of agreement must take account of the marginal frequencies with which two raters use the categories. If observers tend to use the categories with different frequency, this will affect measures of agreement.

Here we describe some simple indices that summarize agreement with a single score (and associated standard errors or confidence intervals). Von Eye and Mun (2006) treat this topic from the perspective of loglinear models.

4.7.1.1   Intraclass correlation

An analysis of variance framework leads to the **intraclass correlation** as a measure of inter-rater reliability, particularly when there are more than two raters. This approach is not covered here, but various applications are described by Shrout and Fleiss (1979), and implemented in R in ICC() in the psych (Revelle, 2015) package.

4.7.1.2   Cohen's Kappa

Cohen's kappa (κ) (Cohen, 1960, 1968) is a commonly used measure of agreement that compares the observed agreement to agreement expected by chance if the two observer's ratings were independent. If p_{ij} is the probability that a randomly selected subject is rated in category i by the first observer and in category j by the other, then the observed agreement is the sum of the diagonal entries, $P_o = \sum_i p_{ii}$. If the ratings were independent, this probability of agreement (by chance) would be $P_c = \sum_i p_{i+} p_{+i}$. Cohen's κ is then the ratio of the difference between actual agreement and chance agreement, $P_o - P_c$, to the maximum value this difference could obtain:

$$\kappa = \frac{P_o - P_c}{1 - P_c} . \tag{4.6}$$

When agreement is perfect, $\kappa = 1$; when agreement is no better than would be obtained from statistically independent ratings, $\kappa = 0$. κ could conceivably be negative, but this rarely occurs in practice. The minimum possible value depends on the marginal totals.

For large samples (n_{++}), κ has an approximate normal distribution when $H_0 : \kappa = 0$ is true and its standard error (Fleiss, 1973, Fleiss et al., 1969) is given by

$$\hat{\sigma}(\kappa) = \frac{P_c + P_c^2 - \sum_i p_{i+}p_{+i}(p_{i+} + p_{+i})}{n_{++}(1 - P_c)^2}.$$

Hence, it is common to conduct a test of $H_0 : \kappa = 0$ by referring $z = \kappa/\hat{\sigma}(\kappa)$ to a unit normal distribution. The hypothesis of agreement no better than chance is rarely of much interest, however. It is preferable to estimate and report a confidence interval for κ.

4.7.1.3 Weighted Kappa

The original (unweighted) κ only counts strict agreement (the same category is assigned by both observers). A weighted version of κ (Cohen, 1968) may be used when one wishes to allow for *partial* agreement. For example, exact agreements might be given full weight, while a one-category difference might be given a weight of 1/2. This typically makes sense only when the categories are *ordered*, as in severity of diagnosis.

Weighted κ uses weights, $0 \le w_{ij} \le 1$ for each cell in the table, with $w_{ii} = 1$ for the diagonal cells. In this case P_o and P_c are defined as weighted sums

$$P_o = \sum_i \sum_j w_{ij}p_{ij}$$

$$P_c = \sum_i \sum_j w_{ij}p_{i+}p_{+j}$$

and these weighted sums are used in Eqn. (4.6).

For an $R \times R$ table, two commonly used patterns of weights are those based on equal spacing of weights (Cicchetti and Allison, 1971) for a near-match, and *Fleiss-Cohen weights* (Fleiss and Cohen, 1972), based on an inverse-square spacing,

$$w_{ij} = 1 - \frac{|i-j|}{R-1} \qquad \text{equal spacing}$$

$$w_{ij} = 1 - \frac{|i-j|^2}{(R-1)^2} \qquad \text{Fleiss-Cohen}$$

The Fleiss-Cohen weights attach greater importance to near disagreements, as you can see below for a 4×4 table. These weights also provide a measure equivalent to the intraclass correlation.

Integer Spacing Cicchetti Allison weights				Inverse Square Spacing Fleiss-Cohen weights			
1	2/3	1/3	0	1	8/9	5/9	0
2/3	1	2/3	1/3	8/9	1	8/9	5/9
1/3	2/3	1	2/3	5/9	8/9	1	8/9
0	1/3	2/3	1	0	5/9	8/9	1

4.7.1.4 Computing Kappa

The function `Kappa()` in vcd calculates unweighted and weighted Kappa. The `weights` argument can be used to specify the weighting scheme as either `"Equal-Spacing"` or `"Fleiss-Cohen"`. The function returns a **"Kappa"** object, for which there is a `confint.Kappa()` method, providing confidence intervals. The `summary.Kappa()` method also prints the weights.

The lines below illustrate Kappa for the *SexualFun* data.

```
> Kappa(SexualFun)

           value      ASE      z Pr(>|z|)
Unweighted 0.129 0.0686 1.89   0.05939
Weighted   0.237 0.0783 3.03   0.00244

> confint(Kappa(SexualFun))


Kappa                  lwr       upr
  Unweighted −0.0051204 0.26378
  Weighted    0.0838834 0.39088
```

4.7.2   Observer agreement chart

The observer agreement chart proposed by Bangdiwala (1985, 1987) provides a simple graphic representation of the strength of agreement in a contingency table, and alternative measures of strength of agreement with an intuitive interpretation. More importantly, it shows the *pattern* of disagreement when agreement is less than perfect.

4.7.2.1   Construction of the basic plot

Given a $k \times k$ contingency table, the agreement chart is constructed as an $n \times n$ square, where $n = n_{++}$ is the total sample size. Black squares, each of size $n_{ii} \times n_{ii}$, show observed agreement. These are positioned within k larger rectangles, each of size $n_{i+} \times n_{+i}$ as shown in the left panel of Figure 4.18. Each rectangle is subdivided by the row/column frequencies n_{ij} of row i/column j, where cell (i, i) is filled black. The large rectangle shows the maximum possible agreement, given the marginal totals. Thus, a visual impression of the strength of agreement is given by

$$B = \frac{\text{area of dark squares}}{\text{area of rectangles}} = \frac{\sum_i^k n_{ii}^2}{\sum_i^k n_{i+} n_{+i}} . \tag{4.7}$$

When there is perfect agreement, the k rectangles determined by the marginal totals are all squares, completely filled by the shaded squares reflecting the diagonal n_{ii} entries, and $B = 1$.

```
> agreementplot(SexualFun, main = "Unweighted", weights = 1)
> agreementplot(SexualFun, main = "Weighted")
```

4.7.2.2   Partial agreement

Partial agreement is allowed by including a weighted contribution from off-diagonal cells, b steps from the main diagonal. For a given cell frequency, n_{ij}, a pattern of weights, $w_1, w_2, \ldots, w_b$ is applied to the cell frequencies as shown schematically below:

$$
\begin{array}{ccccc}
 & n_{i-b,i} & & & w_b \\
 & \vdots & & & \vdots \\
n_{i,i-b} \cdots & n_{i,i} & \cdots n_{i,i+b} & \Leftarrow & w_b \cdots 1 \cdots w_b \\
 & \vdots & & & \vdots \\
 & n_{i+b,i} & & & w_b
\end{array}
$$

These weights are incorporated in the agreement chart (right panel of Figure 4.18) by successively lighter shaded rectangles whose size is proportional to the sum of the cell frequencies, denoted

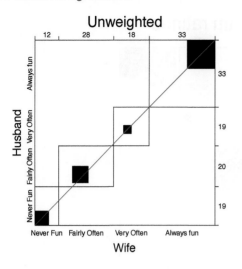
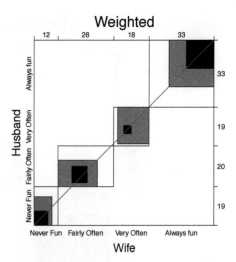

Figure 4.18: Agreement charts for husbands' and wives' sexual fun. Left: unweighted chart, showing only exact agreement; right: weighted chart, using weight $w_1 = 8/9$ for a one-step disagreement.

A_{bi}, shown above. A_{1i} allows 1-step disagreements, using weights 1 and w_1; A_{2i} includes 2-step disagreements, etc. From this, one can define a weighted measure of agreement, B^w, analogous to weighted κ:

$$B^w = \frac{\text{weighted sum of areas of agreement}}{\text{area of rectangles}} = 1 - \frac{\sum_i^k \left[n_{i+} n_{+i} - n_{ii}^2 - \sum_{b=1}^q w_b A_{bi} \right]}{\sum_i^k n_{i+} n_{+i}}$$

where w_b is the weight for A_{bi}, the shaded area b steps away from the main diagonal, and q is the furthest level of partial disagreement to be considered.

The function `agreementplot()` actually calculates both B and B^w and returns them invisibly as the result of the call. The results, $B = 0.146$, and $B^w = 0.498$, indicate a stronger degree of agreement when 1-step disagreements are included.

```
> B <- agreementplot(SexualFun)
> unlist(B)[1 : 2]

        Bangdiwala Bangdiwala_Weighted
          0.14646             0.49817
```

EXAMPLE 4.18: Mammogram ratings

The *Mammograms* data in **vcdExtra** gives a 4×4 table of (probably contrived) ratings of 110 mammograms by two raters from Kundel and Polansky (2003), used to illustrate the calculation and interpretation of agreement measures in this context.[9]

```
> data("Mammograms", package = "vcdExtra")
> B <- agreementplot(Mammograms, main = "Mammogram ratings")
```

The agreement plot in Figure 4.19 shows substantial agreement among the two raters, particularly when one-step disagreements are taken into account. Careful study of this graph shows that

---

[9]In practice, of course, rater agreement on severity of diagnosis from radiology images varies with many factors. See Antonio and Crespi (2010) for a meta-analytic study concerning agreement in breast cancer diagnosis.

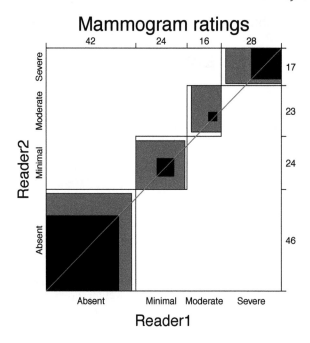

Figure 4.19: Agreement plot for the Mammograms data.

the two raters more often agree exactly for the extreme categories of "Absent" and "Severe." The amounts of unweighted and weighted agreement are shown numerically in the B and B^w statistics.

```
> unlist(B)[1 : 2]

        Bangdiwala Bangdiwala_Weighted
         0.42721            0.83665
```

$\triangle$

4.7.3   Observer bias in agreement

With an ordered scale, it may happen that one observer consistently tends to classify the objects into higher or lower categories than the other, perhaps due to using stricter thresholds for the boundaries between adjacent categories. This bias produces differences in the marginal totals, (n_{i+} and n_{+i}), and decreases the maximum possible agreement. While special tests exist for ***marginal homogeneity***, the observer agreement chart shows this directly by the relation of the dark squares to the diagonal line: When the marginal totals are the same, the squares fall along the diagonal. The measures of agreement, κ and B, cannot determine whether lack of agreement is due to such bias, but the agreement chart can detect this.

EXAMPLE 4.19: Diagnosis of MS patients

Agreement charts for both patient samples in the `MSPatients` data are shown in Figure 4.20. The `agreementplot()` function only handles two-way tables, so we use `cotabplot()` with the `agreementplot` panel function to handle the `Patients` stratum:

```
> cotabplot(MSPatients, cond = "Patients", panel = cotab_agreementplot,
+          text_gp = gpar(fontsize = 18), xlab_rot=20)
```

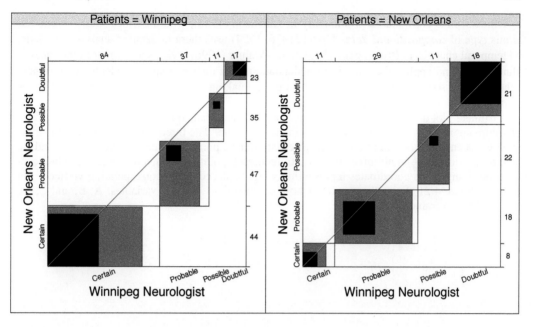

Figure 4.20: Weighted agreement charts for both patient samples in the MSPatients data. Departure of the middle rectangles from the diagonal indicates lack of marginal homogeneity.

It can be seen that, for both groups of patients, the rectangles for the two intermediate categories lie largely below the diagonal line (representing equality). This indicates that the Winnipeg neurologist tends to classify patients into more severe diagnostic categories. The departure from the diagonal is greater for the Winnipeg patients, for whom the Winnipeg neurologist uses the two most severe diagnostic categories very often, as can also be seen from the marginal totals printed in the plot margins.

Nevertheless, there is a reasonable amount of agreement if one-step disagreements are allowed, as can be seen in Figure 4.20 and quantified in the B^w statistics below. The agreement charts also serve to explain why the B measures for exact agreement are so much lower.

```
> agr1 <- agreementplot(MSPatients[, , "Winnipeg"])
> agr2 <- agreementplot(MSPatients[, , "New Orleans"])
> rbind(Winnipeg = unlist(agr1), NewOrleans = unlist(agr2))[, 1 : 2]

           Bangdiwala Bangdiwala_Weighted
Winnipeg      0.27210             0.73808
NewOrleans    0.28537             0.82231
```

$\triangle$

4.8 Trilinear plots

The ***trilinear plot*** (also called a *ternary diagram* or *trinomial plot*) is a specialized display for a 3-column contingency table or for three variables whose relative proportions are to be displayed. Individuals may be assigned to one of three diagnostic categories, for example, or a chemical process may yield three constituents in varying proportions, or we may look at the division of votes among three parties in a parliamentary election. This display is useful, therefore, for both frequencies and proportions.

Trilinear plots are featured prominently in Aitchison (1986), who describes statistical models for this type of ***compositional data***. Upton (1976, 1994) uses them in detailed analyses of spatial and temporal changes in British general elections. Wainer (1996) reviews a variety of other uses of trilinear plots and applies them to aid in understanding the distributions of students' achievement in the National Assessment of Educational Progress, making some aesthetic improvements to the traditional form of these plots along the way.

A trilinear plot displays each observation as a point inside an equilateral triangle whose coordinates correspond to the relative proportions in each column. The three vertices represent the three extremes when 100% occurs in one of the three columns; a point in the exact center corresponds to equal proportions of $\frac{1}{3}$ in all three columns. In fact, each point represents the (weighted) barycenter of the triangle, the coordinates representing weights placed at the corresponding vertices. For instance, Figure 4.21 shows three points whose compositions of three variables, A, B, and C, are given in the data frame DATA below.

```
> library(ggtern)
> DATA <- data.frame(
+    A = c(40, 20, 10),
+    B = c(30, 60, 10),
+    C = c(30, 20, 80),
+    id = c("1", "2", "3"))
>
> aesthetic_mapping <- aes(x = C, y = A, z = B, colour = id)
> ggtern(data = DATA, mapping = aesthetic_mapping) +
+     geom_point(size = 4) +
+     theme_rgbw()
```

(The plot shown requires some more cosmetic parameters not shown for simplicity).

Note that each apex corresponds to 100% of the labeled variable, and the percentage of this variable decreases linearly along a line to the midpoint of the opposite baseline. The grid lines in the figure show the percentage value along each axis.

The construction of trilinear plots is described in detail in http://en.wikipedia.org/

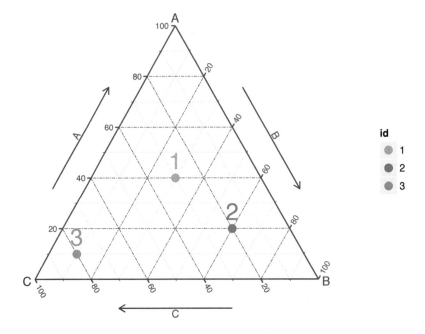

Figure 4.21: A trilinear plot showing three points, for variables A, B, C.

`wiki/Ternary_plot`. Briefly, let $P(a, b, c)$ represent the three components normalized so that $a + b + c = 1.0$. If the apex corresponding to Point A in Figure 4.21 is given (x, y) coordinates of $(x_A, y_A) = (0, 0)$, and those at apex B are $(x_B, y_B) = (100, 0)$, then the coordinates of apex C are $(x_C, y_C) = (50, 50\sqrt{3})$. The cartesian coordinates (x_P, y_P) of point P are then calculated as

$$y_P = c\, y_C$$
$$x_P = y_P \left(\frac{y_C - y_B}{x_C - x_B} \right) + \frac{\sqrt{3}}{2} y_C (1 - a).$$

In R, trilinear plots are implemented in the `triplot()` function in the **TeachingDemos** (Snow, 2013) package, and also in the **ggtern** (Hamilton, 2014) package, an extension of the **ggplot2** framework. The latter is much more flexible, because it inherits all of the capabilities of **ggplot2** for plot annotations, faceting, and layers. In essence, the function `ggtern()` is just a wrapper for `ggplot(...)`, which adds a change in the coordinate system from cartesian (x, y) coordinates to the ternary coordinate system with `coord_tern()`.

EXAMPLE 4.20: Lifeboats on the *Titanic*

We examine the question of who survived and why in the sinking of the *RMS Titanic* in Chapter 5 (Example 5.19, Example 5.21, Exercise 5.12), where we analyze a four-way table, `Titanic`, of the 2, 201 people on board (1, 316 passengers and 885 crew), classified by `Class`, `Sex`, `Age`, and `Survival`. A related data set, `Lifeboats` in **vcd**, tabulates the survivors according to the lifeboats on which they were loaded. This data sheds some additional light on the issue of survival and provides a nice illustration of trilinear plots.

A bit of background: after the disaster, the British Board of Trade launched several inquiries, the most comprehensive of which resulted in the *Report on the Loss of the "Titanic" (S.S.)* by Lord Mersey (Mersey, 1912).[10] The data frame `Lifeboats` in **vcd** contains the data listed on page 38 of that report.[11]

Of interest here is the composition of the boats by the three categories: men, women and children, and crew, and according to the launching of the boats from the port or starboard side. This can be shown in a trilinear display using the following statements. The plot, shown in Figure 4.22, has most of the points near the bottom left, corresponding to a high percentage of women and children. We create a variable, `id`, used to label those boats with more than 10% male passengers. In the **ggplot2** framework, plot aesthetics such as color and shape can be mapped to variables in the data set, and here we map these both to `side` of the boat.

```
> data("Lifeboats", package = "vcd")
> # label boats with more than 10% men
> Lifeboats$id <- ifelse(Lifeboats$men / Lifeboats$total > .1,
+                        as.character(Lifeboats$boat), "")
>
> AES <- aes(x = women, y = men, z = crew, colour = side, shape = side,
+            label = id)
> ggtern(data = Lifeboats, mapping = AES) +
+       geom_text() +
+       geom_point(size=2) +
+       geom_smooth_tern(method = "lm", alpha = 0.2)
```

---

[10]The *Titanic* was outfitted with 20 boats, half on each of the port and starboard sides, of which 14 were large lifeboats with a capacity of 65, two were emergency boats designed for 40 persons, and the remaining four were collapsible boats capable of holding 47, a total capacity of 1, 178 (considered adequate at that time). Two of the collapsible boats, lashed to the roof of the officers' quarters, were ineffectively launched and utilized as rafts after the ship sunk. The report lists the time of launch and composition of the remaining 18 boats according to male passengers, women and children, and "men of crew," as reported by witnesses.

[11]The "data" lists a total of 854 in 18 boats, although only 712 were in fact saved. Mersey notes "it is obvious that these figures are quite unreliable."

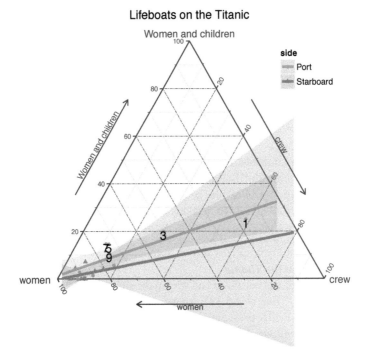

Figure 4.22: Lifeboats on the *Titanic*, showing the composition of each boat. Boats with more than 10% male passengers are labeled.

The resulting plot in Figure 4.22 (for which some more cosmetic parameters than shown in the code above have been used) makes it immediately apparent that many of the boats launched from the port side differ substantially from the starboard boats, whose passengers were almost entirely women and children. Boat 1 had only 20% (2 out of 10) women and children, while the percentage for boat 3 was only 50% (25 out of 50). We highlight the difference in composition of the boats launched from the two sides by adding seperate linear regression lines for the relation men ~ women.

The trilinear plot scales the numbers for each observation to sum to 1.0, so differences in the total number of people on each boat cannot be seen in Figure 4.22. The total number reported loaded is plotted against launch time in Figure 4.23, with a separate regression line and loess smooth fit to the data for the port and starboard sides (code again simplified for clarity):

```
> AES <- aes(x = launch, y = total, colour = side, label = boat)
> ggplot(data = Lifeboats, mapping = AES) +
+       geom_text() +
+       geom_smooth(method = "lm", aes(fill = side), size = 1.5) +
+       geom_smooth(method = "loess", aes(fill = side), se = FALSE,
+               size = 1.2)
```

From the linear regression lines in Figure 4.23, it seems that the rescue effort began in panic on the port side, with relatively small numbers loaded, and (from Figure 4.22), small proportions of women and children. But the loading regime on that side improved steadily over time. The procedures began more efficiently on the starboard side but the numbers loaded increased only slightly. The smoothed loess curves indicate that over time, for each side, there was still a large variability from boat to boat.

△

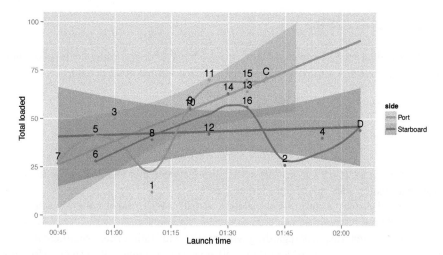

Figure 4.23: Number of people loaded on lifeboats on the Titanic vs. time of launch, by side of boat. The plot annotations show the linear regression and loess smooth.

4.9 Chapter summary

- A contingency table gives the frequencies of observations cross-classified by two or more categorical variables. With such data we are typically interested in testing whether associations exist, quantifying the strength of association, and understanding the nature of the association among these variables.

- For 2×2 tables, association is easily summarized in terms of the odds ratio or its logarithm. This measure can be extended to stratified $2 \times 2 \times k$ tables, where we can also assess whether the odds ratios are equal across strata or how they vary.

- For $R \times C$ tables, measures and tests of general association between two categorical variables are most typically carried out using the Pearson's chi-squared or likelihood-ratio tests provided by `assocstats()`. Stratified tests controlling for one or more background variables, and tests for ordinal categories, are provided by the Cochran–Mantel–Haenszel tests given by `CMHtest()`.

- For 2×2 tables, the fourfold display provides a visualization of the association between variables in terms of the odds ratio. Confidence rings provide a visual test of whether the odds ratio differs significantly from 1. Stratified plots for $2 \times 2 \times k$ tables are also provided by `fourfold()`.

- Sieve diagrams and association plots provide other useful displays of the pattern of association in $R \times C$ tables. These also extend to higher-way tables as part of the strucplot framework.

- When the row and column variables represent different observers rating the same subjects, interest is focused on agreement rather than mere association. Cohen's κ is one measure of strength of agreement. The observer agreement chart provides a visual display of how the observers agree and disagree.

- Another specialized display, the trilinear plot, is useful for three-column frequency tables or compositional data.

4.10   Lab exercises

Exercise 4.1   The data set `fat`, created below, gives a 2×2 table recording the level of cholesterol in diet and the presence of symptoms of heart disease for a sample of 23 people.

```
> fat <- matrix(c(6, 4, 2, 11), 2, 2)
> dimnames(fat) <- list(diet = c("LoChol", "HiChol"),
+                        disease = c("No", "Yes"))
```

(a) Use `chisq.test(fat)` to test for association between diet and disease. Is there any indication that this test may not be appropriate here?

(b) Use a fourfold display to test this association visually. Experiment with the different options for standardizing the margins, using the `margin` argument to `fourfold()`. What evidence is shown in different displays regarding whether the odds ratio differs significantly from 1?

(c) `oddsratio(fat, log = FALSE)` will give you a numerical answer. How does this compare to your visual impression from fourfold displays?

(d) With such a small sample, Fisher's exact test may be more reliable for statistical inference. Use `fisher.test(fat)`, and compare these results to what you have observed before.

(e) Write a one-paragraph summary of your findings and conclusions for this data set.

Exercise 4.2   The data set *Abortion* in **vcdExtra** gives a $2 \times 2 \times 2$ table of opinions regarding abortion in relation to sex and status of the respondent. This table has the following structure:

```
> data("Abortion", package = "vcdExtra")
> str(Abortion)

 table [1:2, 1:2, 1:2] 171 152 138 167 79 148 112 133
 - attr(*, "dimnames")=List of 3
  ..$ Sex             : chr [1:2] "Female" "Male"
  ..$ Status          : chr [1:2] "Lo" "Hi"
  ..$ Support_Abortion: chr [1:2] "Yes" "No"
```

(a) Taking support for abortion as the outcome variable, produce fourfold displays showing the association with sex, stratified by status.

(b) Do the same for the association of support for abortion with status, stratified by sex.

(c) For each of the problems above, use `oddsratio()` to calculate the numerical values of the odds ratio, as stratified in the question.

(d) Write a brief summary of how support for abortion depends on sex and status.

Exercise 4.3   The *JobSat* table on income and job satisfaction created in Example 2.5 is contained in the **vcdExtra** package.

(a) Carry out a standard χ^2 test for association between income and job satisfaction. Is there any indication that this test might not be appropriate? Repeat this test using `simulate.p.value = TRUE` to obtain a Monte Carlo test that does not depend on large sample size. Does this change your conclusion?

(b) Both variables are ordinal, so CMH tests may be more powerful here. Carry out that analysis. What do you conclude?

Exercise 4.4   The *Hospital* data in **vcd** gives a 3×3 table relating the length of stay (in years) of 132 long-term schizophrenic patients in two London mental hospitals with the frequency of visits by family and friends.

(a) Carry out a χ^2 test for association between the two variables.

(b) Use `assocstats()` to compute association statistics. How would you describe the strength of association here?

(c) Produce an association plot for these data, with visit frequency as the vertical variable. Describe the pattern of the relation you see here.

(d) Both variables can be considered ordinal, so `CMHtest()` may be useful here. Carry out that analysis. Do any of the tests lead to different conclusions?

Exercise 4.5 Continuing with the `Hospital` data:

(a) Try one or more of the following other functions for visualizing two-way contingency tables with this data: `plot()`, `tile()`, `mosaic()`, and `spineplot()`. [For all except `spineplot()`, it is useful to include the argument `shade=TRUE`].

(b) Comment on the differences among these displays for understanding the relation between visits and length of stay.

Exercise 4.6 The two-way table `Mammograms` in vcdExtra gives ratings on the severity of diagnosis of 110 mammograms by two raters.

(a) Assess the strength of agreement between the raters using Cohen's κ, both unweighted and weighted.

(b) Use `agreementplot()` for a graphical display of agreement here.

(c) Compare the Kappa measures with the results from `assocstats()`. What is a reasonable interpretation of each of these measures?

Exercise 4.7 Agresti and Winner (1997) gave the data in Table 4.8 on the ratings of 160 movies by the reviewers Gene Siskel and Roger Ebert for the period from April 1995 through September 1996. The rating categories were Con ("thumbs down"), Mixed, and Pro ("thumbs up").

Table 4.8: Movie ratings by Siskel & Ebert, April 1995–September 1996. *Source*: Agresti and Winner (1997)

		Ebert			
		Con	Mixed	Pro	Total
	Con	24	8	13	45
Siskel	Mixed	8	13	11	32
	Pro	10	9	64	83
	Total	42	30	88	160

(a) Assess the strength of agreement between the raters using Cohen's κ, both unweighted and weighted.

(b) Use `agreementplot()` for a graphical display of agreement here.

(c) Assess the hypothesis that the ratings are *symmetric* around the main diagonal, using an appropriate χ^2 test. *Hint*: Symmetry for a square table T means that $t_{ij} = t_{ji}$ for $i \neq j$. The expected frequencies under the hypothesis of symmetry are the average of the off-diagonal cells, $E = (T + T^\mathsf{T})/2$.

(d) Compare the results with the output of `mcnemar.test()`.

Exercise 4.8 For the `VisualAcuity` data set:

(a) Use the code shown in the text to create the table form, `VA.tab`.

(b) Perform the CMH tests for this table.

(c) Use the `woolf_test()` described in Section 4.3.2 to test whether the association between left and right eye acuity can be considered the same for men and women.

Exercise 4.9 The graph in Figure 4.23 may be misleading, in that it doesn't take into account of the differing capacities of the 18 life boats on the *Titanic*, given in the variable `cap` in the *Lifeboats* data.

(a) Calculate a new variable, `pctloaded`, as the percentage loaded relative to the boat capacity.
(b) Produce a plot similar to Figure 4.23, showing the changes over time in this measure.

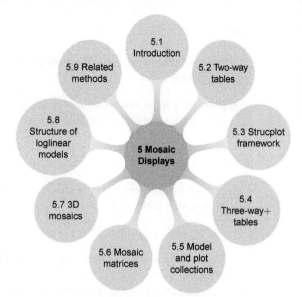

5

Mosaic Displays for n-Way Tables

Mosaic displays help to visualize the pattern of associations among variables in two-way and larger tables. Extensions of this technique can reveal partial associations, marginal associations, and shed light on the structure of loglinear models themselves.

5.1 Introduction

Little boxes on the hillside, little boxes made of ticky tacky,
Little boxes on the hillside, little boxes all the same.
There's a green one and a pink one, and a blue one and a yellow one,
And they're all made out of ticky tacky, and they all look just the same.

Words and music by Malvina Reynolds, ©Schroder Music Company 1962, 1990;
recorded by Pete Seeger

In Chapter 4, we described a variety of graphical techniques for visualizing the pattern of association in simple contingency tables. These methods are somewhat specialized for particular sizes and shapes of tables: 2×2 tables (fourfold display), $R \times C$ tables (tile plot, sieve diagram), square tables (agreement charts), $R \times 3$ tables (trilinear plots), and so forth.

This chapter describes the *mosaic display* and related graphical methods for n-way frequency tables, designed to show various aspects of high-dimensional contingency tables in a hierarchical way. These methods portray the frequencies in an n-way contingency table by a collection of rectangular "tiles" whose size (area) is proportional to the cell frequency. In this respect, the mosaic display is similar to the sieve diagram (Section 4.5). However, mosaic plots and related methods described here:

- generalize more readily to *n*-way tables. One can usefully examine 3-way, 4-way, and even larger tables, subject to the limitations of resolution in any graph;

- are intimately connected to loglinear models, generalized linear models, and generalized non-linear models for frequency data;

- provide a method for fitting a series of sequential loglinear models to the various marginal totals of an *n*-way table; and

- can be used to illustrate the relations among variables that are fitted by various loglinear models.

The basic ideas behind these graphical methods are explained for two-way tables in Section 5.2; the *strucplot framework* on which these are based is described in Section 5.3. The graphical extension of mosaic plots to three-way and large tables (Section 5.4) is quite direct. However, the details require a brief introduction to loglinear models and some terminology for different types of "independence" in such tables, also described in this section. Mosaic methods are further extended to all-pairwise plots in Section 5.6 and 3D plots in Section 5.7.

5.2   Two-way tables

The mosaic display (Friendly, 1992, 1994b, 1997, Hartigan and Kleiner, 1981, 1984) is like a grouped barchart, where the heights (or widths) of the bars show the relative frequencies of one variable, and widths (heights) of the sections in each bar show the conditional frequencies of the second variable, given the first. This gives an area-proportional visualization of the frequencies composed of tiles corresponding to the cells created by successive vertical and horizontal splits of rectangle, representing the total frequency in the table. The construction of the mosaic display, and what it reveals, are most easily understood for two-way tables.

EXAMPLE 5.1:   Hair color and eye color

Consider the data shown earlier in Table 4.2, showing the relation between hair color and eye color among students in a statistics course. The basic mosaic display for this 4×4 table is shown in Figure 5.1.

```
> data("HairEyeColor", package = "datasets")
> haireye <- margin.table(HairEyeColor, 1 : 2)
> mosaic(haireye, labeling = labeling_values)
```

For such a two-way table, the mosaic in Figure 5.1 is constructed by first dividing a unit square in proportion to the marginal totals of one variable, say, hair color.

For these data, the marginal frequencies and proportions of hair color are calculated below:

```
> (hair <- margin.table(haireye, 1))

Hair
Black Brown   Red Blond
  108    286    71   127

> prop.table(hair)

Hair
  Black   Brown     Red   Blond
0.18243 0.48311 0.11993 0.21453
```

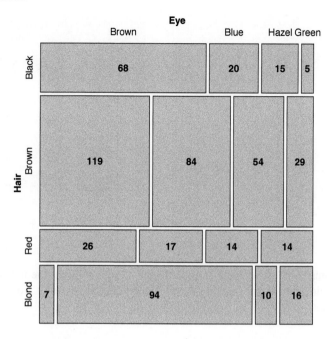

Figure 5.1: Basic mosaic display for hair color and eye color data. The area of each rectangle is proportional to the observed frequency in that cell, shown as numbers.

These frequencies can be shown as the mosaic for the first variable (hair color), with the unit square split according to the marginal proportions as in Figure 5.2 (left). The rectangular tiles are then shaded to show the residuals (deviations) from a particular model as shown in the right panel of Figure 5.2. The details of the calculations for shading are:

- The one-way table of marginal totals can be fit to a model, in this case, the (implausible) model that all hair colors are equally probable. This model has expected frequencies $m_i = 592/4 = 148$:

```
> expected <- rep(sum(hair) / 4, 4)
> names(expected) <- names(hair)
> expected

Black Brown   Red Blond
  148   148   148    148
```

- The Pearson residuals from this model, $r_i = (n_i - m_i)/\sqrt{m_i}$, are:

```
> (residuals <- (hair - expected) / sqrt(expected))

Hair
   Black    Brown      Red    Blond
 -3.2880  11.3435  -6.3294  -1.7262
```

and these values are shown by color and shading as shown in the legend in Figure 5.3. The high positive value for brown hair indicates that people with brown hair are much more frequent in this sample than the equiprobability model would predict; the large negative residual for red hair shows that redheads are much less common. Further details of the schemes for shading are described below, but essentially we use increasing intensities of blue (red) for positive (negative) residuals.

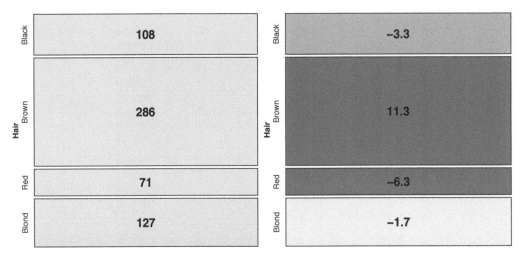

Figure 5.2: First step in constructing a mosaic display. Left: splitting the unit square according to frequencies of hair color; right: shading the tiles according to residuals from a model of equal marginal probabilities.

In the next step, the rectangle for each hair color is subdivided in proportion to the *relative* (conditional) frequencies of the second variable—eye color, giving the following conditional row proportions:

```
> round(addmargins(prop.table(haireye, 1), 2), 3)

        Eye
Hair    Brown  Blue Hazel Green   Sum
  Black 0.630 0.185 0.139 0.046 1.000
  Brown 0.416 0.294 0.189 0.101 1.000
  Red   0.366 0.239 0.197 0.197 1.000
  Blond 0.055 0.740 0.079 0.126 1.000
```

The proportions in each row determine the width of the tiles in the second mosaic display in Figure 5.3.

- Again, the cells are shaded in relation to standardized Pearson residuals, $r_{ij} = (n_{ij} - m_{ij})/\sqrt{m_{ij}}$, from a model. For a two-way table, the model is that hair color and eye color are independent in the population from which this sample was drawn. These residuals are calculated as shown below using `independence_table()` to calculate the expected values m_{ij} under this model ($m_{ij} = n_{++}\pi_{i+}\pi_{+j}$):

```
> exp <- independence_table(haireye)
> resids <- (haireye - exp) / sqrt(exp)
> round(resids, 2)

        Eye
Hair    Brown  Blue Hazel Green
  Black  4.40 -3.07 -0.48 -1.95
  Brown  1.23 -1.95  1.35 -0.35
  Red   -0.07 -1.73  0.85  2.28
  Blond -5.85  7.05 -2.23  0.61
```

- Thus, in Figure 5.3, the two tiles shaded deep blue correspond to the two cells, (Black, Brown) and (Blond, Blue), whose residuals are greater than +4, indicating much greater frequency in

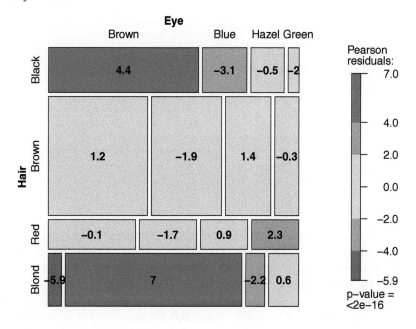

Figure 5.3: Second step in constructing the mosaic display. Each rectangle for hair color is subdivided in proportion to the relative frequencies of eye color, and the tiles are shaded in relation to residuals from the model of independence.

those cells than would be found if hair color and eye color were independent. The tile shaded deep red, (Blond, Brown), corresponds to the largest negative residual $= -5.85$, indicating this combination is extremely rare under the hypothesis of independence.

- The overall Pearson χ^2 statistic for the independence model is just the sum of squares of the residuals, with degrees of freedom $(r-1) \times (c-1)$.

```
> (chisq <- sum(resids ^ 2))

[1] 138.29

> (df <- prod(dim(haireye) - 1))

[1] 9

> pchisq(chisq, df, lower.tail = FALSE)

[1] 2.3253e-25
```

- These results are of course identical to what `chisq.test()` provides. Note that the latter can be used to retrieve the residuals:

```
> chisq.test(haireye)


Pearson's Chi-squared test

data:  haireye
X-squared = 138, df = 9, p-value <2e-16
```

```
> round(residuals(chisq.test(haireye)), 2)

          Eye
Hair    Brown  Blue Hazel Green
  Black  4.40 -3.07 -0.48 -1.95
  Brown  1.23 -1.95  1.35 -0.35
  Red   -0.07 -1.73  0.85  2.28
  Blond -5.85  7.05 -2.23  0.61
```

△

5.2.1 Shading levels

A variety of schemes for shading the tiles are available in the strucplot framework (Section 5.3), but the simplest (and default) shading patterns for the tiles are based on the sign and magnitude of the standardized Pearson residuals, using shades of blue for positive residuals and red for negative residuals, and two threshold values for their magnitudes, $|r_{ij}| > 2$ and $|r_{ij}| > 4$.

Because the standardized residuals are approximately unit-normal $N(0, 1)$ values, this corresponds to highlighting cells whose residuals are *individually* significant at approximately the .05 and .0001 level, respectively. Other shading schemes described later provide tests of significance, but the main purpose of highlighting cells is to draw attention to the *pattern* of departures of the data from the assumed model of independence.

5.2.2 Interpretation and reordering

To interpret the association between hair color and eye color, consider the pattern of positive (blue) and negative (red) tiles in the mosaic display. We interpret positive values as showing cells whose observed frequency is substantially greater than would be found under independence; negative values indicate cells that occur less often than under independence.

The interpretation can often be enhanced by reordering the rows or columns of the two-way table so that the residuals have an *opposite corner* pattern of signs. This usually helps us interpret any systematic patterns of association in terms of the ordering of the row and column categories.

In this example, a more direct interpretation can be achieved by reordering the eye colors as shown in Figure 5.4. Note that in this rearrangement both hair colors and eye colors are ordered from dark to light, suggesting an overall interpretation of the association between hair color and eye color.

```
> # re-order eye colors from dark to light
> haireye2 <- as.table(haireye[, c("Brown", "Hazel", "Green", "Blue")])
> mosaic(haireye2, shade = TRUE)
```

In general, the levels of a factor in mosaic displays are often best reordered by arranging them according to their scores on the first (largest) *correspondence analysis* dimension (Friendly, 1994b); see Chapter 6 for details. Friendly and Kwan (2003) use this as one example of ***effect ordering*** for data displays, illustrated in Chapter 1.

Thus, the mosaic in Figure 5.4 shows that the association between hair and eye color is essentially that:

- people with dark hair tend to have dark eyes,
- those with light hair tend to have light eyes,
- people with red hair and green eyes do not quite fit this pattern.

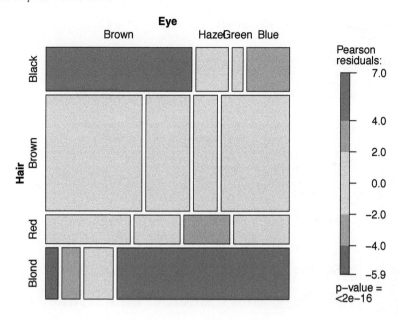

Figure 5.4: Two-way mosaic for hair color and eye color, reordered. The eye colors were reordered from dark to light, enhancing the interpretation.

5.3 The strucplot framework

Mosaic displays have much in common with sieve plots and association plots described in Chapter 4 and with related graphical methods such as ***doubledecker plots*** described later in this chapter. The main idea is to visualize a contingency table of frequencies by "tiles" corresponding to the table cells arranged in rectangular form. For multiway tables with more than two factors, the variables are nested into rows and columns using recursive conditional splits, given the table margins. The result is a "flat" representation that can be visualized in ways similar to a two-dimensional representation of a table. The `structable()` function described in Section 2.5 gives the tabular version of a strucplot. The description below follows Meyer et al. (2006), also included as a vignette (accessible from R as `vignette("strucplot", pkg = "vcd")`), in vcd.

Rather than implementing each of these methods separately, the ***strucplot framework*** in the vcd package provides a general class of methods of which these are all instances. This framework defines a class of conditional displays, which allows for granular control of graphical appearance aspects, including:

- the content of the tiles, e.g., observed or expected frequencies
- the split direction for each dimension, horizontal or vertical
- the graphical parameters of the tiles' content, e.g., color or other visual attributes
- the spacing between the tiles
- the labeling of the tiles

5.3.1 Components overview

The strucplot framework is highly modularized: Figure 5.5 shows the hierarchical relationship between the various components. For the most part, you will directly use the convenience and related

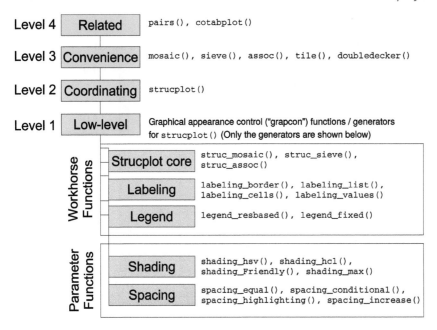

Figure 5.5: Components of the strucplot framework. High-level functions use those at lower levels to provide a general system for tile-based plots of frequency tables.

functions at the top of the diagram, but it is more convenient to describe the framework from the bottom up.

1. On the lowest level, there are several groups of workhorse and parameter functions that directly or indirectly influence the final appearance of the plot (see Table 5.1 for an overview). These are examples of *graphical appearance control* functions (called **grapcon functions**). They are created by generating functions (*grapcon generators*), allowing flexible parameterization and extensibility (Figure 5.5 only shows the generators). The generator names follow the naming convention `group_foo()`, where `group` reflects the group the generators belong to (strucplot core, labeling, legend, shading, or spacing).

   - The workhorse functions (created by `struc_foo()`) are `labeling_foo()`, and `legend_foo()`. These functions directly produce graphical output (i.e., "add ink to the canvas"), for labels and legends respectively.
   - The parameter functions (created by `spacing_foo()` and `shading_foo()`) compute graphical parameters used by the others. The grapcon functions returned by `struc_foo()` implement the core functionality, creating the tiles and their content.

2. On the second level of the framework, a suitable combination of the low-level grapcon functions (or, alternatively, corresponding generating functions) is passed as "hyperparameters" to `strucplot()`. This central function sets up the graphical layout using grid viewports, and coordinates the specified core, labeling, shading, and spacing functions to produce the plot.

3. On the third level, vcd provides several convenience functions such as `mosaic()`, `sieve()`, `assoc()`, `tile()`, and `doubledecker()` which interface to `strucplot()` through sensible parameter defaults and support for model formulae.

4. Finally, on the fourth level, there are "related" vcd functions (such as `cotabplot()` and the

Table 5.1: Available graphical appearance control (grapcon) generators in the strucplot framework

Group	Grapcon generator	Description
strucplot core	`struc_assoc()`	core function for association plots
	`struc_mosaic()`	core function for mosaic plots (also used for tile plots)
	`struc_sieve()`	core function for sieve plots
labeling	`labeling_border()`	border labels
	`labeling_cboxed()`	centered labels with boxes, all labels clipped, and on top and left border
	`labeling_cells()`	cell labels
	`labeling_conditional()`	border labels for conditioning variables and cell labels for conditioned variables
	`labeling_doubledecker()`	draws labels for doubledecker plot
	`labeling_lboxed()`	left-aligned labels with boxes
	`labeling_left()`	left-aligned border labels
	`labeling_left2()`	left-aligned border labels, all labels on top and left border
	`labeling_list()`	draws a list of labels under the plot
	`labeling_residuals()`	show residuals in cells
	`labeling_value()`	show values (observed, expected) in cells
shading	`shading_binary()`	visualizes the sign of the residuals
	`shading_Friendly()`	implements Friendly shading (based on HSV colors)
	`shading_hcl()`	shading based on HCL colors
	`shading_hsv()`	shading based on HSV colors
	`shading_max()`	shading visualizing the maximum test statistic (based on HCL colors)
	`shading_sieve()`	implements Friendly shading customized for sieve plots (based on HCL colors)
spacing	`spacing_conditional()`	increasing spacing for conditioning variables, equal spacing for conditioned variables
	`spacing_dimequal()`	equal spacing for each dimension
	`spacing_equal()`	equal spacing for all dimensions
	`spacing_highlighting()`	increasing spacing, last dimension set to zero
	`spacing_increase()`	increasing spacing
legend	`legend_fixed()`	creates a fixed number of bins (similar to `mosaicplot()`)
	`legend_resbased()`	suitable for an arbitrary number of bins (also for continuous shadings)

`pairs()` methods for table objects) arranging collections of plots of the strucplot framework into more complex displays (e.g., by means of panel functions).

5.3.2 Shading schemes

Unlike other graphics functions in base R, the strucplot framework allows almost full control over the graphical parameters of all plot elements. In particular, in association plots, mosaic plots, and sieve plots, you can modify the graphical appearance of each tile individually.

Built on top of this functionality, the framework supplies a set of shading functions choosing colors appropriate for the visualization of loglinear models. The tiles' graphical parameters are set using the `gp` argument of the functions of the strucplot framework. This argument basically expects an object of class **"gpar"** whose components are arrays of the same shape (length and dimensionality) as the data table.

For added generality, however, you can also supply a `grapcon` function that computes such an object given a vector of residuals, or, alternatively, a *generating function* that takes certain arguments and returns such a grapcon function (see Table 5.1). vcd provides several shading functions, including support for both HSV (hue-saturation-value) and HCL (hue-chroma-luminance) colors, and visualization of significance tests.

5.3.2.1   Specifying graphical parameters for strucplot displays

Strucplot displays in vcd are built using the grid graphics package (Murrell, 2011). There are many graphical parameters that can be set using gp = gpar(...) in a call to a high-level strucplot function. Among these, the following are often most useful to control the drawing components:

col	Color for lines and borders.
fill	Color for filling rectangles, polygons, ...
alpha	Alpha channel for transparency of fill color.
lty	Line type for lines and borders.
lwd	Line width for lines and borders.

In addition, a number of parameters control the display of text labels in these displays:

fontsize	The size of text (in points)
cex	Multiplier applied to fontsize
fontfamily	The font family (serif, sans, mono, ...)
fontface	The font face (**bold**, *italic*, ...)

See help(gpar) for a complete list and help(par) for further details.

We illustrate this capability below using the hair color and eye color data as reordered in Figure 5.4. The following example produces a ***Marimekko chart***, or a "poor-man's mosaic display" as shown in the left panel of Figure 5.6. This is essentially a divided bar chart where the eye colors within each horizontal bar for the hair color group are all given the same color. In the example, the matrix fill_colors is constructed to conform to the haireye2 table, using color values that approximate the eye colors.[1]

```
> # color by hair color
> fill_colors <- c("brown4", "#acba72", "green", "lightblue")
> (fill_colors_mat <- t(matrix(rep(fill_colors, 4), ncol = 4)))

     [,1]     [,2]       [,3]    [,4]
[1,] "brown4" "#acba72" "green" "lightblue"
[2,] "brown4" "#acba72" "green" "lightblue"
[3,] "brown4" "#acba72" "green" "lightblue"
[4,] "brown4" "#acba72" "green" "lightblue"

> mosaic(haireye2, gp = gpar(fill = fill_colors_mat, col = 0))
```

Alternatively, we could have used the convenience function shading_Marimekko() in vcd:

```
> mosaic(haireye2, gp = shading_Marimekko(haireye2))
```

Note that because the hair colors and eye colors are both ordered, this shows the decreasing prevalence of light hair color amongst those with brown eyes and the increasing prevalence of light hair with blue eyes.

Alternatively, for some purposes,[2] we might like to use color to highlight the pattern of diagonal

---

[1]Actually, the fill_colors vector could be directly used since values are recycled as needed by mosaic().

[2]For example, this would be appropriate for a square table, showing agreement between row and column categories, as in Section 4.7.

cells, and the off-diagonals 1, 2, 3 steps removed. The R function `toeplitz()` returns such a patterned matrix, and we can use this to calculate the `fill_colors` by indexing the result of the `rainbow_hcl()` palette function in colorspace (Ihaka et al., 2015) (generating better colors than `palette()`). The code below produces the right panel in Figure 5.6.

```
> # toeplitz designs
> library(colorspace)
> toeplitz(1 : 4)

     [,1] [,2] [,3] [,4]
[1,]    1    2    3    4
[2,]    2    1    2    3
[3,]    3    2    1    2
[4,]    4    3    2    1

> fill_colors <- rainbow_hcl(8)[1 + toeplitz(1 : 4)]
> mosaic(haireye2, gp = gpar(fill = fill_colors, col = 0))
```

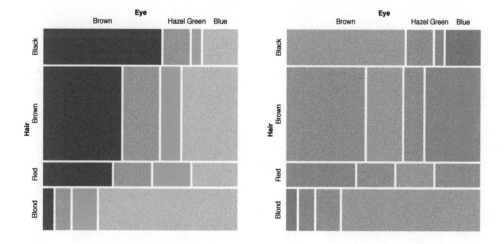

Figure 5.6: Mosaic displays for the `haireye2` data, using custom colors to fill the tiles. Left: Marimekko chart, using colors to reflect the eye colors; right: Toeplitz-based colors, reflecting the diagonal strips in a square table.

Again, vcd offers a convenience function for this purposes, `shading_diagonal()`:

```
> mosaic(haireye2, gp = shading_diagonal(haireye2))
```

More simply, to shade a mosaic according to the levels of one variable (typically a response variable), you can use the `highlighting` arguments of `mosaic()`. The first call below gives a result similar to the left panel of Figure 5.6. Alternatively, using the formula method for `mosaic()`, specify the response variable as the left-hand side.

```
> mosaic(haireye2, highlighting = "Eye", highlighting_fill = fill_colors)
> mosaic(Eye ~ Hair, data = haireye2, highlighting_fill = fill_colors)
```

5.3.2.2   Residual-based shading

The important idea that differentiates mosaic and other strucplot displays from the "poor-man's," Marimekko versions (Figure 5.6) often shown in other software, is that rather than just using shading color to *identify* the cells, we can use these attributes to show something more—*residuals* from some model, whose pattern helps to explain the association between the table variables.

As described above, the strucplot framework includes a variety of `shading_` functions, and these can be customized with optional arguments. Zeileis et al. (2007) describe a general approach to residual-based shadings for area-proportional visualizations, used in the development of the strucplot framework in `vcd`.

EXAMPLE 5.2: Interpolation options

One simple thing to do is to modify the `interpolate` option passed to the default `shading_hcl` function, as shown in Figure 5.7.

```
> # more shading levels
> mosaic(haireye2, shade = TRUE, gp_args = list(interpolate = 1 : 4))
>
> # continuous shading
> interp <- function(x) pmin(x / 6, 1)
> mosaic(haireye2, shade = TRUE, gp_args = list(interpolate = interp))
```

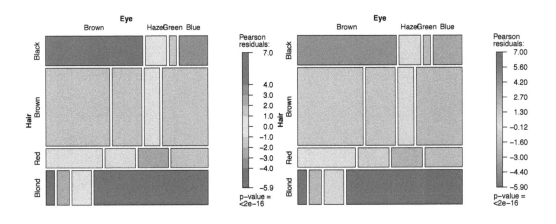

Figure 5.7: Interpolation options for shading levels in mosaic displays. Left: four shading levels; right: continuous shading.

For the left panel of Figure 5.7, a numeric vector is passed as `interpolate=1:4`, defining the boundaries of a step function mapping the absolute values of residuals to saturation levels in the HCL color scheme. For the right panel, a user-defined function, `interp()`, is created which maps the absolute residuals to saturation values in a continuous way (up to a maximum of 6).

Note that these two interpolation schemes produce quite similar results, differing mainly in the shading level of residuals within ± 1 and in the legend. In practice, the default discrete interpolation, using cutoffs of $\pm 2, \pm 4$ usually works quite well.   △

EXAMPLE 5.3: Shading functions

Alternatively, the names of shading functions can be passed as the `gp` argument, as shown below, producing Figure 5.8. Two shading function are illustrated here:

• The left panel of Figure 5.8 uses the classical Friendly (1994b) shading scheme, `shading_Friendly` with HSV colors[3] of blue and red and default cutoffs for absolute residuals, $\pm 2, \pm 4$, corresponding to `interpolate = c(2, 4)`. In this shading scheme, all tiles use an outline color (`col`) corresponding to the sign of the residual. As well, the border line type (`lty`) distinguishes positive and negative residuals, which is useful if a mosaic plot is printed in black and white.

- The right panel uses the `shading_max()` function, based on the ideas of Zeileis et al. (2007) on residual-based shadings for area-proportional visualizations. Instead of using the cutoffs 2 and 4, it employs the critical values, M_α, for the maximum absolute Pearson residual statistic,

$$M = \max_{i,j} |r_{ij}|,$$

by default at $\alpha = 0.10$ and 0.01.[4] Only those residuals with $|r_{ij}| > M_\alpha$ are colored in the plot, using two levels for Value ("lightness") in HSV color space. Consequently, all color in the plot signals a significant departure from independence at 90% or 99% significance level, respectively.[5]

```
> mosaic(haireye2, gp = shading_Friendly, legend = legend_fixed)
> set.seed(1234)
> mosaic(haireye2, gp = shading_max)
```

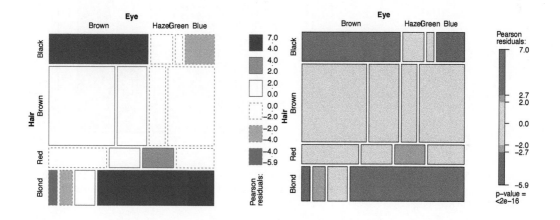

Figure 5.8: Shading functions for mosaic displays. Left: `shading_Friendly` using fixed cutoffs and the "Friendly" color scheme and an alternative legend style (`legend_fixed`); right: `shading_max`, using a permutation-based test to determine significance of residuals.

In this example, the difference between these two shading schemes is largely cosmetic, in that the pattern of association is similar in the two panels of Figure 5.8, and the interpretation would be the same. This is not always the case, as we will see in the next example. △

---

[3]`shading_Friendly2()` is a variant based on HCL colors.

[4]These default significance levels were chosen because this leads to displays where fully colored cells are clearly significant ($p < 0.01$), cells without color are clearly non-significant ($p > 0.1$), and cells in between can be considered to be weakly significant ($0.01 \leq p \leq 0.1$).

[5]This computation uses the vcd function `coindep_test()` to calculate generalized tests of (conditional) independence by simulation from the marginal distribution of the input table under (conditional) independence. In these examples using `shading_max`, the function `set.seed()` is used to initialize the random number generators to a given state for reproducibility.

EXAMPLE 5.4: Arthritis treatment

This example uses the `Arthritis` data, illustrated earlier (Example 2.2), on the relation between treatment and outcome for rheumatoid arthritis. To confine this example to a two-way table, we use only the (larger) female patient group.

```
> art <- xtabs(~ Treatment + Improved, data = Arthritis,
+               subset = Sex == "Female")
> names(dimnames(art))[2] <- "Improvement"
```

The calls to `mosaic()` below compare `shading_Friendly` and `shading_max`, giving the plots shown in Figure 5.9.

```
> mosaic(art, gp = shading_Friendly, margin = c(right = 1),
+        labeling = labeling_residuals, suppress = 0, digits = 2)
> set.seed(1234)
> mosaic(art, gp = shading_max, margin = c(right = 1))
```

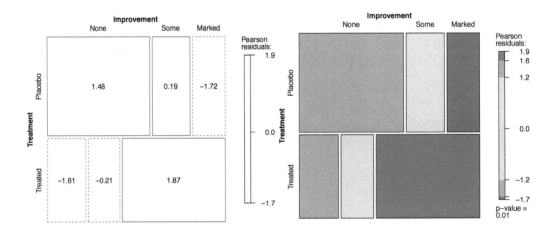

Figure 5.9: Mosaic plots for the female patients in the `Arthritis` data. Left: Fixed shading levels via `shading_Friendly`; right: shading levels determined by significant maximum residuals via `shading_max`.

This data set is somewhat paradoxical, in that the standard `chisq.test()` for association with these data gives a highly significant result, $\chi^2(2) = 11.3, p = 0.0035$, while the shading pattern using `shading_Friendly` in the left panel of Figure 5.9 shows all residuals within ± 2, and thus unshaded.

On the other hand, the `shading_max` shading in the right panel of Figure 5.9 shows that significant deviations from independence occur in the four corner cells, corresponding to more of the treated group showing marked improvement, and more of the placebo group showing no improvement.

Some details behind the `shading_max` method are shown below. The Pearson residuals for this table are calculated as:

```
> residuals(chisq.test(art))

          Improvement
Treatment     None      Some     Marked
  Placebo  1.47752   0.19267  -1.71734
  Treated -1.60852  -0.20975   1.86960
```

The `shading_max()` function then calls `coindep_test(art)` to generate $n = 1000$ random tables with the same margins, and computes the maximum residual statistic for each. This gives a non-parametric p-value for the test of independence, $p = 0.011$ shown in the legend.

```
> set.seed(1243)
> art_max <- coindep_test(art)
> art_max


Permutation test for conditional independence

data:  art
f(x) = 1.87, p-value = 0.011
```

Finally, the 0.90 and 0.99 quantiles of the simulation distribution are used as shading levels, passed as the value of the `interpolate` argument.

```
> art_max$qdist(c(0.90, 0.99))

   90%    99%
1.2393 1.9167
```

$\triangle$

The converse situation can also arise in practice. An overall test for association using Pearson's χ^2 may not be significant, but the maximum residual test may highlight one or more cells worthy of greater attention, as illustrated in the following example.

EXAMPLE 5.5: UK soccer scores

In Example 3.9, we examined the distribution of goals scored by the home team and the away team in 380 games in the 1995/96 season by the 20 teams in the UK Football Association, Premier League. The analysis there focused on the distribution of the total goals scored, under the assumption that the number of goals scored by the home team and the away team were independent.

Here, the rows and columns of the table *UKSoccer* are both ordered, so it is convenient and compact to carry out all the CMH tests taking ordinality into account.

```
> data("UKSoccer", package = "vcd")
> CMHtest(UKSoccer)

Cochran-Mantel-Haenszel Statistics for Home by Away

                 AltHypothesis Chisq Df   Prob
cor          Nonzero correlation  1.01  1 0.315
rmeans   Row mean scores differ  5.63  4 0.229
cmeans   Col mean scores differ  7.42  4 0.115
general     General association 18.65 16 0.287
```

All of these are non-significant, so that might well be the end of the story, as far as independence of goals in home and away games is concerned. Yet, one residual, $r_{42} = 3.08$ stands out, corresponding to 4 or more goals by the home team and only 2 goals by the away team, which accounts for nearly half of the $\chi^2(16) = 18.7$ for general association. The mosaic plot is shown in Figure 5.10.

```
> set.seed(1234)
> mosaic(UKSoccer, gp = shading_max, labeling = labeling_residuals,
+        digits = 2)
```

This occurrence may or may not turn out to have some explanation, but at least the mosaic plot draws it to our attention, and is consistent with the (significant) result from `coindep_test()`.

$\triangle$

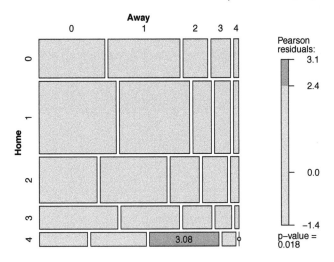

Figure 5.10: Mosaic display for UK soccer scores, highlighting one cell that stands out for further attention.

5.4   Three-way and larger tables

The mosaic displays and other graphical methods within the strucplot framework extend quite naturally to three-way and higher-way tables. The essential idea is that for the variables in a multiway table in a given order, each successive variable is used to subdivide the tile(s) in proportion to the relative (conditional) frequencies of that variable, given all previous variables. This process continues recursively until all table variables have been included.

For simplicity, we continue with the running example of hair color and eye color. Imagine that each cell of the two-way table for hair and eye color is further classified by one or more additional variables—sex and level of education, for example. Then each rectangle can be subdivided horizontally to show the proportion of males and females in that cell, and each of those horizontal portions can be subdivided vertically to show the proportions of people at each educational level in the hair–eye–sex group.

EXAMPLE 5.6: Hair color, eye color, and sex

Figure 5.11 shows the mosaic for the three-way table, with hair and eye color groups divided according to the proportions of males and females. As explained in the next section (Section 5.4.2) there are now different models for "independence" we could investigate, not just the (mutual) independence of all factors. Here, for example, we could examine whether the additional variable (Sex) is independent from the *joint* relationship between Hair and Eye.

```
> HEC <- HairEyeColor[, c("Brown", "Hazel", "Green", "Blue"),]
> mosaic(HEC, rot_labels = c(right = -45))
```

In Figure 5.11 it is easy to see that there is no systematic association between sex and the combinations of hair and eye color—the proportion of male/female students is roughly the same in almost all hair/eye color groups. Yet, among blue-eyed blonds, there seems to be an overabundance of females, and the proportion of blue-eyed males with brown hair also looks suspicious.   △

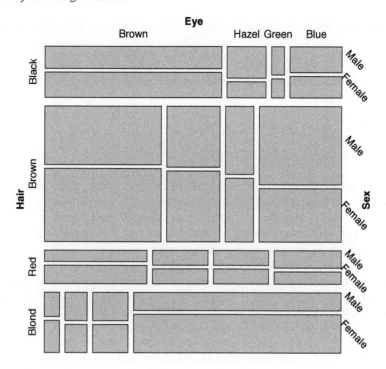

Figure 5.11: Three-way mosaic for hair color, eye color, and sex.

These and other hypotheses are best tested within the framework of ***loglinear model***s, allowing you to flexibly specify various independence models for any number of variables, and analyze them similarly to classical ANOVA models. This general topic is discussed in detail in Chapter 9. For the present purposes, we give a short introduction in the following section.

5.4.1   A primer on loglinear models

The essential idea behind loglinear models is that the multiplicative relationships among expected frequencies under independence (shown as areas in sieve diagrams and mosaic plots) become *additive* models when expressed as models for log frequency, and we briefly explain this connection here for two-way tables.

To see this, consider two discrete variables, A and B, with n_{ij} observations in each cell i, j of an $R \times C$ contingency table, and use $n_{i+} = \Sigma_j n_{ij}$ and $n_{+j} = \Sigma_i n_{ij}$ for the row and column marginal totals, respectively. The total frequency is $n_{++} = \Sigma_{ij} n_{ij}$. Analogously, we use m_{ij} for the expected frequency under any model and also use a subscript $+$ to represent summation over that dimension.

Then, the hypothesis of independence means that the expected frequencies, m_{ij}, obey

$$m_{ij} = \frac{m_{i+} \, m_{+j}}{m_{++}} \, .$$

This multiplicative model can be transformed to an additive (linear) model by taking logarithms of both sides:

$$\log(m_{ij}) = \log(m_{i+}) + \log(m_{+j}) - \log(m_{++}) \, .$$

This is usually re-expressed in an equivalent form in terms of model parameters μ, λ_i^A and λ_j^B

$$\log(m_{ij}) = \mu + \lambda_i^A + \lambda_j^B \equiv [A][B] . \tag{5.1}$$

Model Eqn. (5.1) asserts that the row and column variables are independent because there is no term that depends on both A and B.

In contrast, a model for a two-way table that allows an arbitrary association between the variables is the **saturated model**, including an additional term, λ_{ij}^{AB}:

$$\log(m_{ij}) = \mu + \lambda_i^A + \lambda_j^B + \lambda_{ij}^{AB} \equiv [AB] . \tag{5.2}$$

Except for the difference in notation, model Eqn. (5.2) is formally the same as a two-factor ANOVA model with an interaction, typically expressed as $E(y_{ij}) = \mu + \alpha_i + \beta_j + (\alpha\beta)_{ij}$. Hence, associations between variables in loglinear models are analogous to interactions in ANOVA models.[6] In contrast to ANOVA, the "main effects," λ_i^A and λ_j^B, are rarely of interest—a typical log-linear analysis focuses only on the interaction (association) terms.

Models such as Eqn. (5.1) and Eqn. (5.2) are examples of **hierarchical models**. This means that the model must contain all lower-order terms contained within any high-order term in the model.

Thus, the saturated model, Eqn. (5.2) contains λ_{ij}^{AB}, and therefore *must* contain λ_i^A and λ_j^B. As a result, hierarchical models may be identified by the shorthand notation that lists only the high-order terms: model Eqn. (5.2) is denoted $[AB]$, while model Eqn. (5.1) is $[A][B]$.

In R, the most basic function for fitting loglinear models is `loglin()` in the stats package. It is designed to work with the frequency data in table form, and a model specified in terms of the (high-order) table margins to be fitted. For example, the independence model Eqn. (5.1) is specified as

```
> loglin(mytable, margin = list(1, 2))
```

meaning that variables 1 and 2 are independent, whereas the saturated model Eqn. (5.2) would be specified as

```
> loglin(mytable, margin = list(c(1, 2)))
```

The function `loglm()` in MASS provides a more convenient front-end to `loglin()` to allow loglinear models to be specified using a model formula. With table variables A and B, the independence model can be fit using `loglm()` as

```
> loglm(~ A + B, data = mytable)
```

and the saturated model in either of the following equivalent forms:

```
> loglm(~ A + B + A : B, data = mytable)
> loglm(~ A * B, data = mytable)
```

In such model formulas, A:B indicates an interaction term λ_{ij}^{AB}, while A*B is expanded to also include the terms A + B.

EXAMPLE 5.7: Hair color, eye color, and sex

Getting back to our running example of hair and eye color, we start casting the classical test of independence used in Section 5.2 as log-linear model analysis. Using the `haireye` two-way table, the independence of `Hair` and `Eye` is equivalent to the model [Hair][Eye] and formulated in R using `loglm()` as:

---

[6]The use of superscripted symbols, $\lambda_i^A, \lambda_j^B, \lambda_{ij}^{AB}$, rather than separate Greek letters is a convention in loglinear models, and useful mainly for multiway tables.

```
> loglm(~ Hair + Eye, data = haireye)

Call:
loglm(formula = ~Hair + Eye, data = haireye)

Statistics:
                        X^2 df P(> X^2)
Likelihood Ratio 146.44   9         0
Pearson          138.29   9         0
```

The output includes both the χ^2 and the deviance test statistics, both significant, indicating strong lack of fit. We now extend the analysis by including Sex, i.e., use the full *HairEyeColor* data set. In the section's introductory example, this was visualized using a mosaic plot, leading to the hypothesis whether Hair and Eye were jointly independent of Sex. To test this formally, we fit the corresponding model [HairEye][Sex] to the data:

```
> HE_S <- loglm(~ Hair * Eye + Sex, data = HairEyeColor)
> HE_S

Call:
loglm(formula = ~Hair * Eye + Sex, data = HairEyeColor)

Statistics:
                        X^2 df P(> X^2)
Likelihood Ratio 19.857  15   0.17750
Pearson          19.567  15   0.18917
```

giving a non-significant Pearson $\chi^2(15) = 19.567, p = 0.189$. The residuals from this model could be retrieved using

```
> residuals(HE_S, type = "pearson")
```

for further inspection. Mosaic plots can conveniently be used for this purpose, either by specifying the residuals with the residuals= argument, or by providing the loglm model formula as the expected= argument, letting mosaic() calculate them by calling loglm(). In the call to mosaic() below, the model of joint independence is specified as expected = ~ Hair * Eye + Sex, giving Figure 5.12. The strucplot labeling function labeling_residuals is used to display the residuals in the highlighted cells.

```
> HEC <- HairEyeColor[, c("Brown", "Hazel", "Green", "Blue"),]
> mosaic(HEC, expected = ~ Hair * Eye + Sex,
+        labeling = labeling_residuals,
+        digits = 2, rot_labels = c(right = -45))
```

Although non-significant, the two largest residuals highlighted in the plot account for nearly half $(-2.15^2 + 2.03^2 = 8.74)$ of the lack of fit, and so are worthy of attention here. An easy (probably facile) interpretation is that among the blue-eyed blonds, some of the females benefited from hair products.   △

5.4.2 Fitting models

When three or more variables are represented in a table, we can fit several different models of types of "independence" and display the residuals from each model. We treat these models as null or **baseline models**, which may not fit the data particularly well. The deviations of observed frequencies from expected ones, displayed by shading, will often suggest terms to be added to an explanatory model that achieves a better fit.

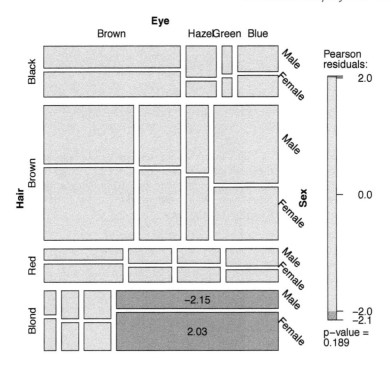

Figure 5.12: Three-way mosaic for hair color, eye color, and sex. Residuals from the model of joint independence, [HE][S] are shown by shading.

For a three-way table, with variables A, B and C, some of the hypothesized models that can be fit are described below and summarized in Table 5.2. Here we use [•] notation to list the **high-order terms** in a hierarchical loglinear model; these correspond to the margins of the table that are fitted exactly, and which translate directly into R formulas used in `loglm()` and `mosaic(..., expected=)`.

The notation [AB][AC], for example, is shorthand for the model `loglm(~ A*B + A*C)` that implies

$$\log(m_{ijk}) = \mu + \lambda_i^A + \lambda_j^B + \lambda_k^C + \lambda_{ij}^{AB} + \lambda_{ik}^{AC} , \qquad (5.3)$$

and reproduces the $\{AB\}$ and $\{AC\}$ marginal subtables.[7] That is, the calculated expected frequencies in these margins are always equal to the corresponding observed frequencies, $m_{ij+} = n_{ij+}$ and $m_{i+k} = n_{i+k}$.

In this table, $A \perp B$ is read, "A is independent of B." The independence interpretation of the model Eqn. (5.3) is $B \perp C \mid A$, which can be read as "B is independent of C, given (conditional on) A." Table 5.2 also depicts the relations among variables as an ***association graph***, where associated variables are connected by an edge and variables that are asserted to be independent are unconnected. In mosaic-like displays, other associations present in the data will appear in the pattern of residuals.

For a three-way table, there are four general classes of independence models illustrated in Table 5.2, as described below.[8] Not included here is the ***saturated model***, [ABC], which fits the observed data exactly.

---

[7]The notation here uses curly braces, {•} to indicate a marginal subtable summed over all other variables.
[8]For H_2 and H_3, permutation of the variables A, B, and C gives other members of each class.

Table 5.2: Fitted margins, model symbols, and interpretations for some hypotheses for a three-way table

Hypothesis	Fitted margins	Model symbol	Independence interpretation	Association graph
H_1	$n_{i++}, n_{+j+}, n_{++k}$	[A][B][C]	$A \perp B \perp C$	
H_2	n_{ij+}, n_{++k}	[AB][C]	$(A, B) \perp C$	
H_3	n_{i+k}, n_{+jk}	[AC][BC]	$A \perp B \mid C$	
H_4	$n_{ij+}, n_{i+k}, n_{+jk}$	[AB][AC][BC]	NA	

H_1: **Complete independence.** The model of complete (mutual) independence, symbolized $A \perp B \perp C$, with model formula ~ A + B + C, asserts that all joint probabilities are products of the one-way marginal probabilities:

$$\pi_{ijk} = \pi_{i++} \, \pi_{+j+} \, \pi_{++k} \, ,$$

for all i, j, k in a three-way table. This corresponds to the log-linear model [A][B][C]. Fitting this model puts all higher terms, and hence all association among the variables, into the residuals.

H_2: **Joint independence.** Another possibility is to fit the model in which variable C is jointly independent of variables A and B, ($\{A, B\} \perp C$), with model formula ~ A*B + C, where

$$\pi_{ijk} = \pi_{ij+} \, \pi_{++k} \, .$$

This corresponds to the loglinear model [AB][C]. Residuals from this model show the extent to which variable C is related to the combinations of variables A and B but they do not show any association between A and B, since that association is fitted exactly. For this model, variable C is also independent of A and B in the marginal $\{AC\}$ table (collapsing over B) and in the marginal $\{BC\}$.

H_3: **Conditional independence.** Two variables, say A and B, are conditionally independent given the third (C) if A and B are independent when we control for C, symbolized as $A \perp B \mid C$, and model formula ~ A*C + B*C (or ~ (A + B) * C). This means that conditional probabilities, $\pi_{ij|k}$ obey

$$\pi_{ij|k} = \pi_{i+|k} \, \pi_{+j|k} \, ,$$

where $\pi_{ij|k} = \pi_{ijk}/\pi_{ij+}$, $\pi_{i+|k} = \pi_{i+k}/\pi_{i++}$, and $\pi_{+j|k} = \pi_{+jk}/\pi_{+j+}$. The corresponding loglinear models is denoted [AC][BC]. When this model is fit, the mosaic display shows the conditional associations between variables A and B, controlling for C, but does not show the associations between A and C, or B and C.

H_4: **No three-way interaction.** For this model, no pair is marginally or conditionally independent, so there is *no* independence interpretation. Nor is there a closed-form expression for the cell probabilities. However, the association between any two variables is the same at each level of the third variable. The corresponding loglinear model formula is [AB][AC][BC], indicating that all two-way margins are fit exactly and so only the three-way association is shown in the mosaic residuals.

EXAMPLE 5.8: Hair color, eye color, and sex

We continue with the analysis of the `HairEyeColor` data from Example 5.6 and Example 5.7. Figure 5.12 showed the fit of the joint-independence model [HairEye][Sex], testing whether the joint distribution of hair color and eye color is associated with sex.

Any other model fit to this table will have the same size tiles in the mosaic since the areas depend on the observed frequencies; the residuals, and hence the shading of the tiles will differ. Figure 5.13 shows mosaics for two other models. Shading in the left panel shows residuals from the model of mutual independence, [Hair][Eye][Sex], and so includes all sources of association among these three variables. The right panel shows the conditional independence model, [HairSex][EyeSex], testing whether hair color and eye color are independent, given sex. Note that the pattern of residuals here is similar to that in the two-way display, Figure 5.4, that collapsed over sex.

```
> abbrev <- list(abbreviate = c(FALSE, FALSE, 1))
> mosaic(HEC, expected = ~ Hair + Eye + Sex, labeling_args = abbrev,
+    main = "Model: ~ Hair + Eye + Sex")
> mosaic(HEC, expected = ~ Hair * Sex + Eye * Sex, labeling_args = abbrev,
+         main="Model: ~ Hair*Sex + Eye*Sex")
```

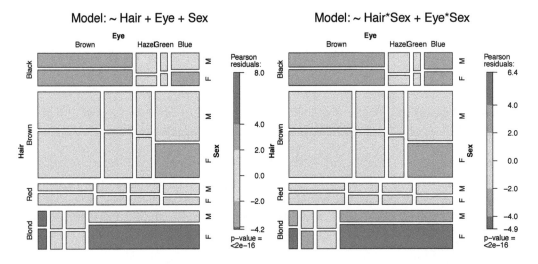

Figure 5.13: Mosaic displays for other models fit to the data on hair color, eye color, and sex. Left: Mutual independence model; right: Conditional independence of hair color and eye color given sex.

Compared with Figure 5.12 for the joint independence model, [HairEye][Sex], it is easy to see that both of these models fit very poorly.

We consider loglinear models in more detail in Chapter 9, but for now note that these models are fit using `loglm()` in the **MASS** package, with the model formula given in the `expected` argument. The details of these models can be seen by fitting these models explicitly, and the fit of several models can be summarized compactly using `LRstats()` in **vcdExtra**.

```
> library(MASS)
> # three types of independence:
> mod1 <- loglm(~ Hair + Eye + Sex, data = HEC)        # mutual
> mod2 <- loglm(~ Hair * Sex + Eye * Sex, data = HEC)  # conditional
> mod3 <- loglm(~ Hair * Eye + Sex, data = HEC)        # joint
> LRstats(mod1, mod2, mod3)

Likelihood summary table:
      AIC BIC LR Chisq Df Pr(>Chisq)
mod1  321 333    166.3 24     <2e-16 ***
mod2  324 344    156.7 18     <2e-16 ***
mod3  193 218     19.9 15       0.18
---
Signif. codes:  0 '***' 0.001 '**' 0.01 '*' 0.05 '.' 0.1 ' ' 1
```

Alternatively, you can get the Pearson and likelihood ratio (LR) tests for a given model using `anova()`, or compare a set of models using LR tests on the *difference* in LR χ^2 from one model to the next, when a list of models is supplied to `anova()`.

```
> anova(mod1)

Call:
loglm(formula = ~Hair + Eye + Sex, data = HEC)

Statistics:
                      X^2 df P(> X^2)
Likelihood Ratio 166.30 24        0
Pearson          164.92 24        0

> anova(mod1, mod2, mod3, test = "chisq")

LR tests for hierarchical log-linear models

Model 1:
 ~Hair + Eye + Sex
Model 2:
 ~Hair * Sex + Eye * Sex
Model 3:
 ~Hair * Eye + Sex

          Deviance df Delta(Dev) Delta(df) P(> Delta(Dev))
Model 1    166.300 24
Model 2    156.678 18     9.6222         6         0.14149
Model 3     19.857 15   136.8213         3         0.00000
Saturated    0.000  0    19.8566        15         0.17750
```

$\triangle$

5.5 Model and plot collections

This section describes a few special circumstances in which a collection of mosaic plots and related loglinear models can be used in a complementary fashion to understand the nature of associations in three-way and larger tables.

5.5.1   Sequential plots and models

As described in Section 5.2, we can think of the mosaic display for an *n*-way table as being constructed in stages, with the variables listed in a given order, and the unit tile decomposed recursively as each variable is entered in turn. This process turns out to have the useful property that it provides an additive (hierarchical) decomposition of the total association in a table, in a way analogous to sequential fitting with Type I sum of squares in regression models.

Typically, we just view the mosaic and fit models to the full *n*-way table, but it is useful to understand the connection with models for the marginal subtables, defined by summing over all variables not yet entered. For example, for a three-way table with variables A, B, C, the marginal subtables $\{A\}$ and $\{AB\}$ are calculated in the process of constructing the three-way mosaic. The $\{A\}$ marginal table can be fit to a model where the categories of variable A are equiprobable as shown in Figure 5.2 (or some other discrete distribution); the independence model can be fit to the $\{AB\}$ subtable as in Figure 5.2, and so forth.

This connection can be seen in the following formula that decomposes the joint cell probability in an *n*-way table with variables $v_1, v_2, \ldots v_n$ as a sequential product of conditional probabilities,

$$p_{ijk\ell\cdots} = \overbrace{p_i \times p_{j|i}}^{\{v_1 v_2\}} \times \underbrace{p_{k|ij}}_{\{v_1 v_2 v_3\}} \times p_{\ell|ijk} \times \cdots \times p_{n|ijk\cdots} \tag{5.4}$$

In Eqn. (5.4), the first term corresponds to the one-way mosaic for v_1, the first two terms to the mosaic for v_1 and v_2, the first three terms to the mosaic for v_1, v_2, and v_2, and so forth.

It can be shown (Friendly, 1994b) that this sequential product of probabilities corresponds to a set of sequential models of *joint independence*, whose likelihood ratio G^2 statistics provide an additive decomposition of the total association, $G^2_{[v_1][v_2]\ldots[v_n]}$ for the mutual independence model in the full table:

$$G^2_{[v_1][v_2]\ldots[v_n]} = G^2_{[v_1][v_2]} + G^2_{[v_1 v_2][v_3]} + G^2_{[v_1 v_2 v_3][v_4]} + \cdots + G^2_{[v_1\ldots v_{n-1}][v_n]} . \tag{5.5}$$

For example, for the hair-eye data, the mosaic displays for the [Hair] [Eye] marginal table (Figure 5.4) and the [HairEye] [Sex] table (Figure 5.12) can be viewed as representing the partition of G^2 shown as a table below:

Model	Model symbol	df	G^2
Marginal	[Hair] [Eye]	9	146.44
Joint	[Hair, Eye] [Sex]	15	19.86
Mutual	[Hair] [Eye] [Sex]	24	166.30

The decomposition in this table reflecting Eqn. (5.5) is shown as a visual equation in Figure 5.14. You can see from the shading how the two sequential submodels contribute to overall association in the model of mutual independence.

Although sequential models of joint independence have the nice additive property illustrated above, other classes of sequential models are possible, and sometimes of substantive interest. The main types of these models are illustrated in Table 5.3 for 3-, 4-, and 5-way tables, with variables A, B, ..., E. In all cases, the natural model for the one-way margin is the equiprobability model, and that for the two-way margin is [A][B].

The vcdExtra package provides a collection of convenience functions that generate the loglinear model formulae symbolically, as indicated in the **function** column. The functions `mutual()`, `joint()`, `conditional()`, `markov()` and so forth simply generate a list of terms suitable for a model formula for `loglin()`. See `help(loglin-utilities)` for further details.

Wrapper functions `loglin2string()` and `loglin2formula()` convert these to character strings or model formulae, respectively, for use with `loglm()` and `mosaic()`-related functions in vcdExtra. Some examples are shown below.

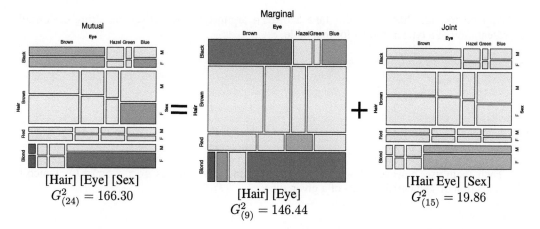

[Hair] [Eye] [Sex]
$G^2_{(24)} = 166.30$

[Hair] [Eye]
$G^2_{(9)} = 146.44$

[Hair Eye] [Sex]
$G^2_{(15)} = 19.86$

Figure 5.14: Visual representation of the decomposition of the G^2 for mutual independence (total) as the sum of marginal and joint independence.

Table 5.3: Classes of sequential models for *n*-way tables

function	3-way	4-way	5-way
mutual	[A] [B] [C]	[A] [B] [C] [D]	[A] [B] [C] [D] [E]
joint	[AB] [C]	[ABC] [D]	[ABCE] [E]
joint (with=1)	[A] [BC]	[A] [BCD]	[A] [BCDE]
conditional	[AC] [BC]	[AD] [BD] [CD]	[AE] [BE] [CE] [DE]
conditional (with=1)	[AB] [AC]	[AB] [AC] [AD]	[AB] [AC] [AD] [AE]
markov (order=1)	[AB] [BC]	[AB] [BC] [CD]	[AB] [BC] [CD] [DE]
markov (order=2)	[A] [B] [C]	[ABC] [BCD]	[ABC] [BCD] [CDE]
saturated	[ABC]	[ABCD]	[ABCDE]

```
> for(nf in 2 : 5) {
+    print(loglin2string(joint(nf, factors = LETTERS[1:5])))
+ }

[1] "[A] [B]"
[1] "[A,B] [C]"
[1] "[A,B,C] [D]"
[1] "[A,B,C,D] [E]"

> for(nf in 2 : 5) {
+    print(loglin2string(conditional(nf, factors = LETTERS[1:5]),
+                 sep = ""))
+ }

[1] "[A] [B]"
[1] "[AC] [BC]"
[1] "[AD] [BD] [CD]"
[1] "[AE] [BE] [CE] [DE]"

> for(nf in 2 : 5) {
+    print(loglin2formula(conditional(nf, factors = LETTERS[1:5])))
+ }

~A + B
~A:C + B:C
~A:D + B:D + C:D
~A:E + B:E + C:E + D:E
```

Applied to data, these functions take a `table` argument, and deliver the string or formula representation of a type of model for that table:

```
> loglin2formula(joint(3, table = HEC))

~Hair:Eye + Sex

> loglin2string(joint(3, table = HEC))

[1] "[Hair,Eye] [Sex]"
```

Their main use, however, is within higher-level functions, such as `seq_loglm()`, which fit the collection of sequential models of a given type.

```
> HEC.mods <- seq_loglm(HEC, type = "joint")
> LRstats(HEC.mods)

Likelihood summary table:
        AIC BIC LR Chisq Df Pr(>Chisq)
model.1 194 194   165.6  3    <2e-16 ***
model.2 241 246   146.4  9    <2e-16 ***
model.3 193 218    19.9 15      0.18
---
Signif. codes:  0 '***' 0.001 '**' 0.01 '*' 0.05 '.' 0.1 ' ' 1
```

In this section we have described a variety of models that can be fit to higher-way tables, some relations among those models, and the aspects of lack of fit that are revealed in the mosaic displays. The following discussion illustrates the process of model fitting, using the mosaic as an interpretive guide to the nature of associations among the variables. In general, we start with a minimal baseline model.[9] The pattern of residuals in the mosaic will suggest associations to be added to an adequate explanatory model. As the model achieves better fit to the data, the degree of shading decreases, so we may think of the process of model fitting as "cleaning the mosaic."

5.5.2 Causal models

The sequence of models of joint independence has another interpretation when the ordering of the variables is based on a set of ordered hypotheses involving causal relationships among variables (Goodman, 1973, Fienberg, 1980, Section 7.2). Suppose, for example, that the causal ordering of four variables is $A \to B \to C \to D$, where the arrow means "is antecedent to." Goodman suggests that the conditional joint probabilities of B, C, and D given A can be characterized by a set of recursive logit models that treat (a) B as a response to A, (b) C as a response to A and B jointly, and (c) D as a response to A, B and C.

These are equivalent to the loglinear models that we fit as the sequential baseline models of joint independence, namely [A][B], [AB][C], and [ABC][D]. The combination of these models with the marginal probabilities of A gives a characterization of the joint probabilities of all four variables, as in Eqn. (5.4). In application, residuals from each submodel show the associations that remain unexplained.

EXAMPLE 5.9: Marital status and pre- and extramarital sex

A study of divorce patterns in relation to premarital and extramarital sex by Thornes and Collard (1979) reported the 2^4 table shown below, and included in **vcd** as `PreSex`.

---

[9]When one variable, R, is a response, this normally is the model of joint independence, $[E_1 E_2 \ldots][R]$, where $E_1, E_2, \ldots$ are the explanatory variables. Better-fitting models will often include associations of the form $[E_i R]$, $[E_i E_j R] \ldots$.

```
> data("PreSex", package = "vcd")
> structable(Gender + PremaritalSex + ExtramaritalSex ~ MaritalStatus,
+               data = PreSex)

                Gender          Women              Men
                PremaritalSex   Yes       No       Yes       No
                ExtramaritalSex Yes No  Yes No   Yes No  Yes No
MaritalStatus
Divorced                        17  54  36 214   28  60  17  68
Married                          4  25   4 322   11  42   4 130
```

These data were analyzed by Agresti (2013, Section 6.1.7) and by Friendly (1994b, 2000), from which this account draws. A sample of about 500 people who had petitioned for divorce, and a similar number of married people, were asked two questions regarding their pre- and extramarital sexual experience: (1) "Before you married your (former) husband/wife, had you ever made love with anyone else?," (2) "During your (former) marriage (did you) have you had any affairs or brief sexual encounters with another man/woman?" The table variables are thus gender (G), reported premarital (P) and extramarital (E) sex, and current marital status (M).

In this analysis we consider the variables in the order G, P, E, and M, and first reorder the table variables for convenience.

```
> PreSex <- aperm(PreSex, 4 : 1)    # order variables G, P, E, M
```

That is, the first stage treats P as a response to G and examines the [Gender][Pre] mosaic to assess whether gender has an effect on premarital sex. The second stage treats E as a response to G and P jointly; the mosaic for [Gender, Pre] [Extra] shows whether extramarital sex is related to either gender or premarital sex. These are shown in Figure 5.15.

```
> # (Gender Pre)
> mosaic(margin.table(PreSex, 1 : 2), shade = TRUE,
+                main = "Gender and Premarital Sex")
>
> ## (Gender Pre)(Extra)
> mosaic(margin.table(PreSex, 1 : 3),
+          expected = ~ Gender * PremaritalSex + ExtramaritalSex,
+          main = "Gender*Pre + ExtramaritalSex")
```

Finally, the mosaic for [Gender, Pre, Extra] [Marital] is examined for evidence of the dependence of marital status on the three previous variables jointly. As noted above, these models are equivalent to the recursive logit models whose path diagram is $G \rightarrow P \rightarrow E \rightarrow M$.[10] The G^2 values for these models shown below provide a decomposition of the G^2 for the model of complete independence fit to the full table.

Model	df	G^2
[G] [P]	1	75.259
[GP] [E]	3	48.929
[GPE] [M]	7	107.956
[G] [P] [E] [M]	11	232.142

The [Gender] [Pre] mosaic in the left panel of Figure 5.15 shows that men are much more likely to report premarital sex than are women; the sample odds ratio is 3.7. We also see that women are about twice as prevalent as men in this sample. The mosaic for the model of joint independence, [Gender Pre] [Extra] in the right panel of Figure 5.15, shows that extramarital sex depends on gender

---

[10]Agresti (2013, Figure 6.1) considers a slightly more complex, but more realistic model in which premarital sex affects both the propensity to have extramarital sex and subsequent marital status.

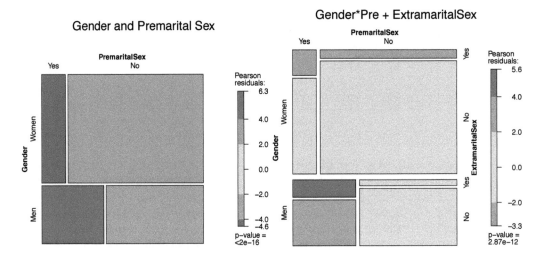

Figure 5.15: Mosaic displays for the first two marginal tables in the PreSex data. Left: Gender and premarital sex; right: fitting the model of joint independence with extramarital sex, [GP][E].

and premarital sex jointly. From the pattern of residuals in Figure 5.15 we see that men and women who have reported premarital sex are far more likely to report extramarital sex than those who have not. In this three-way marginal table, the conditional odds ratio of extramarital sex given premarital sex is nearly the same for both genders (3.61 for men and 3.56 for women). Thus, extramarital sex depends on premarital sex, but not on gender.

```
> loddsratio(margin.table(PreSex, 1 : 3), stratum = 1, log = FALSE)

odds ratios for Gender and PremaritalSex by ExtramaritalSex

      Yes        No
0.28269  0.28611
```

Four-way mosaic plots for two models are shown in Figure 5.16.

```
> ## (Gender Pre Extra)(Marital)
> mosaic(PreSex,
+        expected = ~ Gender * PremaritalSex * ExtramaritalSex
+                   + MaritalStatus,
+        main = "Gender*Pre*Extra + MaritalStatus")
> ## (GPE)(PEM)
> mosaic(PreSex,
+        expected = ~ Gender * PremaritalSex * ExtramaritalSex
+                   + MaritalStatus * PremaritalSex * ExtramaritalSex,
+        main = "G*P*E + P*E*M")
```

△

5.5.3 Partial association

In a three-way (or larger) table it may be that two variables, say A and B, are associated at some levels of the third variable, C, but not at other levels of C. More generally, we may wish to explore whether and how the association among two (or more) variables in a contingency table varies over

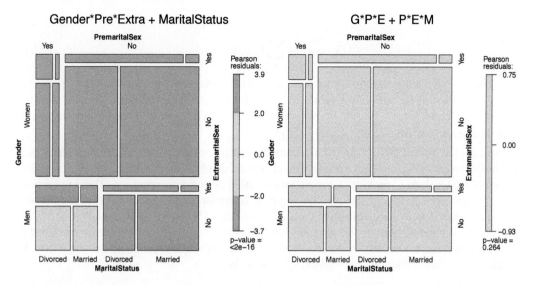

Figure 5.16: Four-way mosaics for the PreSex data. The left panel fits the model [GPE][M]. The pattern of residuals suggests other associations with marital status. The right panel fits the model [GPE][PEM].

the levels of the remaining variables. The term ***partial association*** refers to the association among some variables within the levels of the other variables.

Partial association represents a useful "divide and conquer" statistical strategy: it allows you to refine the question you want to answer for complex relations by breaking it down to smaller, easier questions.[11] It is a statistically happy fact that an answer to the larger, more complex question can be expressed as an algebraic sum of the answers to the smaller questions, just as was the case with sequential models of joint independence.

For concreteness, consider the case where you want to understand the relationship between *attitude* toward corporal punishment of children by parents or teachers (Never, Moderate use OK) and *memory* that the respondent had experienced corporal punishment as a child (Yes, No). But you also have measured other variables on the respondents, including their level of *education* and *age* category. In this case, the question of association among all the table variables may be complex, but we can answer a highly relevant, specialized question precisely, "Is there an association between attitude and memory, *controlling for education and age?*" The answer to this question can be thought of as the sum of the answers to the simpler question of association between attitude and memory across all combinations of the education and age categories.

A simpler version of this idea is considered first below (Example 5.10): among workers who were laid off due to either the closure of a plant or business vs. replacement by another worker, the (conditional) relationship of employment status (new job vs. still unemployed) and duration of unemployment can be studied as a sum of the associations between these focal variables over the separate tables for cause of layoff.

To make this idea precise, consider for example the model of conditional independence, $A \perp B \mid C$ for a three-way table. This model asserts that A and B are independent within *each* level of C. Denote the hypothesis that A and B are independent at level $C(k)$ by $A \perp B \mid C(k), k = 1, \ldots K$. Then one can show (Andersen, 1991) that

---

[11] This is an analog, for categorical data, of the ANOVA strategy for "probing interactions" by testing ***simple effects*** at the levels of one or more of the factors involved in a two- or higher-way interaction.

$$G^2_{A \perp B \mid C} = \sum_k^K G^2_{A \perp B \mid C(k)} \, . \tag{5.6}$$

That is, the overall likelihood ratio G^2 for the conditional independence model with $(I-1)(J-1)K$ degrees of freedom is the sum of the values for the ordinary association between A and B over the levels of C (each with $(I-1)(J-1)$ degrees of freedom). The same additive relationship holds for the Pearson χ^2 statistics: $\chi^2_{A \perp B \mid C} = \sum_k^K \chi^2_{A \perp B \mid C(k)}$.

Thus, (a) the overall G^2 (χ^2) may be decomposed into portions attributable to the AB association in the layers of C, and (b) the collection of mosaic displays for the dependence of A and B for each of the levels of C provides a natural visualization of this decomposition. These provide an analog, for categorical data, of the conditioning plot, or *coplot*, that Cleveland (1993b) has shown to be an effective display for quantitative data. See Friendly (1999a) for further details.

Mosaic and other displays in the strucplot framework for partial association can be produced in several different ways. One way is to use a model formula in the call to `mosaic()` which lists the conditioning variables after the " | " (given) symbol, as in

```
~ Memory + Attitude | Age + Education
```

Another way is to use `cotabplot()`. This takes the same kind of conditioning model formula, but presents each panel for the conditioning variables in a separate frame within a trellis-like grid.[12]

EXAMPLE 5.10: Employment status data

Data from a 1974 Danish study of 1,314 employees who had been laid off are given in the data table *Employment* in **vcd** (from Andersen (1991, Table 5.12)). The workers are classified by: (a) their employment status, on January 1, 1975 ("NewJob" or still "Unemployed"), (b) the length of their employment at the time of layoff, (c) the cause of their layoff ("Closure", etc., or "Replaced").

```
> data("Employment", package = "vcd")
> structable(Employment)
```

EmploymentStatus	LayoffCause	EmploymentLength	<1Mo	1-3Mo	3-12Mo	1-2Yr	2-5Yr	>5Yr
NewJob	Closure		8	35	70	62	56	38
	Replaced		40	85	181	85	118	56
Unemployed	Closure		10	42	86	80	67	35
	Replaced		24	42	41	16	27	10

In this example, it is natural to regard `EmploymentStatus` (variable A) as the response variable, and `EmploymentLength` (B) and `LayoffCause` (C) as predictors. In this case, the minimal baseline model is the joint independence model, $[A][BC]$, which asserts that employment status is independent of both length and cause. This model fits quite poorly, as shown in the output from `loglm()` below.

```
> loglm(~ EmploymentStatus + EmploymentLength * LayoffCause,
+          data = Employment)

Call:
loglm(formula = ~EmploymentStatus + EmploymentLength * LayoffCause,
     data = Employment)

Statistics:
                  X^2 df P(> X^2)
Likelihood Ratio 172.28 11        0
Pearson          165.70 11        0
```

---

[12]Depending on your perspective, this has the advantage of adjusting for the total frequency in each conditional panel, or the disadvantage of ignoring these differences.

The residuals, shown in Figure 5.17, indicate an opposite pattern for the two categories of `LayoffCause`: those who were laid off as a result of a closure are more likely to be unemployed, regardless of length of time they were employed. Workers who were replaced, however, apparently are more likely to be employed, particularly if they were employed for 3 months or more.

```
> # baseline model [A][BC]
> mosaic(Employment, shade = TRUE,
+        expected = ~ EmploymentStatus + EmploymentLength * LayoffCause,
+        main = "EmploymentStatus + Length * Cause")
```

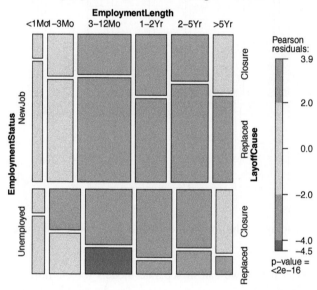

Figure 5.17: Mosaic display for the employment status data, fitting the baseline model of joint independence.

Beyond this baseline model, it is substantively more meaningful to consider the conditional independence model, $A \perp B \mid C$, (or [AC][BC] in shorthand notation), which asserts that employment status is independent of length of employment, given the cause of layoff. We fit this model as shown below:

```
> loglm(~ EmploymentStatus * LayoffCause + EmploymentLength * LayoffCause,
+       data = Employment)

Call:
loglm(formula = ~EmploymentStatus * LayoffCause + EmploymentLength *
    LayoffCause, data = Employment)

Statistics:
                   X^2 df  P(> X^2)
Likelihood Ratio 24.630 10 0.0060927
Pearson          26.072 10 0.0036445
```

This model fits far better ($G^2(10) = 24.63$), but the lack of fit is still significant. The residuals, shown in Figure 5.18, still suggest that the pattern of association between employment and length is different for replaced workers and those laid off due to closure of their workplace.

```
> mosaic(Employment, shade = TRUE, gp_args = list(interpolate = 1 : 4),
+          expected = ~ EmploymentStatus * LayoffCause +
+                        EmploymentLength * LayoffCause,
+          main = "EmploymentStatus * Cause + Length * Cause")
```

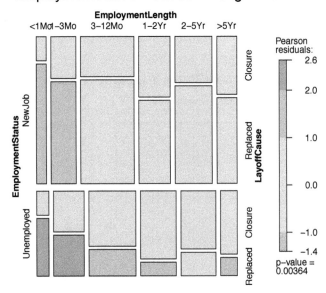

Figure 5.18: Mosaic display for the employment status data, fitting the model of conditional independence, [AC][BC].

To explain this result better, we can fit separate models for the *partial* relationship between `EmploymentStatus` and `EmploymentLength` for the two levels of `LayoffCause`. In R, with the *Employment* data as in table form, this is easily done using `apply()` over the `LayoffCause` margin, giving a list containing the two `loglm()` models.

```
> mods.list <-
+   apply(Employment, "LayoffCause",
+         function(x) loglm(~ EmploymentStatus + EmploymentLength,
+                            data = x))
> mods.list

$Closure
Call:
loglm(formula = ~EmploymentStatus + EmploymentLength, data = x)

Statistics:
                 X^2 df P (> X^2)
Likelihood Ratio 1.4786  5   0.91553
Pearson          1.4835  5   0.91497

$Replaced
Call:
loglm(formula = ~EmploymentStatus + EmploymentLength, data = x)

Statistics:
                  X^2 df   P (> X^2)
Likelihood Ratio 23.151  5 0.00031578
Pearson          24.589  5 0.00016727
```

Extracting the model fit statistics for these partial models and adding the fit statistics for the overall model of conditional independence, [AC][BC], gives the table below, illustrating the additive property of G^2 and χ^2 (Eqn. (5.6)).

Model	df	G^2	χ^2
$A \perp B \mid C_1$	5	1.49	1.48
$A \perp B \mid C_2$	5	23.15	24.59
$A \perp B \mid C$	10	24.63	26.07

One simple way to visualize these results is to call `mosaic()` separately for each of the layers corresponding to `LayoffCause`. The result is shown in Figure 5.19.

```
> mosaic(Employment[,,"Closure"], shade = TRUE,
+        gp_args = list(interpolate = 1 : 4),
+        margin = c(right = 1), main = "Layoff: Closure")
> mosaic(Employment[,,"Replaced"], shade = TRUE,
+        gp_args = list(interpolate = 1 : 4),
+        margin = c(right = 1), main = "Layoff: Replaced")
```

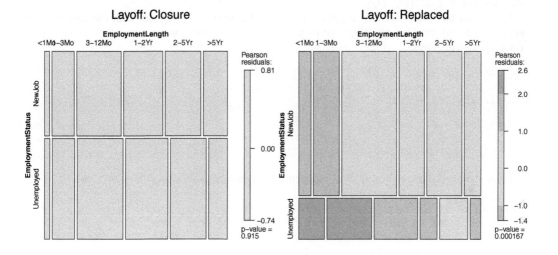

Figure 5.19: Mosaic displays for the employment status data, with separate panels for cause of layoff.

The simple summary from this example is that for workers laid off due to closure of their company, length of previous employment is unrelated to whether or not they are re-employed. However, for workers who were replaced, there is a systematic pattern: those who had been employed for three months or less are likely to remain unemployed, while those with longer job tenure are somewhat more likely to have found a new job.   △

The statistical methods and R techniques described above for three-way tables extend naturally to higher-way tables, as can be seen in the next example.

EXAMPLE 5.11: Corporal punishment data

Here we use the *Punishment* data from vcd, which contains the results of a study by the Gallup Institute in Denmark in 1979 about the attitude of a random sample of 1,456 persons towards corporal punishment of children (Andersen, 1991, pp. 207–208). As shown below, this data set is a frequency data frame representing a $2 \times 2 \times 3 \times 3$ table, with table variables (a) `attitude` toward use of corporal punishment (approve of "moderate" use or "no" approval); (b) `memory` of whether the respondent had experienced corporal punishment as a child (yes/no); (c) `education` level of respondent (elementary, secondary, high); and (d) `age` category of respondent.

```
> data("Punishment", package = "vcd")
> str(Punishment, vec.len = 2)

'data.frame': 36 obs. of  5 variables:
 $ Freq     : num  1 3 20 2 8 ...
 $ attitude : Factor w/ 2 levels "no","moderate": 1 1 1 1 1 ...
 $ memory   : Factor w/ 2 levels "yes","no": 1 1 1 1 1 ...
 $ education: Factor w/ 3 levels "elementary","secondary",..: 1 1 1 2 2 ...
 $ age      : Factor w/ 3 levels "15-24","25-39",..: 1 2 3 1 2 ...
```

Of main interest here is the association between attitude toward corporal punishment as an adult (A) and memory of corporal punishment as a child (B), controlling for age (C) and education (D); that is, the model $A \perp B \mid (C, D)$, or [ACD][BCD] in shorthand notation.

As noted above, this conditional independence hypothesis can be decomposed into the 3×3 partial tests of $A \perp B \mid (C_k, D_\ell)$.

These tests and the associated graphics are somewhat easier to carry out with the data in table form (`pun`) constructed below. While we're at it, we recode the variable names and factor levels for nicer graphical displays.

```
> pun <- xtabs(Freq ~ memory + attitude + age + education,
+              data = Punishment)
> dimnames(pun) <- list(
+    Memory = c("yes", "no"),
+    Attitude = c("no", "moderate"),
+    Age = c("15-24", "25-39", "40+"),
+    Education = c("Elementary", "Secondary", "High"))
```

Then, the overall test of conditional independence can be carried using `loglm()` out as

```
> (mod.cond <- loglm(~ Memory * Age * Education +
+                      Attitude * Age * Education, data = pun))

Call:
loglm(formula = ~Memory * Age * Education + Attitude * Age *
    Education, data = pun)

Statistics:
                   X^2 df   P(> X^2)
Likelihood Ratio 39.679  9 8.6851e-06
Pearson          34.604  9 6.9964e-05
```

Alternatively, `coindep_test()` in **vcd** provides tests of conditional independence of two variables in a contingency table by simulation from the marginal permutation distribution of the input table. The version reporting a Pearson χ^2 statistic is given by

```
> set.seed(1071)
> coindep_test(pun, margin = c("Age", "Education"),
+              indepfun = function(x) sum(x ^ 2), aggfun = sum)


Permutation test for conditional independence

data:  pun
f(x) = 34.6, p-value <2e-16
```

These tests all show substantial association between attitude and memory of corporal punishment. How can we understand and explain this?

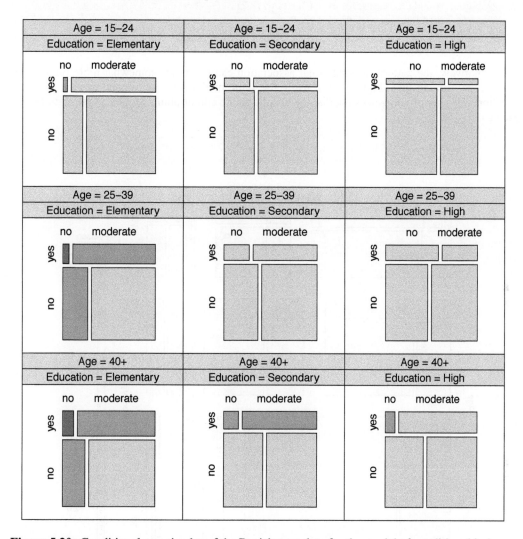

Figure 5.20: Conditional mosaic plot of the Punishment data for the model of conditional independence of attitude and memory, given age and education. Shading of tiles is based on the sum of squares statistic.

As in Example 5.10, we can partition the overall G^2 or χ^2 to show the contributions to this association from the combinations of age and education. The call to `apply()` below fits an independence model for `Memory` and `Attitude` for each stratum defined by the combinations of `Age` and `Education`, and extracts the Pearson χ^2 statistics. The result is returned as a 3×3 matrix.

```
> mods.list <- apply(pun, c("Age", "Education"),
+        function(x) loglm(~ Memory + Attitude, data = x)$pearson)
```

One visual analog of this table of χ^2 statistics is a `cotabplot()` of the (conditional) association of attitude and memory over the age and education cells, shown in Figure 5.20. `cotabplot()` is very general, allowing a variety of functions of the residuals to be used for shading (Zeileis et al., 2007). Here we use the (Pearson) sum of squares statistic, $\sum_{k,\ell} \chi^2_{k,\ell}$.

```
> set.seed(1071)
> pun_cotab <- cotab_coindep(pun, condvars = 3 : 4, type = "mosaic",
+    varnames = FALSE, margins = c(2, 1, 1, 2),
+    test = "sumchisq", interpolate = 1 : 2)
> cotabplot(~ Memory + Attitude | Age + Education,
+           data = pun, panel = pun_cotab)
```

Alternatively, the pattern of conditional association can be shown somewhat more directly in a conditional mosaic plot (Figure 5.21), using the same model formula to condition on age and education. This simply organizes the display to split on the conditioning variables first, with larger spacings.

```
> mosaic(~ Memory + Attitude | Age + Education, data = pun,
+        shade = TRUE, gp_args = list(interpolate = 1 : 4))
```

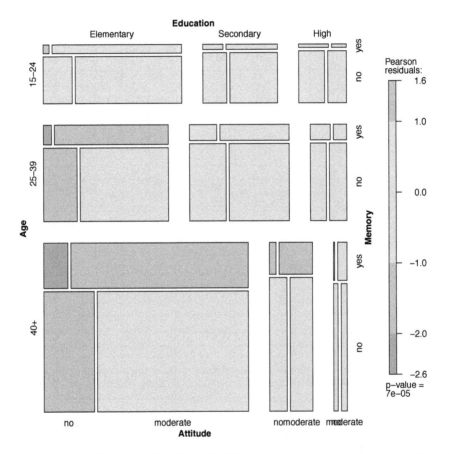

Figure 5.21: Conditional mosaic plot of the Punishment data for the model of conditional independence of attitude and memory, given age and education. This plot explicitly shows the total frequencies in the cells of age and education by the areas of the main blocks for these variables.

Both Figure 5.20 and Figure 5.21 reveal that the association between attitude and memory becomes stronger with increasing age among those with the lowest education (first column). Among those in the highest age group (bottom row), the strength of association *decreases* with increasing education. These two displays differ in that in the cotabplot() of Figure 5.20, the marginal frequencies of age and education are not shown, whereas in the mosaic() of Figure 5.21 they determine the relative sizes of the tiles for the combinations of age and education.

The divide-and-conquer strategy of partial association using statistical tests and visual displays now provides a simple, coherent explanation for this table: memory of experienced violence as a child tends to engender a more favorable attitude toward corporal punishment as an adult, but this association varies directly with both age and education. △

5.6 Mosaic matrices for categorical data

One reason for the wide usefulness of graphs of quantitative data has been the development of effective, general techniques for dealing with high-dimensional data sets. The ***scatterplot matrix*** shows all pairwise (marginal) views of a set of variables in a coherent display, whose design goal is to show the interdependence among the collection of variables as a whole. It combines multiple views of the data into a single display, which allows detection of patterns that could not readily be discerned from a series of separate graphs. In effect, a multivariate data set in p dimensions (variables) is shown as a collection of $p(p-1)$ two-dimensional scatterplots, each of which is the projection of the cloud of points on two of the variable axes. These ideas can be readily extended to categorical data.

A multiway contingency table of p categorical variables, $A, B, C, \ldots$, contains the interdependence among the collection of variables as a whole. The saturated loglinear model, $[ABC\ldots]$ fits this interdependence perfectly, but is often too complex to describe or understand.

By summing the table over all variables except two, A and B, say, we obtain a two-variable (marginal) table, showing the bivariate relationship between A and B, which is also a projection of the p-variable relation into the space of two (categorical) variables. If we do this for all $p(p-1)$ unordered pairs of categorical variables and display each two-variable table as a mosaic, we have a categorical analog of the scatterplot matrix, called a ***mosaic matrix***. Like the scatterplot matrix, the mosaic matrix can accommodate any number of variables in principle, but in practice is limited by the resolution of our display to three or four variables.

In R, the main implementation of this idea is in the generic function `pairs()`. The vcd package extends this to mosaic matrices with methods for "table" and "structable" objects. The gpairs package provides a ***generalized pairs plot***, with appropriate graphics for a mixture of quantitative and categorical variables.

5.6.1 Mosaic matrices for pairwise associations

EXAMPLE 5.12: Bartlett data on plum-root cuttings

The simplest example of what you can see in a mosaic matrix is provided by the $2 \times 2 \times 2$ table used by Bartlett (1935) to illustrate a method for testing for no three-way interaction in a contingency table (hypothesis H_4 in Table 5.2).

The data set `Bartlett` in vcdExtra gives the result of an agricultural experiment to investigate the survival of plum-root cuttings (`Alive`) in relation to two factors: `Time` of planting and the `Length` of the cutting. In this experiment, 240 cuttings were planted for each of the 2×2 combinations of these factors, and their survival (Alive, Dead) was later recorded.

```
> pairs(Bartlett, gp = shading_Friendly2)
```

The mosaic matrix for these data, showing all two-way marginal relations, is shown in Figure 5.22. It can immediately be seen that `Time` and `Length` are independent by the design of the experiment; we use `gp=shading_Friendly` here to emphasize this.

The top row and left column show the relation of survival to each of time of planting and cutting length. It is easily seen that greater survival is associated with cuttings taken now (vs. spring) and

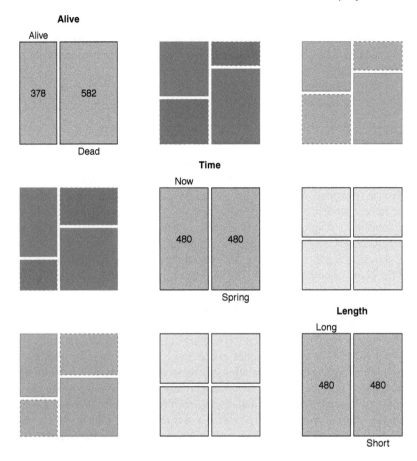

Figure 5.22: Mosaic pairs plot for the Bartlett data. Each panel shows the bivariate marginal relation between the row and column variables.

those cut long (vs. short), and the degree of association is stronger for planting time than for cutting length.   △

EXAMPLE 5.13: Marital status and pre- and extramarital sex

In Example 5.9 we examined a series of models relating marital status to reported premarital and extramarital sexual activity and gender in the `PreSex` data. Figure 5.23 shows the mosaic matrix for these data. The diagonal panels show the labels for the category levels as well as the one-way marginal totals.

```
> data("PreSex", package = "vcd")
> pairs(PreSex, gp = shading_Friendly2, space = 0.25,
+      gp_args = list(interpolate = 1 : 4),
+      diag_panel_args = list(offset_varnames = -0.5))
```

If we view gender, premarital sex, and extramarital sex as explanatory, and marital status (Divorced vs. still Married) as the response, then the mosaics in row 1 (and in column 1)[13] show how marital status depends on each predictor marginally. The remaining panels show the relations within the set of explanatory variables.

---

[13]Rows and columns in a mosaic matrix are identified as in a table or numerical matrix, with row 1, column 1 in the upper left corner.

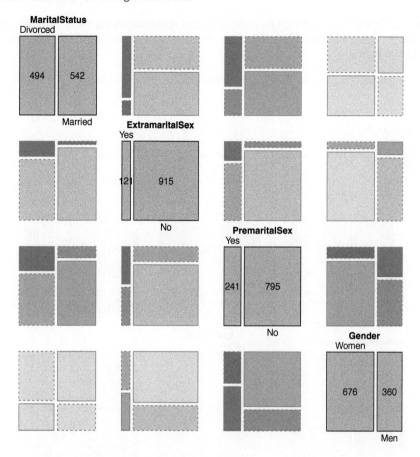

Figure 5.23: Mosaic pairs plot for the PreSex data. Each panel shows the bivariate marginal relation between the row and column variables.

Thus, we see in row 1, column 4, that marital status is independent of gender (all residuals equal zero, here), by design of the data collection. In the (1, 3) panel, we see that reported premarital sex is more often followed by divorce, while non-report is more prevalent among those still married. The (1, 2) panel shows a similar, but stronger relation between extramarital sex and marriage stability. These effects pertain to the associations of P and E with marital status (M)—the terms [PM] and [EM] in the loglinear model. We saw earlier that an interaction of P and E (the term [PEM]) is required to fully account for these data. This effect is not displayed in Figure 5.23.

Among the background variables (the loglinear term [GPE]), the (2, 3) panel shows a strong relation between premarital sex and subsequent extramarital sex, while the (2, 4) and (3, 4) panels show that men are far more likely to report premarital sex than women in this sample, and also more likely to report extramarital sex.

Even though the mosaic matrix shows only pairwise, bivariate associations, it provides an integrated view of all of these together in a single display.

△

EXAMPLE 5.14: Berkeley admissions

In Chapter 4 we examined the relations among the variables Admit, Gender, and Department in the Berkeley admissions data (Example 4.1, Example 4.11, Example 4.15) using fourfold displays (Figure 4.5 and Figure 4.6) and sieve diagrams (Figure 4.13). These displays showed either a marginal relation (e.g., Admit, Gender) or the full three-way table.

In contrast, Figure 5.24 shows all pairwise marginal relations among these variables, produced using `pairs()`. Some additional arguments are used to control the details of labels for the diagonal and off-diagonal panels.

```
> largs <- list(labeling = labeling_border(varnames = FALSE,
+                  labels = c(T, T, F, T), alternate_labels = FALSE))
> dargs <- list(gp_varnames = gpar(fontsize = 20), offset_varnames = -1,
+                  labeling = labeling_border(alternate_labels = FALSE))
> pairs(UCBAdmissions, shade = TRUE, space = 0.25,
+        diag_panel_args = dargs,
+        upper_panel_args = largs, lower_panel_args = largs)
```

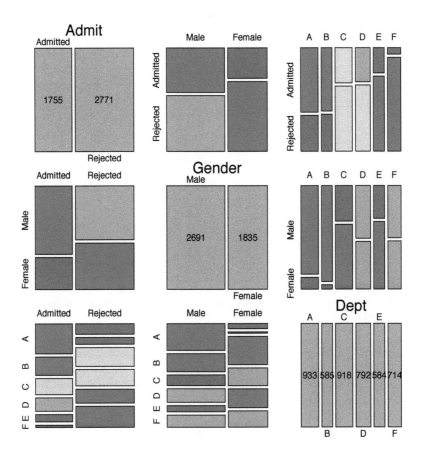

Figure 5.24: Mosaic matrix of the UCBAdmissions data showing bivariate marginal relations.

The panel in row 2, column 1 shows that Admission and Gender are strongly associated marginally, as we saw in Figure 4.5, and overall, males are more often admitted. The diagonally opposite panel (row 1, column 2) shows the same relation, splitting first by gender.[14]

The panels in the third column (and third row) provide the explanation for the paradoxical result (see Figure 4.6) that, within all but department A, the likelihood of admission is equal for men and women, yet, overall, there appears to be a bias in favor of admitting men (see Figure 4.5). The (1,

---

[14]Note that this is different than just the transpose or interchange of horizontal and vertical dimensions as in a scatterplot matrix, because the mosaic display splits the total frequency first by the horizontal variable and then (conditionally) by the vertical variable. The areas of all corresponding tiles are the same in each diagonally opposite pair, however, as are the residuals shown by color and shading.

3) and (3, 1) panels show the marginal relation between Admission and Department; that is, how admission rate varies across departments. Departments A and B have the greatest overall admission rate, departments E and F the least. The (2, 3) and (3, 2) panels show how men and women apply differentially to the various departments. It can be seen that men apply in much greater numbers to departments A and B, with higher admission rates, while women apply in greater numbers to the departments C–F, with the lowest overall rate of admission.

△

5.6.2 Generalized mosaic matrices and pairs plots

We need not show only the marginal relation between each pair of variables in a mosaic matrix. Friendly (1999b) describes the extension of this idea to conditional, partial, and other views of a contingency table.

In `pairs.table()`, different *panel functions* can be used to specify what is displayed in the upper, lower, and diagonal panels. For the off-diagonal panels, a `type` argument can be used to plot mosaics showing various kinds of independence relations:

`type = "pairwise"` – Shows bivariate marginal relations, collapsed over all other variables.
`type = "total"` – Shows mosaic plots for mutual independence.
`type = "conditional"` – Shows mosaic plots for conditional independence given all other variables.
`type = "joint"` – Shows mosaic plots for joint independence of all pairs of variables from the others.

EXAMPLE 5.15: Berkeley admissions

Figure 5.25 shows the generalized mosaic matrix for the `UCBAdmissions` data, using 3-way mosaics for all the off-diagonal cells. The observed frequencies, of course, are the same in all these cells. However, in the lower panels, the tiles are shaded according to models of joint independence, while in the upper panels, they are shaded according to models of mutual independence.

```
> pairs(UCBAdmissions, space = 0.2,
+        lower_panel = pairs_mosaic(type = "joint"),
+        upper_panel = pairs_mosaic(type = "total"))
```

In this example, it is more useful to fit and display the models of conditional independence for each pair of row, column variables given the remaining one, as shown in Figure 5.26.

```
> pairs(UCBAdmissions, type = "conditional", space = 0.2)
```

Thus, the shading in the (1, 2) and (2, 1) panels shows the fit of the model [Admit, Dept] [Gender, Dept], which asserts that Admission and Gender are independent, given (controlling for) Department. Except for Department A, this model fits quite well, again indicating lack of gender bias. The (1, 3) and (3, 1) panels show the relation between admission and department controlling for gender, highlighting the differential admission rates across departments.

△

Beyond this, the framework of pairs plots can be further generalized to *mixtures* of quantitative and categorical variables, as first described in Friendly (2003) and then in a wider context by Emerson et al. (2013) and Friendly (2013). The essential idea is to consider the combination of two variables, each of which can be either categorical (**C**) or quantitative (**Q**), and various ways to *render* that combination in a graphical display:

CC: mosaic display, sieve diagram, doubledecker plot, faceted or divided bar chart;

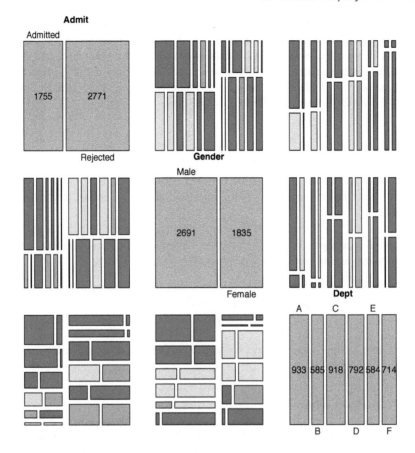

Figure 5.25: Generalized mosaic matrix of the UCBAdmissions data. The above-diagonal plots fit models of joint independence; below-diagonal plots fit models of mutual independence.

CQ: side-by-side boxplots, stripplots, faceted histograms, aligned density plots;
QQ: scatterplot, corrgram, data ellipses, etc.

In R some of these possibilities are provided in the **gpairs** package (using grid graphics and the **vcd** strucplot framework), and the **GGally** (Schloerke et al., 2014) package (an extension to **ggplot2**).

EXAMPLE 5.16: Arthritis treatment

We illustrate these ideas with the *Arthritis* data using the **gpairs** package in Figure 5.27. In this data, the variables Treatment, Sex, and Improved are categorical, and Age is quantitative. The call to gpairs() below reorders the variables to put the response variable Improved in row 1, column 1. Various options can be passed to mosaic() using the mosaic.pars argument.

```
> library(gpairs)
> data("Arthritis", package = "vcd")
> gpairs(Arthritis[,c(5, 2, 3, 4)],
+        diag.pars = list(fontsize = 20),
+        mosaic.pars = list(gp = shading_Friendly,
+                           gp_args = list(interpolate = 1 : 4)))
```

gpairs() provides a variety of options for the **CQ** and **QQ** combinations, as well as the diagonal cells, but only the defaults are used here. The bottom row, corresponding to Age, uses boxplots to show the distributions of age for each of the categorical variables. The last column

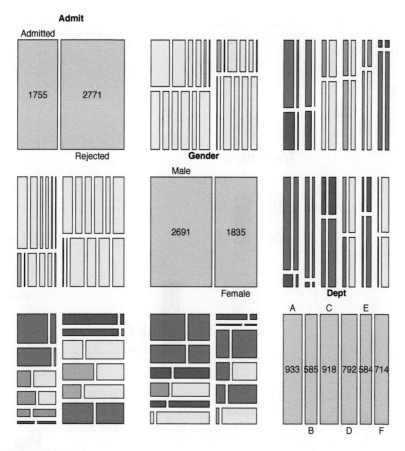

Figure 5.26: Generalized mosaic matrix of the UCBAdmissions data. The off-diagonal plots fit models of conditional indpendence.

shows these same variables as stripplots (or "barcodes"), which show all the individual observations. In the (1, 4) and (4, 1) panels, it can be seen that younger patients are more likely to report no improvement. The other panels in the first row (and column) show that improvement is more likely in the treated condition and greater among women than men. △

5.7 3D mosaics

Mosaic-like displays use the idea of recursive partitioning of a unit square to portray the frequencies in an n-way table by the area of rectangular tiles with (x, y) coordinates. The same idea extends naturally to a 3D graphic. This starts with a unit cube, which is successively subdivided into 3D cuboids along (x, y, z) dimensions, and the frequency in a table cell is then represented by volume.

As in the 2D versions, each cuboid can be shaded to represent some other feature of the data, typically the residual from some model of independence. In principle, the display can accommodate more than 3 variables by using a sequence of split directions along the (x, y, z) axes.

One difficulty in implementing this method is that, short of using a 3D printer, the canvas for a 3D plot on a screen or printer is still projected on a two-dimensional surface, and graphical elements (volumes, lines, text) toward the front of the view will obscure those in the back. In R, a major advance in 3D graphics is available in the rgl (Adler and Murdoch, 2014) package, which mitigates

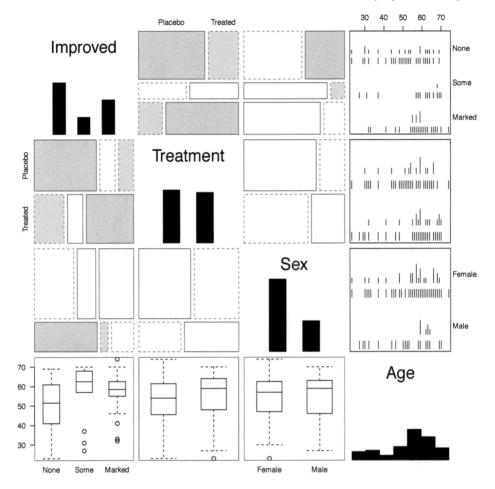

Figure 5.27: Generalized pairs plot of the Arthritis data. Combinations of categorical and quantitative variables can be rendered in various ways.

these problems by: (a) providing an interactive graphic window that can be zoomed and rotated manually with the mouse; (b) allowing dynamic graphics under program control, for example to animate a plot or make a movie; (c) providing control of the details of 3D rendering, including transparency of shapes, surface shading, lighting, and perspective.

The **vcdExtra** package implements 3D mosaics using rgl graphics. mosaic3d() provides methods for "loglm" as well as "table" (or "structable") objects. At the time of writing, only some features of 2D mosaics are available.

EXAMPLE 5.17: Bartlett data on plum-root cuttings
In Example 5.12 we showed the mosaic matrix for the *Bartlett*, fitting the model of mutual independence to show all associations among the table variables, Alive, Time of planting, and Length of cutting. Figure 5.28 shows the 3D version, produced using mosaic3d():

```
> mosaic3d(Bartlett)
```

In the view of this figure, it can be seen that cuttings are more likely to be alive when planted Now and when cut Long. These relations can more easily be appreciated by rotating the 3D display.

△

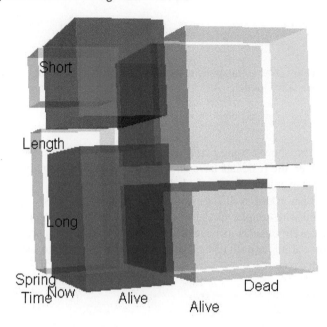

Figure 5.28: 3D mosaic plot of the Bartlett data, according to the model of mutual independence.

5.8 Visualizing the structure of loglinear models

For quantitative response data, it is easy to visualize a fitted model—for linear regression, this is just a plot of the fitted line; for multiple regression or nonlinear regression with two predictors, this is a plot of the fitted response surface. For a categorical response variable, an analog of such plots is provided by effect plots, described later in this book.

For contingency table data, mosaic displays can be used in a similar manner to illuminate the relations among variables in a contingency table represented in various loglinear models, a point described by Theus and Lauer (1999). In fact, each of the model types depicted in Table 5.2 has a characteristic shape and structure in a mosaic display. This, in turn, leads to a clearer understanding of the structure that appears in real data when a given model fits, the relations among the models, and the use of mosaic displays. The essential idea is a simple extension of what we do for more traditional models: show the *expected* (fitted) frequencies under a given model rather than observed frequencies in a mosaic-like display.

To illustrate, we use some artificial data on the relations among age, sex, and symptoms of some disease shown in the $2 \times 2 \times 2$ table `struc` below.

```
> struc <- array(c(6, 10, 312, 44,
+                  37, 31, 192, 76),
+   dim = c(2, 2, 2),
+   dimnames = list(Age = c("Young", "Old"),
+                   Sex = c("F", "M"),
+                   Disease = c("No", "Yes")))
+ )
> struc <- as.table(struc)
> structable(struc)

              Sex   F    M
Age    Disease
Young  No           6  312
       Yes         37  192
```

```
Old    No              10   44
       Yes             31   76
```

First, note that there are substantial associations in this table, as shown in Figure 5.29, fitting the (default) mutual independence model.

```
> mosaic(struc, shade = TRUE)
```

The first split by `Age` shows strong partial associations between `Sex` and `Disease` for both young and old. However, the residuals have an opposite pattern for young and old, suggesting a more complex relationship among these variables.

In this section we are asking a different question: what would mosaic displays look like if the data were in accord with simpler models? One way to do this is simply to use the expected frequencies to construct the tiles, as in sieve diagrams. The result, in Figure 5.30, shows that the tiles for sex and disease align for each of the age groups, but it is harder to see the relations among all three variables in this plot.

```
> mosaic(struc, type = "expected")
```

We can visualize the model-implied relations among all variables together more easily using mosaic matrices.

5.8.1 Mutual independence

For example, to show the structure of a table that exactly fits the model of mutual independence, H_1, use the `loglm()` to find the fitted values, `fit`, as shown below. The function `fitted()` extracts these from the `"loglm"` object.

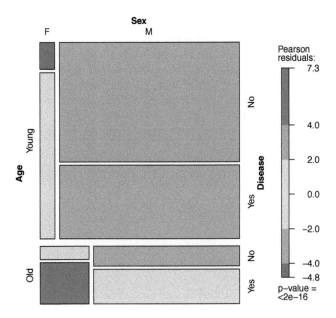

Figure 5.29: Mosaic display for the data on age, sex, and disease. Observed frequencies are shown in the plot, and residuals reflect departure from the model of mutual independence.

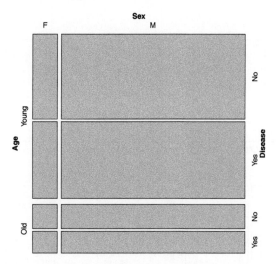

Figure 5.30: Mosaic display for the data on age, sex, and disease, using expected frequencies under mutual independence.

```
> mutual <- loglm(~ Age + Sex + Disease, data = struc, fitted = TRUE)
> fit <- as.table(fitted(mutual))
> structable(fit)

            Sex       F        M
Age    Disease
Young  No         34.0991 253.3077
       Yes        30.7992 228.7940
Old    No         10.0365  74.5567
       Yes         9.0652  67.3416
```

These fitted frequencies then have the same one-way margins as the data in *struc*, but have no two-way or higher associations. Then `pairs()` for this table, using `type="total"`, shows the three-way mosaic for each pair of variables, giving the result in Figure 5.31. We use `gp=shading_Friendly` to explicitly indicate the zero residuals in the display,

```
> pairs(fit, gp = shading_Friendly2, type = "total")
```

In this figure the same data are shown in all the off-diagonal panels and the mutual independence model was fitted in each case, but with the table variables permuted. All residuals are exactly zero in all cells, by construction. We see that in each view, the four large tiles corresponding to the first two variables align, indicating that these two variables are marginally independent. For example, in the (1, 2) panel, age and sex are independent, collapsed over disease.

Moreover, comparing the top half to the bottom half in any panel we see that the divisions by the third variable are the same for both levels of the second variable. In the (1, 2) panel, for example, age and disease are independent for both males and females. This means that age and sex are conditionally independent given disease (age $\perp$ sex | disease).

Because this holds in all six panels, we see that mutual independence implies that *all pairs* of variables are conditionally independent, given the remaining one, $(X \perp Y \mid Z)$ for all permutations of variables. A similar argument can be used to show that joint independence also holds, i.e., $((X, Y) \perp Z)$ for all permutations of variables.

Alternatively, you can also visualize these relationships interactively in a 3D mosaic using `mosaic3d()` that allows you to rotate the mosaic to see all views. In Figure 5.32, all of the 3D

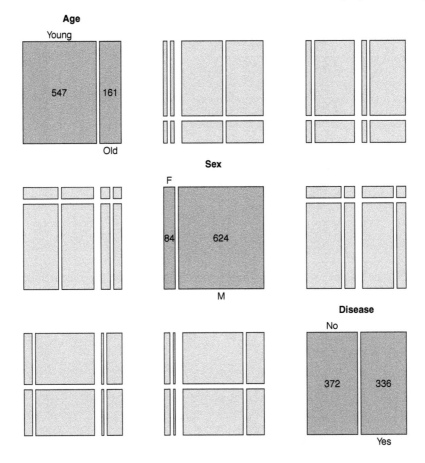

Figure 5.31: Mosaic matrix for fitted values under mutual independence. In all panels the joint frequencies conform to the one-way margins.

tiles are unshaded and you can see that the 3D unit cube has been sliced according to the marginal frequencies.

```
> mosaic3d(fit)
```

5.8.2 Joint independence

The model of joint independence, $H_2 : (A, B) \perp C$, or equivalently, the loglinear model $[AB][C]$ may be visualized similarly by a mosaic matrix in which the data are replaced by fitted values under this model. We illustrate this for the model [Age Sex][Disease], calculating the fitted values in a similar way as before.

```
> joint <- loglm(~ Age * Sex + Disease, data = struc, fitted = TRUE)
> fit <- as.table(fitted(joint))
> structable(fit)

          Sex        F        M
Age  Disease
Young No         22.593  264.814
     Yes         20.407  239.186
```

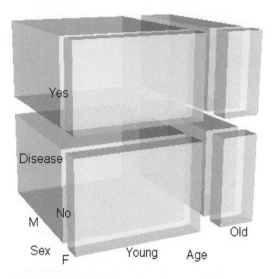

Figure 5.32: 3D mosaic plot of frequencies according to the model of mutual independence. The one-way margins are slices through the unit cube.

```
Old    No              21.542   63.051
       Yes             19.458   56.949
```

The `pairs.table()` plot, now using simpler pairwise plots (`type="pairwise"`), is shown in Figure 5.33.

```
> pairs(fit, gp = shading_Friendly2)
```

This shows, in row 3 and column 3, the anticipated independence of both age and sex with disease, collapsing over the remaining variable. The (1, 2) and (2, 1) panels show that age and sex are still associated when disease is ignored.

5.9 Related visualization methods

A variety of other graphical methods provide the means for visualizing relationships in multiway frequency tables. We briefly describe a few of these here, without much detail, to give a sense of some alternatives.

5.9.1 Doubledecker plots

Doubledecker plots visualize the dependence of one categorical (typically binary) variable on further categorical variables. Formally, they are mosaic plots with vertical splits for all dimensions (predictors) except the last one, which represents the dependent variable (outcome). The last variable is visualized by horizontal splits, no space between the tiles, and separate colors for the levels.

They have the advantage of making it easier to "read" the differences among the conditional response proportions in relation to combinations of the explanatory variables. Moreover, for a binary response, the difference in these conditional proportions for any two columns has a direct relation to the odds ratio for a positive response in relation to those predictor levels (Hofmann, 2001).

The `doubledecker()` function in **vcd** takes a formula argument of the form R ~ E1 + E2

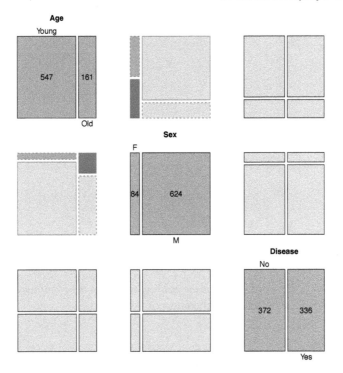

Figure 5.33: Mosaic matrix for fitted values under joint independence for the model [Age Sex][Disease].

+ ... where R is the response variable and E1, E2, ... are the predictors in the contingency table in array form. The shorthand notation, R ~ . means that all variables other than R are taken as predictors, in their order in the array.

EXAMPLE 5.18: Berkeley admissions

Figure 5.34 shows the doubledecker plot for the *UCBAdmissions* data. By default, the levels of the response (Admit) are taken in their order in the array and shaded to highlight the *last* level (Rejected). We want to highlight Admitted, so we reverse this dimension in the call below.

```
> doubledecker(Admit ~ Dept + Gender, data = UCBAdmissions[2:1, , ])
```

In Figure 5.34, it is easy to see the effects of both Dept and Gender on Admit. Admission rate declines across departments A–E, and within departments, the proportion admitted is roughly the same, except for department A, where more female applicants are admitted.   △

EXAMPLE 5.19: Titanic data

Figure 5.35 shows the doubledecker plot for the *Titanic* data. The levels of the response (Survived) are shaded in increasing grey levels, highlighting the proportions of survival.

```
> doubledecker(Survived ~ Class + Age + Sex, Titanic)
```

This order of variables makes it easiest to compare survival of men and women within each age–class combination, but you can also see that survival of adult women decreases with class, and survival among men was greatest in first class. Some additional visualizations of these relationships are illustrated using the next topic in Example 5.21.

    △

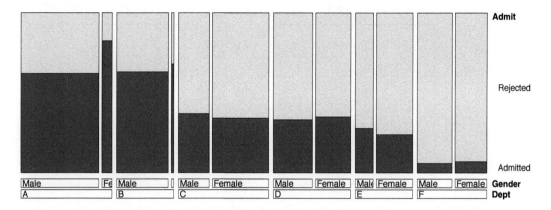

Figure 5.34: Doubledecker plot for the UCBAdmissions data.

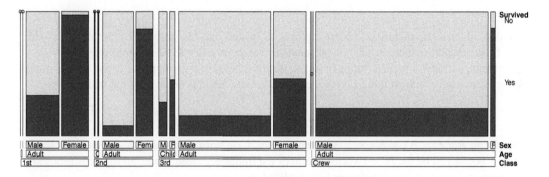

Figure 5.35: Doubledecker plot for the Titanic data.

5.9.2 Generalized odds ratios*

In Example 4.12, we used fourfold displays (Figure 4.7) to analyze the odds ratio between breathlessness and wheeze in coal miners as a function of age. Figure 4.8 showed that a plot of the odds ratio directly against age gave a simplified description of this three-way relationship.

Odds ratios for 2×2 tables can be generalized to $R \times C$ tables in a variety of ways, and these can also be calculated for n-way tables by treating all but the first two dimensions as strata. Plots of these generalized odds ratios can be quite informative, perhaps more so than in the $2 \times 2 \times k$ case.

Consider an $R \times C$ table with frequencies n_{ij}. Then a set of $(R-1) \times (C-1)$ **local odds ratios**, $\theta_{i,j}$, can be calculated as the odds ratios for adjacent pairs of rows and columns as shown in the left panel of Figure 5.36.

$$\theta_{ij} = \frac{n_{ij}/n_{i+1,j}}{n_{i,j+1}/n_{i+1,j+1}} = \frac{n_{ij} \times n_{i+1,j+1}}{n_{i+1,j} \times n_{i,j+1}}, \quad \begin{array}{l} i = 1, 2, \ldots, R-1 \\ j = 1, 2, \ldots, C-1 \end{array}.$$

These odds ratios correspond to "profile contrasts" (or sequential contrasts or successive differences) for ordered categories. Similarly, if one row category and one column category (say, the last) are considered baseline or reference categories, odds ratios with respect to contrasts with those categories (Figure 5.36, right panel) are defined as

$$\theta_{ij} = \frac{n_{i,j} \times n_{R,C}}{n_{i,C} \times n_{R,j}}, \quad \begin{array}{l} i = 1, 2, \ldots, R-1 \\ j = 1, 2, \ldots, C-1 \end{array}.$$

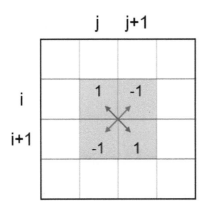 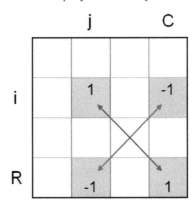

Figure 5.36: Generalized odds ratios for an $R \times C$ table. Left: local odds ratios for adjacent categories. Right: odds ratios with respect to a reference category (the last). Each log odds ratio is a contrast of the log frequencies, shown by the cell weights.

Note that all such parameterizations are equivalent, in that one can derive all other possible odds ratios from any non-redundant set, but substance-driven contrasts will be easier to interpret.

This calculation is simple in terms of log odds ratios, because it corresponds to a contrast among the log frequencies, with weights ± 1 for the four relevant cells. For local odds ratios, these are

$$\log(\theta_{ij}) = \begin{pmatrix} 1 & -1 & -1 & 1 \end{pmatrix} \log \begin{pmatrix} n_{ij} & n_{i+1,j} & n_{i,j+1} & n_{i+1,j+1} \end{pmatrix}^{\mathsf{T}}.$$

Consider an $R \times C \times K_1 \times K_2 \times \dots$ frequency table $n_{ij\dots}$, with factors $K_1, K_2 \dots$ taken as strata. Let $n = \text{vec}(n_{ij\dots})$ be the $N \times 1$ vectorization of the frequency table. Then, all log odds ratios and their asymptotic covariance matrix can be calculated as:

$$\log(\widehat{\boldsymbol{\theta}}) = \boldsymbol{C}\log(\boldsymbol{n})$$
$$\boldsymbol{S} \equiv \mathcal{V}[\log(\boldsymbol{\theta})] = \boldsymbol{C}\,\text{diag}\,(\boldsymbol{n})^{-1}\,\boldsymbol{C}^{\mathsf{T}}$$

where $\boldsymbol{C}$ is an N-column matrix containing all zeros, except for two $+1$ elements and two -1 elements in each row that select the four cells involved in each log lodds ratio.[15]

The function `loddsratio()` in **vcd** calculates these values for the categories of the first two dimensions of an n-way table, together with their asymptotic covariance matrix. Additional dimensions are treated as strata. The `as.array()` and `as.data.frame()` methods can be used to convert a `loddsratio` object to a form suitable for plotting or further analysis.

EXAMPLE 5.20: Corporal punishment data

Example 5.11 used mosaic displays to describe the relationship between attitude toward corporal punishment of children in relationship to memory of having experienced that as a child and education and age of the respondent. Given that `attitude` is the response, we could examine the odds ratios among this variable and any one predictor, treating the other variables as strata. Continuing the analysis of Example 5.11, we calculate log odds ratios for the association of `attitude` and `memory`, stratified by `age` and `education`.

```
> data("Punishment", package = "vcd")
> pun_lor <- loddsratio(Freq ~ memory + attitude | age + education,
+                       data = Punishment)
```

---

[15] Some additional theory and applications of generalized odds ratios for ordered variables is given by Goodman (1983). Hofmann (2001) describes some connections between odds ratios, loglinear models, and visual modeling using doubledecker plots and mosaic plots.

The `as.data.frame()` method converts this to a data frame, and adds standard errors (ASE).

```
> pun_lor_df <- as.data.frame(pun_lor)
```

The plot method for `loddsratio` objects conveniently plots the log odds ratio (LOR) against the strata variables, `age` or `education`, and by default also adds error bars. The result is shown in Figure 5.37.

```
> plot(pun_lor)
```

log odds ratios for memory and attitude by age, education

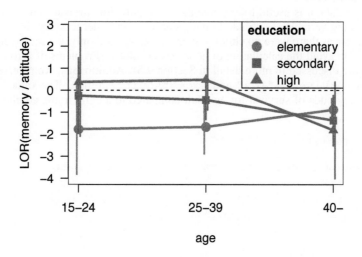

Figure 5.37: Log odds ratio for the association between attitude and memory of corporal punishment, stratified by age and education. Error bars show ±1 standard error.

Compared to Figure 5.20, the differences among the age and education groups are now clear. For respondents less than age 40, increasing education increases the association (log odds ratio) between attitude and memory: those who remembered corporal punishment as a child are more likely to approve of it as their education increases. This result is reversed for those over 40, where all log odds ratios are negative: memory of corporal punishment makes it *less* likely to approve, and this effect becomes stronger with increased education.

Because log odds ratios have an approximate normal distribution under the null hypothesis that all $\log\theta_{ij} = 0$, you can treat these values as data, and carry out a rough analysis of the effects of the stratifying variables using ANOVA, with weights inversely proportional to the estimated sampling variances.[16] In the analysis shown below, we have treated age and education as ordered (numeric) variables.

```
> pun_mod <- lm(LOR ~ age * education, data = pun_lor_df,
+              weights = 1 / ASE^2)
> anova(pun_mod)
```

---

[16]This ignores the covariances among the log odds ratios, which are not independent. A proper analysis uses generalized least squares with a weight matrix S^{-1}, where $S = \mathcal{V}[\log(\theta)]$ is the covariance matrix.

```
Analysis of Variance Table

Response: LOR
              Df Sum Sq Mean Sq F value Pr(>F)
age            1   1.04    1.04    2.72  0.160
education      1   1.84    1.84    4.79  0.080 .
age:education  1   5.04    5.04   13.13  0.015 *
Residuals      5   1.92    0.38
---
Signif. codes:  0 '***' 0.001 '**' 0.01 '*' 0.05 '.' 0.1 ' ' 1
```

This confirms the interaction of age and education on the association between attitude and memory that we described from visual inspection of Figure 5.37.   △

EXAMPLE 5.21: Titanic data

For the `Titanic` data, it is useful to examine the odds ratios for survival in relation to age or sex, using the remaining variables as strata. Some preprocessing is nececessary first: These data contain **structural zeros** as there were no children in the crew. Accordingly, we set the corresponding cell entries to NA to avoid the calculation of nonsensical values. (Problems of zero frequencies in frequency tables are discussed in more detail in Section 9.5). Additionally, we reverse the order of the levels so that `Survived=="Yes"` and `Age=="Adult"` are first. The values calculated below then give the log odds of survival for an adult compared to a child in the combinations sex and class.

```
> Titanic2 <- Titanic[, , 2:1, 2:1]
> Titanic2["Crew", , "Child", ] <- NA
> titanic_lor1 <- loddsratio(~ Survived + Age | Class + Sex,
+                            data = Titanic2)
> titanic_lor1

log odds ratios for Survived and Age by Class, Sex

       Sex
Class       Male     Female
  1st   -3.12102   2.342518
  2nd   -5.50154  -1.510269
  3rd   -0.66874   0.031104
  Crew        NA         NA
```

Similarly, for survival and sex, we obtain the log odds ratios of survival for males versus females, for the combinations of age and class.

```
> titanic_lor2 <- loddsratio(~ Survived + Sex | Class + Age,
+                            data = Titanic2)
> titanic_lor2

log odds ratios for Survived and Sex by Class, Age

       Age
Class      Adult      Child
  1st    -4.1643    1.29928
  2nd    -4.1516   -0.16034
  3rd    -1.4786   -0.77879
  Crew   -3.0156         NA
```

The plots for both tables are shown in Figure 5.38.

In the left panel of Figure 5.38 you can see that the odds ratio of survival for adults relative to children was always greater for females as compared to males, but much less so in 3rd class. In the right panel, the odds ratio of survival for males versus females was always greater for children than adults, again less so in 3rd class.   △

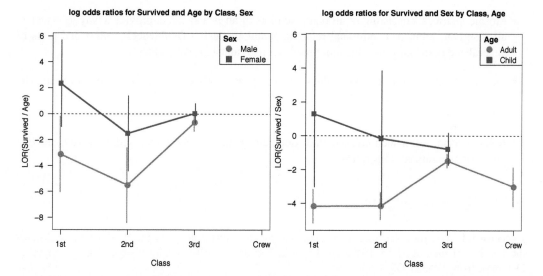

Figure 5.38: Log odds ratio plots for the Titanic data. Left: Odds ratios for survival and age, by sex and class. Right: for survival and sex, by age and class. Error bars show ±1 standard error.

Other examples and plots for log odds ratios are shown in `help(loddsratio)`.

5.10   Chapter summary

- The mosaic display depicts the frequencies in a contingency table by a collection of rectangular "tiles" whose area is proportional to the cell frequency. The residual from a specified model is portrayed by shading the tile to show the sign and magnitude of the deviation from the model.

- For two-way tables, the tiles for the second variable align at each level of the first variable when the two variables are independent (see Figure 5.10).

- The perception and understanding of *patterns of association* (deviations from independence) are enhanced by reordering the rows or columns to give the shading of the residuals a more coherent pattern. An opposite-corner pattern "explains" the association in terms of the ordering of the factor levels.

- For three-way and larger tables, a variety of models can be fit and visualized. Starting with a minimal baseline model, the pattern of residuals will often suggest additional terms that must be added to "clean the mosaic."

- It is often useful to examine the *sequential* mosaic displays for the marginal subtables with the variables in a given order. Sequential models of joint independence provide a breakdown of the total association in the full table, and are particularly appropriate when the last variable is a response.

- Partial association, which refers to the associations among a subset of variables, within the levels of other variables, may be easily studied by constructing separate mosaics for the subset variables for the levels of the other, "given" variables. These displays provide a breakdown of a model of conditional association for the whole table, and serve as an analog of coplots for quantitative data.

- Mosaic matrices, consisting of all pairwise plots of an n-way table, provide a way to visualize all marginal, joint, or conditional relations simultaneously. Doubledecker plots and plots of generalized odds ratios provide other methods to visualize n-way tables.

- The structural relations among model terms in various loglinear models themselves can also be visualized by mosaic matrices showing the expected, rather than observed, frequencies under different models.

- Related visualization techniques include doubledecker plots for binary response models and line plots for generalized odds ratios.

5.11   Lab exercises

Exercise 5.1 The data set *criminal* in the package logmult gives the 4×5 table below of the number of men aged 15–19 charged with a criminal case for whom charges were dropped in Denmark from 1955–1958.

```
> data("criminal", package = "logmult")
> criminal

      Age
Year    15   16   17   18   19
  1955 141  285  320  441  427
  1956 144  292  342  441  396
  1957 196  380  424  462  427
  1958 212  424  399  442  430
```

(a) Use loglm() to test whether there is an association between Year and Age. Is there evidence that dropping of charges in relation to age changed over the years recorded here?

(b) Use mosaic() with the option shade=TRUE to display the pattern of signs and magnitudes of the residuals. Compare this with the result of mosaic() using "Friendly shading," from the option gp=shading_Friendly. Describe verbally what you see in each regarding the pattern of association in this table.

Exercise 5.2 The data set *AirCrash* in vcdExtra gives a database of all crashes of commercial airplanes between 1993–2015, classified by Phase of the flight and Cause of the crash. How can you best show is the nature of the association between these variables in a mosaic plot? Start by making a frequency table, aircrash.tab:

```
> data("AirCrash", package = "vcdExtra")
> aircrash.tab <- xtabs(~ Phase + Cause, data = AirCrash)
```

(a) Make a default mosaic display of the data with shade=TRUE and interpret the pattern of the high-frequency cells.

(b) The default plot has overlapping labels due to the uneven marginal frequencies relative to the lengths of the category labels. Experiment with some of the labeling_args options (abbreviate, rot_labels, etc.) to see if you can make the plot more readable. *Hint*: a variety of these are illustrated in Section 4.1 of vignette("strucplot")

(c) The levels of Phase and Cause are ordered alphabetically (because they are factors). Experiment with other orderings of the rows/columns to make interpretation clearer, e.g., ordering Phase temporally or ordering both factors by their marginal frequency.

Exercise 5.3 The Lahman package contains comprehensive data on baseball statistics for Major League Baseball from 1871 through 2012. For all players, the `Master` table records the handedness of players, in terms of throwing (L, R) and batting (B, L, R), where B indicates "both." The table below was generated using the following code:

```
> library(Lahman)
> data("Master", package = "Lahman")
> basehands <- with(Master, table(throws, bats))
```

	Bats		
Throws	B	L	R
L	177	2640	527
R	924	1962	10442

- Use the code above, or else enter these data into a frequency table in R.
- Construct mosaic displays showing the relation of batting and throwing handedness, split first by batting and then by throwing.
- From these displays, what can be said about players who throw with their left or right hands in terms of their batting handedness?

Exercise 5.4 * A related analysis concerns differences in throwing handedness among baseball players according to the fielding position they play. The following code calculates such a frequency table.

```
> library(Lahman)
> MasterFielding <- data.frame(merge(Master, Fielding, by = "playerID"))
> throwPOS <- with(MasterFielding, table(POS, throws))
```

(a) Make a mosaic display of throwing hand vs. fielding position.
(b) Calculate the percentage of players throwing left-handed by position. Make a sensible graph of this data.
(c) Re-do the mosaic display with the positions sorted by percentage of left-handers.
(d) Is there anything you can say about positions that have very few left-handed players?

Exercise 5.5 For the `Bartlett` data described in Example 5.12, fit the model of no three-way association, H_4 in Table 5.2.

(a) Summarize the goodness of fit for this model, and compare to simpler models that omit one or more of the two-way terms.
(b) Use a mosaic-like display to show the lack of fit for this model.

Exercise 5.6 Red core disease, caused by a fungus, is not something you want if you are a strawberry. The data set `jansen.strawberry` from the agridat package gives a frequency data frame of counts of damage from this fungus from a field experiment reported by Jansen (1990). See the help file for details. The following lines create a $3 \times 4 \times 3$ table of crossings of 3 male parents with 4 (different) female parents, recording the number of plants in four blocks of 9 or 10 plants each showing red core disease in three ordered categories, C1, C2, or C3.

```
> data("jansen.strawberry", package = "agridat")
>
> dat <- jansen.strawberry
> dat <- transform(dat, category = ordered(category,
```

```
+                                            levels = c('C1','C2','C3')))
> levels(dat$male) <- paste0("M", 1:3)
> levels(dat$female) <- paste0("F", 1:4)
>
> jansen.tab <- xtabs(count ~ male + female + category, data = dat)
> names(dimnames(jansen.tab)) <- c("Male parent", "Female parent",
+                                  "Disease category")
> ftable(jansen.tab)
```

(a) Use `pairs(jansen.tab, shade=TRUE)` to display the pairwise associations among the three variables. Describe how disease category appears to vary with male and female parent. Why is there no apparent association between male and female parent?

(b) As illustrated in Figure 5.6, use `mosaic()` to prepare a 3-way mosaic plot with the tiles colored in increasing shades of some color according to disease category. Describe the pattern of category C3 in relation to male and female parent. (Hint: the `highlighting` arguments are useful here.)

(c) With `category` as the response variable, the minimal model for association is [MF][C], or `~ 1*2 + 3`. Fit this model using `loglm()` and display the residuals from this model with `mosaic()`. Describe the pattern of lack of fit of this model.

Exercise 5.7 The data set `caith` in **MASS** gives another classic 4×5 table tabulating hair color and eye color, this for people in Caithness, Scotland, originally from Fisher (1940). The data is stored as a data frame of cell frequencies, whose rows are eye colors and whose columns are hair colors.

```
> data("caith", package = "MASS")
> caith

       fair red medium dark black
blue    326  38    241  110     3
light   688 116    584  188     4
medium  343  84    909  412    26
dark     98  48    403  681    85
```

(a) The `loglm()` and `mosaic()` functions don't understand data in this format, so use `Caith <- as.matrix(caith)` to convert to array form. Examine the result, and use `names(dimnames(Caith))<-c()` to assign appropriate names to the row and column dimensions.

(b) Fit the model of independence to the resulting matrix using `loglm()`.

(c) Calculate and display the residuals for this model.

(d) Create a mosaic display for this data.

Exercise 5.8 The `HairEyePlace` data in **vcdExtra** gives similar data on hair color and eye color, for both Caithness and Aberdeen as a $4 \times 5 \times 2$ table.

(a) Prepare separate mosaic displays, one for each of Caithness and Aberdeen. Comment on any difference in the pattern of residuals.

(b) Construct conditional mosaic plots, using the formula `~ Hair + Eye | Place` and both `mosaic()` and `cotabplot()`. It is probably more useful here to suppress the legend in these plots. Comment on the difference in what is shown in the two displays.

Exercise 5.9 Bertin (1983, pp. 30–31) used a 4-way table of frequencies of traffic accident victims in France in 1958 to illustrate his scheme for classifying data sets by numerous variables, each of which could have various types and could be assigned to various visual attributes. His data are contained in *Accident* in **vcdExtra**, a frequency data frame representing his $5 \times 2 \times 4 \times 2$ table of the variables age, result (died or injured), mode of transportation, and gender.

```
> data("Accident", package = "vcdExtra")
> str(Accident, vec.len=2)

'data.frame': 80 obs. of  5 variables:
 $ age    : Ord.factor w/ 5 levels "0-9"<"10-19"<..: 5 5 5 5 5 ...
 $ result : Factor w/ 2 levels "Died","Injured": 1 1 1 1 1 ...
 $ mode   : Factor w/ 4 levels "4-Wheeled","Bicycle",..: 4 4 2 2 3 ...
 $ gender : Factor w/ 2 levels "Female","Male": 2 1 2 1 2 ...
 $ Freq   : int  704 378 396 56 742 ...
```

(a) Use loglm() to fit the model of mutual independence, Freq ~ age+mode+gender+result to this data set.

(b) Use mosaic() to produce an interpretable mosaic plot of the associations among all variables under the model of mutual independence. Try different orders of the variables in the mosaic. (*Hint*: the abbreviate component of the labeling_args argument to mosaic() will be useful to avoid some overlap of the category labels.)

(c) Treat result ("Died" vs. "Injured") as the response variable, and fit the model Freq ~ age*mode*gender + result that asserts independence of result from all others jointly.

(d) Construct a mosaic display for the residual associations in this model. Which combinations of the predictor factors are more likely to result in death?

Exercise 5.10 The data set *Vietnam* in **vcdExtra** gives a $2 \times 5 \times 4$ contingency table in frequency form reflecting a survey of student opinion on the Vietnam War at the University of North Carolina in May 1967. The table variables are sex, year in school, and response, which has categories: (A) Defeat North Vietnam by widespread bombing and land invasion; (B) Maintain the present policy; (C) De-escalate military activity, stop bombing and begin negotiations; (D) Withdraw military forces immediately. How does the chosen response vary with sex and year?

```
> data("Vietnam", package = "vcdExtra")
> str(Vietnam)

'data.frame': 40 obs. of  4 variables:
 $ sex     : Factor w/ 2 levels "Female","Male": 1 1 1 1 1 1 1 1 1 1 ...
 $ year    : int  1 1 1 1 2 2 2 2 3 3 ...
 $ response: Factor w/ 4 levels "A","B","C","D": 1 2 3 4 1 2 3 4 1 2 ...
 $ Freq    : int  13 19 40 5 5 9 33 3 22 29 ...
```

(a) With response (R) as the outcome variable and year (Y) and sex (S) as predictors, the minimal baseline loglinear model is the model of joint independence, [R][YS]. Fit this model, and display it in a mosaic plot.

(b) Construct conditional mosaic plots of the response versus year separately for males and females. Describe the associations seen here.

(c) Follow the methods shown in Example 5.10 to fit separate models of independence for the levels of sex, and the model of conditional independence, $R \perp Y \mid S$. Verify that the decomposition of G^2 in Eqn. (5.6) holds for these models.

(d) Construct a useful 3-way mosaic plot of the data for the model of conditional independence.

Exercise 5.11 Consider the models for 4-way tables shown in Table 5.3.

(a) For each model, give an independence interpretation. For example, the model of mutual independence corresponds to $A \perp B \perp C \perp D$.

(b) Use the functions shown in the table together with `loglin2formula()` to print the corresponding model formulas for each.

Exercise 5.12 The dataset `Titanic` classifies the 2,201 pasengers and crew of the *Titanic* by `Class` (1st, 2nd, 3rd, Crew), `Sex`, `Age`, and `Survived`. Treating `Survived` as the response variable,

(a) Fit and display a mosaic plot for the baseline model of joint independence, [CGA][S]. Describe the remaining pattern of associations.
(b) Do the same for a "main effects" model that allows two-way associations between each of C, G, and A with S.
(c) What three-way association term should be added to this model to allow for greater survival among women and children? Does this give an acceptable fit?
(d) Test and display models that allow additional three-way associations until you obtain a reasonable fit.

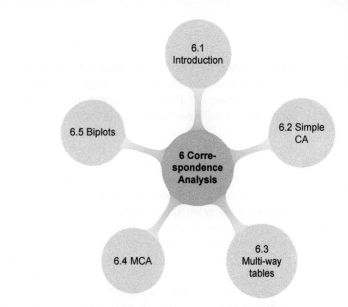

6

Correspondence Analysis

Correspondence analysis provides visualizations of associations in a two-way contingency table in a small number of dimensions. Multiple correspondence analysis extends this technique to *n*-way tables. Other graphical methods, including mosaic matrices and biplots, provide complementary views of loglinear models for two-way and *n*-way contingency tables, but correspondence analysis methods are particularly useful for a simple visual analysis.

6.1 Introduction

> Whenever a large sample of chaotic elements is taken in hand and marshalled in the order of their magnitude, an unsuspected and most beautiful form of regularity proves to have been latent all along.
>
> Sir Francis Galton, *Natural Inheritance*, London: Macmillan, 1889.

Correspondence analysis (CA) is an exploratory technique that displays the row and column categories in a two-way contingency table as points in a graph, so that the positions of the points represent the associations in the table. Mathematically, correspondence analysis is related to the *biplot*, to *canonical correlation*, and to *principal component analysis*.

This technique finds scores for the row and column categories on a small number of dimensions that account for the greatest proportion of the χ^2 for association between the row and column categories, just as principal components account for maximum variance of quantitative variables. But CA does more—the scores provide a quantification of the categories, and have the property that they maximize the correlation between the row and column variables. For graphical display two or three dimensions are typically used to give a reduced rank approximation to the data.

Correspondence analysis has a very large, multi-national literature and was rediscovered several times in different fields and different countries. The method, in slightly different forms, is also discussed under the names **dual scaling**, **optimal scaling**, **reciprocal averaging**, **homogeneity analysis**, and **canonical analysis of categorical data**.

See Greenacre (1984) and Greenacre (2007) for an accessible introduction to CA methodology, or Gifi (1981) and Lebart et al. (1984) for a detailed treatment of the method and its applications from the Dutch and French perspectives. Greenacre and Hastie (1987) provide an excellent discussion of the geometric interpretation, while van der Heijden and de Leeuw (1985) and van der Heijden et al. (1989) develop some of the relations between correspondence analysis and log-linear methods for three-way and larger tables. Correspondence analysis is usually carried out in an exploratory, graphical way. Goodman (1981, 1985, 1986) has developed related inferential models, the RC model (see Section 10.1.3) and the canonical correlation model, with close links to CA.

One simple development of CA is as follows: For a two-way table the scores for the row categories, namely $X = \{x_{im}\}$, and column categories, $Y = \{y_{jm}\}$, on dimension $m = 1, \ldots, M$ are derived from a (generalized) **singular value decomposition** of (Pearson) residuals from independence, expressed as $d_{ij}/\sqrt{n}$, to account for the largest proportion of the χ^2 in a small number of dimensions. This decomposition may be expressed as

$$\frac{d_{ij}}{\sqrt{n}} = \frac{n_{ij} - m_{ij}}{\sqrt{n\, m_{ij}}} = X\, D_\lambda\, Y^{\mathsf{T}} = \sum_{m=1}^{M} \lambda_m\, x_{im}\, y_{jm} \,, \tag{6.1}$$

where m_{ij} is the expected frequency and where D_λ is a diagonal matrix with elements $\lambda_1 \geq \lambda_2 \geq \cdots \geq \lambda_M$, and $M = \min(I-1, J-1)$. In M dimensions, the decomposition Eqn. (6.1) is exact. For example, an $I \times 3$ table can be depicted exactly in two dimensions when $I \geq 3$. The useful result for visualization purposes is that a rank-d approximation in d dimensions is obtained from the first d terms on the right side of Eqn. (6.1). The proportion of the Pearson χ^2 accounted for by this approximation is

$$n \sum_{m}^{d} \lambda_m^2 / \chi^2 \,.$$

The quantity $\chi^2/n = \sum_i \sum_j d_{ij}^2/n$ is called the total **inertia** and is identical to the measure of association known as Pearson's mean-square contingency, the square of the ϕ coefficient.

Thus, correspondence analysis is designed to show how the data deviate from expectation when the row and column variables are independent, as in the sieve diagram, association plot, and mosaic display. However, the sieve, association, and mosaic plots depict every *cell* in the table, and for large tables it may be difficult to see patterns. Correspondence analysis shows only row and column *categories* as points in the two (or three) dimensions that account for the greatest proportion of deviation from independence. The pattern of the associations can then be inferred from the positions of the row and column points.

6.2 Simple correspondence analysis

6.2.1 Notation and terminology

Because Correspondence analysis grew up in so many homes, the notation, formulae, and terms used to describe the method vary considerably. The notation used here generally follows Greenacre (1984, 1997, 2007).

The descriptions here employ the following matrix and vector definitions:

- $N = \{n_{ij}\}$ is the $I \times J$ contingency table with row and column totals n_{i+} and n_{+j}, respectively. The grand total n_{++} is also denoted by n for simplicity.

- $P = \{p_{ij}\} = N/n$ is the matrix of joint cell proportions, called the ***correspondence matrix***.

- $r = \sum_j p_{ij} = P\mathbf{1}$ is the row margin of P; $c = \sum_i p_{ij} = P^\mathsf{T}\mathbf{1}$ is the column margin. r and c are called the *row masses* and *column masses*.

- D_r and D_c are diagonal matrices with r and c on their diagonals, used as weights.

- $R = D_r^{-1}P = \{n_{ij}/n_{+j}\}$ is the matrix of row conditional probabilities, called *row profiles*. Similarly, $C = D_c^{-1}P^\mathsf{T} = \{n_{ij}/n_{i+}\}$ is the matrix of column conditional probabilities or *column profiles*.

- $S = D_r^{-1/2}(P - rc^\mathsf{T})D_c^{-1/2}$ is the matrix of standardized Pearson residuals from independence (denoted d_{ij} in the introduction).

Two types of coordinates, X, Y for the row and column categories are defined, based on the singular value decomposition (SVD) of S,

$$S = UD_\lambda V^\mathsf{T} \quad \text{where} \quad U^\mathsf{T}U = V^\mathsf{T}V = I,$$

and D_λ is the diagonal matrix of singular values $\lambda_1 \geq \lambda_2 \geq \cdots \geq \lambda_M$. U is the orthonormal $I \times M$ matrix of left singular vectors, and V is the $J \times M$ matrix of right singular vectors.

The SVD of S is related to the eigenvalue–eigenvector decomposition of a square symmetric matrix, in that $SS^\mathsf{T} = UD_\lambda^2 U$ and $S^\mathsf{T}S = VD_\lambda^2 V$, so the values λ^2 are the eigenvalues in both cases and the singular vectors are the corresponding eigenvectors. In correspondence analysis, these eigenvalues (squares of the singular values) are called the ***principal inertia***s, and are the values used in the decomposition of the Pearson χ^2 for the dimensions, $\chi^2 = n \sum_m \lambda_m^2$.

principal coordinates: The coordinates of the row (F) and column (G) profiles with respect to their own principal axes are defined so that the inertia along each axis is the corresponding eigenvalue value, λ_m,

$$
\begin{align}
F &= D_r^{-1/2}UD_\lambda^2 \quad \text{scaled so that} \quad F^\mathsf{T}D_rF = D_\lambda^2, \tag{6.2}\\
G &= D_c^{-1/2}VD_\lambda^2 \quad \text{scaled so that} \quad G^\mathsf{T}D_cG = D_\lambda^2. \tag{6.3}
\end{align}
$$

The joint plot in principal coordinates, F and G, is called the ***symmetric map*** because both row and column profiles are overlaid in the same coordinate system.

standard coordinates: The standard coordinates (Φ, Γ) are a rescaling of the principal coordinates to unit inertia along each axis,

$$
\begin{align}
\Phi &= D_r^{-1}U \quad \text{scaled so that} \quad \Phi^\mathsf{T}D_r\Phi = I, \tag{6.4}\\
\Gamma &= D_c^{-1}V \quad \text{scaled so that} \quad \Gamma^\mathsf{T}D_c\Gamma = I. \tag{6.5}
\end{align}
$$

These differ from the principal coordinates in Eqn. (6.2) and Eqn. (6.3) simply by the absence of the scaling factors, D_λ^2. An ***asymmetric map*** shows one set of points (say, the rows) in principal coordinates and the other set in standard coordinates.

Thus, the weighted average of the squared principal coordinates for the rows or columns on a principal axis equals the squared singular value, λ^2 for that axis, whereas the weighted average of the squared standard coordinates equals 1. The relative positions of the row or column points along any axis is the same under either scaling, but the distances between points differ, because the axes are weighted differentially in the two scalings.

6.2.2 Geometric and statistical properties

We summarize here some geometric and statistical properties of the Correspondence analysis solutions that are useful in interpretation.

nested solutions: Because they use successive terms of the SVD Eqn. (6.1), correspondence analysis solutions are *nested*, meaning that the first two dimensions of a three-dimensional solution will be identical to the two-dimensional solution.

centroids at the origin: In both principal coordinates and standard coordinates the points representing the row and column profiles have their centroids (weighted averages) at the origin. Thus, in CA plots, the origin represents the (weighted) average row profile and column profile.

reciprocal averages: CA assigns scores to the row and column categories such that the column scores are proportional to the weighted averages of the row scores, and vice-versa.

chi-square distances: In principal coordinates, the row coordinates may be shown equal to the row profiles $D_r^{-1} P$, rescaled by the inverse by the square-root of the column masses, $D_c^{-1/2}$. Distances between two row profiles, R_i and $R_{i'}$, are most sensibly defined as χ^2 distances, where the squared difference $[R_{ij} - R_{i'j}]^2$ is inversely weighted by the column frequency, to account for the different relative frequency of the column categories. The rescaling by $D_c^{-1/2}$ transforms this weighted χ^2 metric into ordinary Euclidean distance. The same is true of the column principal coordinates.

interpretation of distances: In principal coordinates, the distance between two row points may be interpreted as described above, and so may the distance between two column points. The distance between a row and column point, however, does not have a clear distance interpretation.

residuals from independence: The distance between a row and column point do have a rough interpretation in terms of residuals or the difference between observed and expected frequencies, $n_{ij} - m_{ij}$. Two row (or column) points deviate from the origin (the average profile) when their profile frequencies have similar values. A row point appears in a similar direction away from the origin as a column point when $n_{ij} - m_{ij} > 0$, and in an opposite different direction from that column point when the residual is negative.

Because of these differences in interpretations of distances, there are different possibilities for graphical display. A joint display of principal coordinates for the rows and standard coordinates for the columns (or vice-versa), sometimes called an ***asymmetric map***, is suggested by Greenacre and Hastie (1987) and by Greenacre (1989) as the plot with the most coherent geometric interpretation (for the points in principal coordinates) and is sometimes used in the French literature.

Another common joint display is the ***symmetric map*** of the principal coordinates in the same plot. This is the default in the ca package described below. In the authors' opinion, this produces better graphical displays, because both sets of coordinates are scaled with the same weights for each axis. Symmetric plots are used exclusively in this book, but that should not imply that these plots are universally preferred. Another popular choice is to avoid the possibility of misinterpretation by making separate plots of the row and column coordinates.

6.2.3 R software for correspondence analysis

Correspondence analysis methods for computation and plotting are available in a number of R packages including:

MASS: `corresp()`; the plot method calls `biplot()` for a 2-factor solution, using a a symmetric biplot factorization that scales the row and column points by the square roots of the the singular values. There is also an `mca()` function for multiple correspondence analysis.

ca: `ca()`; provides 2D plots via the `plot.ca()` method and interactive (rgl) 3D plots via `plot3d.ca()`. This package is the most comprehensive in terms of plotting options for various coordinate types, plotting supplementary points (see Section 6.3.2), and other features. It also provides `mjca()` for multiple and joint correspondence analysis of higher-way tables.

FactoMineR (Husson et al., 2015): `CA()`; provides a wide variety of measures for the quality of the CA representation and many options for graphical display

These methods also differ in terms of the types of input they accept. For example, MASS::`corresp()` handles matrices, data frames, and "xtabs" objects, but not "table" objects. `ca()` is the most general, with methods for two-way tables, matrices, data frames, and "xtabs" objects. In the following, we largely use the ca package.

EXAMPLE 6.1: Hair color and eye color

The script below uses the two-way table `haireye` from the *HairEyeColor* data, collapsed over `Sex`. In this table, `Hair` colors form the rows, and `Eye` colors form the columns. By default, `ca()` produces a two-dimensional solution. In this example, the complete, exact solution would have $M = \min((I - 1), (J - 1)) = 3$ dimensions, and you could obtain this using the argument nd=3 in the call to `ca()`.

```
> haireye <- margin.table(HairEyeColor, 1 : 2)
> library(ca)
> (haireye.ca <- ca(haireye))


Principal inertias (eigenvalues):
            1         2         3
Value    0.208773  0.022227  0.002598
Percentage 89.37%    9.52%     1.11%


Rows:
            Black     Brown      Red   Blond
Mass      0.18243   0.48311   0.1199  0.2145
ChiDist   0.55119   0.15946   0.3548  0.8384
Inertia   0.05543   0.01228   0.0151  0.1508
Dim. 1   -1.10428  -0.32446  -0.2835  1.8282
Dim. 2    1.44092  -0.21911  -2.1440  0.4667


Columns:
            Brown     Blue     Hazel     Green
Mass      0.37162   0.3632   0.15710   0.10811
ChiDist   0.50049   0.5537   0.28865   0.38573
Inertia   0.09309   0.1113   0.01309   0.01608
Dim. 1   -1.07713   1.1981  -0.46529   0.35401
Dim. 2    0.59242   0.5564  -1.12278  -2.27412
```

In the printed output, the table labeled "Principal inertias (eigenvalues)" indicates that nearly 99% of the Pearson χ^2 for association is accounted for by two dimensions, with most of that attributed to the first dimension.

The `summary` method for "ca" objects gives a more nicely formatted display, showing a *scree plot* of the eigenvalues, a portion of which is shown below.

```
> summary(haireye.ca)


Principal inertias (eigenvalues):

 dim    value       %    cum%    scree plot
```

```
1        0.208773  89.4   89.4  ***********************
2        0.022227   9.5   98.9  **
3        0.002598   1.1  100.0
        --------  -----
Total:   0.233598 100.0
...
```

The Pearson χ^2 for this table (given by `chisq.test(haireye)`) is 138.29. This value is n (592) times the sum of the eigenvalues (0.2336) shown above.

The result returned by `ca()` can be plotted using the `plot.ca()` method. However, it is useful to understand that `ca()` returns the CA solution in terms of *standard coordinates*, Φ (`rowcoord`) and Γ (`colcoord`). We illustrate Eqn. (6.4) and Eqn. (6.5) using the components of the `"ca"` object `haireye.ca`.

```
> # standard coordinates Phi (Eqn 6.4) and Gamma (Eqn 6.5)
> (Phi <- haireye.ca$rowcoord)

          Dim1      Dim2      Dim3
Black -1.10428   1.44092  -1.08895
Brown -0.32446  -0.21911   0.95742
Red   -0.28347  -2.14401  -1.63122
Blond  1.82823   0.46671  -0.31809

> (Gamma <- haireye.ca$colcoord)

          Dim1      Dim2       Dim3
Brown -1.07713   0.59242  -0.423960
Blue   1.19806   0.55642   0.092387
Hazel -0.46529  -1.12278   1.971918
Green  0.35401  -2.27412  -1.718443

> # demonstrate orthogonality of std coordinates
> Dr <- diag(haireye.ca$rowmass)
> zapsmall(t(Phi) %*% Dr %*% Phi)

     Dim1 Dim2 Dim3
Dim1    1    0    0
Dim2    0    1    0
Dim3    0    0    1

> Dc <- diag(haireye.ca$colmass)
> zapsmall(t(Gamma) %*% Dc %*% Gamma)

     Dim1 Dim2 Dim3
Dim1    1    0    0
Dim2    0    1    0
Dim3    0    0    1
```

These standard coordinates are transformed internally within the plot function according to the `map` argument, which defaults to `map="symmetric"`, giving principal coordinates. The following call to `plot.ca()` produces Figure 6.1.

```
> res <- plot(haireye.ca)
```

For use in further customizing such plots (as we will see in the next example), the function `plot.ca()` returns (invisibly) the coordinates for the row and column points actually plotted, which we saved above as `res`:

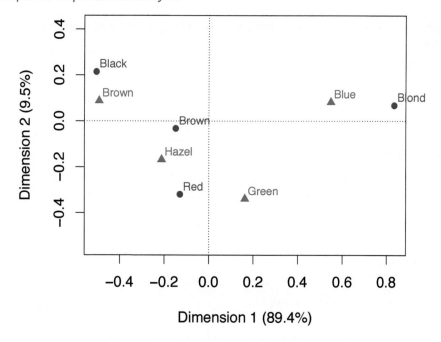

Figure 6.1: Correspondence analysis solution for the hair color and eye color data.

```
> res

$rows
          Dim1        Dim2
Black  -0.50456   0.214820
Brown  -0.14825  -0.032666
Red    -0.12952  -0.319642
Blond   0.83535   0.069579

$cols
          Dim1        Dim2
Brown  -0.49216   0.088322
Blue    0.54741   0.082954
Hazel  -0.21260  -0.167391
Green   0.16175  -0.339040
```

It is important to understand that in CA plots (and related biplots, Section 6.5), the interpretation of distances between points (and angles between vectors) is meaningful. In order to achieve this, the axes in such plots must be *equated*, meaning that the two axes are scaled so that the number of data units per inch are the same for both the horizontal and vertical axes, or an *aspect ratio* = 1.[1]

The interpretation of the CA plot in Figure 6.1 is then as follows:

- Dimension 1, accounting for nearly 90% of the association between hair and eye color corresponds to dark (left) vs. light (right) on both variables.
- Dimension 2 largely contrasts red hair and green eyes with the remaining categories, accounting for an additional 9.5% of the Pearson χ^2.
- With equated axes, and a symmetric map, the distances between row points and distances between column points are meaningful. Along Dimension 1, the eye colors could be considered

---

[1]In base R graphics, this is achieved with the plot() option asp=1.

roughly equally spaced, but for the hair colors, Blond is quite different in terms of its frequency profile.

△

EXAMPLE 6.2: Mental impairment and parents' SES

In Example 4.3 we introduced the data set *Mental*, relating mental health status to parents' SES. As in Example 4.7, we convert this to a two-way table, `mental.tab`, to conduct a correspondence analysis.

```
> data("Mental", package="vcdExtra")
> mental.tab <- xtabs(Freq ~ ses + mental, data = Mental)
```

We calculate the CA solution, and save the result in `mental.ca`:

```
> mental.ca <- ca(mental.tab)
> summary(mental.ca)
```

```
Principal inertias (eigenvalues):

 dim    value       %    cum%    scree plot
 1      0.026025   93.9  93.9    *************************
 2      0.001379    5.0  98.9    *
 3      0.000298    1.1 100.0
        --------  -----
 Total: 0.027702 100.0
...
```

The scree plot produced by `summary(mental.ca)` shows that the association between mental health and parents' SES is almost entirely 1-dimensional, with 94% of the χ^2 (45.98, with 15 df) accounted for by Dimension 1.

We then plot the solution as shown below, giving Figure 6.2. For this example, it is useful to connect the row points and the column points by lines, to emphasize the pattern of these ordered variables.

```
> res <- plot(mental.ca,  ylim = c(-.2, .2))
> lines(res$rows, col = "blue", lty = 3)
> lines(res$cols, col = "red", lty = 4)
```

The plot of the CA scores in Figure 6.2 shows that diagnostic mental health categories are well-aligned with Dimension 1. The mental health scores are approximately equally spaced, except that the two intermediate categories are a bit closer on this dimension than the extremes. The SES categories are also aligned with Dimension 1, and approximately equally spaced, with the exception of the highest two SES categories, whose profiles are extremely similar, suggesting that these two categories could be collapsed.

Because both row and column categories have the same pattern on Dimension 1, we may interpret the plot as showing that the profiles of both variables are ordered, and their relation can be explained as a positive association between high parents' SES and higher mental health status of children. A mosaic display of these data (Exercise 6.5) would show a characteristic opposite corner pattern of association.

From a modeling perspective, we might ask how strong is the evidence for the spacing of categories noted above. For example, we might ask whether assigning integer scores to the levels of SES and mental impairment provides a simpler, but satisfactory account of their association. Questions of this type can be explored in connection with loglinear models in Chapter 9.

△

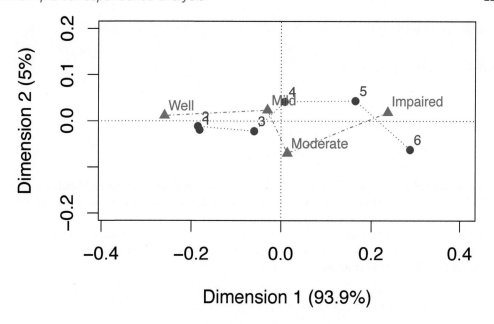

Figure 6.2: Correspondence analysis solution for the Mental health data.

EXAMPLE 6.3: Repeat victimization

The data set *RepVict* in the **vcd** package gives an 8×8 table (from Fienberg (1980, Table 2-8)) on repeat victimization for various crimes among respondents to a U.S. National Crime Survey. A special feature of this data set is that row and column categories reflect the *same* crimes, so substantial association is expected. Here we examine correspondence analysis results in a bit more detail and also illustrate how to customize the displays created by plot(ca(...)).

```
> data("RepVict", package = "vcd")
> victim.ca <- ca(RepVict)
> summary(victim.ca)


Principal inertias (eigenvalues):

 dim    value      %    cum%   scree plot
 1    0.065456   33.8   33.8   ********
 2    0.059270   30.6   64.5   ********
 3    0.029592   15.3   79.8   ****
 4    0.016564    8.6   88.3   **
 5    0.011140    5.8   94.1   *
 6    0.007587    3.9   98.0   *
 7    0.003866    2.0  100.0
      --------  -----
 Total: 0.193474 100.0
...
```

The results above show that, for this 8×8 table, 7 dimensions are required for an exact solution, of which the first two account for 64.5% of the Pearson χ^2. The lines below illustrate that the Pearson χ^2 is n times the sum of the squared singular values, $n \sum \lambda_i^2$.

```
> chisq.test(RepVict)


Pearson's Chi-squared test

data:  RepVict
X-squared = 11100, df = 49, p-value <2e-16

> (chisq <- sum(RepVict) * sum(victim.ca$sv^2))

[1] 11131
```

The default plot produced by `plot.ca(victim.ca)` plots both points and labels for the row and column categories. However, what we want to emphasize here is the relation between the *same* crimes on the first and second occurrence.

To do this, we label each crime just once (using `labels=c(2,0)`) and connect the two points for each crime by a line, using `segments()`, as shown in Figure 6.3. The addition of a `legend()` makes the plot more easily readable.

```
> res <- plot(victim.ca, labels = c(2, 0))
> segments(res$rows[,1], res$rows[,2], res$cols[,1], res$cols[,2])
> legend("topleft", legend = c("First", "Second"), title = "Occurrence",
+        col = c("blue", "red"), pch = 16 : 17, bg = "gray90")
```

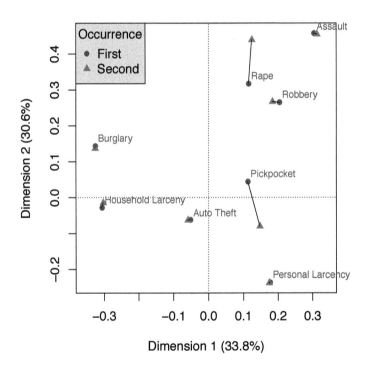

Figure 6.3: 2D CA solution for the repeat victimization data. Lines connect the category points for first and second occurrence to highlight these relations.

In Figure 6.3 it may be seen that most of the points are extremely close for the first and second occurrence of a crime, indicating that the row profile for a crime is very similar to its corresponding column profile, with Rape and Pickpocket as exceptions.

In fact, if the table was symmetric, the row and column points in Figure 6.3 would be identical, as can be easily demonstrated by analyzing a symmetric version.

```
> RVsym <- (RepVict + t(RepVict)) / 2
> RVsym.ca <- ca(RVsym)
> res <- plot(RVsym.ca)
> all.equal(res$rows, res$cols)

[1] TRUE
```

The first dimension appears to contrast crimes against the person (right) with crimes against property (left), and it may be that the second dimension represents degree of violence associated with each crime. The latter interpretation is consistent with the movement of Rape towards a higher position and Pickpocket towards a lower one on this dimension.

△

6.2.4 Correspondence analysis and mosaic displays

For a two-way table, CA and mosaic displays give complementary views of the pattern of association between the row and column variables, but both are based on the (Pearson) residuals from independence. CA shows the row and column categories as points in a 2D (or 3D) space accounting for the largest proportion of the Pearson χ^2, while mosaics show the association by the pattern of shading in the mosaic tiles. It is useful to compare them directly to see how associations can be interpreted from these graphs.

EXAMPLE 6.4: TV viewing data
The data on television viewership from Hartigan and Kleiner (1984) was used as an example of manipulating complex categorical data in Section 2.9. The main association here concerns how viewership across days of the week varies by TV network, so we first collapse the *TV* data to a 5×3 two-way table.

```
> data("TV", package = "vcdExtra")
> TV2 <- margin.table(TV, c(1, 3))
> TV2

          Network
Day          ABC  CBS  NBC
  Monday    2847 2923 2629
  Tuesday   3110 2403 2568
  Wednesday 2434 1283 2212
  Thursday  1766 1335 5886
  Friday    2737 1479 1998
```

In this case, the 2D CA solution is exact, meaning that two dimensions account for 100% of the association.

```
> TV.ca <- ca(TV2)
> TV.ca


 Principal inertias (eigenvalues):
            1        2
Value    0.081934 0.010513
Percentage 88.63%   11.37%
...
```

The plot of this solution is shown in the left panel of Figure 6.4, using lines from the origin to the category points for the networks.

```
> res <- plot(TV.ca)
> segments(0, 0, res$cols[,1], res$cols[,2], col = "red", lwd = 2)
```

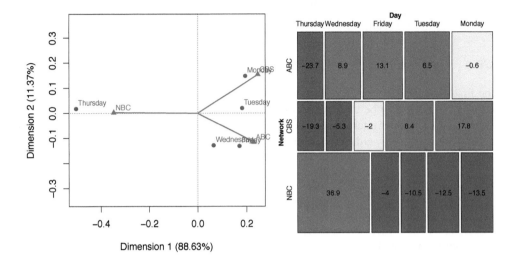

Figure 6.4: CA plot and mosaic display for the TV viewing data. The days of the week in the mosaic plot were permuted according to their order in the CA solution.

An analogous mosaic display, informed by the CA solution, is shown in the right panel of Figure 6.4. Here, the days of the week are reordered according to their positions on the first CA dimension, another example of effect ordering.

```
> days.order <- order(TV.ca$rowcoord[,1])
> mosaic(t(TV2[days.order,]), shade = TRUE, legend = FALSE,
+        labeling = labeling_residuals, suppress=0)
```

In the CA plot, you can see that the dominant dimension separates viewing on Thursday, with the largest share of viewers watching NBC, from the other weekdays. In the mosaic plot, Thursday stands out as the only day with a higher than expected frequency for NBC, and this is the largest residual in the entire table. The second dimension in the CA plot separates CBS, with its greatest proportion of viewers on Monday, from ABC, with greater viewership on Wednesday and Friday.

Emerson (1998, Fig. 2) gives a table listing the shows in each half-hour time slot. Could the overall popularity of NBC on Thursday have been due to *Friends* or *Seinfeld*? An answer to this and similar questions requires analysis of the three-way table (Exercise 6.9) and model-based methods for polytomous outcome variables described in Section 8.3.

△

6.3   Multi-way tables: Stacking and other tricks

A three- or higher-way table can be analyzed by correspondence analysis in several ways. Multiple correspondence analysis (MCA), described in Section 6.4, is an extension of simple correspondence analysis that analyzes simultaneously all possible two-way tables contained within a multiway table. Another approach, described here, is called ***stacking*** or ***interactive coding***. This is a bit of a trick, to force a multiway table into a two-way table for a standard correspondence analysis, but it is a useful one.

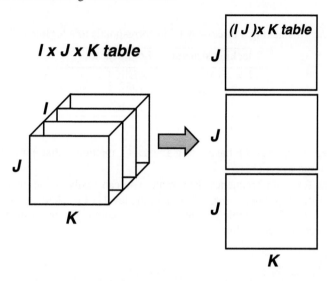

Figure 6.5: Stacking approach for a three-way table. Two of the table variables are combined interactively to form the rows of a two-way table.

A three-way table of size $I \times J \times K$ can be sliced into I two-way tables, each $J \times K$. If the slices are concatenated vertically, the result is one two-way table, of size $(I \times J) \times K$, as illustrated in Figure 6.5. In effect, the first two variables are treated as a single composite variable with IJ levels, which represents the main effects and interaction between the original variables that were combined. Van der Heijden and de Leeuw (1985) discuss this use of correspondence analysis for multi-way tables and show how *each* way of slicing and stacking a contingency table corresponds to the analysis of a specified loglinear model. Like the mosaic display, this provides another way to visualize the relations in a loglinear model.

In particular, for the three-way table with variables A, B, C that is reshaped as a table of size $(I \times J) \times K$, the correspondence analysis solution analyzes residuals from the log-linear model $[AB][C]$. That is, for such a table, the $I \times J$ rows represent the joint combinations of variables A and B. The expected frequencies under independence for this table are

$$m_{[ij]k} = \frac{n_{ij+} \, n_{++k}}{n} \, , \tag{6.6}$$

which are the ML estimates of expected frequencies for the log-linear model $[AB][C]$. The χ^2 that is decomposed by correspondence analysis is the Pearson χ^2 for this log-linear model. When the table is stacked as $I \times (J \times K)$ or $J \times (I \times K)$, correspondence analysis decomposes the residuals from the log-linear models $[A][BC]$ and $[B][AC]$, respectively, as shown in Table 6.1. In this approach, only the associations in separate [] terms are analyzed and displayed in the correspondence analysis maps. Van der Heijden and de Leeuw (1985) show how a generalized form of correspondence analysis can be interpreted as decomposing the difference between two specific loglinear models, so their approach is more general than is illustrated here.

6.3.1 Interactive coding in R

In the general case of an n-way table, the stacking approach is similar to that used by `ftable()` and `structable()` in vcd as described in Section 2.5 to flatten multiway tables to a two-way, printable form, where some variables are assigned to the rows and the others to the columns. Both

Table 6.1: Each way of stacking a three-way table corresponds to a loglinear model

Stacking structure	Loglinear model
$(I \times J) \times K$	$[AB][C]$
$I \times (J \times K)$	$[A][BC]$
$J \times (I \times K)$	$[B][AC]$

ftable() and structable() have as.matrix() methods[2] that convert their result into a matrix suitable as input to ca().

With data in the form of a frequency data frame, you can easily create interactive coding using interaction() or simply use paste() to join the levels of stacked variables together.

To illustrate, create a 4-way table of random Poisson counts (with constant mean, $\lambda = 15$) of types of Pet, classified by Age, Color, and Sex.

```
> set.seed(1234)
> dim <- c(3, 2, 2, 2)
> tab <- array(rpois(prod(dim), 15), dim = dim)
> dimnames(tab) <- list(Pet = c("dog", "cat", "bird"),
+                       Age = c("young", "old"),
+                       Color = c("black", "white"),
+                       Sex = c("male", "female"))
```

You can use ftable() to print this, with a formula that assigns Pet and Age to the columns and Color and Sex to the rows.

```
> ftable(Pet + Age ~ Color + Sex, tab)

             Pet    dog        cat        bird
             Age young old young old young old
Color Sex
black male          10   12   16   16   16   12
      female         8   12   13   15   11   13
white male          18   11   12   18   13   20
      female        13   13   16   15   12   15
```

Then, as.matrix() creates a matrix with the levels of the stacked variables combined with some separator character. Using ca(pet.mat) would then calculate the CA solution for the stacked table, analyzing only the associations in the loglinear model [Pet Age][Color Sex].[3]

```
> (pet.mat <- as.matrix(ftable(Pet + Age ~ Color + Sex, tab), sep = '.'))

                 Pet.Age
Color.Sex        dog.young dog.old cat.young cat.old bird.young bird.old
   black.male           10      12        16      16         16       12
   black.female          8      12        13      15         11       13
   white.male           18      11        12      18         13       20
   white.female         13      13        16      15         12       15
```

With data in a frequency data frame, a similar result (as a frequency table) can be obtained using interaction() as shown below. The result of xtabs() looks the same as pet.mat.

```
> tab.df <- as.data.frame(as.table(tab))
> tab.df <- within(tab.df,
+   {Pet.Age = interaction(Pet, Age)
```

---

[2]This requires at least R version 3.1.0 or vcd 1.3-2 or later.
[3]The result would not be at all interesting here. Why?

```
+    Color.Sex = interaction(Color, Sex)
+    })
> xtabs(Freq ~ Color.Sex + Pet.Age, data = tab.df)
```

EXAMPLE 6.5: Suicide rates in Germany

To illustrate the use of correspondence analysis for the analysis for three-way tables, we use data on suicide rates in West Germany classified by sex, age, and method of suicide used. The data, from Heuer (1979, Table 1) have been discussed by Friendly (1991, 1994b), van der Heijden and de Leeuw (1985), and others.

The original $2 \times 17 \times 9$ table contains 17 age groups from 10 to 90 in 5-year steps and 9 categories of suicide method, contained in the frequency data frame *Suicide* in vcd, with table variables sex, age, and method. To avoid extremely small cell counts and cluttered displays, this example uses a reduced table in which age groups are combined in the variable age.group, a factor with 15-year intervals except for the last interval, which includes ages 70–90; the methods "toxic gas" and "cooking gas" were collapsed (in the variable method2) giving the $2 \times 5 \times 8$ table shown in the output below. These changes do not affect the general nature of the data or conclusions drawn from them.

In this example, we decided to stack the combinations of age and sex, giving an analysis of the loglinear model $[AgeSex][Method]$, to show how the age–sex categories relate to method of suicide.

In the case of a frequency data frame, it is quite simple to join two or more factors to form the rows of a new two-way table. Here we use paste() to form a new, composite factor, called age_sex here, abbreviating sex for display purposes.

```
> data("Suicide", package = "vcd")
> # interactive coding of sex and age.group
> Suicide <- within(Suicide, {
+    age_sex <- paste(age.group, toupper(substr(sex, 1, 1)))
+    })
```

Then, use xtabs() to construct the two-way table suicide.tab:

```
> suicide.tab <- xtabs(Freq ~ age_sex + method2, data = Suicide)
> suicide.tab
            method2
age_sex    poison  gas hang drown  gun knife jump other
  10-20 F     921   40  212    30   25    11  131   100
  10-20 M    1160  335 1524    67  512    47  189   464
  25-35 F    1672  113  575   139   64    41  276   263
  25-35 M    2823  883 2751   213  852   139  366   775
  40-50 F    2224   91 1481   354   52    80  327   305
  40-50 M    2465  625 3936   247  875   183  244   534
  55-65 F    2283   45 2014   679   29   103  388   296
  55-65 M    1531  201 3581   207  477   154  273   294
  70-90 F    1548   29 1355   501    3    74  383   106
  70-90 M     938   45 2948   212  229   105  268   147
```

The results of the correspondence analysis of this table are shown below:

```
> suicide.ca <- ca(suicide.tab)
> summary(suicide.ca)


Principal inertias (eigenvalues):
```

```
dim     value        %    cum%    scree plot
1       0.096151    57.2   57.2   **************
2       0.059692    35.5   92.6   *********
3       0.008183     4.9   97.5   *
4       0.002158     1.3   98.8
5       0.001399     0.8   99.6
6       0.000557     0.3  100.0
7       6.7e-050     0.0  100.0
        --------   -----
Total:  0.168207   100.0
...
```

It can be seen that 92.6% of the χ^2 for this model is accounted for in the first two dimensions. Plotting these gives the display shown in Figure 6.6.

```
> plot(suicide.ca)
```

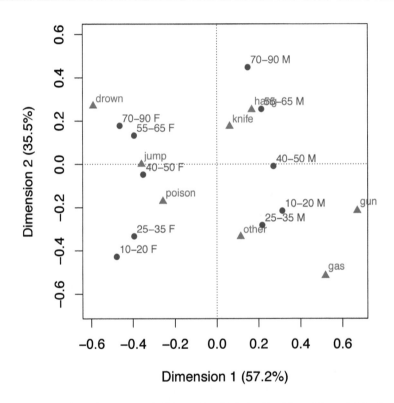

Figure 6.6: 2D CA solution for the stacked [AgeSex][Method] table of the suicide data.

Dimension 1 in the plot separates males (right) and females (left), indicating a large difference between suicide profiles of males and females with respect to methods of suicide. The second dimension is mostly ordered by age with younger groups at the bottom and older groups at the top. Note also that the positions of the age groups are roughly parallel for the two sexes. Such a pattern indicates that sex and age do not interact in this analysis.

The relation between the age–sex groups and methods of suicide can be approximately interpreted in terms of similar distance and direction from the origin, which represents the marginal row and column profiles. Young males are more likely to commit suicide by gas or a gun, older males by hanging, while young females are more likely to ingest some toxic agent and older females by jumping or drowning.   △

EXAMPLE 6.6: Suicide rates in Germany — mosaic plot

For comparison, it is useful to see how to construct a mosaic display showing the same associations for the loglinear model $[AS][M]$ as in the correspondence analysis plot. To do this, we first construct the three-way table, `suicide.tab3`,

```
> suicide.tab3 <- xtabs(Freq ~ sex + age.group + method2, data = Suicide)
```

As discussed in Chapter 5, mosaic plots are sensitive both to the order of variables used in successive splits, and to the order of levels within variables and are most effective when these orders are chosen to reflect the some meaningful ordering.

In the present example, `method2` is an unordered table factor, but Figure 6.6 shows that the methods of suicide vary systematically with both sex and age, corresponding to dimensions 1 and 2, respectively. Here we choose to reorder the table according to the coordinates on Dimension 1. We also delete the low-frequency `"other"` category to simplify the display.

```
> # methods, ordered as in the table
> suicide.ca$colnames

[1] "poison" "gas"    "hang"   "drown"  "gun"    "knife"
[7] "jump"   "other"

> # order of methods on CA scores for Dim 1
> suicide.ca$colnames[order(suicide.ca$colcoord[,1])]

[1] "drown"  "jump"   "poison" "knife"  "other"  "hang"
[7] "gas"    "gun"

> # reorder methods by CA scores on Dim 1
> suicide.tab3 <- suicide.tab3[, , order(suicide.ca$colcoord[,1])]
> # delete "other"
> suicide.tab3 <- suicide.tab3[,, -5]
> ftable(suicide.tab3)

               method2 drown jump poison knife hang  gas  gun
sex    age.group
male   10-20             67  189   1160    47 1524  335  512
       25-35            213  366   2823   139 2751  883  852
       40-50            247  244   2465   183 3936  625  875
       55-65            207  273   1531   154 3581  201  477
       70-90            212  268    938   105 2948   45  229
female 10-20             30  131    921    11  212   40   25
       25-35            139  276   1672    41  575  113   64
       40-50            354  327   2224    80 1481   91   52
       55-65            679  388   2283   103 2014   45   29
       70-90            501  383   1548    74 1355   29    3
```

To construct the mosaic display for the same model analyzed by correspondence analysis, we use the argument `expected=~age.group*sex + method2` to supply the model formula. For this large table, it is useful to tweak the labels for the `method2` variable to reduce overplotting; the `labeling_args` argument provides many options for customizing `strucplot` displays.

```
> library(vcdExtra)
> mosaic(suicide.tab3, shade = TRUE, legend = FALSE,
+        expected = ~ age.group * sex + method2,
+        labeling_args = list(abbreviate_labs = c(FALSE, FALSE, 5)),
+                            rot_labels = c(0, 0, 0, 90))
```

This figure (Figure 6.7) again shows the prevalence of `gun` and `gas` among younger males and decreasing with age, whereas use of `hang` increases with age. For females, these three methods are used less frequently, whereas `poison`, `jump`, and `drown` occur more often. You can also see that

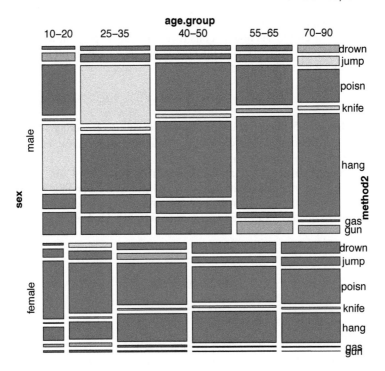

Figure 6.7: Mosaic display showing deviations from the model [AgeSex][Method] for the suicide data.

for females the excess prevalence of these high-frequency methods varies somewhat less with age than it does for males.

$\triangle$

6.3.2 Marginal tables and supplementary variables

An *n*-way table in frequency form or case form is automatically collapsed over factors that are not listed in the call to xtabs() when creating the table input for ca(). The analysis gives a ***marginal model*** for the categorical variables that *are* listed.

The positions of the categories of the omitted variables may nevertheless be recovered, by treating them as ***supplementary variables***, given as additional rows or columns in the two-way table. A supplementary variable is ignored in finding the CA solution, but its categories are then projected into that space. This is another useful trick to extend traditional CA to higher-way tables.

To illustrate, the code below lists only the age and method2 variables, and hence produces an analysis collapsed over sex. This ignores not only the effect of sex itself, but also all associations of age and method with sex, which are substantial. We don't show the ca() result or the plot yet.

```
> # two way, ignoring sex
> suicide.tab2 <- xtabs(Freq ~ age.group + method2, data = Suicide)
> suicide.tab2
           method2
age.group  poison  gas hang drown  gun knife jump other
    10-20    2081  375 1736    97  537    58  320   564
```

25-35	4495	996	3326	352	916	180	642	1038
40-50	4689	716	5417	601	927	263	571	839
55-65	3814	246	5595	886	506	257	661	590
70-90	2486	74	4303	713	232	179	651	253

```
> suicide.ca2 <- ca(suicide.tab2)
```

To treat the levels of sex as supplementary points, we calculate the two-way table of sex and method, and append this to the suicide.tab2 as additional rows:

```
> # relation of sex and method
> suicide.sup <- xtabs(Freq ~ sex + method2, data = Suicide)
> suicide.tab2s <- rbind(suicide.tab2, suicide.sup)
```

In the call to ca(), we then indicate these last two rows as supplementary:

```
> suicide.ca2s <- ca(suicide.tab2s, suprow = 6 : 7)
> summary(suicide.ca2s)


Principal inertias (eigenvalues):

  dim    value      %    cum%   scree plot
  1      0.060429  93.9  93.9   ************************
  2      0.002090   3.2  97.1   *
  3      0.001479   2.3  99.4   *
  4      0.000356   0.6 100.0
         --------  -----
  Total: 0.064354 100.0

...
```

This CA analysis has the same total Pearson chi-square, $\chi^2(28) = 3422.5$, as the result of chisq.test(suicide.tab2). However, the scree plot display above shows that the association between age and method is essentially one-dimensional, but note also that dimension 1 ("age–method") in this analysis has nearly the same inertia (0.0604) as the second dimension (0.0596) in the analysis of the stacked table. We plot the CA results as shown below (see Figure 6.8), and add a line connecting the supplementary points for sex.

```
> res <- plot(suicide.ca2s, pch = c(16, 15, 17, 24))
> lines(res$rows[6 : 7,])
```

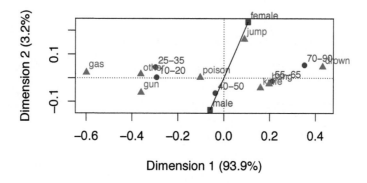

Figure 6.8: 2D CA solution for the [Age] [Method] marginal table. Category points for Sex are shown as supplementary points.

Comparing this graph with Figure 6.6, you can see that ignoring sex has collapsed the differences between males and females, which were the dominant feature of the analysis including sex. The dominant feature in Figure 6.8 is the Dimension 1 ordering of both age and method. However, as in Figure 6.6, the supplementary points for sex point toward the methods that are more prevalent for females and males.

6.4 Multiple correspondence analysis

Multiple correspondence analysis (MCA) is designed to display the relationships of the categories of two or more discrete variables, but it is best used for multiway tables where the extensions of classical CA described in Section 6.3 do not suffice. Again, this is motivated by the desire to provide an *optimal scaling* of categorical variables, giving scores for the discrete variables in an n-way table with desirable properties, and which can be plotted to visualize the relations among the category points.

The most typical development of MCA starts by defining indicator ("dummy") variables for each category and reexpresses the n-way contingency table in the form of a cases-by-variables indicator matrix, Z. Simple correspondence analysis for a two-way table can, in fact, be derived as the canonical correlation analysis of the indicator matrix.

Unfortunately, the generalization to more than two variables follows a somewhat different path, so that simple CA does not turn out to be precisely a special case of MCA in some respects, particularly in the decomposition of an interpretable χ^2 over the dimensions in the visual representation.

Nevertheless, MCA does provide a useful graphic portrayal of the *bivariate* relations among any number of categorical variables, and has close relations to the mosaic matrix (Section 5.6). If its limitations are understood, it is helpful in understanding large, multivariate categorical data sets, in a similar way to the use of scatterplot matrices and dimension-reduction techniques (e.g., principal component analysis) for quantitative data.

6.4.1 Bivariate MCA

For the hair color–eye color data, the indicator matrix Z has 592 rows and $4 + 4 = 8$ columns. The columns refer to the eight categories of hair color and eye color and the rows to the 592 students in Snee's 1974 sample.

For simplicity, we show the calculation of the indicator matrix below in frequency form, using `model.matrix()` to compute the dummy (0/1) variables for the levels of hair color (`Hair1`–`Hair4`) and eye color (`Eye1`–`Eye4`).

```
> haireye.df <- cbind(
+     as.data.frame(haireye),
+     model.matrix(Freq ~ Hair + Eye, data=haireye,
+         contrasts.arg=list(Hair=diag(4), Eye=diag(4)))[,-1]
+     )
> haireye.df
```

	Hair	Eye	Freq	Hair1	Hair2	Hair3	Hair4	Eye1	Eye2	Eye3	Eye4
1	Black	Brown	68	1	0	0	0	1	0	0	0
2	Brown	Brown	119	0	1	0	0	1	0	0	0
3	Red	Brown	26	0	0	1	0	1	0	0	0
4	Blond	Brown	7	0	0	0	1	1	0	0	0
5	Black	Blue	20	1	0	0	0	0	1	0	0
6	Brown	Blue	84	0	1	0	0	0	1	0	0
7	Red	Blue	17	0	0	1	0	0	1	0	0
8	Blond	Blue	94	0	0	0	1	0	1	0	0
9	Black	Hazel	15	1	0	0	0	0	0	1	0
10	Brown	Hazel	54	0	1	0	0	0	0	1	0

11	Red	Hazel	14	0	0	1	0	0	0	1	0
12	Blond	Hazel	10	0	0	0	1	0	0	1	0
13	Black	Green	5	1	0	0	0	0	0	0	1
14	Brown	Green	29	0	1	0	0	0	0	0	1
15	Red	Green	14	0	0	1	0	0	0	0	1
16	Blond	Green	16	0	0	0	1	0	0	0	1

Thus, the first row in `haireye.df` represents the 68 individuals having black hair (`Hair1=1`) and brown eyes (`Eye1=1`). The indicator matrix Z is then computed by replicating the rows in `haireye.df` according to the `Freq` value, using the function `expand.dft`. The result has 592 rows and 8 columns.

```
> Z <- expand.dft(haireye.df)[,-(1:2)]
> vnames <- c(levels(haireye.df$Hair), levels(haireye.df$Eye))
> colnames(Z) <- vnames
> dim(Z)

[1] 592    8
```

Note that if the indicator matrix is partitioned as $Z = [Z_1, Z_2]$, corresponding to the two sets of categories, then the contingency table is given by $N = Z_1^{\mathsf{T}} Z_2$.

```
> (N <- t(as.matrix(Z[,1:4])) %*% as.matrix(Z[,5:8]))

      Brown Blue Hazel Green
Black    68   20    15     5
Brown   119   84    54    29
Red      26   17    14    14
Blond     7   94    10    16
```

With this setup, MCA can be described as the application of the simple correspondence analysis algorithm to the indicator matrix Z. This analysis would yield scores for the rows of Z (the cases), usually not of direct interest, and for the columns (the categories of both variables). As in simple CA, each row point is the weighted average of the scores for the column categories, and each column point is the weighted average of the scores for the row observations.[4]

Consequently, the point for any category is the centroid of all the observations with a response in that category, and all observations with the same response pattern coincide. As well, the origin reflects the weighted average of the categories for *each* variable. As a result, category points with low marginal frequencies will be located further away from the origin, while categories with high marginal frequencies will be closer to the origin. For a binary variable, the two category points will appear on a line through the origin, with distances inversely proportional to their marginal frequencies.

EXAMPLE 6.7: Hair color and eye color

For expository purposes, we illustrate the analysis of the indicator matrix below for the hair color–eye color data using `ca()`, rather than the function `mjca()`, which is designed for a more general approach to MCA.

```
> Z.ca <- ca(Z)
> res <- plot(Z.ca, what = c("none", "all"))
```

In the call to `plot.ca`, the argument `what` is used to suppress the display of the row points for the cases. The plot shown in Figure 6.9 is an enhanced version of this basic plot.

---

[4]Note that, in principle, this use of an indicator matrix could be extended to three (or more) variables. That extension is more easily described using an equivalent form, the *Burt matrix*, described in Section 6.4.2.

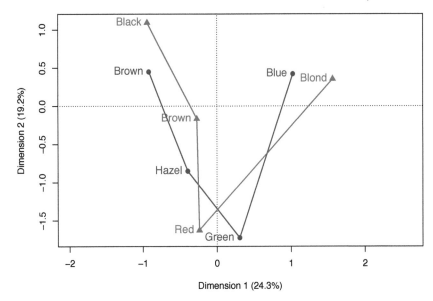

Figure 6.9: Correspondence analysis of the indicator matrix Z for the hair color–eye color data. The category points are joined separately by lines for the hair color and eye color categories.

```
       Dim1      Dim2 factor levels
1  -0.94250   1.09220   Hair  Black
2  -0.27693  -0.16608   Hair  Brown
3  -0.24194  -1.62513   Hair    Red
4   1.56039   0.35376   Hair  Blond
5  -0.91933   0.44905    Eye  Brown
6   1.02254   0.42176    Eye   Blue
7  -0.39712  -0.85105    Eye  Hazel
8   0.30215  -1.72375    Eye  Green
```

Comparing Figure 6.9 with Figure 6.1, we see that the general pattern of the hair color and eye color categories is the same in the analysis of the contingency table (Figure 6.1) and the analysis of the indicator matrix (Figure 6.9), except that the axes are scaled differently—the display has been stretched along the second (vertical) dimension. The interpretation is the same: Dimension 1 reflects a dark–light ordering of both hair and eye colors, and Dimension 2 reflects something that largely distinguishes red hair and green eyes from the other categories.

Indeed, it can be shown (Greenacre, 1984, 2007) that the two displays are identical, except for changes in scales along the axes. There is no difference at all between the displays in standard coordinates. Greenacre (1984, pp. 130–134) describes the precise relations between the geometries of the two analyses.

△

Aside from the largely cosmetic difference in relative scaling of the axes, a major difference between analysis of the contingency table and analysis of the indicator matrix is in the decomposition of principal inertia and corresponding χ^2 contributions for the dimensions. The plot axes in Figure 6.9 indicate 24.3% and 19.2% for the contributions of the two dimensions, whereas Figure 6.1 shows 89.4% and 9.5%. This difference is the basis for the more general development of MCA methods and is reflected in the `mcja()` function illustrated later in this chapter. But first, we describe a second approach to extending simple CA to the multivariate case based on the **Burt matrix**.

6.4.2 The Burt matrix

The same solution for the category points as in the analysis of the indicator matrix may be obtained more simply from the so-called **Burt matrix** (Burt, 1950),

$$ B = Z^\mathsf{T} Z = \begin{bmatrix} N_1 & N \\ N^\mathsf{T} & N_2 \end{bmatrix}, $$

where N_1 and N_2 are diagonal matrices containing the marginal frequencies of the two variables (the column sums of Z_1 and Z_2). In this representation, the contingency table of the two variables, N, appears in the off-diagonal block in this equation. This calculation is shown below.

```
> Burt <- t(as.matrix(Z)) %*% as.matrix(Z)
> rownames(Burt) <- colnames(Burt) <- vnames
> Burt

      Black Brown Red Blond Brown Blue Hazel Green
Black   108     0   0     0    68   20    15     5
Brown     0   286   0     0   119   84    54    29
Red       0     0  71     0    26   17    14    14
Blond     0     0   0   127     7   94    10    16
Brown    68   119  26     7   220    0     0     0
Blue     20    84  17    94     0  215     0     0
Hazel    15    54  14    10     0    0    93     0
Green     5    29  14    16     0    0     0    64
```

The standard coordinates from an analysis of the Burt matrix B are identical to those of Z. (However, the singular values of B are the squares of those of Z.) Then, the following code, using the `Burt` matrix produces the same display of the category points for hair color and eye color as shown for the indicator matrix Z in Figure 6.9.

```
> Burt.ca <- ca(Burt)
> plot(Burt.ca)
```

6.4.3 Multivariate MCA

The coding of categorical variables in an indicator matrix and the relationship to the Burt matrix provides a direct and natural way to extend this analysis to more than two variables. If there are Q categorical variables, and variable q has J_q categories, then the Q-way contingency table, of size $J = \prod_{q=1}^{Q} J_q = J_1 \times J_2 \times \cdots \times J_Q$, with a total of $n = n_{++\ldots}$ observations, may be represented by the partitioned $(n \times J)$ indicator matrix $[Z_1\, Z_2\, \ldots\, Z_Q]$.

Then the Burt matrix is the symmetric partitioned matrix

$$ B = Z^\mathsf{T} Z = \begin{bmatrix} N_1 & N_{12} & \cdots & N_{1Q} \\ N_{21} & N_2 & \cdots & N_{2Q} \\ \vdots & \vdots & \ddots & \vdots \\ N_{Q1} & N_{Q2} & \cdots & N_Q \end{bmatrix}, \tag{6.7} $$

where again the diagonal blocks N_i contain the one-way marginal frequencies. The off-diagonal blocks N_{ij} contain the bivariate marginal contingency tables for each pair (i, j) of variables.

Classical MCA (see, e.g., Greenacre (1984), Gower and Hand (1996)) can then be defined as a singular value decomposition of the matrix B, which produces scores for the categories of *all* variables so that the greatest proportion of the bivariate, pairwise associations in all blocks (including the diagonal blocks) is accounted for in a small number of dimensions.

In this respect, MCA resembles multivariate methods for quantitative data based on the joint bivariate correlation or covariance matrix (Σ) and there is some justification for regarding the Burt matrix as the categorical analog of Σ.[5]

There is a close connection between this analysis and the bivariate mosaic matrix (Section 5.6): The mosaic matrix displays the residuals from independence for each pair of variables, and thus provides a visual representation of the Burt matrix. The one-way margins shown (by default) in the diagonal cells reflect the diagonal matrices N_i in Eqn. (6.7). The total amount of shading in all the individual mosaics portrays the total pairwise associations decomposed by MCA. See Friendly (1999a) for further details.

For interpretation of MCA plots, we note the following relations (Greenacre, 1984, Section 5.2):[6]

- The inertia contributed by a given variable increases with the number of response categories.
- The centroid of the categories for each discrete variable is at the origin of the display.
- For a particular variable, the inertia contributed by a given category increases as the marginal frequency in that category *decreases*. Low frequency points therefore appear further from the origin.
- The category points for a binary variable lie on a line through the origin. The distance of each point to the origin is inversely related to the marginal frequency.

EXAMPLE 6.8: Marital status and pre- and extramarital sex

The data on the relation between marital status and reported premarital and extramarital sex was explored earlier using mosaic displays in Example 5.9 and Example 5.13.

Using the ca package, an MCA analysis of the `PreSex` data is carried out using `mjca()`. This function typically takes a data frame in *case form* containing the factor variables, but converts a table to this form. This example analyzes the Burt matrix calculated from the `PreSex` data, specified as `lambda="Burt"`

```
> data("PreSex", package = "vcd")
> PreSex <- aperm(PreSex, 4:1)    # order variables G, P, E, M
> presex.mca <- mjca(PreSex, lambda = "Burt")
> summary(presex.mca)


Principal inertias (eigenvalues):

  dim    value      %    cum%    scree plot
  1    0.149930   53.6   53.6    *************
  2    0.067201   24.0   77.6    ******
  3    0.035396   12.6   90.2    ***
  4    0.027365    9.8  100.0    **
       --------  -----
  Total: 0.279892 100.0
...
```

The output from `summary()` seems to show that 77.6% of the total inertia is accounted for in two dimensions. A basic, default plot of the MCA solution is provided by the `plot()` method for "mjca" objects.

---

[5]For multivariate normal data, however, the mean vector and covariance matrix are sufficient statistics, so all higher-way relations are captured in the covariance matrix. This is not true of the Burt matrix. Moreover, the covariance matrix is typically expressed in terms of mean-centered variables, while the Burt matrix involves the marginal frequencies. A more accurate statement is that the uncentered covariance matrix is analogous to the Burt matrix.

[6]This book, now out of print, is available for free download at http://www.carme-n.org/.

```
> plot(presex.mca)
```

This plotting method is not very flexible in terms of control of graphical parameters or the ability to add additional annotations (labels, lines, legend) to ease interpretation. Instead, we use the plot method to create an empty plot (with no points or labels), and return the calculated plot coordinates (res) for the categories. A bit of processing of the coordinates provides the customized display shown in Figure 6.10.

```
> # plot, but don't use point labels or points
> res <- plot(presex.mca, labels = 0, pch = ".", cex.lab = 1.2)
>
> # extract factor names and levels
> coords <- data.frame(res$cols, presex.mca$factors)
> nlev <- presex.mca$levels.n
> fact <- unique(as.character(coords$factor))
>
> cols <- c("blue", "red", "brown", "black")
> points(coords[,1:2], pch=rep(16:19, nlev), col=rep(cols, nlev), cex=1.2)
> text(coords[,1:2], label=coords$level, col=rep(cols, nlev), pos=3,
+       cex=1.2, xpd=TRUE)
> lwd <- c(2, 2, 2, 4)
> for(i in seq_along(fact)) {
+    lines(Dim2 ~ Dim1, data = coords, subset = factor==fact[i],
+          lwd = lwd[i], col = cols[i])
+ }
>
> legend("bottomright",
+        legend = c("Gender", "PreSex", "ExtraSex", "Marital"),
+        title = "Factor", title.col = "black",
+        col = cols, text.col = cols, pch = 16:19,
+        bg = "gray95", cex = 1.2)
```

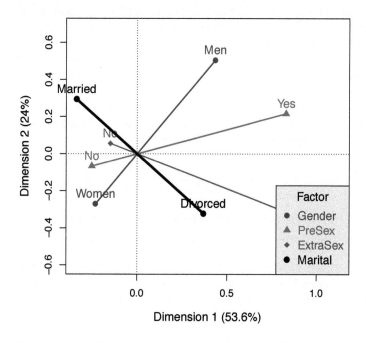

Figure 6.10: MCA plot of the Burt matrix for the PreSex data. The category points are joined separately by lines for the factor variables.

As indicated above, the category points for each factor appear on lines through the origin, with distances inversely proportional to their marginal frequencies. For example, the categories for No premarital and extramarital sex are much larger than the corresponding Yes categories, so the former are positioned closer to the origin. In contrast, the categories of gender and marital status are more nearly equal marginally.

Another aspect of interpretation of Figure 6.10 concerns the alignment of the lines for different factors. The positions of the category points on Dimension 1 suggest that women are less likely to have had pre-marital and extra-marital sex and that still being married is associated with the absence of pre- and extra-marital sex. As well, the lines for gender and marital status are nearly at right angles, suggesting that these variables are unassociated. This interpretation is more or less correct, but it is only approximate in this MCA scaling of the coordinate axes. An alternative scaling, based on a *biplot* representation is described in Section 6.5.

If you compare the MCA result in Figure 6.10 with the mosaic matrix in Figure 5.23, you will see that they are both showing the bivariate pairwise associations among these variables, but in different ways. The mosaic plots show the details of marginal and joint frequencies together with residuals from independence for each 2×2 marginal subtable. The MCA plot using the Burt matrix summarizes each category point in terms of a 2D representation of contributions to total inertia (association).   △

6.4.3.1   Inertia decomposition

The transition from simple CA to MCA is straightforward in terms of the category scores derived from the indicator matrix Z or the Burt matrix, B. It is less so in terms of the calculation of total inertia, and therefore in the chi-square values and corresponding percentages of association accounted for in some number of dimensions.

In simple CA, the total inertia is χ^2/n, and it therefore makes sense to talk of percentage of association accounted for by each dimension. But in MCA of the indicator matrix, the total inertia, $\sum \lambda^2$, is simply $(J - Q)/Q$, because the inertia of each subtable, Z_i, is equal to its dimensionality, $J_i - 1$, and the total inertia of an indicator matrix is the average of the inertias of its subtables. Consequently, the average inertia per dimension is $1/Q$, and it is common to interpret only those dimensions that exceed this average (analogous to the use of 1 as a threshold for eigenvalues in principal components analysis).

To more adequately reflect the percentage of association in MCA, Greenacre (1990), revising an earlier proposal by Benzécri (1977), suggested the calculation of *adjusted inertia*, which ignores the contributions of the diagonal blocks in the Burt matrix,

$$(\lambda_i^\star)^2 = \left[\frac{Q}{Q-1}(\lambda_i^Z - \frac{1}{Q}) \right]^2 \tag{6.8}$$

as the principal inertia due to the dimensions with $(\lambda^Z)^2 > 1/Q$. This adjustment expresses the contribution of each dimension as $(\lambda_i^\star)^2/\sum(\lambda_i^\star)^2$, with the summation over only dimensions with $(\lambda^Z)^2 > 1/Q$.

A related method, also handled by `mjca()`, is *joint correspondence analysis* (Greenacre, 1994, Greenacre, 2007, Chapter 19), an iterative method that replaces the diagonal blocks of the Burt matrix with values that minimize their impact on inertia. Unlike MCA, solutions in JCA are not nested, however.

EXAMPLE 6.9: Survival on the *Titanic*

An MCA analysis of the `Titanic` data is carried out using `mjca()` as shown below.

```
> titanic.mca <- mjca(Titanic)
```

mjca() allows different scaling methods for the contributions to inertia of the different dimensions. The default (lambda="adjusted"), used here, is the adjusted inertias as in Eqn. (6.8).

```
> summary(titanic.mca)

Principal inertias (eigenvalues):

dim    value       %    cum%   scree plot
1      0.067655   76.8  76.8   ***********************
2      0.005386    6.1  82.9   **
3      00000000    0.0  82.9
       --------   -----
Total: 0.088118
...
```

Using similar code to that used in Example 6.8, Figure 6.11 shows an enhanced version of the default plot that connects the category points for each factor by lines using the result returned by the plot() function.

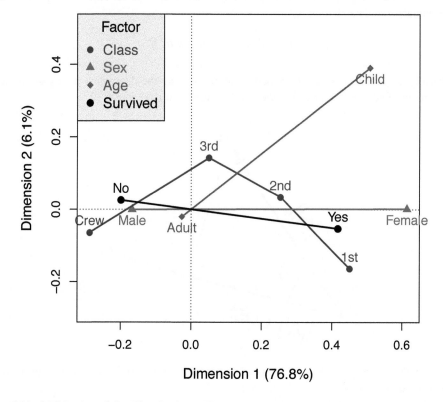

Figure 6.11: MCA plot of the Titanic data. The category points are joined separately by lines for the factor variables.

In this plot, the points for each factor have the property that the sum of coordinates on each dimension, weighted inversely by the marginal proportions, equals zero. Thus high-frequency categories (e.g., Adult and Male) are close to the origin.

The first dimension is perfectly aligned with gender, and also strongly aligned with Survival. The second dimension pertains mainly to Class and Age effects. Consider those points that differ from the origin most similarly (in distance and direction) to the point for Survived ("Yes"); this gives the interpretation that survival was associated with being female or upper class or (to a lesser degree) being a child.

$\triangle$

6.5 Biplots for contingency tables

Like correspondence analysis, the ***biplot*** (Bradu and Gabriel, 1978, Gabriel, 1971, 1980, 1981, Gower et al., 2011) is a visualization method that uses the SVD to display a matrix in a low-dimensional (usually 2-dimensional) space. They differ in the relationships in the data that are portrayed, however:

- In correspondence analysis the (weighted, χ^2) *distances* between row points and distances between column points are designed to reflect *differences* between the row profiles and column profiles.

- In the biplot, on the other hand, row and column points are represented by *vectors* from the origin such that the projection (inner product) of the vector $\boldsymbol{a}_i$ for row i on $\boldsymbol{b}_j$ for column j approximates the data element y_{ij},

$$\boldsymbol{Y} \approx \boldsymbol{A}\boldsymbol{B}^{\mathsf{T}} \iff y_{ij} \approx \boldsymbol{a}_i^{\mathsf{T}}\boldsymbol{b}_j \,. \tag{6.9}$$

Geometrically, Eqn. (6.9) may be described as approximating the data value y_{ij} by the projection of the end point of vector $\boldsymbol{a}_i$ on $\boldsymbol{b}_j$ (and vice-versa), as shown in Figure 6.12.

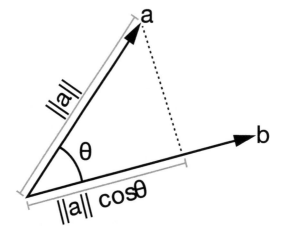

Figure 6.12: The scalar product of vectors of two points from the origin is the length of the projection of one vector on the other.

6.5.1 CA bilinear biplots

As in CA, there are a number of different representations of coordinates for row and column points for a contingency table within a biplot framework. One set of connections between CA and the biplot can be seen through the *reconstitution formula*, giving the decomposition of the correspondence matrix $P = N/n$ in terms of the standard coordinates Φ and Γ, defined in Eqn. (6.4) and Eqn. (6.5) as:

$$p_{ij} = r_i c_j \left(1 + \sum_{m=1}^{M} \sqrt{\lambda_m} \phi_{im} \gamma_{jm} \right) , \qquad (6.10)$$

or, in matrix terms,

$$P = D_r (11^{\mathsf{T}} + \Phi D_\lambda^{1/2} \Gamma^{\mathsf{T}}) D_c . \qquad (6.11)$$

The CA solution approximates this by a sum over $d \ll M$ dimensions, or by using only the first d (usually 2) columns of Φ and Γ.

Eqn. (6.10) can be re-written in biplot scalar form as

$$\left(\frac{p_{ij}}{r_i c_j} \right) - 1 \approx \sum_{m=1}^{d} (\sqrt{\lambda_m} \phi_{im}) \gamma_{jm} = \sum_{m=1}^{d} f_{im} \gamma_{jm} \qquad (6.12)$$

where $f_{im} = (\sqrt{\lambda_m} \phi_{im})$ gives the principal coordinates of the row points. The left-hand side of Eqn. (6.12) contains the **contingency ratios**, $p_{ij}/r_i c_j$, of the observed cell probabilities to their expected values under independence. This shows that an **asymmetric CA plot** of row principal coordinates F and the column standard coordinates Γ is a biplot that approximates the deviations of the contingency ratios from their values under independence.

In the ca package, this plot is obtained by specifying map="rowprincipal" in the call to plot(), or map="colprincipal" to plot the column points in principal coordinates. It is typical in such biplots to display one set of coordinates as points and the other as vectors from the origin, as controlled by the arrows argument, so that one can interpret the data values represented as approximated by the projections of the points on the vectors.

Two other types of asymmetric "maps" are also defined with different scalings that turn out to have better visual properties in terms of representing the relations between the row and column categories, particularly when the strength of association (inertia) in the data is low.

- The option map="rowgab" (or map="colgab") gives a biplot form proposed by Gabriel and Odoroff (1990) with the rows (columns) shown in principal coordinates and the columns (rows) in standard coordinates multiplied by the mass c_j (r_i) of the corresponding point.
- The *contribution biplot* for CA (Greenacre, 2013), with the option map="rowgreen" (or map="colgreen") provides a reconstruction of the standardized residuals from independence, using the points in standard coordinates multiplied by the square root of the corresponding masses. This has the nice visual property of showing more directly the contributions of the vectors to the low-dimensional solution.

EXAMPLE 6.10: Suicide rates in Germany — biplot

To illustrate the biplot representation, we continue with the data on suicide rates in Germany from Example 6.5, using the stacked table suicide.tab comprised of the age–sex combinations as rows and methods of suicide as columns.

```
> suicide.tab <- xtabs(Freq ~ age_sex + method2, data = Suicide)
> suicide.ca <- ca(suicide.tab)
```

Using this result, suicide.ca, in the call to plot() below, we use map="colgreen", and vectors represent the methods of suicide, as shown in Figure 6.13.

```
> plot(suicide.ca, map = "colgreen", arrows = c(FALSE, TRUE))
```

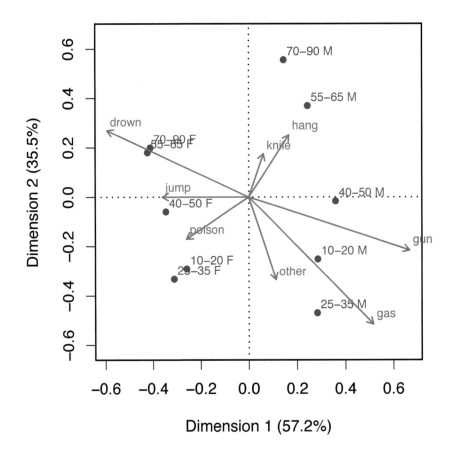

Figure 6.13: CA biplot of the suicide data using the contribution biplot scaling. Associations between the age–sex categories and the suicide methods can be read as the projections of the points on the vectors. The lengths of the vectors for the suicide categories reflect their contributions to this representation in a 2D plot.

The interpretation of the row points for the age–sex categories is similar to what we saw earlier in Figure 6.6. But now, the vectors for the suicide categories reflect the contributions of those methods to the representation of association. Thus, the methods `drown`, `gun`, and `gas` have large contributions, while `knife`, `hang`, and `poison` are relatively small. Moreover, the projections of the points for the age–sex combinations on the method vectors reflect the standardized residuals from independence.

The most comprehensive modern treatment of biplot methodology is the book *Understanding Biplots* (Gower et al., 2011). Together with the book, they provide an R package, UBbipl (le Roux and Lubbe, 2013), that is capable of producing an astounding variety of high-quality plots. Unfortunately, that package is only available on their publisher's web site,[7] and you need the book to be able to use it because all the documentation is in the book. Nevertheless, we illustrate the use of the `cabipl()` function to produce the version of the CA biplot shown in Figure 6.14.

---

[7]http://www.wiley.com/legacy/wileychi/gower/material.html.

```
> library(UBbipl)
> cabipl(as.matrix(suicide.tab),
+       axis.col = gray(.4), ax.name.size = 1,
+       ca.variant = "PearsonResA",
+       markers = FALSE,
+       row.points.size = 1.5,
+       row.points.col = rep(c("red", "blue"), 4),
+       plot.col.points = FALSE,
+       marker.col = "black", marker.size = 0.8,
+       offset = c(2, 2, 0.5, 0.5),
+       offset.m = rep(-0.2, 14),
+       output = NULL)
```

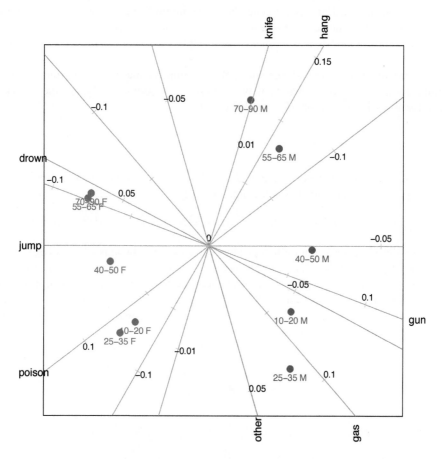

Figure 6.14: CA biplot of the suicide data, showing calibrated axes for the suicide methods.

This plot uses ca.variant = "PearsonResA" to specify that the biplot is to approximate the standardized Pearson residuals by the inner product of each row point on the vector for the column point for the suicide methods, as also in Figure 6.13. However, Figure 6.14 represents the methods as calibrated axis lines, designed to be read as scales for the projections of the row points (age–sex) on the methods. The UBbipl package has a huge number of options for controlling the details of the biplot display. See Gower et al. (2011, Ch. 2) for all the details.

△

6.5.2 Biadditive biplots

A different use of biplots for contingency tables stems from the close analogy between additive relations for a quantitative response when there is no interaction between factors, and the multiplicative relations for a contingency table when there is no association.

For quantitative data, Bradu and Gabriel (1978) show how the biplot can be used to diagnose additive relations among rows and columns. For example, when a two-way table is well-described by a two-factor ANOVA model with no interaction,

$$y_{ij} = \mu + \alpha_i + \beta_j + \epsilon_{ij} \iff Y \approx a1^\mathsf{T} + 1b^\mathsf{T} ,$$

then the row points, a_i, and the column points, b_j, will fall on two straight lines at right angles to each other in the biplot. For a contingency table, the multiplicative relations among frequencies under independence become additive relations in terms of log frequency, and Gabriel et al. (1997) illustrate how biplots of log frequency can be used to explore associations in two-way and three-way tables.

That is, for a two-way table, independence, $A \perp B$, implies that ratios of frequencies should be proportional for any two rows, i, i' and any two columns, j, j'. Equivalently, this means that the log odds ratio for all such sets of four cells should be zero:

$$A \perp B \iff \log \theta_{ii',jj'} = \log \left(\frac{n_{ij} n_{i'j'}}{n_{i'j} n_{ij'}} \right) = 0 .$$

Now, if the log frequencies have been centered by subtracting the grand mean, Gabriel et al. (1997) show that $\log \theta_{ii',jj'}$ is approximated in the biplot (of $\log(n_{ij}) - \overline{\log(n_{ij})}$),

$$\log \theta_{ii',jj'} \approx a_i^\mathsf{T} b_j - a_{i'}^\mathsf{T} b_j - a_i^\mathsf{T} b_{j'} + a_i^\mathsf{T} b_{j'} = (a_i - a_{i'})^\mathsf{T} (b_j - b_{j'}) .$$

Therefore, this biplot criterion for independence in a two-way table is whether $(a_i - a_{i'})^\mathsf{T} (b_j - b_{j'}) \approx 0$ for all pairs of rows, i, i', and all pairs of columns, j, j'. But $(a_i - a_{i'})$ is the vector connecting a_i to $a_{i'}$ and $(b_j - b_{j'})$ is the vector connecting b_j to $b_{j'}$, as shown in Figure 6.15, and the inner product of any two vectors equals zero *iff* they are orthogonal. Hence, this criterion implies that all lines connecting pairs of row points are orthogonal to lines connecting pairs of column points, as illustrated in Figure 6.15.

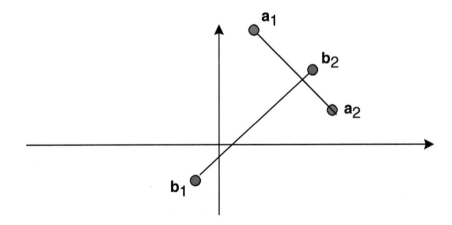

Figure 6.15: Independence implies orthogonal vector differences in a biplot of log frequency. The line joining a_1 to a_2 represents $(a_1 - a_2)$. This line is perpendicular to the line $(b_1 - b_2)$ under independence.

EXAMPLE 6.11: UK soccer scores

We examined the data on UK soccer scores in Example 5.5 and saw that the number of goals scored by the home and away teams were largely independent (see Figure 5.10). This data set provides a good test of the ability of the biplot to diagnose independence.

```
> data("UKSoccer", package = "vcd")
> dimnames(UKSoccer) <- list(Home = paste0("H", 0:4),
+                            Away = paste0("A", 0:4))
```

Basic biplots in R are provided by `biplot()`, which works mainly with the result calculated by `prcomp()` or `princomp()`. Here, we use `prcomp()` on the log frequencies in the *UKSoccer* table, adding 1, because there is one cell with zero frequency.

```
> soccer.pca <- prcomp(log(UKSoccer + 1), center = TRUE, scale. = FALSE)
```

The result is plotted using a customized plot based on `biplot()`, as shown in Figure 6.16.

```
> biplot(soccer.pca, scale = 0, var.axes = FALSE,
+   col = c("blue", "red"), cex = 1.2, cex.lab = 1.2,
+   xlab = "Dimension 1", ylab = "Dimension 2")
```

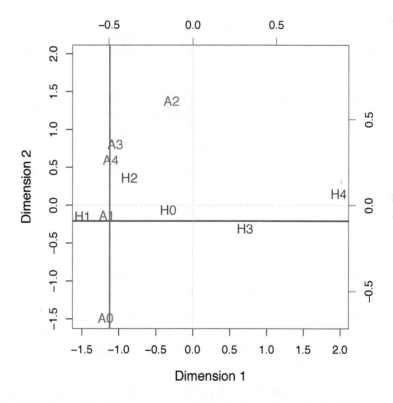

Figure 6.16: Biplot for the biadditive representation of independence for the UK soccer scores. The row and column categories are independent in this plot when they appear as points on approximately orthogonal lines.

To supplement this plot and illustrate the orthogonality of row and column category points under independence, we added horizontal and vertical lines as calculated below, using the results returned by `prcomp()`. The initial version of this plot showed that two points, A2 and H2, did not align with the others, so these were excluded from the calculations.

```
> # get the row and column scores
> rscores <- soccer.pca$x[, 1 : 2]
> cscores <- soccer.pca$rotation[, 1 : 2]
> # means, excluding A2 and H2
> rmean <- colMeans(rscores[-3,])[2]
> cmean <- colMeans(cscores[-3,])[1]
>
> abline(h = rmean, col = "blue", lwd = 2)
> abline(v = cmean, col = "red", lwd = 2)
> abline(h = 0, lty = 3, col = "gray")
> abline(v = 0, lty = 3, col = "gray")
```

You can see that all the A points (except for A2) and all the H points (except for H2) lie along straight lines, and these lines are indeed at right angles, signifying independence. The fact that these straight lines are parallel to the coordinate axes is incidental, and unrelated to the independence interpretation.

$\triangle$

6.6 Chapter summary

- Correspondence analysis is an exploratory technique, designed to show the row and column categories in a two- (or three-) dimensional space. These graphical displays, and various extensions, provide ways to interpret the patterns of association and visually explore the adequacy of certain loglinear models.

- The scores assigned to the categories of each variable are optimal in several equivalent ways. Among other properties, they maximize the (canonical) correlations between the quantified variables (weighted by cell frequencies), and make the regressions of each variable on the other most nearly linear, for each CA dimension.

- Multi-way tables may be analyzed in several ways. In the "stacking" approach, two or more variables may be combined interactively in the rows and/or columns of an n-way table. Simple CA of the restructured table reveals associations between the row and column categories of the restructured table, but hides associations between the variables combined interactively. Each way of stacking corresponds to a particular loglinear model for the full table.

- Multiple correspondence analysis is a generalization of CA to two or more variables based on representing the data as an indicator matrix, or the Burt matrix. The usual MCA provides an analysis of the joint, bivariate relations between all pairs of variables.

- The biplot is a related technique for visualizing the elements of a data array by points or vectors in a joint display of their row and column categories. A standard CA biplot represents the contributions to lack of independence as the projection of the points for rows (or columns) on vectors for the other categories.

- Another application of the biplot to contingency table data is described, based on analysis of log frequency. This analysis also serves to diagnose patterns of independence and partial independence in two-way and larger tables.

6.7 Lab exercises

Exercise 6.1 The *JobSat* data in **vcdExtra** gives a 4×4 table recording job satisfaction in relation to income.

(a) Carry out a simple correspondence analysis on this table. How much of the inertia is accounted for by a one-dimensional solution? How much by a two-dimensional solution?

(b) Plot the 2D CA solution. To what extent can you consider the association between job satisfaction and income "explained" by the ordinal nature of these variables?

Exercise 6.2 Refer to Exercise 5.1 in Chapter 5. Carry out a simple correspondence analysis on the 4×5 table `criminal` from the **logmult** package.

(a) What percentages of the Pearson χ^2 for association are explained by the various dimensions?

(b) Plot the 2D correspondence analysis solution. Describe the pattern of association between year and age.

Exercise 6.3 Refer to Exercise 5.2 for a description of the `AirCrash` data from the **vcdExtra** package. Carry out a simple correspondence analysis on the 5×5 table of `Phase` of the flight and `Cause` of the crash.

(a) What percentages of the Pearson χ^2 for association are explained by the various dimensions?

(b) Plot the 2D correspondence analysis solution. Describe the pattern of association between phase and cause. How would you interpret the dimensions?

(c) The default plot method uses `map="symmetric"` with points for both rows and columns. Try using `map="symbiplot"` with vectors (`arrows=`) for either rows or columns. (Read `help(plot.ca)` for a description of these options.)

Exercise 6.4 The data set `caith` in **MASS** gives a classic table tabulating hair color and eye color of people in Caithness, Scotland, originally from Fisher (1940).

(a) Carry out a simple correspondence analysis on this table. How many dimensions seem necessary to account for most of the association in the table?

(b) Plot the 2D solution. The interpretation of the first dimension should be obvious; is there any interpretation for the second dimension?

Exercise 6.5 The same data, plus a similar table for Aberdeen, are given as a three-way table as `HairEyePlace` in **vcdExtra**.

(a) Carry out a similar correspondence analysis to the last exercise for the data from Aberdeen. Comment on any differences in the placement of the category points.

(b) Analyze the three-way table, stacked to code hair color and place interactively, i.e., for the loglinear model [Hair Place][Eye]. What does this show?

Exercise 6.6 The data set `Gilby` in **vcdExtra** gives a classic (but now politically incorrect) 6×4 table of English schoolboys classified according to their clothing and their teacher's rating of "dullness" (lack of intelligence).

(a) Compute and plot a correspondence analysis for this data. Write a brief description and interpretation of these results.

(b) Make an analogous mosaic plot of this table. Interpret this in relation to the correspondence analysis plot.

Exercise 6.7 For the mental health data analyzed in Example 6.2, construct a shaded sieve diagram and mosaic plot. Compare these with the correspondence analysis plot shown in Figure 6.2. What features of the data and the association between SES and mental health status are shown in each?

Exercise 6.8 Simulated data are often useful to help understand the connections between data, analysis methods, and associated graphic displays. Section 6.3.1 illustrated interactive coding in R, using a simulated 4-way table of counts of pets, classified by age, color, and sex, but with no associations because the counts had a constant Poisson mean, $\lambda = 15$.

  (a) Re-do this example, but in the call to `rpois()`, specify a non-negative vector of Poisson means to create some associations among the table factors.
  (b) Use CA methods to determine if and how the structure you created in the data appears in the results.

Exercise 6.9 The `TV` data was analyzed using CA in Example 6.4, ignoring the variable `Time`. Carry out analyses of the 3-way table, reducing the number of levels of `Time` to three hourly intervals as shown below.

```
> data ("TV", package="vcdExtra")
> # reduce number of levels of Time
> TV.df <- as.data.frame.table(TV)
> levels(TV.df$Time) <- rep(c("8", "9", "10"), c(4, 4, 3))
> TV3 <- xtabs(Freq ~ Day + Time + Network, TV.df)
> structable(Day ~ Network + Time, TV3)
```

	Day	Monday	Tuesday	Wednesday	Thursday	Friday
Network	Time					
ABC	8	536	861	744	735	1119
	9	1401	1205	1022	682	907
	10	910	1044	668	349	711
CBS	8	1167	646	550	680	509
	9	967	959	409	385	544
	10	789	798	324	270	426
NBC	8	858	1090	512	1927	823
	9	946	890	831	1858	590
	10	825	588	869	2101	585

  (a) Use the stacking approach (Section 6.3) to perform a CA of the table with `Network` and `Time` coded interactively. You can create this using the `as.matrix()` method for a "structable" object.

```
> TV3S <- as.matrix(structable(Day ~ Network + Time, TV3), sep=":")
```

  (b) What loglinear model is analyzed by this approach?
  (c) Plot the 2D solution. Compare this to the CA plot of the two-way table in Figure 6.4.
  (d) Carry out an MCA analysis using `mjca()` of the three-way table `TV3`. Plot the 2D solution, and compare this with both the CA plot and the solution for the stacked three-way table.

Exercise 6.10 Refer to the MCA analysis of the `PreSex` data in Example 6.8. Use the stacking approach to analyze the stacked table with the combinations of premarital and extramarital sex in the rows and the combinations of gender and marital status in the columns. As suggested in the exercise above, you can use `as.matrix(structable())` to create the stacked table.

  (a) What loglinear model is analyzed by this approach? Which associations are included and which are excluded in this analysis?
  (b) Plot the 2D CA solution for this analysis. You might want to draw lines connecting some of the row points or column points to aid in interpretation.
  (c) How does this analysis differ from the MCA analysis shown in Figure 6.10?

Exercise 6.11 Refer to Exercise 5.10 for a description of the `Vietnam` data set in vcdExtra.

(a) Using the stacking approach, carry out a correspondence analysis corresponding to the loglinear model [R][YS], which asserts that the response is independent of the combinations of year an sex.

(b) Construct an informative 2D plot of the solution, and interpret in terms of how the response varies with year for males and females.

(c) Use mjca() to carry out an MCA on the three-way table. Make a useful plot of the solution and interpret in terms of the relationship of the response to year and sex.

Exercise 6.12 Refer to Exercise 5.9 for a description of the *Accident* data set in vcdExtra. The data set is in the form of a frequency data frame, so first convert to table form.

```
> accident.tab <- xtabs(Freq ~ age + result + mode + gender, data=Accident)
```

(a) Use mjca() to carry out an MCA on the four-way table accident.tab.

(b) Construct an informative 2D plot of the solution, and interpret in terms of how the variable result varies in relation to the other factors.

Exercise 6.13 The *UCBAdmissions* data was featured in numerous examples in Chapter 4 (e.g., Example 4.11, Example 4.15) and Chapter 5 (e.g., Example 5.14, Example 5.18).

(a) Use mjca() to carry out an MCA on the three-way table UCBAdmissions.

(b) Plot the 2D MCA solution in a style similar to that shown in Figure 6.10 and Figure 6.11

(c) Interpret the plot. Is there some interpretation for the first dimension? What does the plot show about the relation of admission to the other factors?

Part III

Model-Building Methods

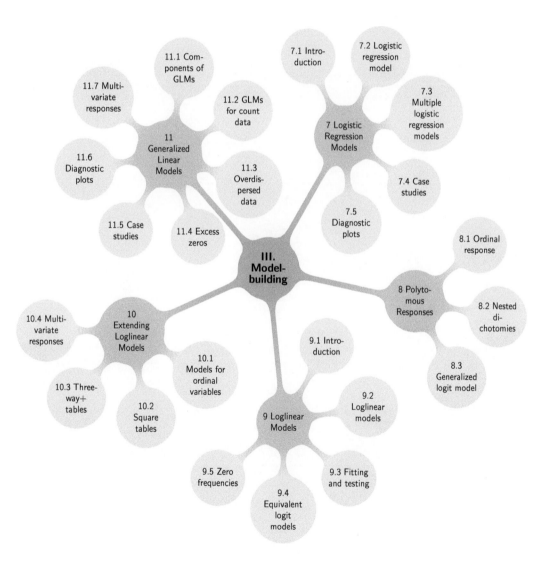

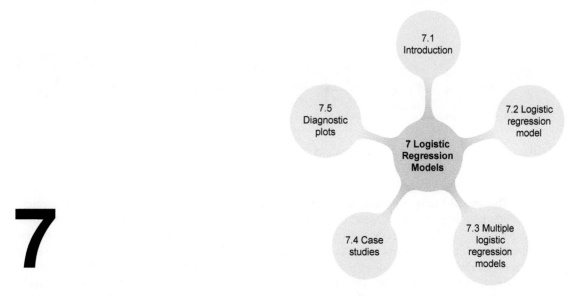

7

Logistic Regression Models

This chapter introduces the modeling framework for categorical data in the simple situation where we have a categorical response variable, often binary, and one or more explanatory variables. A fitted model provides both statistical inference and prediction, accompanied by measures of uncertainty. Data visualization methods for discrete response data must often rely on smoothing techniques, including both direct, non-parametric smoothing and the implicit smoothing that results from a fitted parametric model. Diagnostic plots help us to detect influential observations that may distort our results.

7.1 Introduction

All models are wrong, but some are useful.

George E. P. Box, (Box and Draper, 1987, p. 424)

Chapters 4–6 have been concerned primarily with simple exploratory methods for studying the relations among categorical variables and with testing hypotheses about their associations through non-parametric tests and with overall goodness-of-fit statistics.

This chapter begins our study of model-based methods for the analysis of discrete data. These models differ from those we have examined earlier primarily in that they consider *explicitly* an assumed probability distribution for the observations, and make clear distinctions between the systematic component, which is explained by the model, and the random component, which is not. More importantly, the model-based approach allows a compact summary of categorical data in terms of a (hopefully) small number of parameters accompanied by measures of uncertainty (standard errors), and the ability to estimate predicted values over the range of explanatory variables.

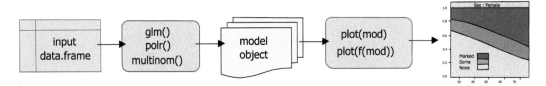

Figure 7.1: Overview of fitting and graphing for model-based methods in R.

This model-fitting approach has several advantages: (a) Inferences for the model parameters include both hypothesis tests and confidence intervals. (b) The former help us to assess which explanatory variables affect the outcome; the size of the estimated parameters and the widths of their confidence intervals help us to assess the strength and importance of these effects. (c) There are a variety of methods for model selection, designed to help determine a favorable trade-off between goodness-of-fit and parsimony. (d) Finally, the predicted values obtained from the model effectively smooth the discrete responses, allow predictions for unobserved values of the explanatory variables, and provide important means to interpret the fitted relationship graphically.

Figure 7.1 provides a visual overview of the steps for fitting and graphing with model-based methods in R. (a) A modeling function such as `glm()` is applied to an input data frame. The result is a ***model object*** containing all the information from the fitting process. (b) As is standard in R, `print()` and `summary()` methods give, respectively, basic and detailed printed output. (c) Many modeling functions have `plot()` methods that produce different types of summary and diagnostic plots. (d) For visualizing the fitted model, most model methods provide a `predict()` method that can be used to plot the fitted values from the model over the ranges of the predictors. Such plots can be customized by the addition of points (showing the observations), lines, confidence bands, and so forth.

In this chapter we consider models for a ***binary response***, such as "success" or "failure," or the number of "successes" in a fixed number of "trials," where we might reasonably assume a binomial distribution for the random component. As we will see in Chapter 8, these methods extend readily to a ***polytomous response*** with more than two outcome categories, such as improvement in therapy, with categories "none," "some," and "marked."

These models can be seen as simple extensions of familiar ANOVA and regression models for quantitative data. They are also important special cases of a more general approach, the ***generalized linear model*** that subsumes a wide variety of families of techniques within a single, unified framework. However, rather than starting at the top with the fully general version, this chapter details the important special cases of models for discrete outcomes, beginning with binary responses.

This chapter proceeds as follows: in Section 7.2 we introduce the simple logistic regression model for a binary response and a single quantitative predictor. This model extends directly to models for grouped, binomial data (Section 7.2.4) and to models with any number of regressors (Section 7.3), which can be quantitative, discrete factors, and more general forms.

For interpreting and understanding the results of a fitted model, we emphasize plotting predicted probabilities and predicted log odds in various ways, for which effect plots (Section 7.3.3) are particularly useful for complex models.

Section 7.4 presents several case studies to highlight issues of data analysis, model building, and visualization in the context of constructing and interpreting multiple logistic regression models. These focus on the combination of exploratory plots to see the data, modeling steps, and graphs to interpret a given model. Individual observations sometimes exert great influence on a fitted model. Some measures of influence and diagnostic plots are illustrated in Section 7.5.

7.2 The logistic regression model

The logistic regression model describes the relationship between a discrete outcome variable, the "response," and a set of explanatory variables. The response variable is often *dichotomous*, although extensions to the model permit multi-category, *polytomous* outcomes, discussed in Chapter 8. The explanatory variables may be continuous or (with factor variables) discrete.

For a binary response, Y, and a continuous explanatory variable, X, we may be interested in modeling the probability of a successful outcome, which we denote $\pi(x) \equiv \Pr(Y = 1 \mid X = x)$. That is, at a given value $X = x$, you can imagine that there is a binomial distribution of the responses, $\text{Bin}(\pi(x), n_x)$.

The simplest naive model, called the *linear probability model*, supposes that this probability, $\pi(x)$, varies linearly with the value of x,

$$E(Y \mid x) = \pi(x) = \alpha + \beta x \,, \tag{7.1}$$

where the notation $E(Y \mid x)$ indicates that the probability $\pi(x)$ represents the population conditional average of the 1s and 0s for all observations with a fixed value of x. For binary observations, this is simply the proportion of 1s.

Figure 7.2 illustrates the basic setup for modeling a binary outcome using the `Arthritis` data, and described more fully in Example 7.1–Example 7.3: The "Better" response represents a positive effect of some arthritis medicament, given age. The 0/1 observations are shown as (jittered) points. The predicted values under the linear probability model (Eqn. (7.1)) are shown as the red lines in both panels. As you can see, this model cannot be right, because it predicts a probability less than 0 for small values of Age, and would also predict probabilities greater than 1 for larger values of Age.

The linear probability model is also wrong because it assumes that the distribution of residuals, $Y_i - \hat{\pi}(x_i)$, is normal, with mean 0 and constant variance. However, because Y is dichotomous, the residuals are also dichotomous, and have variance $\pi(x_i)(1 - \pi(x_i))$, which is maximal for $\pi = 0.5$ and decreases as π goes toward 0 or 1.

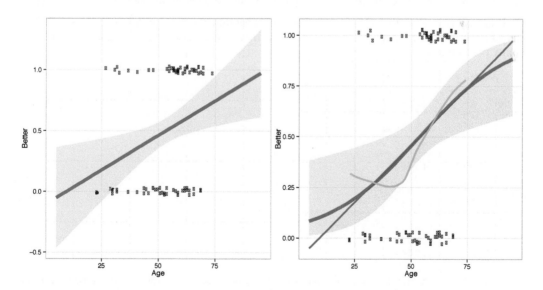

Figure 7.2: Arthritis treatment data, for the relationship of the binary response "Better" to Age, shown as jittered points. The left panel shows the predicted values and 95% confidence envelope under the linear probability model. The right panel shows the fitted logistic regression, together with the simple linear regression (red) and a non-parametric (loess) smoothed curve (green).

One way around the difficulty of needing to constrain the predicted values to the interval $[0, 1]$ is to re-specify the model so that a *transformation* of π has a linear relation to x, and that transformation keeps $\hat{\pi}$ between 0 and 1 for all x. This idea of modeling a transformation of the response that has desired statistical properties is one of the fundamental ones that led to the development of **generalized linear models**, which we treat more fully later in Chapter 11.

A particularly convenient choice of the transformation gives the **linear logistic regression model** (or **linear logit model**[1]), which posits a linear relation between the **log odds** (or **logit**) of this probability and x,

$$\text{logit}[\pi(x)] \equiv \log\left(\frac{\pi(x)}{1 - \pi(x)}\right) = \alpha + \beta x \,. \tag{7.2}$$

When $\beta > 0$, $\pi(x)$ and the log odds increase as X increases; when $\beta < 0$ they decrease with X.

This model can also be expressed as a model for the probabilities $\pi(x)$ in terms of the *inverse* of the logit transformation used in Eqn. (7.2),

$$\pi(x) = \text{logit}^{-1}[\pi(x)] = \frac{1}{1 + \exp[-(\alpha + \beta x)]} \,. \tag{7.3}$$

This transformation uses the cumulative distribution function of the logistic distribution, $\Lambda(p) = \frac{1}{1+exp(-p)}$, giving rise to the term *logistic regression*.[2]

From Eqn. (7.2) we see that the odds of a success response can be expressed as

$$\text{odds}(Y = 1) \equiv \frac{\pi(x)}{1 - \pi(x)} = \exp(\alpha + \beta x) = e^{\alpha}(e^{\beta})^x \,, \tag{7.4}$$

which is a multiplicative model for the odds.

So, under the logistic model,

- β is the change in the log odds associated with a unit increase in x. The odds are multiplied by e^{β} for each unit increase in x.
- α is log odds at $x = 0$; e^{α} is the odds of a favorable response at this x-value (which may not have a reasonable interpretation if $X = 0$ is far from the range of the data).

It is easy to explore the relationships among probabilities, odds, and log odds using R, as we show below, using the function `fractions()` in MASS to print the odds corresponding to probability p as a fraction.

```
> library(MASS)
> p <- c(.05, .10, .25, .50, .75, .90, .95)
> odds <- p / (1 - p)
> data.frame(p,
+                 odds = as.character(fractions(odds)),
+                 logit = log(odds))

     p odds   logit
1 0.05 1/19 -2.9444
2 0.10  1/9 -2.1972
3 0.25  1/3 -1.0986
4 0.50    1  0.0000
5 0.75    3  1.0986
6 0.90    9  2.1972
7 0.95   19  2.9444
```

---

[1]Some writers use the term *logit model* to refer to those using only categorical predictors; we use the terms logistic regression and logit regression interchangeably.

[2]Any other cumulative probability transformation serves the purpose of constraining the probabilities to the interval $[0, 1]$. The cumulative normal transformation $\pi(x) = \Phi(\alpha + \beta x)$ gives the **linear probit regression** model. We don't treat probit models here because: (a) The logistic and probit models give results so similar that it is hard to distinguish them in practice; (b) The logistic model is simpler to interpret as a linear model for the log odds or a multiplicative model for the odds.

Thus, a probability of $\pi = 0.25$ represents an odds of 1 to 3, or 1/3, while a probability of $\pi = 0.75$ represents an odds of 3 to 1, or 3. The logits are symmetric around 0, so $\text{logit}(.25) = -\text{logit}(.75)$.

Another simple way to interpret the parameter β in the logistic regression model is to consider the relationship between the probability $\pi(x)$ and x. From Eqn. (7.3) it can be shown that the fitted curve (the blue line in Figure 7.2) has slope equal to $\beta\pi(1 - \pi)$. This has a maximum value of $\beta/4$ when $\pi = \frac{1}{2}$, so taking $\beta/4$ gives a quick estimate of the maximum effect of x on the probability scale.

In Figure 7.2 and other plots later in this chapter we try to show the binary responses (as jittered points or a rug plot) to help you appreciate how the fitted logistic curve arises from their distribution across the range of a predictor. For didactic purposes this can be seen more readily by plotting the conditional distributions of $f(x \mid y), y \in \{0, 1\}$ as a histogram, boxplot, or density plot. The function `logi.hist.plot()` in the **popbio** (Stubben et al., 2012) package is a nice implementation of this idea (de la Cruz Rot, 2005). The call below produces Figure 7.3, and it is easy to see how increasing age produces a greater probability of a Better response.

```
> with(Arthritis,
+      logi.hist.plot(Age, Improved > "None", type = "hist",
+                     counts = TRUE, ylabel = "Probability (Better)",
+                     xlab = "Age", col.cur = "blue",
+                     col.hist = "lightblue", col.box = "lightblue")
+    )
```

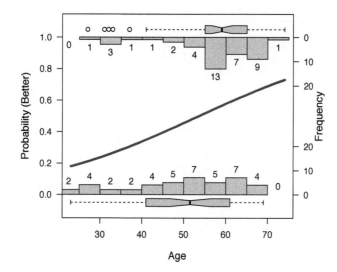

Figure 7.3: Plot of the Arthritis treatment data, showing the conditional distributions of the 0/1 observations of the Better response by histograms and boxplots.

7.2.1 Fitting a logistic regression model

Logistic regression models are the special case of generalized linear models fit in R using `glm()` for a binary response using `family=binomial`. We first illustrate how simple models can be fit and interpreted.

EXAMPLE 7.1: Arthritis treatment

In Chapter 4 we examined the data on treatment for rheumatoid arthritis in relation to two categorical predictors, sex of patient and treatment. In addition, the `Arthritis` data gives the age of each patient in this study, and we focus here on the relationship between `Age` and the outcome, `Improved`. This response variable has three categories (none, some, or marked improvement), but for now we consider whether the patient showed any improvement at all, defining the event `Better` to be some or marked improvement.

```
> data("Arthritis", package = "vcd")
> Arthritis$Better <- as.numeric(Arthritis$Improved > "None")
```

The logistic regression model is fit using `glm()` as shown below, specifying `family=binomial` for a binary response.

```
> arth.logistic <- glm(Better ~ Age, data = Arthritis, family = binomial)
```

As usual for **R** modeling functions, the `print()` method for `"glm"` objects gives brief printed output, while the `summary()` method is more verbose, and includes standard errors and hypothesis tests for the model coefficients. To save some space, it is convenient to use the generic function `coeftest()` from the **lmtest** (Hothorn et al., 2014) package. Then, we can use this instead of the more detailed `summary()`:

```
> library(lmtest)
> coeftest(arth.logistic)


z test of coefficients:

            Estimate Std. Error z value Pr(>|z|)
(Intercept)  -2.6421     1.0732   -2.46    0.014 *
Age           0.0492     0.0194    2.54    0.011 *
---
Signif. codes:  0 '***' 0.001 '**' 0.01 '*' 0.05 '.' 0.1 ' ' 1
```

In the output above, the parameter estimates are $\alpha = -2.642$, and $\beta = 0.0492$. So, the estimated odds of a better response are multiplied by $e^{\beta} = \exp(0.0492) = 1.05$ for each one-year increase in age. Equivalently, you can think of this as a 5% increase per year (using $100(e^{\beta} - 1)$ to convert). Over 10 years, the odds are multiplied by $\exp(10 \times 0.0492) = 1.64$, a 64% increase, a substantial effect in the range for these data. You can do these calculations in **R** using the `coef()` method for the `"glm"` object.

```
> exp(coef(arth.logistic))

(Intercept)        Age
   0.071214   1.050482

> exp(10 * coef(arth.logistic)["Age"])

    Age
1.6364
```

For comparison with the logistic model, we could fit the linear probability model Eqn. (7.1) using either `lm()` or `glm()` with the default `family=gaussian` argument.

```
> arth.lm <- glm(Better ~ Age, data = Arthritis)
> coef(arth.lm)

(Intercept)        Age
  -0.107170   0.011379
```

The coefficient for age can be interpreted to indicate that the probability of a better response increases by 0.011 for each one-year increase in age. You can compare this with the $\beta/4$ rule of thumb, which gives $0.0492/4 = 0.0123$. Even though the linear probability model is inappropriate theoretically, you can see in Figure 7.2 (the black line) that it gives similar predicted probabilities to those of the logistic model between age 25–75, where most of the data points are located.

△

7.2.2 Model tests for simple logistic regression

There are two main types of hypothesis tests one might want to perform for a logistic regression model. We postpone general discussion of this topic until Section 7.3, but introduce the main ideas here using the analysis of the `Arthritis` data.

- The most basic test answers the question, "How much better is the fitted model, $\text{logit}(\pi) = \alpha + \beta x$, than the null model $\text{logit}(\pi) = \alpha$, which includes only the regression intercept?" One answer to this question is given by the (Wald) test of the coefficient for age, testing the hypothesis $H_0 : \beta = 0$ that appeared in the output from `summary(arth.logistic)` shown above. The more direct test compares the deviance[3] of the fitted model to the deviance of the null model, and can be obtained using the `anova()` function:

```
> anova(arth.logistic, test = "Chisq")

Analysis of Deviance Table

Model: binomial, link: logit

Response: Better

Terms added sequentially (first to last)


      Df Deviance Resid. Df Resid. Dev Pr(>Chi)
NULL                     83          116
Age    1     7.29        82          109     0.007 **
---
Signif. codes:  0 '***' 0.001 '**' 0.01 '*' 0.05 '.' 0.1 ' ' 1
```

- A second question is, "How bad is this model, compared to a model (the *saturated model*) that fits the data perfectly?" This is a test of the size of the residual deviance, which is given by the function `LRstats()` in **vcdExtra**.

```
> library(vcdExtra)
> LRstats(arth.logistic)

Likelihood summary table:
               AIC BIC LR Chisq Df Pr(>Chisq)
arth.logistic 113 118      109 82      0.024 *
---
Signif. codes:  0 '***' 0.001 '**' 0.01 '*' 0.05 '.' 0.1 ' ' 1
```

The summary of these tests is that linear logistic model Eqn. (7.2) fits significantly better than the null model, but that model also shows significant lack of fit.

---

[3]The deviance is basically defined as -2 times the log-likelihood ratio of some reduced model to the full model. Two nested models can thus be compared by computing the difference of the corresponding deviances. If the larger model has k more parameters than the reduced one, this difference follows a chi-squared distribution with k degrees of freedom.

7.2.3 Plotting a binary response

It is often difficult to understand how a binary response can give rise to a smooth, continuous relation between the predicted response, usually the probability of an event, and a continuous explanatory variable. Beyond this, plots of the data together with fitted models help you to interpret what these models imply.

We illustrate two approaches below using the `Arthritis` data shown in Figure 7.2, first using R base graphics, and then with the ggplot2 package that makes such graphs somewhat easier to do.

That plot, which was designed for didactic purposes, has the following features:

- It shows the *data*, that is, the 0/1 observations of the `Better` response in relation to age. To do this effectively and avoid over-plotting, the binary responses are jittered.
- It plots the predicted (fitted) logistic regression relationship on the scale of probability, together with a 95% confidence band.
- It also plots the predicted probabilities from the linear probability model.
- A smoothed, non-parametric regression curve for the binary observations is also added to the plot to give some indication of possible nonlinearity in the relationship of Better to age.

EXAMPLE 7.2: Arthritis treatment — Plotting logistic regression with base graphics

Here we explain how plots similar to Figure 7.2 can be constructed using R base graphics. We describe the steps needed to calculate predicted values and confidence bands and how to add these to a basic plot. These ideas are the basis for the higher-level and more convenient plotting methods illustrated later in this chapter (Section 7.3.2) The steps detailed below give the plot shown in Figure 7.4.

First, we set up the basic plot of the jittered values of `Better` vs. `Age`, setting `xlim` to a larger range than that in the data, only to emphasize where the logistic and linear probability models diverge.

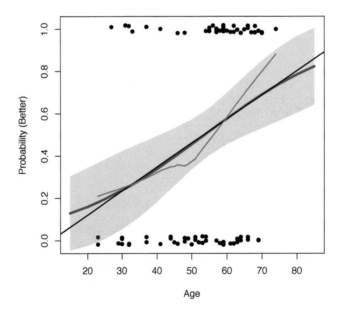

Figure 7.4: A version of plot of the Arthritis treatment data (Figure 7.2) produced with R base graphics, showing logistic, linear regression and lowess fits.

```
> plot(jitter(Better, .1) ~ Age, data = Arthritis,
+        xlim = c(15, 85), pch = 16,
+        ylab="Probability (Better)")
```

The fitted logistic curve can be obtained using the `predict()` method for the `"glm"` object `arth.logistic`. For this example, we wanted to get fitted values for the range of Age from 15–85, which is specified in the `newdata` argument.[4] The argument `type="response"` gives fitted values of the probabilities. (The default, `type="link"` would give predicted logits.) Standard errors of the fitted values are not calculated by default, so we set `se.fit=TRUE`.

```
> xvalues <- seq(15, 85, 5)
> pred.logistic <- predict(arth.logistic,
+                          newdata = data.frame(Age = xvalues),
+                          type = "response", se.fit = TRUE)
```

When `se.fit=TRUE`, the `predict()` function returns its result in a list, with components `fit` for the fitted values and `se.fit` for the standard errors. From these, we can calculate 95% pointwise prediction intervals using the standard normal approximation.

```
> upper <- pred.logistic$fit + 1.96 * pred.logistic$se.fit
> lower <- pred.logistic$fit - 1.96 * pred.logistic$se.fit
```

We can then plot the confidence band using `polygon()` and the fitted logistic curve using `lines`. A graphics trick is to use a transparent color for the confidence band using `rgb(r, g, b, alpha)`, where `alpha` is the transparency value.

```
> polygon(c(xvalues, rev(xvalues)),
+         c(upper, rev(lower)),
+         col = rgb(0, 0, 1, .2), border = NA)
> lines(xvalues, pred.logistic$fit, lwd=4 , col="blue")
```

This method, using `predict()` for calculations and `polygon()` and `lines()` for plotting can be used to display the predicted relationships and confidence bands under other models. Here, we simply used `abline()` to plot the fitted line for the linear probability model `arth.lm` and `lowess()` to calculate a smoothed, non-parametric curve.

```
> abline(arth.lm, lwd = 2)
> lines(lowess(Arthritis$Age, Arthritis$Better, f = .9),
+       col = "red", lwd = 2)
```

$\triangle$

EXAMPLE 7.3: Arthritis treatment — Plotting logistic regression with ggplot2

Model-based plots such as Figure 7.2 are relatively more straightforward to produce using gg-plot2. The basic steps here are to:

- set up the plot frame with `ggplot()` using Age and Better as (x, y) coordinates;
- use `geom_point()` to plot the observations, whose positions are jittered with `position_jitter()`;
- use `stat_smooth()` with `method = "glm"` and `family = binomial` to plot the predicted probability curve and confidence band. By default, `stat_smooth()` calculates and plots 95% confidence bands on the response (probability) scale.

---

[4]Omitting the `newdata` argument would give predicted values using the linear predictors in the data used for the fitted model. Some care needs to be taken if the predictor(s) contain missing values.

```
> library(ggplot2)
> # basic logistic regression plot
> gg <- ggplot(Arthritis, aes(x = Age, y = Better)) +
+   xlim(5, 95) +
+   geom_point(position = position_jitter(height = 0.02, width = 0)) +
+   stat_smooth(method = "glm", family = binomial,
+               alpha = 0.1, fill = "blue", size = 2.5, fullrange = TRUE)
```

Finally, we can add other smoothers to the plot, literally by using + to add these to the "ggplot" object.

```
> # add linear model and loess smoothers
> gg <- gg + stat_smooth(method = "lm", se = FALSE,
+                        size = 1.2, color = "black", fullrange = TRUE)
> gg <- gg + stat_smooth(method = "loess", se = FALSE,
+                        span = 0.95, colour = "red", size = 1.2)
> gg   # show the plot
```

△

7.2.4   Grouped binomial data

A related case occurs with grouped data, where rather than binary observations, $y_i \in \{0, 1\}$ in case form, the data is given in what is called ***events/trials form*** that records the number of successes, y_i that occurred in n_i trials associated with each setting of the explanatory variable(s) x_i.[5] Case form, with binary observations, is the special case where $n_i = 1$.

Data in events/trials form often arises from contingency table data with a binary response. For example, in the *UCBAdmissions* data, the response variable Admit with levels "Admitted", "Rejected" could be treated in this way using the number of applicants as the number of trials.

As before, we can consider y_i/n_i to estimate the probability of success, π_i, and the distribution of Y to be binomial, $\text{Bin}(\pi_i, n_i)$ at each x_i.

In practical applications, there are two main differences between the cases of ungrouped, case form data and grouped, event/trials form.

- In fitting models using glm(), the model formula, response ~ terms, can be given using a response consisting of a two-column matrix, whose columns contain the numbers of successes y_i and failures $n_i - y_i$. Alternatively, the response can be given as the proportion of successes, y_i/n_i, but then it is necessary to specify the number of trials as a weight.

- In plotting the fitted model on the scale of probability, you usually have to explicitly plot the fraction of successes, y_i/n_i.

EXAMPLE 7.4: Space shuttle disaster

In Example 1.2 and Example 1.10 we described the background behind the post-mortem examination of the evidence relating to the disastrous launch of the space shuttle *Challenger* on January 28, 1986. Here we consider a simple, but proper analysis of the data available at the time of launch. We also use this example to illustrate some details of the fitting and plotting of grouped binomial data. As well, we describe some of the possibilities for dealing with missing data.

The data set *SpaceShuttle* in **vcd** contains data on the failures of the O-rings in 24 NASA launches preceding the launch of *Challenger*, as given by Dalal et al. (1989) and Tufte (1997), also analyzed by Lavine (1991).

---

[5]Alternatively, the data may record the number of successes, y_i, and number of failures, $n_i - y_i$.

Each launch used two booster rockets with a total of six O-rings, and the data set records as nFailures the number of these that were considered damaged after the rockets were recovered at sea. In one launch (flight # 4), the rocket was lost at sea, so the relevant response variables are missing.

In this example, we focus on the variable nFailures as a binomial with $n_i = 6$ trials. The missing data for flight 4 can be handled in several ways in the call to glm()

```
> data("SpaceShuttle", package = "vcd")
> shuttle.mod <- glm(cbind(nFailures, 6 - nFailures) ~ Temperature,
+            data = SpaceShuttle, na.action = na.exclude,
+            family = binomial)
```

Alternatively, we can add an explicit trials variable, represent the response as the proportion nFailures/trials, and use weight = trials to indicate the total number of observations.

```
> SpaceShuttle$trials <- 6
> shuttle.modw <- glm(nFailures / trials ~ Temperature, weight = trials,
+            data = SpaceShuttle, na.action = na.exclude,
+            family = binomial)
```

These two approaches give identical results for all practical purposes:

```
> all.equal(coef(shuttle.mod), coef(shuttle.modw))

[1] TRUE
```

As before, we can test whether temperature significantly improves prediction of failure probability using anova():

```
> # testing, vs. null model
> anova(shuttle.mod, test = "Chisq")

Analysis of Deviance Table

Model: binomial, link: logit

Response: cbind(nFailures, 6 - nFailures)

Terms added sequentially (first to last)


             Df Deviance Resid. Df Resid. Dev Pr(>Chi)
NULL                        22        24.2
Temperature   1     6.14   21        18.1      0.013 *
---
Signif. codes:  0 '***' 0.001 '**' 0.01 '*' 0.05 '.' 0.1 ' ' 1
```

The code below gives a **ggplot2** version in Figure 7.5 of the plot we showed earlier in Example 1.2 (Figure 1.2). The relevant details here are:

- We specify y = nFailures / trials to calculate the failure probabilities.
- Points are jittered in the call to geom_point() to prevent overplotting.
- In the call to geom_smooth(), we need to use weight = trials, just as in the call to glm() above.
- fullrange = TRUE makes the fitted regression curve and confidence band extend across the entire plot.

```
> library(ggplot2)
> ggplot(SpaceShuttle, aes(x = Temperature, y = nFailures / trials)) +
+    xlim(30, 81) +
+    xlab("Temperature (F)") +
+    ylab("O-Ring Failure Probability") +
+    geom_point(position=position_jitter(width = 0, height = 0.01),
+               aes(size = 2)) +
+    theme(legend.position = "none") +
+    geom_smooth(method = "glm", family = binomial, fill = "blue",
+                aes(weight = trials), fullrange = TRUE, alpha = 0.2,
+                size = 2)
```

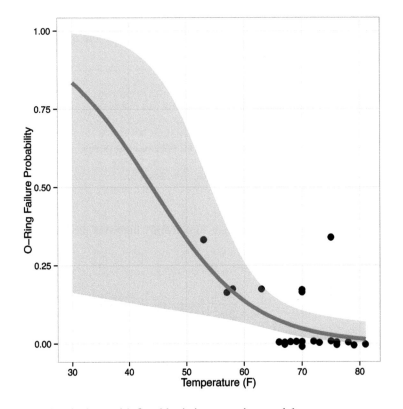

Figure 7.5: Space shuttle data, with fitted logistic regression model.

$\triangle$

7.3 Multiple logistic regression models

As is the case in classical regression, generalizing the simple logistic regression to an arbitrary number of explanatory variables is quite straightforward. We let $x_i = (x_{i1}, x_{i2}, \ldots, x_{ip})$ denote the vector of p explanatory variables for case or cluster i. Then the general logistic regression model can be expressed as

$$\begin{aligned}
\mathrm{logit}(\pi_i) \equiv \log \frac{\pi_i}{1 - \pi_i} &= \alpha + x_i^{\mathsf{T}} \beta \\
&= \alpha + \beta_1 x_{i1} + \beta_2 x_{i2} + \cdots + \beta_p x_{ip} .
\end{aligned} \tag{7.5}$$

Equivalently, we can represent this model in terms of probabilities as the logistic transformation of the **linear predictor**, $\eta_i = \alpha + \boldsymbol{x}_i^\mathsf{T} \boldsymbol{\beta}$,

$$\pi_i = \Lambda(\eta_i) \;\; = \;\; \Lambda(\alpha + \boldsymbol{x}_i^\mathsf{T} \boldsymbol{\beta}) \tag{7.6}$$

$$= \frac{1}{1 + \exp(\alpha + \beta_1 x_{i1} + \beta_2 x_{i2} + \cdots + \beta_p x_{ip})} \; .$$

The xs can include any of the following sorts of regressors, as in the general linear model:

- **quantitative** variables (e.g., age, income)
- **polynomial** powers of quantitative variables (e.g., age, age^2, age^3)
- **transformations** of quantitative variables (e.g., log salary)
- factors, represented as **dummy** variables for qualitative predictors (e.g., P_1, P_2, P_3 for four political party affiliations)
- **interaction** terms (e.g., sex $\times$ age, or age $\times$ income)

EXAMPLE 7.5: Arthritis treatment

We continue with the analysis of the `Arthritis` data, fitting a model containing the main effects of Age, Sex, and Treatment, with Better as the response. This model has the form

$$\text{logit}(\pi_i) = \alpha + \beta_1 x_{i1} + \beta_2 x_{i2} + \beta_3 x_{i3}$$

where x_1 is Age and x_2 and x_3 are the factors representing Sex and Treatment, respectively.

Using the default (0/1) dummy coding that R uses ("treatment" contrasts against the lowest factor level),[6] they are defined as:

$$x_2 = \begin{cases} 0 & \text{if Female} \\ 1 & \text{if Male} \end{cases} \qquad x_3 = \begin{cases} 0 & \text{if Placebo} \\ 1 & \text{if Treatment} \end{cases}$$

In this model,

- α doesn't have a sensible interpretation here, but formally it would be the log odds of improvement for a person at age $x_1 = 0$ in the baseline or reference group, with $x_2 = 0$ and $x_3 = 0$—that is, females receiving the placebo. To make the intercept interpretable, we will fit the model centering age near the mean, by using $x_1 - 50$ as the first regressor.

- β_1 is the increment in log odds of improvement for each one-year increase in age.

- β_2 is the increment in log odds for male as compared to female. Therefore, e^{β_2} gives the odds of improvement for males relative to females.

- β_3 is the increment in log odds for being in the treated group. e^{β_3} gives the odds of improvement for the active treatment group relative to placebo.

We fit the model as follows. In `glm()` model formulas, "$-$" has a special meaning, so we use the identity function, `I(Age-50)` to center age.

```
> arth.logistic2 <- glm(Better ~ I(Age-50) + Sex + Treatment,
+                        data = Arthritis,
+                        family = binomial)
```

---

[6]For factor variables with the default treatment contrasts, you can change the reference level using `relevel()`. In this example, you could make male the baseline category using `Arthritis$Sex <- relevel(Arthritis$Sex, ref = "Male")`.

The parameters defined here are *incremental effects*. The intercept corresponds to a baseline group (50-year-old females given the placebo); the other parameters are incremental effects for the other groups compared to the baseline group. Thus, when α, β_1, β_2, and β_3 have been estimated, the fitted logits and predicted odds at `Age==50` are:

Sex	Treatment	Logit	Odds Improved
Female	Placebo	α	e^{α}
Female	Treated	$\alpha + \beta_3$	$e^{\alpha+\beta_3}$
Male	Placebo	$\alpha + \beta_2$	$e^{\alpha+\beta_2}$
Male	Treated	$\alpha + \beta_2 + \beta_3$	$e^{\alpha+\beta_2+\beta_3}$

We first focus on the interpretation of the coefficients estimated for this model shown below.

```
> coeftest(arth.logistic2)


z test of coefficients:

                  Estimate Std. Error z value Pr(>|z|)
(Intercept)        -0.5781     0.3674   -1.57    0.116
I(Age - 50)         0.0487     0.0207    2.36    0.018 *
SexMale            -1.4878     0.5948   -2.50    0.012 *
TreatmentTreated    1.7598     0.5365    3.28    0.001 **
---
Signif. codes:  0 '***' 0.001 '**' 0.01 '*' 0.05 '.' 0.1 ' ' 1
```

To interpret these in terms of odds ratios and also find confidence intervals, just use `exp()` and `confint()`.

```
> exp(cbind(OddsRatio = coef(arth.logistic2),
+           confint(arth.logistic2)))

                  OddsRatio    2.5 %   97.5 %
(Intercept)          0.5609  0.26475   1.1323
I(Age - 50)          1.0500  1.01000   1.0963
SexMale              0.2259  0.06524   0.6891
TreatmentTreated     5.8113  2.11870  17.7266
```

Here,

- $\alpha = -0.578$: At age 50, females given the placebo have an odds of improvement of $\exp(-0.578) = 0.56$.
- $\beta_1 = 0.0487$: Each year of age multiplies the odds of improvement by $\exp(0.0487) = 1.05$, or a 5% increase.
- $\beta_2 = -1.49$: Males are only $\exp(-1.49) = 0.26$ times as likely to show improvement relative to females. (Or, females are $\exp(1.49) = 4.437$ times more likely than males to improve.)
- $\beta_3 = 1.76$: People given the active treatment are $\exp(1.76) = 5.8$ times more likely to show improvement compared to those given the placebo.

As you can see, the interpretation of coefficients in multiple logistic models is straightforward, though a bit cumbersome. This becomes more difficult in larger models, particularly when there are interactions. In these cases, you can understand (and explain) a fitted model more easily through plots of predicted values, either on the scale of response probability or on the logit scale of the linear predictor. We describe these methods in Section 7.3.1–Section 7.3.3 below.

$\triangle$

7.3.1 Conditional plots

The simplest kind of plots display the data together with a representation of the fitted relationship (predicted values, confidence bands) separately for subsets of the data defined by one or more of the predictors. Such plots can show the predicted values for the response variable on the ordinate against one chosen predictor on the abscissa, and can use multiple curves and multiple panels to represent other predictors.

However, these plots are ***conditional plots***, meaning that the data shown in each panel and used in each fitted curve are limited to the subset of the observations defined by the curve and panel variables. As well, predictors that are not shown in a given plot are effectively ignored (or marginalized), as was the case in Figure 7.2 that showed only the effect of age in the `Arthritis` data.

EXAMPLE 7.6: Arthritis treatment — conditional plots

For the `Arthritis` data, a basic conditional plot of `Better` vs. `Age`, showing the observations as jittered points (with `geom_point()`) and the fitted logistic curves (with `stat_smooth()` using `method="glm"`), can be produced with **ggplot2** as shown below, giving Figure 7.6.

```
> library(ggplot2)
> gg <- ggplot(Arthritis, aes(Age, Better, color = Treatment)) +
+    xlim(5, 95) + theme_bw() +
+    geom_point(position = position_jitter(height = 0.02, width = 0)) +
+    stat_smooth(method = "glm", family = binomial, alpha = 0.2,
+               aes(fill = Treatment), size = 2.5, fullrange = TRUE)
> gg    # show the plot
```

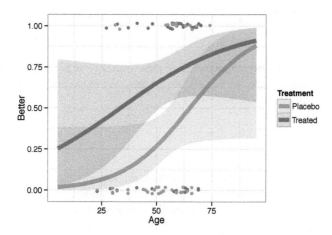

Figure 7.6: Conditional plot of Arthritis data showing separate points and fitted curves stratified by Treatment. A separate fitted curve is shown for the two treatment conditions, ignoring Sex.

In this call to `ggplot()`, specifying `color=Treatment` gives different point and line colors, but also automatically stratifies the fitted curves using the levels of this variable.

With such a plot, it is easy to add further stratifying variables in the data using *facets* to produce separate panels (functions `facet_wrap()` or `facet_grid()`, with different options to control the details). The following line further stratifies by `Sex`, producing Figure 7.7.

```
> gg + facet_wrap(~ Sex)
```

However, you can see from this plot how this method breaks down when the sample size is small in some of the groups defined by the stratifying factors. The panel for males shows a paradoxical

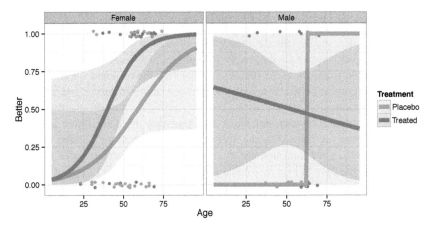

Figure 7.7: Conditional plot of Arthritis data, stratified by Treatment and Sex. The unusual patterns in the panel for Males signals a problem with this data.

negative relation with age for the treated group and a step function for the placebo group. The explanation for this is shown in the two-way frequency table of the sex and treatment combinations:

```
> addmargins(xtabs(~Sex + Treatment, data = Arthritis), 2)

          Treatment
Sex      Placebo Treated Sum
  Female      32      27  59
  Male        11      14  25
```

Less than 1/3 of the sample were males, and of these only 11 were in the placebo group. `glm()` cannot estimate the fitted relationship against `Age` here—the slope coefficient is infinite, and the fitted probabilities are all either 0 or 1.[7]

△

7.3.2   Full-model plots

For a model with two or more explanatory variables, *full-model plots* display the fitted response surface for all predictors together, rather than stratified by conditioning variables. Such plots show the predicted values for the response variable on the ordinate against one chosen predictor on the abscissa, and can use multiple curves and multiple panels to represent other predictors.

The programming steps used to plot a fitted logistic regression with base graphics and ggplot2 in the style of earlier examples (Examples 7.2, 7.2, and 7.4) become more tedious with multiple predictors. The vcd package provides the function `binreg_plot()` designed to plot the predicted response surface for a binary outcome directly from a fitted model object. At the time of writing, this function does not yet handle multiple panels or facets, but separate plots for panel variables can be produced using the `subset` argument as illustrated in the next example.

EXAMPLE 7.7:   Arthritis treatment — full-model plots
   This example shows how to plot the fitted main effects model using `binreg_plot()`. These plots can be shown either on the logit scale (with `type = "link"`) or the probability scale (`type = "response"`, the default).

---
[7]This is called *complete separation*, and occurs whenever the responses have no overlap on the predictor variable(s) used in fitting the logistic regression model.

This plot method is designed to use a numeric predictor (Age here) as the horizontal axis, and show separate point symbols and curves for the levels of the combinations of factors (if any). A basic plot on the logit scale (not included here) showing both factors (Sex, Treatment) can be produced using:

```
> library(vcd)
> binreg_plot(arth.logistic2, type = "link")
```

With two or more factors, such plots are often easier to read when the main factor(s) to be compared appear (Treatment here) as lines or curves within a plot, and other factors (Sex) are shown in separate panels. Figure 7.8 does this in two plots, using the subset argument to select the appropriate data and predicted values for males and females. When this is done, it is important to include the same xlim and ylim arguments so that the scales of all plots are identical.

```
> binreg_plot(arth.logistic2, type = "link", subset = Sex == "Female",
+             main = "Female", xlim=c(25, 75), ylim = c(-3, 3))
> binreg_plot(arth.logistic2, type = "link", subset = Sex == "Male",
+             main = "Male", xlim=c(25, 75), ylim = c(-3, 3))
```

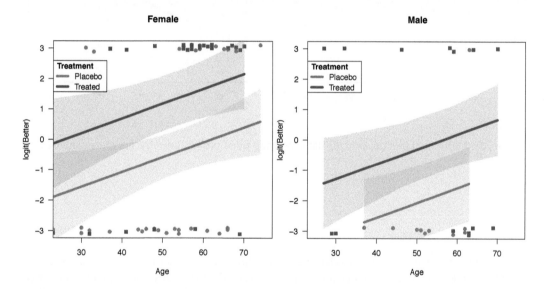

Figure 7.8: Full-model plot of Arthritis data, showing fitted logits by Treatment and Sex.

This plot method has several nice features:

- Plotting on the logit scale shows the additive linear effects of all predictors (parallel lines for the combinations of Sex and Treatment).
- It provides a visual representation of the information contained in the table of coefficients.
- The choice to display Treatment within each panel makes it easier to judge the size of this effect, compared to the effect of Sex, which must be judged across the panels.
- It shows the data as points, and the fitted lines and confidence bands are restricted to the range of the data in each. You can easily see the reason for the unusual pattern in the conditional plot for Males shown in Figure 7.7.
- It generalizes directly to any fitted model, because the predicted values are obtained from the model object. For example, you could easily add the interaction term Age:Sex and plot the result.

While plots on the logit scale have a simpler form, many people find it easier to think about such relationships in terms of probabilities, as we have done in earlier plots in this chapter. Figure 7.9 shows these plots using the default `type = "response"`.

```
> binreg_plot(arth.logistic2, subset = Sex == "Female",
+              main = "Female", xlim = c(25, 75))
> binreg_plot(arth.logistic2, subset = Sex == "Male",
+              main = "Male", xlim = c(25, 75))
```

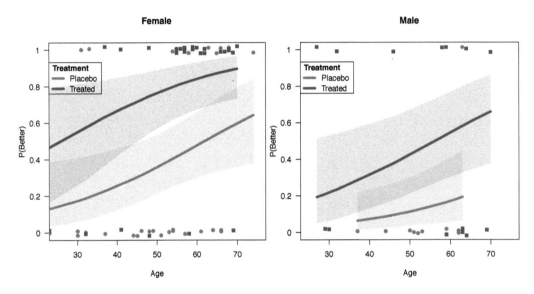

Figure 7.9: Full-model plot of Arthritis data, showing fitted probabilities by Treatment and Sex.

$\triangle$

7.3.3   Effect plots

For more than two variables, full-model plots of the fitted response surface can be cumbersome, particularly when the model contains interactions or when the main substantive interest is focused on a given main effect or interaction, controlling for all other explanatory variables. The method of **effect displays** (tables and graphs), developed by John Fox (1987, 2003) and implemented in the effects package, is a useful solution to these problems.

The idea of effect plots is quite simple but very general and handles models of arbitrary complexity:[8] consider a particular subset of predictors (*focal predictors*) we wish to visualize in a given linear model or generalized linear model. The essence is to calculate fitted values (and standard errors) for the model terms involving these variables and all low-order relatives (e.g., main effects that are marginal to an interaction), as these variables are allowed to vary over their range.

All other variables are "controlled" by being fixed at typical values. For example, a quantitative covariate could be fixed at its mean or median; a factor could be fixed at equal proportions of its levels or its proportions in the data. The result, when plotted, shows all effects of the focal predictors and their low-order relatives, but with all other variables controlled (or "adjusted for").

---

[8]Less general expression of these ideas include the use of **adjusted means** in analysis of covariance, and **least squares means** or **population marginal means** (Searle et al., 1980) in analysis of variance; for example, see the lsmeans (Lenth and Hervé, 2015) package for classical linear models.

7.3.3.1 The score model matrix*

More formally, assume we have fit a model with a linear predictor $\eta_i = \alpha + x_i^{\mathsf{T}}\beta$ (on the logit scale, for logistic regression). Letting $\beta_0 = \alpha$ and $x_0 = 1$, we can rewrite this in matrix form as $\eta = X\beta$ where X is the model matrix constructed by the modeling function, such as `glm()`. Fitting the model gives the estimated coefficients b and its estimated covariance matrix $\widehat{\mathcal{V}}(b)$.

The `Effect()` function constructs an analogous *score model matrix*, X^*, where the focal variables have been varied over their range, and all other variables represented as constant, typical values. Using this as input (the `newdata` argument) to the `predict()` function then gives the fitted values $\eta^* = X^*b$. Standard errors used for confidence intervals are calculated by `predict()` (when `se.fit=TRUE`) as the square roots of $\text{diag}\left(X^*\widehat{\mathcal{V}}(b)X^{*\mathsf{T}}\right)$. Note that these ideas work not only for `glm()` models, but potentially for any modeling function that has a `predict()` and `vcov()` method.[9]

These results are calculated on the scale of the linear predictor η (logits, for logistic regression) when the `type` argument to `predict()` is `type="link"` or on the response scale (probabilities, here) when `type="response"`. The latter makes use of the inverse transformation, Eqn. (7.6).

There are two main calculation functions in the **effects** package:

- `Effect()` takes a character vector of the names of a subset of focal predictors and constructs the score matrix X^* by varying these over their ranges, while holding all other predictors constant at "typical" values. There are many options that control these calculations. For example, `xlevels` can be used to specify the values of the focal predictors; `typical` or `given.values`, respectively, can be used to specify either a function (`mean`, `median`) or a list of specific typical values used for the variables that are controlled. The result is an object of class "eff", for which there are `print()`, `summary()`, and (most importantly) `plot()` methods. See `help(Effect)` for a complete description.

- `allEffects()` takes a model object, and calculates the effects for each high-order term in the model (including their low-order) relatives. Similar optional arguments control the details of the computation. The result is an object of class "efflist".

In addition, the plotting methods for "eff" and "efflist" objects offer numerous options to control the plot details, only a few of which are used in the examples below. For logistic regression models, they also solve the problem of the trade-off between plots on the logit scale, which have a simple representation in terms of additive effects, and plots on the probability scale which are usually simpler to understand. By default, the fitted model effects are plotted on the logit scale, but the response y axis is labeled with the corresponding probability values.

7.3.3.2 Partial residuals

We noted earlier that for discrete response data, it is usually important to display the *data* in some fashion, along with the fitted relationship. Conditional and full-model plots do this by jittering the binary values at 0 and 1 so you can see where the data exists.

The **effects** package takes this idea further, by allowing the display of ***partial residuals***. Letting r denote the vector of residuals for a given model (see Section 7.5.1 for details), the partial residuals r_j pertaining to predictor x_j are defined as

$$r_j = r + \widehat{\beta}_j x_j .$$

---

[9]For example, the **effects** package presently provides methods for models fit by `lm()` (including multivariate linear response models), `glm()`, `gls()`, multinomial (`multinom()` in the **nnet** (Ripley, 2015b) package) and proportional odds models (`polr()` in **MASS**), polytomous latent class models (**poLCA** (Linzer and Lewis., 2014) package), as well as a variety of multi-level and mixed-effects linear models fit with `lmer()` from the **lme4** (Bates et al., 2014) package, or with `lme()` from the **nlme** (Pinheiro et al., 2015) package.

These are a natural extension of residuals in simple regression to the multiple regression setting, in that the slope of a simple regression of r on x is equal to the value of $\widehat{\beta}_j$ in the full multiple regression model (Cook, 1993). Adding partial residuals to an effect plot (together with a non-parametric smoothing) can help to visualize lack of fit or misspecification of the response mean attributable to continuous predictors, such as a nonlinear relation or an omitted interaction (Fox and Weisberg, 2015b).

EXAMPLE 7.8: Arthritis treatment
    Here we illustrate the use of the **effects** package with the simple main effects model that was fit in Example 7.5. `allEffects()` is used to calculate the predicted probabilities of the `Better` response for `Age` and the two factors, `Sex` and `Treatment`. Partial residuals (for quantitative predictors) must be requested in the call to `allEffects()` or `Effect()`.

```
> library(effects)
> arth.eff2 <- allEffects(arth.logistic2, partial.residuals = TRUE)
> names(arth.eff2)

[1] "I(Age-50)"   "Sex"          "Treatment"
```

The result, `arth.eff2`, is a list containing the fitted values (response probabilities, by default) for each of the model terms. For example the main effect for `Sex` is shown below; the associated score `model.matrix` illustrates how `Sex` is varied over its range, while `Age-50` and `Treatment` are fixed at their average values in the data.

```
> arth.eff2[["Sex"]]


 Sex effect
Sex
 Female      Male
0.60932 0.26050

> arth.eff2[["Sex"]]$model.matrix

  (Intercept) I(Age - 50) SexMale TreatmentTreated
1           1      3.3571       0           0.4881
2           1      3.3571       1           0.4881
```

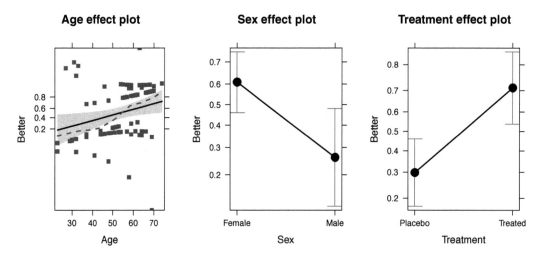

Figure 7.10: Plot of all effects in the main effects model for the Arthritis data. Partial residuals and their loess smooth are also shown for the continuous predictor, Age.

The default plot method for the **"efflist"** object produces one plot for each high-order term, which are just the main effects in this model. The call below produces Figure 7.10.

```
> plot(arth.eff2, rows = 1, cols = 3,
+      type="response", residuals.pch = 15)
```

The smoothed loess curve for the partial residuals with respect to age show a hint of nonlinearity, but perhaps not enough to worry about.

You can quite easily also produce effect plots for several predictors jointly, or full-model plots by using all predictors in the model in a call to `Effect()`, as shown in the call below.

```
> arth.full <- Effect(c("Age", "Treatment", "Sex"), arth.logistic2)
```

Then plotting the result, with some options, gives the plot shown in Figure 7.11.

```
> plot(arth.full, multiline = TRUE, ci.style = "bands",
+      colors = c("red", "blue"), lwd = 3,
+      ticks = list(at = c(.05, .1, .25, .5, .75, .9, .95)),
+      key.args = list(x = .52, y = .92, columns = 1),
+      grid = TRUE)
```

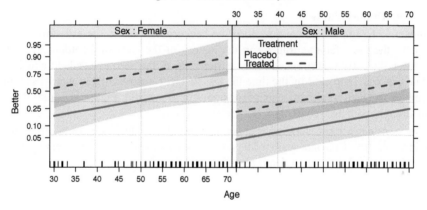

Figure 7.11: Full-model plot of the effects of all predictors in the main effects model for the Arthritis data, plotted on the logit scale.

Alternatively, we can plot these results directly on the scale of probabilities, as shown in Figure 7.12.

```
> plot(arth.full, multiline = TRUE, ci.style = "bands",
+      type="response",
+      colors = c("red", "blue"), lwd = 3,
+      key.args = list(x = .52, y = .92, columns = 1),
+      grid = TRUE)
```

△

7.4   Case studies

The examples below take up some issues of data analysis, model building, and visualization in the context of multiple logistic regression models. We focus on the combination of exploratory plots to see the data, modeling steps, and graphs to interpret a given model.

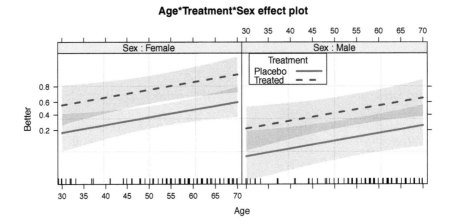

Figure 7.12: Full-model plot of the effects of all predictors in the main effects model for the Arthritis data, plotted on the probability scale.

7.4.1   Simple models: Group comparisons and effect plots

EXAMPLE 7.9: Donner Party

In Chapter 1, Example 1.3, we described the background behind the sad story of the Donner Party, perhaps the most famous tragedy in the history of the westward settlement in the United States. In brief, the party was stranded on the eastern side of the Sierra Nevada mountains by heavy snow in late October 1846, and by the time the last survivor was rescued in April 1847, nearly half of the members had died from famine and exposure to extreme cold. Figure 1.3 showed that survival decreased strongly with age.

Here we consider a more detailed analysis of these data, which are contained in the data set `Donner` in **vcdExtra**. This data set lists 90 people in the Donner Party by name, together with age, sex, survived (0/1), and the date of death for those who died.[10]

```
> data("Donner", package = "vcdExtra")    # load the data
> library(car)                            # for some() and Anova()
> some(Donner, 8)

                        family age     sex survived      death
Breen, Peter              Breen   3    Male        1       <NA>
Donner, Jacob            Donner  65    Male        0 1846-12-21
Foster, Jeremiah       MurFosPik   1    Male        0 1847-03-13
Graves, Nancy            Graves   9 Female         1       <NA>
McCutchen, Harriet    McCutchen   1 Female         0 1847-02-02
Reed, James               Reed  46    Male        1       <NA>
Reinhardt, Joseph        Other  30    Male        0 1846-12-21
Wolfinger, Doris       FosdWolf  20 Female         1       <NA>
```

The main purpose of this example is to try to understand, through graphs and models, how survival was related to age and sex. However, first, we do some data preparation and exploration. The response variable, `survived`, is a 0/1 integer, and it is more convenient for some purposes to make it a factor.

---

[10]Most historical sources count the number in the Donner Party at 87 or 89. An exact accounting of the members of the Donner Party is difficult, because: (a) several people joined the party in mid-route, at Fort Bridger and in the Wasatch Mountains; (b) several rode ahead to search for supplies and one (Charles Stanton) brought two more with him (Luis and Salvador); (c) five people died before reaching the Sierra Nevada mountains. `Donner` incorporates updated information from Kristin Johnson's listing, http://user.xmission.com/~octa/DonnerParty/Roster.htm.

```
> Donner$survived <- factor(Donner$survived, labels = c("no", "yes"))
```

Some historical accounts (Grayson, 1990) link survival in the Donner Party to kinship or family groups, so we take a quick look at this factor here. The variable `family` reflects a recoding of the last names of individuals to reduce the number of factor levels. The main families in the Donner party were: Donner, Graves, Breen, and Reed. The families of Murphy, Foster, and Pike are grouped as `"MurFosPik"`, those of Fosdick and Wolfinger are coded as `"FosdWolf"`, and all others as `"Other"`.

```
> xtabs(~ family, data = Donner)

family
    Breen    Donner      Eddy  FosdWolf     Graves  Keseberg
        9        14         4         4         10         4
McCutchen MurFosPik     Other      Reed
        3        12        23         7
```

For the present purposes, we reduce these 10 family groups further, collapsing some of the small families into `"Other"`, and reordering the levels. Assigning new values to the `levels()` of a factor is a convenient trick for recoding factor variables.

```
> # collapse small families into "Other"
> fam <- Donner$family
> levels(fam)[c(3, 4, 6, 7, 9)] <- "Other"
>
> # reorder, putting Other last
> fam = factor(fam,levels(fam)[c(1, 2, 4:6, 3)])
> Donner$family <- fam
> xtabs(~family, data=Donner)

family
    Breen    Donner    Graves MurFosPik      Reed     Other
        9        14        10        12         7        38
```

`xtabs()` then shows the counts of survival by these family groups:

```
> xtabs(~ survived + family, data = Donner)

        family
survived Breen Donner Graves MurFosPik Reed Other
     no      0      7      3         6    1    25
    yes      9      7      7         6    6    13
```

Plotting this distribution of survival by family with a formula gives a **spineplot**, a special case of the mosaic plot, or a generalization of a stacked bar plot, shown in Figure 7.13. The widths of the bars are proportional to family size, and the shading highlights in light blue the proportion who survived in each family.

```
> plot(survived ~ family, data = Donner, col = c("pink", "lightblue"))
```

A generalized pairs plot (Section 5.6.2), shown in Figure 7.14, gives a visual overview of the data. The diagonal panels here show the marginal distributions of the variables as bar plots, and highlight the skewed distribution of age and the greater number of males than females in the party. The boxplots and barcode plots for survived and age show that those who survived were generally younger than those who perished.

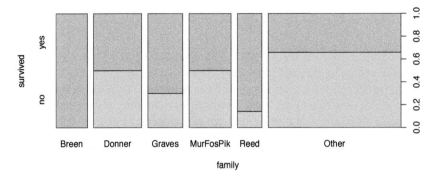

Figure 7.13: Spineplot of survival in the Donner Party by family.

```
> library(gpairs)
> library(vcd)
> gpairs(Donner[,c(4, 2, 3, 1)],
+     diag.pars = list(fontsize = 20, hist.color = "gray"),
+     mosaic.pars = list(gp = shading_Friendly),
+     outer.rot = c(45, 45)
+ )
```

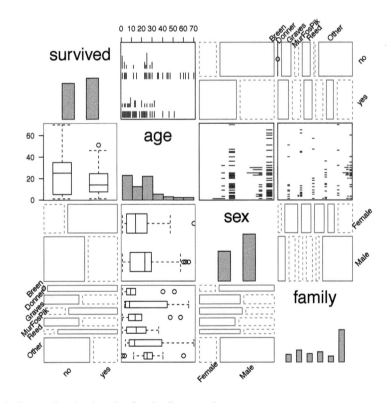

Figure 7.14: Generalized pairs plot for the Donner data.

From an exploratory perspective, we now proceed to examine the relationship of survival to age and sex, beginning with the kind of conditional plots we illustrated earlier (in Example 7.6). Figure 7.15 shows a plot of `survived`, converted back to a 0/1 variable as required by `ggplot()`, together with the binary responses as points and the logistic regressions fitted separately for males and females.

```
> # basic plot: survived vs. age, colored by sex, with jittered points
> gg <- ggplot(Donner, aes(age, as.numeric(survived=="yes"),
+                          color = sex)) +
+   ylab("Survived") + theme_bw() +
+   geom_point(position = position_jitter(height = 0.02, width = 0))
>
> # add conditional linear logistic regressions
> gg + stat_smooth(method = "glm", family = binomial, formula = y ~ x,
+                   alpha = 0.2, size = 2, aes(fill = sex))
```

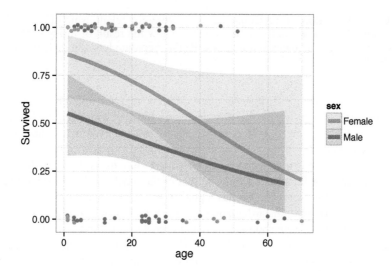

Figure 7.15: Conditional plot of the Donner data, showing the relationship of survival to age and sex. The smoothed curves and confidence bands show the result of fitting separate linear logistic regressions on age for males and females.

It is easy to see that survival among women was greater than for men, perhaps narrowing the gap among the older people, but the data gets thin towards the upper range of age.

The curves plotted in Figure 7.15 assume a linear relationship between the log odds of survival and age (expressed as `formula = y ~ x` in the call to `stat_smooth()`). One simple way to check whether the relationship between survival and age is nonlinear is to re-do this plot, but now allow a quadratic relationship with age, using `formula = y ~ poly(x,2)`. The result is shown in the left panel of Figure 7.16.

```
> # add conditional quadratic logistic regressions
> gg + stat_smooth(method = "glm", family = binomial,
+                   formula = y ~ poly(x,2), alpha = 0.2, size = 2,
+                   aes(fill = sex))
>
> # add loess smooth
> gg + stat_smooth(method = "loess", span=0.9, alpha = 0.2, size = 2,
+                   aes(fill = sex)) +
+   coord_cartesian(ylim = c(-.05,1.05))
```

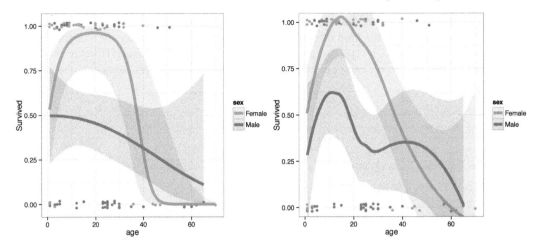

Figure 7.16: Conditional plots of the Donner data, showing the relationship of survival to age and sex. Left: The smoothed curves and confidence bands show the result of fitting separate quadratic logistic regressions on age for males and females. Right: Separate loess smooths are fit to the data for males and females.

This plot is quite surprising. It suggests quite different regimes relating to survival for men and women. Among men, survival probability decreases steadily with age, at least after age 20. For women, those in the age range 10–35 were very likely to have lived, while those over 40 were almost all predicted to perish.

Another simple technique is to fit a non-parametric loess smooth, as shown in the right panel of Figure 7.16.[11] The curve for females is similar to that of the quadratic fit in the left panel, but the curve for males suggests that survival also has a peak around the teenage years. One lesson to be drawn from these graphs is that a linear logistic regression, as shown in Figure 7.16, may tell only part of the story, and, for a binary response, it is not easy to discern whether the true relationship is linear. If it really is, all these graphs would look much more similar. As well, we usually obtain a more realistic smoothing of the data using full-model plots or effect plots.

The suggestions from these exploratory graphs can be used to define and test some models for survival in the Donner Party. The substantive questions of interest are:

- Is the relationship different for men and women? This is, is it necessary to allow for an interaction of age with sex, or separate fitted curves for men and women?

- Is the relationship between survival and age well-represented in a linear logistic regression model?

The first question is the easiest to deal with: we can simply fit a model allowing an interaction of age (or some function of age) and sex,

```
survived ~ age * sex
survived ~ f(age) * sex
```

and compare the goodness of fit with the analogous additive, main-effects models.

From a modeling perspective, there is a wide variety of approaches for testing for nonlinear relationships. We only scratch the surface here, and only for a single quantitative predictor, x, such

---

[11] A technical problem with the use of the loess smoother for binary data is that it can produce fitted values outside the [0–1] interval, as happens in the right panel of this figure. Kernel smoothers, such as the KernSmooth (Wand, 2015) package avoid this problem, but are not available through ggplot2.

as age in this example. One simple approach, illustrated in Figure 7.16, is to allow a quadratic (or higher-power, e.g., cubic) function to describe the relationship between the log odds and x,

$$\begin{aligned}
\text{logit}(\pi_i) &= \alpha + \beta_1 x_i + \beta_2 x_i^2 \\
\text{logit}(\pi_i) &= \alpha + \beta_1 x_i + \beta_2 x_i^2 + \beta_3 x_i^3
\end{aligned}$$

$$\cdots \qquad \cdot$$

In R, these model terms can be fit using `poly(x, 2)`, `poly(x, 3)` ..., which generate orthogonal polynomials for the powers of x. A simple way to test for nonlinearity is a likelihood ratio test comparing the more complex model to the linear one. This method is often sufficient for a hypothesis test, and, if the relationship truly is linear, the fitted logits and probabilities will not differ greatly from what they would be under a linear model. A difficulty with this approach is that polynomial models are often unrealistic, particularly for data that approach an asymptote.

Another simple approach is to use a ***regression spline***, which fits the relationship with x in terms of a set of piecewise polynomials, usually cubic, joined at a collection of points, called *knots*, so that the overall fitted relationship is smooth and continuous. See Fox (2008, Section 17.2) for a cogent, brief description of these methods.

One particularly convenient method is a ***natural spline***, implemented in the splines package in the `ns()` function. This method constrains the fitted cubic spline to be linear at lower and upper limits of x, and, for k knots, fits df $= k + 1$ parameters not counting the intercept. The k knots can be conveniently chosen as k cutpoints in the percentiles of the distribution of x. For example, with $k = 1$, the knot would be placed at the median, or 50th percentile; with $k = 3$, the knots would be placed at the quartiles of the distribution of x; $k = 0$ corresponds to no knots, i.e., a simple linear regression.

In the `ns()` function, you can specify the locations of knots or the number of knots with the `knots` argument, but it is conceptually simpler to specify the number of degrees of freedom used in the spline fit. Thus, `ns(x, 2)` and `poly(x, 2)` both specify a term in x of the same complexity, the former a natural spline with $k = 1$ knot and the later a quadratic function in x.

We illustrate these ideas in the remainder of this example, fitting a 2×2 collection of models to the *Donner* data corresponding to: (a) whether or not age and sex effects are additive; (b) whether the effect is linear on the logit scale or nonlinear (quadratic, here). A brief summary of each model is given using the `Anova()` in the car package, providing Type II tests of each effect. As usual, `summary()` would give more detailed output, including tests for individual coefficients. First, we fit the linear models, without and with an interaction term:

```
> donner.mod1 <- glm(survived ~ age + sex,
+                     data = Donner, family  =binomial)
> Anova(donner.mod1)

Analysis of Deviance Table (Type II tests)

Response: survived
     LR Chisq Df Pr(>Chisq)
age      5.52  1     0.0188 *
sex      6.73  1     0.0095 **
---
Signif. codes:  0 '***' 0.001 '**' 0.01 '*' 0.05 '.' 0.1 ' ' 1

> donner.mod2 <- glm(survived ~ age * sex,
+                     data = Donner, family = binomial)
> Anova(donner.mod2)

Analysis of Deviance Table (Type II tests)
```

```
Response: survived
          LR Chisq Df Pr(>Chisq)
age          5.52   1     0.0188 *
sex          6.73   1     0.0095 **
age:sex      0.40   1     0.5269
---
Signif. codes:  0 '***' 0.001 '**' 0.01 '*' 0.05 '.' 0.1 ' ' 1
```

The main effects of `age` and `sex` are both significant here, but the interaction term, `age:sex`, is not in model `donner.mod2`. Note that the terms tested by `Anova()` in `donner.mod1` are a redundant subset of those in `donner.mod2`.

Next, we fit nonlinear models, representing the linear and nonlinear trends in age by `poly(age, 2)`.[12] The `Anova()` results for terms in both models are contained in the output from `Anova(donner.mod4)`.

```
> donner.mod3 <- glm(survived ~ poly(age, 2) + sex,
+                    data = Donner, family = binomial)
> donner.mod4 <- glm(survived ~ poly(age, 2) * sex,
+                    data = Donner, family = binomial)
> Anova(donner.mod4)

Analysis of Deviance Table (Type II tests)

Response: survived
                 LR Chisq Df Pr(>Chisq)
poly(age, 2)         9.91  2     0.0070 **
sex                  8.09  1     0.0044 **
poly(age, 2):sex     8.93  2     0.0115 *
---
Signif. codes:  0 '***' 0.001 '**' 0.01 '*' 0.05 '.' 0.1 ' ' 1
```

Now, in model `donner.mod4`, the interaction term `poly(age, 2):sex` is significant, indicating that the fitted quadratics for males and females differ in "shape," meaning either their linear (slope) or quadratic (curvature) components.

These four models address the questions posed earlier. A compact summary of these models, giving the likelihood ratio tests of goodness of fit, together with AIC and BIC statistics, are shown below, using the `LRstats()` method in **vcdExtra** for a list of `"glm"` models.

```
> library(vcdExtra)
> LRstats(donner.mod1, donner.mod2, donner.mod3, donner.mod4)

Likelihood summary table:
            AIC BIC LR Chisq Df Pr(>Chisq)
donner.mod1 117 125    111.1 87     0.042 *
donner.mod2 119 129    110.7 86     0.038 *
donner.mod3 115 125    106.7 86     0.064 .
donner.mod4 110 125     97.8 84     0.144
---
Signif. codes:  0 '***' 0.001 '**' 0.01 '*' 0.05 '.' 0.1 ' ' 1
```

By AIC and BIC, `donner.mod4` is best, and it is also the only model with a non-significant LR χ^2 (residual deviance). Because these models comprise a 2×2 set of hypotheses, it is easier to compare models by extracting the LR statistics and arranging these in a table, together with their row and column differences. The entries in the table below are calculated as follows.

---

[12] Alternatively, we could use the term `ns(age, 2)`, or higher-degree polynomials, or natural splines with more knots, but we don't do this here.

```
> mods <- list(donner.mod1, donner.mod2, donner.mod3, donner.mod4)
> LR <- sapply(mods, function(x) x$deviance)
> LR <- matrix(LR, 2, 2)
> rownames(LR) <- c("additive", "non-add")
> colnames(LR) <- c("linear", "non-lin")
> LR <- cbind(LR, diff = LR[,1] - LR[,2])
> LR <- rbind(LR, diff = c(LR[1,1:2] - LR[2,1:2], NA))
```

	linear	non-linear	$\Delta\chi^2$	*p*-value
additive	111.128	106.731	4.396	0.036
non-additive	110.727	97.799	12.928	0.000
$\Delta\chi^2$	0.400	8.932		
p-value	0.527	0.003		

Thus, the answer to our questions seems to be that: (a) there is evidence that the relationship of survival to age differs for men and women in the Donner Party; (b) these relationships are not well-described by a linear logistic regression.

For simplicity, we used a quadratic effect, `poly(age,2)`, to test for nonlinearity here. An alternative test of the same complexity could use a regression spline, `ns(age,2)`, also with 2 degrees of freedom for the main effect and interaction, or allow more knots. To illustrate, we fit two natural spline modes models with 2 and 4 df, and compare these with the quadratic model (`donner.mod4`), all of which include the interaction of age and sex.

```
> library(splines)
> donner.mod5 <- glm(survived ~ ns(age,2) * sex,
+                    data = Donner, family = binomial)
> Anova(donner.mod5)

Analysis of Deviance Table (Type II tests)

Response: survived
              LR Chisq Df Pr(>Chisq)
ns(age, 2)        9.28  2     0.0097 **
sex               7.98  1     0.0047 **
ns(age, 2):sex    8.71  2     0.0129 *
---
Signif. codes:  0 '***' 0.001 '**' 0.01 '*' 0.05 '.' 0.1 ' ' 1

> donner.mod6 <- glm(survived ~ ns(age,4) * sex,
+                    data = Donner, family = binomial)
> Anova(donner.mod6)

Analysis of Deviance Table (Type II tests)

Response: survived
              LR Chisq Df Pr(>Chisq)
ns(age, 4)       22.05  4     0.0002 ***
sex              10.49  1     0.0012 **
ns(age, 4):sex    8.54  4     0.0737 .
---
Signif. codes:  0 '***' 0.001 '**' 0.01 '*' 0.05 '.' 0.1 ' ' 1

> LRstats(donner.mod4, donner.mod5, donner.mod6)

Likelihood summary table:
            AIC BIC LR Chisq Df Pr(>Chisq)
donner.mod4 110 125     97.8 84       0.14
donner.mod5 111 126     98.7 84       0.13
donner.mod6 106 131     86.1 80       0.30
```

With four more parameters, `donner.mod6` fits better and has a smaller AIC.

We conclude this example with an effect plot for the spline model `donner.mod6` shown in Figure 7.17. The complexity of the fitted relationships for men and women is intermediate between the two conditional plots shown in Figure 7.16. (However, note that the fitted effects are plotted on the logit scale in Figure 7.17 and labeled with the corresponding probabilities, whereas the conditional plots are plotted directly on the probability scale.)

```
> library(effects)
> donner.eff6 <- allEffects(donner.mod6, xlevels = list(age=seq(0, 50, 5)))
> plot(donner.eff6, ticks = list(at=c(0.001, 0.01, 0.05, 0.1, 0.25, 0.5,
+                                     0.75, 0.9, 0.95, 0.99, 0.999)))
```

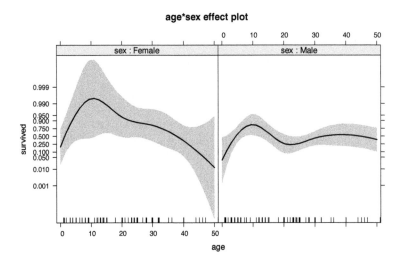

Figure 7.17: Effect plot for the spline model `donner.mod6` fit to the Donner data.

This plot confirms that for women in the Donner Party, survival was greatest for those aged 10–30. Survival among men was overall much less and there is a hint of greater survival for men aged 10–15.

Of course, this statistical analysis does not provide explanations for these effects, and it ignores the personal details of the Donner Party members and the individual causes and circumstances of death, which are generally well-documented in the historical record (Johnson, 1996). See `http://user.xmission.com/~octa/DonnerParty/` for a comprehensive collection of historical sources.

Grayson (1990) attributes the greater survival of women of intermediate age to demographic arguments that women are overall better able to withstand conditions of famine and extreme cold, and high age-specific mortality rates among the youngest and oldest members of human societies. He also concludes (without much analysis) that members with larger social and kinship networks would be more likely to survive.   △

EXAMPLE 7.10: Racial profiling: Arrests for marijuana possession

In the summer of 2002, the *Toronto Star* newspaper launched an investigation on the topic of possible racial profiling by the Toronto police service. Through freedom of information requests, they obtained a data base of over 600,000 arrest records on all potential charges in the period from 1996–2002, the largest data bases on crime arrests and disposition ever assembled in Canada. An initial presentation of this study was given in Example 1.4.

In order to examine the issue of racial profiling (different treatment as a function of race) they excluded all charges such as assault, robbery, speeding, and driving under the influence, where the police have no discretion regarding the laying of a charge. They focused instead on a subset of arrests, where the police had various options.

Among these, for people arrested for a single charge of simple possession of a small amount of marijuana, police have the option of releasing the arrestee, with a summons ("Form 9") to appear in court (similar to a parking ticket), or else the person could be given harsher treatment—brought to a police station or held in jail for a bail hearing ("Show cause"). The main question for the *Toronto Star* was whether the subject's skin color had any influence on the likelihood that the person would be released with a summons.[13]

Their results, published in a week-long series of articles in December 2002, concluded that there was strong evidence that black and white subjects were treated differently. For example, the analysis showed that blacks were 1.5 times more likely than whites to be given harsher treatment than release with a summons; if the subject was taken to the police station, a black was 1.6 times more likely to be held in jail for a bail hearing. An important part of the analysis and the public debate that ensued was to show that other variables that might account for these differences had been controlled or adjusted for.[14]

The data set `Arrests` in the **effects** package gives a simplified version of the *Star* database, containing records for 5,226 cases of arrest on the charge of simple possession of marijuana analyzed by the newspaper. The response variable here is `released` (Yes/No) and the main predictor of interest is skin color of the person arrested, `colour` (Black/White).[15] A random subset of the data set is shown below.

```
> library(effects)
> data("Arrests", package = "effects")
> Arrests[sample(nrow(Arrests), 6),]

     released colour year age  sex employed citizen checks
3768      Yes  Black 2000  23 Male       No     Yes      4
4576      Yes  Black 2001  17 Male      Yes     Yes      0
3976       No  White 2002  20 Male       No     Yes      3
4629      Yes  White 2000  18 Male      Yes     Yes      1
2384       No  Black 2000  19 Male      Yes     Yes      3
869       Yes  White 2001  15 Male      Yes     Yes      1
```

Other available predictors, to be used as control variables, included the `year` of the arrest, `age` and `sex` of the person, and binary indicators of whether the person was `employed` and a `citizen` of Canada. In addition, when someone is stopped by police, his/her name is checked in six police data bases that record previous arrests, convictions, whether on parole, etc. The variable `checks` records the number, 0–6, in which the person's name appeared.

A variety of logistic models were fit to these data including all possible main effects and some two-way interactions. To allow for possible nonlinear effects of `year`, this variable was treated as a factor rather than as a (linear) numeric variable, but the effects of `age` and `checks` were reasonably linear on the logit scale. A reasonable model included the interactions of `colour` with both `year` and `age`, as fit below:

---

[13] Another discretionary charge they investigated was police stops for non-moving violations under the Ontario *Highway Traffic Act*, such as being pulled over for a faulty muffler or having an expired license plate renewal sticker. A disproportionate rate of charges against blacks is sometimes referred to as "driving while black" (DWB). This investigation found that the number of blacks so charged, but particularly young black males, far outweighed their representation in the population.

[14] The Toronto Police Service launched a class-action libel law suit against the *Toronto Star* and the first author of this book, who served as their statistical consultant, claiming damages of $5,000 for every serving police officer in the city, a total of over 20 million dollars. The suit was thrown out of court, and the Toronto police took efforts to enhance training programs to combat the perception of racial profiling.

[15] The original data set also contained the categories Brown and Other, but these appeared with small frequencies.

```
> Arrests$year <- as.factor(Arrests$year)
> arrests.mod <- glm(released ~ employed + citizen + checks
+                    + colour*year + colour*age,
+                    family = binomial, data = Arrests)
```

For such models, significance tests for the model terms are best carried out using the Anova()
function in the car package that uses Type II tests:

```
> library(car)
> Anova(arrests.mod)

Analysis of Deviance Table (Type II tests)

Response: released
            LR Chisq Df Pr(>Chisq)
employed       72.7  1   < 2e-16 ***
citizen        25.8  1   3.8e-07 ***
checks        205.2  1   < 2e-16 ***
colour         19.6  1   9.7e-06 ***
year            6.1  5   0.29785
age             0.5  1   0.49827
colour:year    21.7  5   0.00059 ***
colour:age     13.9  1   0.00019 ***
---
Signif. codes:  0 '***' 0.001 '**' 0.01 '*' 0.05 '.' 0.1 ' ' 1
```

The difficulty in interpreting these results from tables of coefficients can be seen in the output
below:

```
> coeftest(arrests.mod)


z test of coefficients:

                       Estimate Std. Error z value Pr(>|z|)
(Intercept)             0.34443    0.31007    1.11  0.26665
employedYes             0.73506    0.08477    8.67  < 2e-16 ***
citizenYes              0.58598    0.11377    5.15  2.6e-07 ***
checks                 -0.36664    0.02603  -14.08  < 2e-16 ***
colourWhite             1.21252    0.34978    3.47  0.00053 ***
year1998               -0.43118    0.26036   -1.66  0.09770 .
year1999               -0.09443    0.26154   -0.36  0.71805
year2000               -0.01090    0.25921   -0.04  0.96647
year2001                0.24306    0.26302    0.92  0.35541
year2002                0.21295    0.35328    0.60  0.54664
age                     0.02873    0.00862    3.33  0.00086 ***
colourWhite:year1998    0.65196    0.31349    2.08  0.03756 *
colourWhite:year1999    0.15595    0.30704    0.51  0.61152
colourWhite:year2000    0.29575    0.30620    0.97  0.33411
colourWhite:year2001   -0.38054    0.30405   -1.25  0.21073
colourWhite:year2002   -0.61732    0.41926   -1.47  0.14091
colourWhite:age        -0.03737    0.01020   -3.66  0.00025 ***
---
Signif. codes:  0 '***' 0.001 '**' 0.01 '*' 0.05 '.' 0.1 ' ' 1
```

By direct calculation (e.g., using exp(coef(arrests.mod))) you can find that the odds
of a quick release was $\exp(0.735) = 2.08$ times greater for someone employed, $\exp(0.586) = 1.80$
times more likely for a Canadian citizen, and $\exp(1.21) = 3.36$ times more likely for a white than
a black person. It is much more difficult to interpret the interaction terms.

The primary question for the newspaper concerned the overall difference between the the treatment of blacks and whites— the main effect of colour. We plot this as shown below, giving the
plot shown in Figure 7.18. This supports the claim by the *Star* because the 95% confidence limits
for blacks and whites do not overlap, and all other relevant predictors that could account for this
effect have been controlled or adjusted for.

```
> plot(Effect("colour", arrests.mod),
+       lwd = 3, ci.style = "bands", main = "",
+       xlab = list("Skin color of arrestee", cex = 1.25),
+       ylab = list("Probability(released)", cex = 1.25)
+    )
```

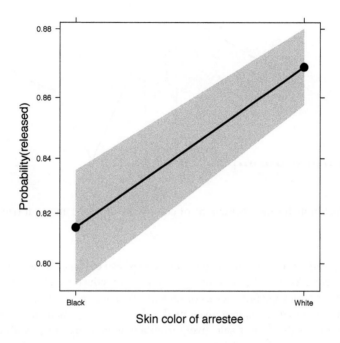

Skin color of arrestee

Figure 7.18: Effect plot for the main effect of skin color in the Arrests data.

Of course, one should be very wary of interpreting main effects when there are important interactions, and the story turned out to be far more nuanced than was reported in the newspaper. In particular, the interactions of color with with age and year provided a more complete account. Effect plots for these interactions are shown in Figure 7.19.

```
> # colour x age interaction
> plot(Effect(c("colour", "age"), arrests.mod),
+       lwd = 3, multiline = TRUE, ci.style = "bands",
+       xlab = list("Age", cex = 1.25),
+       ylab = list("Probability(released)", cex = 1.25),
+       key.args = list(x = .05, y = .99, cex = 1.2, columns = 1)
+    )
> # colour x year interaction
> plot(Effect(c("colour", "year"), arrests.mod),
+       lwd = 3, multiline = TRUE,
+       xlab = list("Year", cex = 1.25),
+       ylab = list("Probability(released)", cex = 1.25),
+       key.args = list(x = .7, y = .99, cex = 1.2, columns = 1)
+    )
```

From the left panel in Figure 7.19, it is immediately apparent that the effect of age was in opposite directions for blacks and whites: Young blacks were indeed treated more severely than young whites; however, for older people, blacks were treated less harshly than whites, controlling for all other predictors.

The right panel of Figure 7.19 shows the changes over time in the treatment of blacks and whites. It can be seen that up to the year 2000 there was strong evidence for differential treatment on these

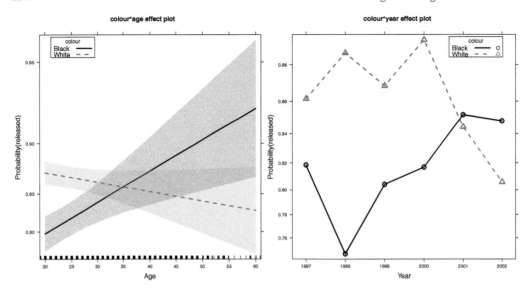

Figure 7.19: Effect plots for the interactions of color with age (left) and year (right) in the Arrests data.

charges, again controlling for other predictors. There was also evidence to support the claim by the police that in the year 2001 they began training of officers to reduce racial effects in treatment.

Finally, the **effects** package provides a convenience function, `allEffects()`, that calculates the effects for all high-order terms in a given model. The `plot()` method for the `"efflist"` object can be used to plot individual terms selectively from a graphic menu, or plot all terms together in one comprehensive display using `ask=FALSE`.

```
> arrests.effects <- allEffects(arrests.mod,
+                               xlevels = list(age = seq(15, 45, 5)))
> plot(arrests.effects,
+      ylab = "Probability(released)", ci.style = "bands", ask = FALSE)
```

The result, shown in Figure 7.20, is a relatively compact and understandable summary of the `arrests.mod` model: (a) people were more likely to be released if they were employed and citizens; (b) each additional police check decreased the likelihood of release with a summons; (c) the effect of skin color varied with age and year of arrest, in ways that tell a far more nuanced story than reported in the newspaper.

Finally, another feature of this plot bears mention: by default, the scales for each effect plot are determined separately for each effect, to maximize use of the plot region. However, you have to read the Y scale values to judge the relative sizes of these effects. An alternative plot, using the *same* scale in each subplot,[16] would show the relative sizes of these effects.

$\triangle$

7.4.2 More complex models: Model selection and visualization

Models with more predictors or more complex terms (interactions, nonlinear terms) present additional challenges for model fitting, summarization, and visualization and interpretation. These problems increase rapidly with the number of potential predictors.

---

[16]With the **effects** package, you can set the `ylim` argument to equate the vertical range for all plots, but this should be done on the logit scale. For this plot, `ylim = plogis(c(0.5, 1))` would work.

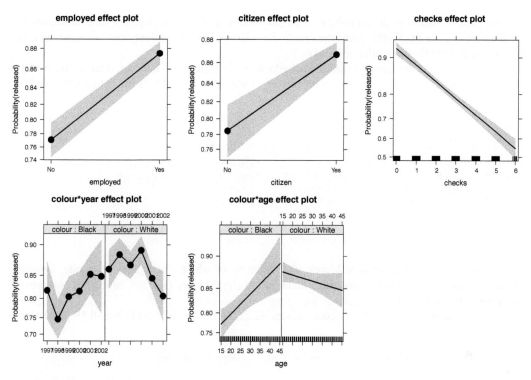

Figure 7.20: Effect plot for all high-order terms in the model for the Arrests data.

A very complicated model, with many terms and interactions, may fit the data at hand quite well. However, because goodness-of-fit is optimized in the sample, terms that appear significant are less likely to be important in a future sample, and we need to worry about inflation of Type I error rates that accompany multiple significance tests. As well, it becomes increasingly difficult to visualize and understand a fitted model as the model becomes increasingly complex. On the other hand, a very simple model may omit important predictors, interactions, or nonlinear relationships with the response and give an illusion of a comfortable interpretation.

Model selection for logistic regression seeks to balance the trade-off between the competing goals of goodness-of-fit and simplicity. A full discussion of this topic is beyond the scope of this book, but is well treated in Agresti (2013, Chapter 6), and extensively in Harrell (2001, Chapters 10–13).

Here, we illustrate some important ideas using the AIC and BIC statistics as parsimony-adjusted measures of goodness-of-fit. These are discussed Section 9.3.2. AIC is defined as

$$\text{AIC} = -2 \log \mathcal{L} + 2k$$

where $\log \mathcal{L}$ is the maximized log likelihood and k is the number of parameters estimated in the model. Better models correspond to *smaller* AIC. BIC is similar, but uses a penalty of $\log(n)k$, and so prefers smaller models as the sample size n increases.

EXAMPLE 7.11: Death in the ICU

In this example we briefly examine some aspects of logistic regression related to model selection and graphical display with a large collection of potential predictors, including both quantitative and discrete variables. We use data from a classic study by Lemeshow et al. (1988) of patients admitted to an intensive care unit at Baystate Medical Center in Springfield, Massachusetts. The major goal of this study was to develop a model to predict the probability of survival (until hospital discharge) of these patients and to study the risk factors associated with ICU mortality. The data, contained in the data set *ICU* in **vcdExtra**, gives the results for a sample of 200 patients that was presented in Hosmer et al. (2013) (and earlier editions).

The *ICU* data set contains 22 variables of which the first, died, is a factor. Among the predictors, two variables (race, coma) were represented initially as 3-level factors, but then recoded to binary variables (white, uncons).

```
> data("ICU", package = "vcdExtra")
> names(ICU)

 [1] "died"    "age"     "sex"     "race"     "service"
 [6] "cancer"  "renal"   "infect"  "cpr"      "systolic"
[11] "hrtrate" "previcu" "admit"   "fracture" "po2"
[16] "ph"      "pco"     "bic"     "creatin"  "coma"
[21] "white"   "uncons"

> ICU <- ICU[,-c(4, 20)]  # remove redundant race, coma
```

Removing the 3-level versions leaves 19 predictors, of which three (age, heart rate, systolic blood pressure) are quantitative and the remainder are either binary (service, cancer) or had previously been dichotomized (e.g., ph<7.25).

As an initial step, and a basis for comparison, we fit the full model containing all 19 predictors.

```
> icu.full <- glm(died ~ ., data = ICU, family = binomial)
> summary(icu.full)


Call:
glm(formula = died ~ ., family = binomial, data = ICU)

Deviance Residuals:
    Min       1Q    Median       3Q       Max
-1.8040  -0.5606  -0.2044  -0.0863    2.9773

Coefficients:
                  Estimate Std. Error z value Pr(>|z|)
(Intercept)       -6.72670    2.38551   -2.82   0.0048 **
age                0.05639    0.01862    3.03   0.0025 **
sexMale            0.63973    0.53139    1.20   0.2286
serviceSurgical   -0.67352    0.60190   -1.12   0.2631
cancerYes          3.10705    1.04585    2.97   0.0030 **
renalYes          -0.03571    0.80165   -0.04   0.9645
infectYes         -0.20493    0.55319   -0.37   0.7110
cprYes             1.05348    1.00661    1.05   0.2953
systolic          -0.01547    0.00850   -1.82   0.0686 .
hrtrate           -0.00277    0.00961   -0.29   0.7732
previcuYes         1.13194    0.67145    1.69   0.0918 .
admitEmergency     3.07958    1.08158    2.85   0.0044 **
fractureYes        1.41140    1.02971    1.37   0.1705
po2<=60            0.07382    0.85704    0.09   0.9314
ph<7.25            2.35408    1.20880    1.95   0.0515 .
pco>45            -3.01844    1.25345   -2.41   0.0160 *
bic<18            -0.70928    0.90978   -0.78   0.4356
creatin>2          0.29514    1.11693    0.26   0.7916
whiteNon-white     0.56573    0.92683    0.61   0.5416
unconsYes          5.23229    1.22630    4.27   2e-05 ***
---
Signif. codes:  0 '***' 0.001 '**' 0.01 '*' 0.05 '.' 0.1 ' ' 1

(Dispersion parameter for binomial family taken to be 1)
```

```
     Null deviance: 200.16  on 199  degrees of freedom
 Residual deviance: 120.78  on 180  degrees of freedom
 AIC: 160.8

 Number of Fisher Scoring iterations: 6
```

You can see that a few predictors are individually significant, but many are not.

However, it is useful to carry out a simultaneous global test of $H_0 : \beta = 0$ that *all* regression coefficients are zero. If this test is not significant, it makes little sense to use selection methods to choose individually significant predictors. For convenience, we define a simple function, LRtest(), to calculate the likelihood ratio test from the model components.

```
> LRtest <- function(model)
+    c(LRchisq = (model$null.deviance - model$deviance),
+       df = (model$df.null - model$df.residual))
>
> (LR <- LRtest(icu.full))

LRchisq       df
 79.383   19.000


> (pvalue <- 1 - pchisq(LR[1], LR[2]))

   LRchisq
2.3754e-09
```

At this point, it is tempting to examine the output from summary(icu.full) shown above and eliminate those predictors that fail significance at some specified level such as the conventional $\alpha = 0.05$. This is generally a bad idea for many reasons.[17]

A marginally better approach is to remove non-significant variables whose coefficients have signs that don't make sense from the substance of the problem. For example, in the full model, both renal (history of chronic renal failure) and infect (infection probable at ICU admission) have negative signs, meaning that their presence *decreases* the odds of death. We remove those variables using update(); as expected they make little difference.

```
> icu.full1 <- update(icu.full, . ~ . - renal - fracture)
> anova(icu.full1, icu.full, test = "Chisq")

Analysis of Deviance Table

Model 1: died ~ age + sex + service + cancer + infect + cpr + systolic +
    hrtrate + previcu + admit + po2 + ph + pco + bic + creatin +
    white + uncons
Model 2: died ~ age + sex + service + cancer + renal + infect + cpr +
    systolic + hrtrate + previcu + admit + fracture + po2 + ph +
    pco + bic + creatin + white + uncons
  Resid. Df Resid. Dev Df Deviance Pr(>Chi)
1       182        122
2       180        121  2      1.7     0.43
```

Before proceeding to consider model selection, it is useful to get a better visual overview of the current model than is available from a table of coefficients and significance tests. Some very useful print(), summary() and plot() methods are available in the rsm (Lenth, 2014) package. Unfortunately, these require that the logistic model is fitted with lrm() in that package rather than with glm(). We pause here to refit the same model as icu.full1 in order to show a plot of odds ratios for the terms in this model.

---

[17]It ignores the facts of (a) an arbitrary cutoff value for significance, (b) the strong likelihood that chance features of the data or outliers influence the result, and (c) problems of collinearity, etc. See Harrell (2001, Section 4.3) for a useful discussion of these issues.

```
> library(rms)
> dd <- datadist(ICU[,-1])
> options(datadist = "dd")
> icu.lrm1 <- lrm(died ~ ., data = ICU)
> icu.lrm1 <- update(icu.lrm1, . ~ . - renal - fracture)
```

The `summary()` method for `"rms"` objects produces a much more detailed descriptive summary of a fitted model, and the `plot()` method for that summary object gives a sensible plot of the odds ratios for the model terms together with confidence intervals, at levels (0.9, 0.95, 0.99) by default. The following lines produce Figure 7.21.

```
> sum.lrm1 <- summary(icu.lrm1)
> plot(sum.lrm1, log = TRUE, main = "Odds ratio for 'died'", cex = 1.25,
+       col = rgb(0.1, 0.1, 0.8, alpha = c(0.3, 0.5, 0.8)))
```

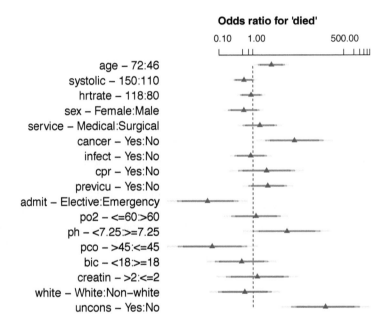

Figure 7.21: Odds ratios for the terms in the model for the ICU data. Each line shows the odds ratio for a term, together with lines for 90, 95, and 99% confidence intervals in progressively darker shades.

In this plot, continuous variables are shown at the top, followed by the discrete predictors. In each line, the range or levels of the predictors are given in the form $a : b$, such that the value a corresponds to the numerator of the odds ratio plotted. Confidence intervals that don't overlap the vertical line for odds ratio = 1 are significant, but this graph shows those at several confidence levels, allowing you to decide what is "significant" visually. As well, the widths of those intervals convey the precision of these estimates.

Among several stepwise selection methods in R for `"glm"` models, `stepAIC()` in the MASS package implements a reasonable collection of methods for forward, backward, and stepwise selection using penalized AIC-like criteria that balance goodness of fit against parsimony. The method takes an argument, `scope`, which is a list of two model formulae: `upper` defines the largest (most complex) model to consider and `lower` defines the smallest (simplest) model, e.g., `lower = ~ 1` is the intercept-only model.

By default, the function produces verbose printed output showing the details of each step, but we suppress that here to save space. It returns the final model as its result, along with an `anova` component that summarises the deviance and AIC from each step.

```
> library(MASS)
> icu.step1 <- stepAIC(icu.full1, trace = FALSE)
> icu.step1$anova

Stepwise Model Path
Analysis of Deviance Table

Initial Model:
died ~ age + sex + service + cancer + infect + cpr + systolic +
    hrtrate + previcu + admit + po2 + ph + pco + bic + creatin +
    white + uncons

Final Model:
died ~ age + cancer + systolic + admit + ph + pco + uncons


           Step Df  Deviance Resid. Df Resid. Dev    AIC
1                                   182     122.48 158.48
2        - po2  1  0.062446         183     122.54 156.54
3    - creatin  1  0.059080         184     122.60 154.60
4    - hrtrate  1  0.072371         185     122.67 152.67
5     - infect  1  0.122772         186     122.79 150.79
6      - white  1  0.334999         187     123.13 149.13
7    - service  1  0.671313         188     123.80 147.80
8        - bic  1  0.377521         189     124.18 146.18
9        - cpr  1  1.148260         190     125.33 145.33
10       - sex  1  1.543523         191     126.87 144.87
11   - previcu  1  1.569976         192     128.44 144.44
```

Alternatively, we can use the BIC criterion, by specifying k=log(n), which generally will select a smaller model when the sample size is reasonably large.

```
> icu.step2 <- stepAIC(icu.full, trace = FALSE, k = log(200))
> icu.step2$anova

Stepwise Model Path
Analysis of Deviance Table

Initial Model:
died ~ age + sex + service + cancer + renal + infect + cpr +
    systolic + hrtrate + previcu + admit + fracture + po2 + ph +
    pco + bic + creatin + white + uncons

Final Model:
died ~ age + cancer + admit + uncons


            Step Df  Deviance Resid. Df Resid. Dev    AIC
1                                    180     120.78 226.74
2     - renal  1  0.0019881          181     120.78 221.45
3       - po2  1  0.0067968          182     120.79 216.16
4   - creatin  1  0.0621463          183     120.85 210.92
5   - hrtrate  1  0.0658870          184     120.92 205.69
6    - infect  1  0.2033221          185     121.12 200.59
7     - white  1  0.3673180          186     121.49 195.66
8       - bic  1  0.6002993          187     122.09 190.96
9   - service  1  0.7676303          188     122.85 186.43
10 - fracture  1  1.3245086          189     124.18 182.46
11      - cpr  1  1.1482598          190     125.33 178.31
```

```
12       - sex  1 1.5435228        191      126.87 174.55
13  - previcu  1 1.5699762        192      128.44 170.83
14       - ph  1 4.4412370        193      132.88 169.97
15      - pco  1 2.7302934        194      135.61 167.40
16 - systolic  1 3.5231028        195      139.13 165.63
```

This model differs from model `icu.step1` selected using AIC in the last three steps, which also removed `ph`, `pco`, and `systolic`.

```
> coeftest(icu.step2)


z test of coefficients:

                Estimate Std. Error z value Pr(>|z|)
(Intercept)      -6.8698     1.3188   -5.21 1.9e-07 ***
age               0.0372     0.0128    2.91 0.00360 **
cancerYes         2.0971     0.8385    2.50 0.01238 *
admitEmergency    3.1022     0.9186    3.38 0.00073 ***
unconsYes         3.7055     0.8765    4.23 2.4e-05 ***
---
Signif. codes:  0 '***' 0.001 '**' 0.01 '*' 0.05 '.' 0.1 ' ' 1
```

These two models are nested, so we can compare them directly using a likelihood ratio test from `anova()`.

```
> anova(icu.step2, icu.step1, test = "Chisq")

Analysis of Deviance Table

Model 1: died ~ age + cancer + admit + uncons
Model 2: died ~ age + cancer + systolic + admit + ph + pco + uncons
  Resid. Df Resid. Dev Df Deviance Pr(>Chi)
1       195        139
2       192        128  3     10.7    0.013 *
---
Signif. codes:  0 '***' 0.001 '**' 0.01 '*' 0.05 '.' 0.1 ' ' 1
```

The larger model is significantly better by this test, but the smaller model is simpler to interpret. We retain these both as "candidate models" to be explored further, but for ease in this example, we do so using the smaller model, `icu.step2`.

Another important step is to check for nonlinearity of quantitative predictors such as `age` and interactions among the predictors. This is easy to do using `update()` and `anova()` as shown below. First, allow a nonlinear term in `age`, and all two-way interactions of the binary predictors.

```
> icu.glm3 <- update(icu.step2, . ~ . - age + ns(age, 3) +
+                   (cancer + admit + uncons) ^ 2)
> anova(icu.step2, icu.glm3, test = "Chisq")

Analysis of Deviance Table

Model 1: died ~ age + cancer + admit + uncons
Model 2: died ~ cancer + admit + uncons + ns(age, 3) + cancer:admit +
    cancer:uncons + admit:uncons
  Resid. Df Resid. Dev Df Deviance Pr(>Chi)
1       195        139
2       191        135  4     3.73     0.44
```

Next, we can check for interactions with `age`:

```
> icu.glm4 <- update(icu.step2, . ~ . + age * (cancer + admit + uncons))
> anova(icu.step2, icu.glm4, test = "Chisq")

Analysis of Deviance Table

Model 1: died ~ age + cancer + admit + uncons
Model 2: died ~ age + cancer + admit + uncons + age:cancer + age:admit +
    age:uncons
  Resid. Df Resid. Dev Df Deviance Pr(>Chi)
1       195        139
2       192        134  3     5.37     0.15
```

None of these additional terms have much effect. △

So, we will tentatively adopt the simple main effects model, `icu.step2`, and consider how to visualize and interpret this result.

EXAMPLE 7.12: Death in the ICU — Visualization

One interesting display is a ***nomogram*** that shows how values on the various predictors translate into a predicted value of the log odds, and the relative strengths of their effects on this prediction. This kind of plot is shown in Figure 7.22, and is produced using `nomogram()` in the **rms** (Harrell, Jr., 2015) package as follows. This only works with models fit using `lrm()`, so we have to refit this model.

```
> icu.lrm2 <- lrm(died ~ age + cancer  + admit + uncons, data = ICU)
> plot(nomogram(icu.lrm2), cex.var = 1.2, lplabel = "Log odds death")
```

In this nomogram, each predictor is scaled according to the size of its effect on a common scale of 0–100 "points." A representative observation is shown by the marked points, corresponding to a person of age 60, without cancer, who was admitted to emergency and was unconscious at that time. Adding the points associated with each variable value gives the result shown on the scale of total points. For this observation, the result is $50 + 0 + 84 + 100 = 234$, for which the scale of log odds at the bottom gives a predicted logit of 2.2, or a predicted probability of death of $1/(1 + \exp(-2.2)) = 0.90$.

This leaves us with the problem of how to visualize the fitted model compactly and comprehensively. Multi-panel full-model plots and effect plots, as we have used them, are somewhat unwieldy with four or more predictors if we want to view all effects simultaneously, because it becomes more difficult to make comparisons across multiple panels (particularly if the vertical scales differ).

One way to reduce the visual complexity of such graphs is to combine some predictors that would otherwise be shown in separate panels into a recoding that can be shown as multiple curves for their combinations in fewer panels. In general, this can be done by combining some predictors *interactively*; for example, with sex and education as factors, their combinations, `M:Hi, M:Lo`, etc., could be used to define a new variable, `group`, used as the curves in one plot, rather than separate panels. This, in fact, is precisely what `binreg_plot()` does when there are two or more factors to be shown in a given plot.

In this case, because age is continuous, it makes sense to plot fitted values against age.[18] With `cancer`, `admit`, and `uncons` as binary factors associated with risk of death, it is also convenient for plotting to represent them in a way that reflects the level assiciated with higher risk. We do this by recoding their levels using `"-"` for low risk.

---

[18] By default, `binreg_plot()` uses the first numeric predictor as the horizontal variable.

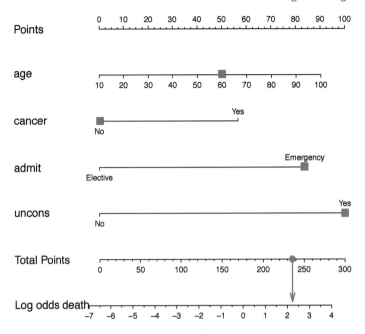

Figure 7.22: Nomogram for predicted values in the simple main effects model for the ICU data. Each predictor is scaled in relation to its effect on the outcome in terms of "points," 0–100. Adding the points for a given case gives total points that have a direct translation to log odds. The marked points show the prediction for someone of age 60, admitted to the emergency ward and unconscious.

```
> levels(ICU$cancer) <- c("-", "Cancer")
> levels(ICU$admit) <- c("-","Emerg")
> levels(ICU$uncons) <- c("-","Uncons")
>
> icu.glm2 <- glm(died ~ age + cancer + admit + uncons,
+                 data = ICU, family = binomial)
```

Then, `binreg_plot()` is called as follows, giving the plot shown in Figure 7.23. Such multiline graphs are more easily read with direct labels on the lines rather than a legend, so the legend is suppressed, and the lines are labeled using `labels = TRUE`. Points along the fitted lines are shown when `point_size>0`.

```
> binreg_plot(icu.glm2, type = "link", conf_level = 0.68,
+             legend = FALSE,
+             labels = TRUE, labels_just = c("right", "bottom"),
+             cex = 0, point_size = 0.8, pch = 15:17,
+             ylab = "Log odds (died)",
+             ylim = c(-7, 4))
```

From Figure 7.23, it is apparent that the log odds of mortality increases with age in all cases. Relative to the line labeled `"-:-:-"` (no risk factors), mortality is higher when any of these risk factors are present, particularly when the patient is admitted to emergency; it is highest when the patient is also unconscious at admission. The vertical gaps between lines that share a common risk (e.g., `Cancer`, `CancerEmerg`) indicate the additional increment from one more risk.

Finally, the plotted points show the number and age distribution of these various combinations. The greatest number of patients have only `Emerg` as a risk factor and only one patient was unconscious with no other risk.

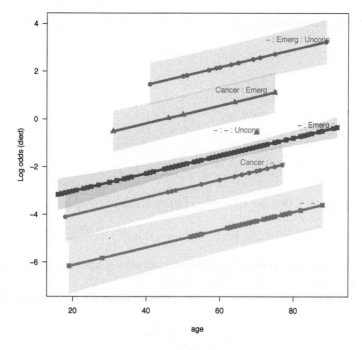

Figure 7.23: Fitted log odds of death in the ICU data for the model `icu.glm2`. Each line shows the relationship with age, for patients having various combinations of risk factors and 1 standard error confidence bands.

Before concluding that this model provides an adequate description of the data, we should examine whether any individual cases are unduly influencing the predicted results, and more importantly, the choice of variables in the model. We examine this question in Section 7.5 where we return to these data (Example 7.14).

△

7.5 Influence and diagnostic plots

In ordinary least squares (OLS) regression, measures of ***influence*** (leverage, Cook's D, DFBETAs, etc.) and associated plots help you to determine whether individual cases (or cells in grouped data) have undue impact on the fitted regression model and the coefficients of individual predictors. Analogs of most of these measures have been suggested for logistic regression and generalized linear models. Pregibon (1981) provided the theoretical basis for these methods, exploiting the relationship between logistic models and weighted least squares. Some additional problems occur in practical applications to logistic regression because the response is discrete, and because the leave-one-out diagnostics are more difficult to compute, but the ideas are essentially the same.

7.5.1 Residuals and leverage

As in ordinary least squares regression, the influence (actual impact) of an observation in logistic models depends multiplicatively on its residual (disagreement between y_i and $\hat{y}_i$) and its leverage (how unusual x_i is in the space of the explanatory variables). A conceptual formula is

$$\text{Influence} = \text{Leverage} \times \text{Residual}$$

This multiplicative definition implies that a case is influential to the extent that it is both poorly fit *and* has unusual values of the predictors.

7.5.1.1 Residuals

In logistic regression, the simple raw residual is just $e_i \equiv y_i - \hat{p}_i$, where $\hat{p}_i = 1/[1 + \exp(-\boldsymbol{x}_i^\mathsf{T}\boldsymbol{b})]$.

The Pearson and deviance residuals are more useful for identifying poorly fitted observations, and are components of overall goodness-of-fit statistics. The (raw) ***Pearson residual*** is defined as

$$r_i \equiv \frac{e_i}{\sqrt{\widehat{p}_i(1 - \widehat{p}_i)}} \tag{7.7}$$

and the Pearson chi-square is therefore $\chi^2 = \sum r_i^2$. The ***deviance residual*** is

$$g_i \equiv \pm -2[y_i \log \widehat{p}_i + (1 - y_i) \log(1 - \widehat{p}_i)]^{1/2} \tag{7.8}$$

where the sign of g_i is the same as that of e_i. Likewise, the sum of squares of the deviance residuals gives the overall deviance, $G^2 = -2 \log \mathcal{L}(\boldsymbol{b}) = \sum g_i^2$.

When y_i is a binomial count based on n_i trials (grouped data), the Pearson residuals Eqn. (7.7) then become

$$r_i \equiv \frac{y_i - n_i \widehat{p}_i}{\sqrt{n_i \widehat{p}_i(1 - \widehat{p}_i)}}$$

with similar modifications made to Eqn. (7.8).

In R, `residuals()` is the generic function for obtaining (raw) residuals from a model fitted with `glm()` (or `lm()`). However, ***standardized residuals***, given by `rstandard()`, and ***studentized residuals***, provided by `rstudent()`, are often more useful because they rescale the residuals to have unit variance. They use, respectively, an overall estimate, $\hat{\sigma}^2$, of error variance, and the leave-one-out estimate, $\hat{\sigma}^2_{(-i)}$, omitting the ith observation; the studentized version is usually to be preferred in model diagnostics because it also accounts for the impact of the observation on residual variance.

7.5.1.2 Leverage

Leverage measures the *potential* impact of an individual case on the results, which is directly proportional to how far an individual case is from the centroid in the space of the predictors. Leverage is defined as the diagonal elements, h_{ii}, of the "Hat" matrix, $\boldsymbol{H}$,

$$\boldsymbol{H} = \boldsymbol{X}^\star(\boldsymbol{X}^{\star\mathsf{T}}\boldsymbol{X}^\star)^{-1}\boldsymbol{X}^{\star\mathsf{T}},$$

where $\boldsymbol{X}^\star = \boldsymbol{V}^{1/2}\boldsymbol{X}$, and $\boldsymbol{V} = \mathrm{diag}\,[\hat{\boldsymbol{p}}(1 - \hat{\boldsymbol{p}})]$.

As in OLS, leverage values are between 0 and 1, and a leverage value, $h_{ii} > \{2 \text{ or } 3\}k/n$, is considered "large;" here, $k = p + 1$ is the number of coefficients including the intercept and n is the number of cases. In OLS, however, the hat values depend only on the Xs, whereas in logistic regression, they also depend on the dependent variable values and the fitted probabilities (through $\boldsymbol{V}$). As a result, an observation may be extremely unusual on the predictors, yet not have a large hat value, if the fitted probability is near 0 or 1. The function `hatvalues()` calculates these values for a fitted "glm" model object.

7.5.2 Influence diagnostics

Influence measures assess the effect that deleting an observation has on the regression parameters, fitted values, or the goodness-of-fit statistics. In OLS, these measures can be computed exactly from a single regression. In logistic regression, the exact effect of deletion requires refitting the model with each observation deleted in turn, a time-intensive computation. Consequently, Pregibon (1981) showed how analogous deletion diagnostics may be approximated by performing one additional step of the iterative procedure. Most modern implementations of these methods for generalized linear models follow Williams (1987).

The simplest measure of influence of observation i is the standardized change in the coefficient for each variable due to omitting that observation, termed **DFBETA**s. From the relation (Pregibon, 1981, p. 716)

$$b - b_{(-i)} = (X^\mathsf{T} V X)^{-1} x_i (y_i - \widehat{p}_i)/(1 - h_{ii}),$$

the estimated standardized change in the coefficient for variable j is

$$\text{DFBETA}ij \equiv \frac{b_{(-i)j} - b_j}{\hat{\sigma}(b_j)}, \tag{7.9}$$

where $\hat{\sigma}(b_j)$ is the estimated standard error of b_j. With k regressors, there are $k + 1$ sets of DF-BETAs, which makes their examination burdensome. Graphical displays ease this burden, as do various summary measures considered below.

The most widely used summary of the overall influence of observation i on the estimated regression coefficients is ***Cook's distance***, which measures the average squared distance between b for all the data and $b_{(-i)}$ estimated without observation i. It is defined as

$$C_i \equiv (b - b_{(-i)})^\mathsf{T} X^\mathsf{T} V X (b - b_{(-i)})/k\hat{\sigma}^2.$$

However, Pregibon (1981) showed that C_i could be calculated simply as

$$C_i = \frac{r_i^2 h_{ii}}{k(1 - h_{ii})^2}, \tag{7.10}$$

where $r_i = y_i - \hat{p}_i / \sqrt{v_{ii}(1 - h_{ii})}$ is the ith standardized Pearson residual and v_{ii} is the ith diagonal element of V. Rules of thumb for noticeably "large" values of Cook's D are only rough indicators, and designed so that only "noteworthy" observations are nominated as unusually influential. One common cutoff for an observation to be treated as influential is $C_i > 1$. Others refer the values of C_i to a χ_k^2 or $F_{k,n-k}$ distribution.

Another commonly used summary statistic of overall influence is the ***DFFITS*** statistic, a standardized measure of the difference between the predicted value $\hat{y}_i$ using all the data and the predicted value $\hat{y}_{(-i)}$ calculated omitting the ith observation.

$$\text{DFFITS}_i = \frac{\hat{y}_i - \hat{y}_{(-i)}}{\hat{\sigma}_{(-i)} \sqrt{h_{ii}}},$$

where $\hat{\sigma}_{(-i)}$ is the estimated standard error with the ith observation deleted. For computation, DFFITS can be expressed in terms of the standardized Pearson residual and leverage as

$$\text{DFFITS}_i = r_i \sqrt{\frac{h_{ii}}{(1 - h_{ii})} \frac{v_{ii}}{v_{(-ii)}}}. \tag{7.11}$$

From Eqn. (7.10) and Eqn. (7.11), it can be shown that Cook's distance is nearly the square of DFFITS divided by k,

$$C_i = \frac{v_{(-ii)}^2}{v_{ii}^2} \frac{\text{DFFITS}_i^2}{k}. \tag{7.12}$$

Noteworthy values of DFFITS are often nominated by the rule-of-thumb $\text{DFFITS}_i > 2$ or $3\sqrt{k/n - k}$.

In R, these influence measures are calculated for a fitted "glm" model using cooks.distance() and dffits(). A convenience function, influence.measures() gives a tabular display showing the DFBETA$_{ij}$ for each model variable, DFFITS, Cook's distances and the diagonal elements of the hat matrix. Cases which are influential with respect to any of these measures are marked with an asterisk.[19]

Beyond printed output of these numerical summaries, plots of these measures can shed light on potential problems due to influential or other noteworthy cases. By highlighting them, such plots provide the opportunity to determine if and how any of these affect your conclusions, or to take some corrective action.

Basic diagnostic plots are provided by the plot() method for a "glm" model object. These are easy to do, but the results for discrete response data are often unsatisfactory. The car package contains a variety of enhanced and extended functions for model diagnostic plots. We illustrate some of these in the examples below.

EXAMPLE 7.13: Donner Party

This example re-visits the data on the Donner Party examined in Example 7.9. For illustrative purposes, we consider the influence measures and diagnostic plots for one specific model, the model donner.mod3, which included a quadratic effect of age and a main effect of sex, but no interaction.

Details of all the diagnostic measures for a given model, including the DFBETAs for individual coefficients, can be obtained using influence.measures. This can be useful for custom plots not provided elsewhere (see Example 7.14).

```
> infl <- influence.measures(donner.mod3)
> names(infl)

[1] "infmat" "is.inf" "call"
```

The summary() method for the "infl" object prints those observations considered noteworthy on one or more of these statistics, as indicated by a "*" next to the value.

```
> summary(infl)

Potentially influential observations of
  glm(formula = survived ~ poly(age, 2) + sex, family = binomial,        data = Donner) :

                       dfb.1_ dfb.p(,2)1 dfb.p(,2)2 dfb.sxM1 dffit   cov.r   cook.d hat
Breen, Patrick          0.08   0.65       0.56       0.23    0.69_*  0.93    0.32   0.09
Donner, Elizabeth      -0.26  -0.34      -0.22       0.12   -0.40    1.15_*  0.03   0.14_*
Graves, Elizabeth C.   -0.24  -0.37      -0.26       0.10   -0.42    1.20_*  0.03   0.16_*
```

The simplest overview of adequacy of a fitted model is provided by the plot() method for a "glm" (or "lm") object, which can produce up to six different diagnostic plots. Among them, we consider the residual-leverage graph (number 5) as being the most useful for assessing influential observations, plotting residuals against leverages. An extended version is produced by the function influencePlot() in the car package, which additionally uses the size (area) of the plotting symbol to also show the value of Cook's D as shown in Figure 7.24. Like other diagnostic plots in car, it is considerably more general than illustrated here, because it allows for different id.methods to label noteworthy points, including id.method = "identify" for interactive point identification by clicking with the mouse. The id.n argument works differently than with plot(), because it selects the most extreme id.n observations on *each* of the studentized residual, hat value, and Cook's D, and labels all of these.

---

[19]See help(influence.measures) for the description of all of these functions for residuals, leverage, and influence diagnostics in generalized linear models.

```
> op <- par(mar = c(5, 4, 1, 1) + .1, cex.lab = 1.2)
> res <- influencePlot(donner.mod3, id.col = "blue", scale = 8, id.n = 2)
> k <- length(coef(donner.mod3))
> n <- nrow(Donner)
> text(x = c(2, 3) * k / n, y = -1.8, c("2k/n", "3k/n"), cex = 1.2)
```

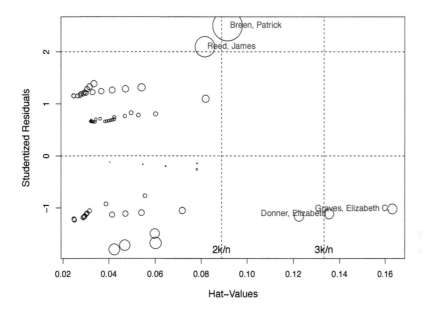

Figure 7.24: Influence plot (residual vs. leverage) for the Donner data model, showing Cook's D as the size of the bubble symbol. Horizontal and vertical reference lines show typical cutoff values for noteworthy residuals and leverage.

Conveniently, `influencePlot()` returns a data frame containing the influence statistics for the points identified in the plot (`res` in the call above). We can combine this with the data values to help learn why these points are considered influential.

```
> # show data together with diagnostics for influential cases
> idx <- which(rownames(Donner) %in% rownames(res))
> cbind(Donner[idx,2:4], res)
```

	age	sex	survived	StudRes	Hat	CookD
Breen, Patrick	51	Male	yes	2.501	0.09148	0.5688
Donner, Elizabeth	45	Female	no	-1.114	0.13541	0.1846
Graves, Elizabeth C.	47	Female	no	-1.019	0.16322	0.1849
Reed, James	46	Male	yes	2.098	0.08162	0.3790

We can see that Patrick Breen and James Reed[20] are unusual because they were both older men who survived, and have large positive residuals; Breen is the most influential by Cook's D, but this value is not excessively large. The two women were among the older women who died. They are selected here because they have the largest hat values, meaning they are unusual in terms of the distribution of age and sex, but they are not particularly influential in terms of Cook's D.

A related graphical display is the collection of index plots provided by `influenceIndexPlot()` in car, which plots various influence diagnostics against the observation numbers in the data. The

---

[20]Breen and Reed, both born in Ireland, were the leaders of their family groups. Among others, both kept detailed diaries of their experiences, from which most of the historical record derives. Reed was also the leader of two relief parties sent out to find rescue or supplies over the high Sierra mountains, so it is all the more remarkable that he survived.

id.n argument here works to label that number of the most extreme observations *individually* for each measure plotted. The following call produces Figure 7.25.

```
> influenceIndexPlot(donner.mod3, vars=c("Cook", "Studentized", "hat"),
+                    id.n=4)
```

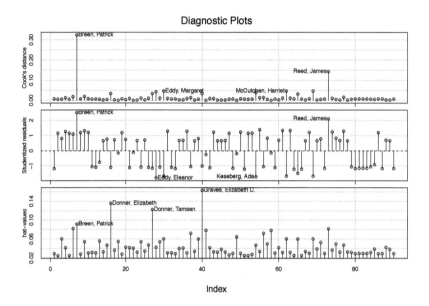

Figure 7.25: Index plots of influence measures for the Donner data model. The four most extreme observations on each measure are labeled.

In our opinion, *separate* index plots are often less useful than combined plots such as the leverage-influence plot that shows residuals, leverage and Cook's D together. However, the car version in Figure 7.25 does that too, and allows us to consider how unusual the labeled observations are both individually and in combination.

$\triangle$

EXAMPLE 7.14: Death in the ICU

In Example 7.11 we examined several models to account for death in the *ICU* data set. We continue this analysis here, with a focus on the simple main effects model, icu.glm2, for which the fitted logits were shown in Figure 7.23. For ease of reference, we restate that model here:

```
> icu.glm2 <- glm(died ~ age + cancer  + admit + uncons,
+                 data = ICU, family = binomial)
```

The plot of residual vs. leverage for this model is shown in Figure 7.26.

```
> library(car)
> res <- influencePlot(icu.glm2, id.col = "red",
+                      scale = 8, id.cex = 1.5, id.n = 3)
```

Details for the cases identified in the figure are shown below, again using rownames(res) to select the relevant observations from the *ICU* data.

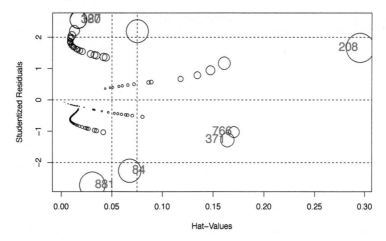

Figure 7.26: Influence plot for the main effects model for the ICU data.

```
> idx <- which(rownames(ICU) %in% rownames(res))
> cbind(ICU[idx, c("died", "age", "cancer", "admit", "uncons")], res)

    died age cancer admit  uncons StudRes     Hat  CookD
84    No  59      - Emerg Uncons  -2.258 0.06781 0.3626
371   No  46 Cancer Emerg      -  -1.277 0.16408 0.2210
766   No  31 Cancer Emerg      -  -1.028 0.17062 0.1719
881   No  89      - Emerg Uncons  -2.718 0.03081 0.4106
127  Yes  19      - Emerg      -   2.565 0.01679 0.2724
208  Yes  70      -     - Uncons   1.662 0.29537 0.4568
380  Yes  20      - Emerg      -   2.548 0.01672 0.2668
```

None of the cases are particularly influential on the model coefficients overall: the largest Cook's D is only 0.45 for case 208. This observation also has the largest hat value. It is unusual on the predictors in this sample: a 70-year-old man without cancer, admitted on an elective basis, who nonetheless died. However, this case is also highly unusual in his having been unconscious on admission for an elective procedure, and signals that there might have been a coding error or other anomaly for this observation.

Another noteworthy observation identified here is case 881, an 89-year-old male, admitted unconscious as an emergency; this case is poorly predicted because he survived. Similarly, two other cases (127, 380) with large studentized residuals are poorly predicted because they died, although they were young, did not have cancer, and were conscious at admission. However, these cases have relatively small Cook's D values. From this evidence we might conclude that, case 208 bears further scrutiny, but none of these cases greatly affects the model, its coefficients, or interpretation.

For comparison with Figure 7.26, the related index plot of these measures is shown in Figure 7.27.

```
> influenceIndexPlot(icu.glm2, vars = c("Cook", "Studentized", "hat"),
+                     id.n = 4)
```

Cook's D and DFFITS are *overall* measures of the total influence that cases have on the regression coefficients and fitted values, respectively. It might be that some cases have a large impact on some individual regression coefficients, but don't appear particularly unusual in these aggregate measures.

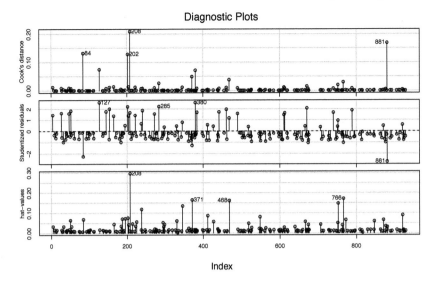

Figure 7.27: Index plots of influence measures for the ICU data model. The four most extreme observations on each measure are labeled.

One way to study this is to make plots of the DFBETA$_{ij}$ statistics. Such plots are not available (as far as we know) in R packages, but it is not hard to construct them from the result returned by `influence.measures()`. To do this, we select the appropriate columns from the `infmat` component returned by that function.

```
> infl <- influence.measures(icu.glm2)
> dfbetas <- data.frame(infl$infmat[,2:5])
> colnames(dfbetas) <- c("dfb.age", "dfb.cancer", "dfb.admit",
+                        "dfb.uncons")
> head(dfbetas)

      dfb.age dfb.cancer dfb.admit dfb.uncons
8    0.047340   0.013418  0.004067   0.009254
12   0.018988   0.018412 -0.004174   0.018106
14  -0.001051   0.014882  0.026278   0.005555
28   0.031562   0.018424 -0.001511   0.016640
32  -0.164084   0.003788 -0.036505   0.023488
38  -0.021525   0.016539 -0.011937   0.020803
```

To illustrate this idea, plotting an individual column of `dfbetas` using `type = "h"` gives an index plot against the observation number. This is shown in Figure 7.28 for the impact on the coefficient for age. The lines and points are colored blue or red according to whether the patient lived or died. Observations for which the |DFBETA$_{\text{age}}$| > 0.2 (an arbitrary value) are labeled.

```
> op <- par(mar = c(5, 5, 1, 1) + .1)
> cols <- ifelse(ICU$died == "Yes", "red", "blue")
> plot(dfbetas[,1], type = "h", col = cols,
+      xlab = "Observation index",
+      ylab = expression(Delta * beta[Age]),
+      cex.lab = 1.3)
> points(dfbetas[,1], col = cols)
> # label some points
> big <- abs(dfbetas[,1]) > .25
```

```
> idx <- 1 : nrow(dfbetas)
> text(idx[big], dfbetas[big, 1], label = rownames(dfbetas)[big],
+        cex = 0.9, pos = ifelse(dfbetas[big, 1] > 0, 3, 1),
+        xpd = TRUE)
> abline(h = c(-.25, 0, .25), col = "gray")
> par(op)
```

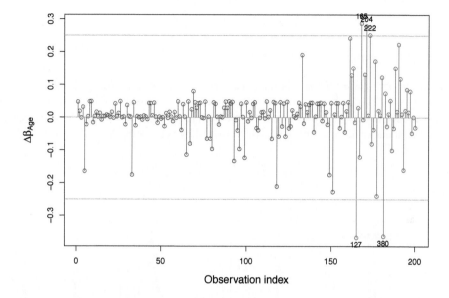

Figure 7.28: Index plot for DFBETA (Age) in the ICU data model. The observations are colored blue or red according to whether the patient lived or died.

None of the labeled points here are a cause for concern, since the standardized DFBETAs are all relatively small. However, the plot shows that patients who died have generally larger impacts on this coefficient.

An interesting alternative to individual index plots is a scatterplot matrix (Figure 7.29) that shows the pairwise changes in the regression coefficients for the various predictors. Here we use scatterplotMatrix() from car, which offers features for additional plot annotations, including identifying the most unusual points in each pairwise plot. In each off-diagonal panel, a 95% data ellipse and linear regression line help to show the marginal relationship between the two measures and highlight why the labeled points are atypical in each plot.[21]

```
> scatterplotMatrix(dfbetas, smooth = FALSE, id.n = 2,
+    ellipse = TRUE, levels = 0.95, robust = FALSE,
+    diagonal = "histogram",
+    groups = ICU$died, col = c("blue", "red"))
```

As Figure 7.29 illustrates, the *joint* effect of observations on *pairs* of coefficients is more complex than is apparent from the univariate views that appear in the plots along the diagonal. The DFBETAs for cancer, admit, and uncons are all extremely peaked, yet the pairwise plots show considerable structure. The points identified would be worthy of further study.

$\triangle$

---

[21]This plot uses the id.method = "mahal" method to label the most extreme observations according to the Mahalanobis distance of each point from the centroid in the plot.

7.5.3   Other diagnostic plots*

The graphical methods described in this section are relatively straightforward indicators of the adequacy of a particular model, with a specified set of predictors, each expressed in a given way. More sophisticated methods have also been proposed, which focus on the need to include a particular predictor and whether its relationship is linear. These include the *component-plus-residual plot*, the *added-variable plot*, and the *constructed variable plot*, which are all analogous to techniques developed in OLS.

7.5.3.1   Component-plus-residual plots

The *component-plus-residual plot* (also called a *partial residual plot*) proposed originally by Larsen and McCleary (1972) is designed to show whether a given quantitative predictor, x_j, included linearly in the model, actually shows a nonlinear relation, requiring transformation. The essential idea is to move the linear term for x_j back into the residual, by calculating the *partial residuals*,

$$r_j^\star = r + \beta_j x_j \ .$$

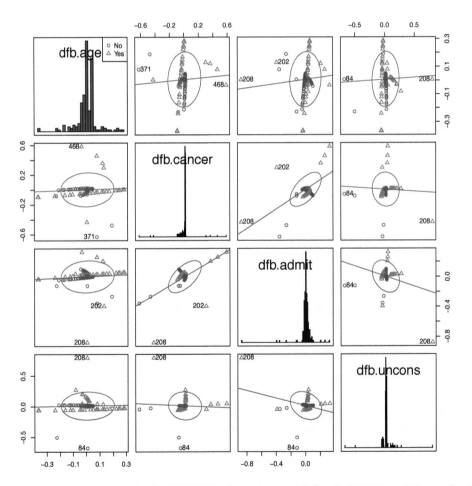

Figure 7.29: Scatterplot matrix for DFBETAs from the model for the ICU data. Those who lived or died are shown with blue circles and red triangles, respectively. The diagonal panels show histograms of each variable.

Then, a plot of $r_j^\star$ against x_j will have the same slope, β_j, as the full model including it among other predictors. However, any nonlinear trend will be shown in the pattern of the points, usually aided by a smoothed non-parametric curve.

As adapted to logistic regression by Landwehr et al. (1984), the partial residual for variable x_j is defined as

$$r_j^\star = V^{-1}r + \beta_j x_j \,.$$

The partial residual plot is then a plot of $r_j^\star$ against x_j, possibly with the addition of a smoothed lowess curve (Fowlkes, 1987) and a linear regression line to aid interpretation. The linear regression of the partial residuals on x_j has the same slope, β_j, as in the full model.

If x_j affects the binary response linearly, the plot should be approximately linear with a slope approximately equal to β_j. A nonlinear plot suggests that x_j needs to be transformed, and the shape of the relation gives a rough guide to the required transformation. For example, a parabolic shape would suggest a term in x_j^2. These plots complement the conditional data plots described earlier (Section 7.3.1), and are most useful when there are several quantitative predictors, so that it is more convenient and sensible to examine their relationships individually.

The car package implements these plots in the crPlots() and crPlot() functions. They also work for models with factor predictors (using parallel boxplots for the factor levels) but not for those with interaction terms.

EXAMPLE 7.15: Donner Party
In Example 7.13, we fit several models for the Donner Party data, and we recall two here to illustrate component-plus-residual plots. Both assert additive effects of age and sex, but the model donner.mod3 allows a quadratic effect of age.

```
> donner.mod1 <- glm(survived ~ age + sex,
+                    data = Donner, family = binomial)
> donner.mod3 <- glm(survived ~ poly(age, 2) + sex,
+                    data = Donner, family = binomial)
```

Had we not made exploratory plots earlier (Example 7.13), and naively fit only the linear model in age, donner.mod1, we could use crPlots() to check for a nonlinear relationship of survival with age as follows, giving Figure 7.30.

```
> crPlots(donner.mod1, ~age, id.n=2)
```

The smoothed loess curve in this plot closely resembles the trend we saw in the conditional plot for age by sex (Figure 7.16), suggesting the need to include a nonlinear term for age. The points identified in this plot, by default, are those with either the most extreme x values (giving them high leverage) or the largest absolute Pearson residuals in the full model. The four structured bands of points in the plot correspond to the combinations of sex and survival.

For comparison, you can see the result of allowing for a nonlinear relationship in age in a partial residual plot for the model donner.mod.3 that includes the effect poly(age, 2) for age. Note that the syntax of the crPlots() function requires that you specify a *term* in the model, rather than just a predictor variable.

```
> crPlots(donner.mod3, ~poly(age,2), id.n=2)
```

Except possibly at the extreme right, this plot (Figure 7.31) shows no indication of a (further) nonlinear relationship.

△

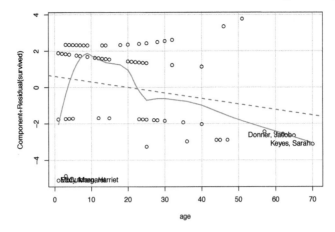

Figure 7.30: Component-plus-residual plot for the simple additive linear model, `donner.mod1`. The dashed red line shows the slope of age in the full model; the smoothed green curve shows a loess fit with span = 0.5.

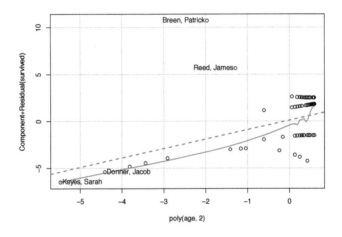

Figure 7.31: Component-plus-residual plot for the nonlinear additive model, `donner.mod3`.

7.5.3.2 Added-variable plots

Added-variable plots (Cook and Weisberg, 1999, Wang, 1985) (also called ***partial-regression plot***s) are another important tool for diagnosing problems in logistic regression and other linear or generalized linear models. These are essentially plots, for each x_i, of an adjusted response, $y_i^\star = y \mid \text{others}_i$, against an adjusted predictor, $x_i^\star = x_i \mid \text{others}_i$, where $\text{others}_i = X \notin x_i \equiv X^{(-i)}$ indicates all other predictors excluding x_i. As such, they show the *conditional* relationship between the response and the predictor x_i, controlling for, or adjusting for, all other predictors. Here, $y_i^\star$ and $x_i^\star$ represent, respectively, the residuals from the regressions of y, and of x_i, on all the other xs excluding x_i.

It might seem from this description that each added-variable plot requires two additional auxiliary logistic regressions to calculate the residuals $y_i^\star$ and $x_i^\star$. However, Wang (1985) showed that the added-variable plot may be constructed by following the logistic regression for the model

$y \sim X^{(-i)}$ with one weighted least-squares regression of x_i on $X^{(-i)}$ to find the residual part, $x_i^{\star}$, of x not predicted by the other regressors.

Let r be the vector of Pearson residuals from the initial logistic fit of y on the variables in $X^{(-i)}$, and let H and $V = \text{diag}\,[\hat{p}(1 - \hat{p})]$ be the hat matrix and V matrix from this analysis. Then, the added-variable plot is a scatterplot of the residuals r against the x_i-residuals,

$$x_i^{\star} = (I - H)V^{1/2}x \,.$$

There are several important uses of added-variable plots:

First, *marginal* plots of the response variable y against the predictor variables x_i can conceal or misrepresent the relationships in a model including several predictors together due to correlations or associations among the predictors. This problem is compounded by the fact that graphical methods for discrete responses (boxplots, mosaic plots) cannot easily show influential observations or nonlinear relationships. Added-variable plots solve this problem by plotting the residuals, $y_i^{\star} = y \,|\, \text{others}_i$, which are less discrete than the marginal responses in y.

Second, the numerical measures and graphical methods for detecting influential observations described earlier in this section are based on the idea of *single-case deletion*, comparing coefficients or fitted values for the full data with those that result from deleting each case in turn. Yet it is well-known (Lawrance, 1995) that sets of two (or more) observations can have **joint influence**, which greatly exceeds their individual influential. Similarly, the influence of one discrepant point can be offset by another influential point in an opposite direction, a phenomenon called **masking**. The main cases of joint influence are illustrated in Figure 7.32. Added-variable plots, showing the partial regression for one predictor controlling all others, can make such cases visually apparent.

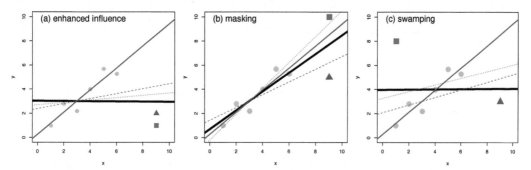

Figure 7.32: Jointly influential points in regression models. In each panel, the thick black line shows the regression of y on x using all the data points. The solid purple line shows the regression deleting *both* the red and blue points and the broken and dotted lines show the regression retaining only the point in its color in addition to the constant gray points. (a) Two points whose joint influence enhance each other; (b) two points where the influence of one is masked by that of the other; (c) two points whose combined influence greatly exceeds the effect of either one individually.

Finally, given a tentative model using predictors x, the added-variable plot for another regressor, z, can provide a useful visual assessment of its additional contribution. An overall test could be based on the difference in G^2 for the enlarged model $\text{logit}(p) = X\beta + \gamma z$, compared to the reduced model $\text{logit}(p) = X\beta$. But the added-variable plot shows whether the evidence for including z is spread throughout the sample or confined to a small subset of observations. The regressor z may be a new explanatory variable, or a higher-order term for variable(s) already in the model.

The `car` package implements these plots with the function `avPlot()` for a single term and `avPlots()` for all terms in a linear or generalized linear model, as shown in the example(s) below. See `http://www.datavis.ca/gallery/animation/duncanAV/` for an animated graphic showing the transition between a marginal plot of the relationship of y to x and the added-variable plot of $y^\star$ to $x^\star$ for the case of multiple linear regression with a quantitative response.

EXAMPLE 7.16: Donner Party

The simple additive model `donner.mod1` for the Donner Party data can be used to illustrate some features of added-variable plots. In the call to `avPlots()` below, we use color for the plotting symbol to distinguish those who survived vs. died, shape to distinguish men from women.

```
> col <- ifelse(Donner$survived == "yes", "blue", "red")
> pch <- ifelse(Donner$sex == "Male", 16, 17)
> avPlots(donner.mod1, id.n = 2,
+         col = col, pch = pch, col.lines = "darkgreen")
```

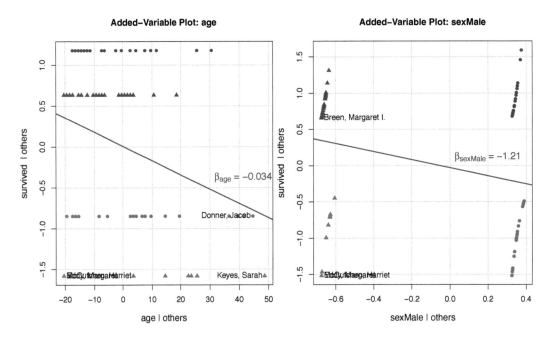

Figure 7.33: Added-variable plots for age (left) and sex (right) in the Donner Party main effects model. Those who survived are shown in blue; those who died in red. Men are plotted with filled circles; women with filled triangles.

These plots have the following properties:

1. The slope in the simple regression of $y_i^\star$ on $x_i^\star$ is the same as the partial coefficient β_i in the full multiple regression model including both predictors here (or all predictors in general).

2. The residuals from this regression line are the same as the residuals in the full model.

3. Because the response, `survived`, is binary, the vertical axis $y_{\text{age}}^\star$ in the left panel for `age` is the part of the logit for survival that cannot be predicted from `sex`. Similarly, the vertical axis in the right panel is the part of survival that cannot be predicted from `age`. This property allows the clusters of points corresponding to discrete variables to be seen more readily, particularly if they are distinguished by visual attributes such as color and shape, as in Figure 7.33.

△

EXAMPLE 7.17: Death in the ICU

We illustrate some of the uses of added-variable plots using the main effects model, `icu.glm2`, predicting death in the ICU from the variables `age`, `cancer`, `admit`, and `uncons`.

To see why marginal plots of the discrete response against each predictor are often unrevealing for the purpose of model assessment, consider the collection of plots in Figure 7.34 showing the default plots (spineplots) for the factor response, `died`, against each predictor. These show the marginal distribution of each predictor by the widths of the bars, and highlight the proportion who died by color. Such plots are useful for some purposes, but not for assessing the adequacy of the fitted model.

```
> op <- par(mfrow = c(2, 2), mar = c(4, 4, 1, 2.5) + .1, cex.lab = 1.4)
> plot(died ~ age, data = ICU, col = c("lightblue", "pink"))
> plot(died ~ cancer, data = ICU, col = c("lightblue", "pink"))
> plot(died ~ admit, data = ICU, col = c("lightblue", "pink"))
> plot(died ~ uncons, data = ICU, col = c("lightblue", "pink"))
```

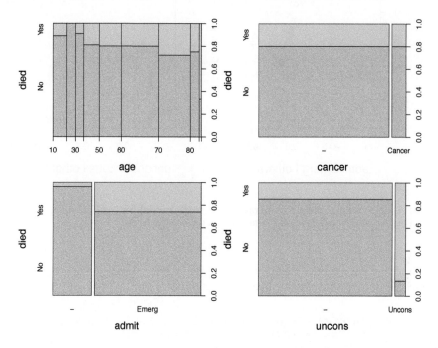

Figure 7.34: Marginal plots of the response `died` against each of the predictors in the model `icu.glm2` for the *ICU* data.

The added-variable plot for this model is shown in Figure 7.35. In each plot, the solid red line shows the partial slope, β_j for the focal predictor, controlling for all others.

```
> pch <- ifelse(ICU$died=="No", 1, 2)
> avPlots(icu.glm2, id.n=2, pch=pch, cex.lab=1.3)
```

The labeled points in each panel use the default `id.method` for `avPlots()`, selecting those with either large absolute model residuals or extreme $x_i^\star$ residuals, given all other predictors. Cases 127 and 881, identified earlier as influential, stand out in all these plots.

Next, we illustrate the use of added-variable plots for checking the effect of influential observations on the decision to include an additional predictor in some given model. In the analysis of the *ICU* data using model selection methods, the variable `systolic` (systolic blood pressure at admission) was nominated by several different procedures. Here we take a closer look at the evidence

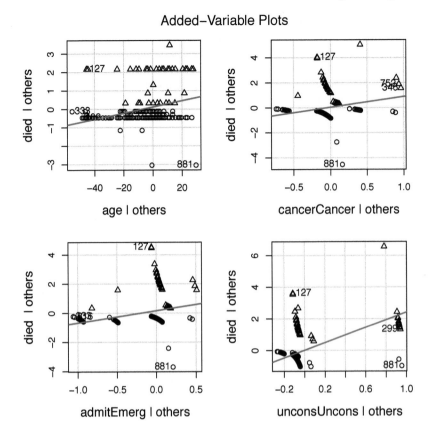

Figure 7.35: Added-variable plots for the predictors in the model for the ICU data. Those who died and survived are shown by triangles ($\triangle$) and circles ($\bigcirc$), respectively.

for inclusion of this variable in a predictive model. We fit a new model adding `systolic` to the others and test the improvement with a likelihood ratio test:

```
> icu.glm2a <- glm(died ~ age + cancer  + admit + uncons + systolic,
+                  data = ICU, family = binomial)
> anova(icu.glm2, icu.glm2a, test = "Chisq")

Analysis of Deviance Table

Model 1: died ~ age + cancer + admit + uncons
Model 2: died ~ age + cancer + admit + uncons + systolic
  Resid. Df Resid. Dev Df Deviance Pr(>Chi)
1       195        139
2       194        136  1     3.52    0.061 .
---
Signif. codes:  0 '***' 0.001 '**' 0.01 '*' 0.05 '.' 0.1 ' ' 1
```

So, the addition of systolic blood pressure is nearly significant at the conventional $\alpha = 0.05$ level. The added-variable plot for this variable in Figure 7.36 shows the strength of evidence for its contribution, above and beyond the other variables in the model, as well as the partial leverage and influence of individual points.

```
> avPlot(icu.glm2a, "systolic", id.n = 3, pch = pch)
```

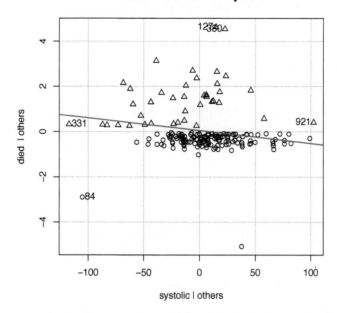

Figure 7.36: Added-variable plot for the effect of adding systolic blood pressure to the main effects model for the ICU data.

In this plot, cases 331 and 921 have high partial leverage, but they are not influential. Case 84, however, has high leverage and a large residual, so it is possibly influential on the evidence for inclusion of systolic in the model. Note also that the partial regression line in this plot nicely separates nearly all the patients who died from those who survived.

△

7.6 Chapter summary

• Model-based methods for categorical data provide confidence intervals for parameters and predicted values for observed and unobserved values of the explanatory variables. Graphical displays of predicted values help us to interpret the fitted relations by smoothing a discrete response.

• The logistic regression model (Section 7.2) describes the relationship between a categorical response variable, usually dichotomous, and a set of one or more quantitative or discrete explanatory variables (Section 7.3) It is conceptually convenient to specify this model as a linear model predicting the log odds (or logit) of the probability of a success from the explanatory variables.

• The relationship between a discrete response and a quantitative predictor may be explored graphically by plotting the binary observations against the predictor with some smoothed curve(s), either parametric or non-parametric, possibly stratified by other predictors.

• For both quantitative and discrete predictors, the results of a logistic regression are most easily interpreted from full-model plots of the fitted values against the predictors, either on the scale of predicted probabilities or log odds (Section 7.3.2). In these plots, confidence intervals provide a visual indication of the precision of the predicted results.

- When there are multiple predictors and/or higher-order interaction terms, effect plots (Section 7.3.3) provide an important method for constructing simplified displays, focusing on the higher-order terms in a given model.

- Influence diagnostics (Section 7.5) assess the impact of individual cases or groups on the fitted model, predicted values, and the coefficients of individual predictors. Among other displays, plots of residuals against leverage showing Cook's D are often most useful.

- Other diagnostic plots (Section 7.5.3) include component-plus-residual plots, which are useful for detecting non-linear relationships for a quantitative predictor, and added-variable plots, which show the partial relations of the response to a given predictor, controlling or adjusting for all other predictors.

7.7   Lab exercises

Exercise 7.1 Arbuthnot's data on the sex ratio of births in London was examined in Example 3.1. Use a binomial logistic regression model to assess whether the proportion of male births varied with the variables `Year`, `Plague`, and `Mortality` in the `Arbuthnot` data set. Produce effect plots for the terms in this model. What do you conclude?

Exercise 7.2 For the Donner Party data in `Donner`, examine Grayson's 1990 claim that survival in the Donner Party was also mediated by the size of the family unit. This takes some care, because the `family` variable in the `Donner` data is a simplified grouping based on the person's name and known alliances among families from the historical record. Use the following code to compute a `family.size variable` from each individual's last name:

```
> data("Donner", package="vcdExtra")
> Donner$survived <- factor(Donner$survived, labels=c("no", "yes"))
> # use last name for family
> lame <- strsplit(rownames(Donner), ",")
> lame <- sapply(lame, function(x) x[[1]])
> Donner$family.size <- as.vector(table(lname)[lname])
```

(a) Choose one of the models (`donner.mod4`, `donner.mod6`) from Example 7.9 that include the interaction of age and sex and nonlinear terms in age. Fit a new model that adds a main effect of `family.size`. What do you conclude about Grayson's claim?

(b) Produce an effect plot for this model.

(c) Continue, by examining whether the effect of family size can be taken as linear, or whether a nonlinear term should be added.

Exercise 7.3 Use component+residual plots (Section 7.5.3) to examine the additive model for the `ICU` data given by

```
> icu.glm2 <- glm(died ~ age + cancer  + admit + uncons,
+                 data=ICU, family=binomial)
```

(a) What do you conclude about the linearity of the (partial) relationship between age and death in this model?

(b) An alternative strategy is to allow some nonlinear relation for age in the model using a quadratic (or cubic) term like `poly(age, 2)` (or `poly(age, 3)`) in the model formula. Do these models provide evidence for a nonlinear effect of age on death in the ICU?

Exercise 7.4 Explore the use of other marginal and conditional plots to display the relationships among the variables predicting death in the ICU in the model `icu.glm2`. For example, you might begin with a marginal `gpairs()` plot showing all bivariate marginal relations, something like this:

```
> library(gpairs)
> gpairs(ICU[,c("died", "age", "cancer", "admit", "uncons")],
+    diag.pars=list(fontsize=16, hist.color="lightgray"),
+    mosaic.pars=list(gp=shading_Friendly,
+                       gp_args=list(interpolate=1:4)))
```

Exercise 7.5 The data set *Caesar* in **vcdExtra** gives a 3×2^3 frequency table classifying 251 women who gave birth by Caesarian section by `Infection` (three levels: none, Type 1, Type2) and `Risk`, whether `Antibiotics` were used, and whether the Caesarian section was `Planned` or not. `Infection` is a natural response variable. In this exercise, consider only the binary outcome of infection vs. no infection.

```
> data("Caesar", package="vcdExtra")
> Caesar.df <- as.data.frame(Caesar)
> Caesar.df$Infect <- as.numeric(Caesar.df$Infection %in%
+                          c("Type 1", "Type 2"))
```

(a) Fit the main-effects logit model for the binary response `Infect`. Note that with the data in the form of a frequency data frame you will need to use `weights=Freq` in the call to `glm()`. (It might also be convenient to reorder the levels of the factors so that `"No"` is the baseline level for each.)

(b) Use `summary()` or `car::Anova()` to test the terms in this model.

(c) Interpret the coefficients in the fitted model in terms of their effect on the odds of infection.

(d) Make one or more effects plots for this model, showing separate terms, or their combinations.

Exercise 7.6 The data set *birthwt* in the MASS package gives data on 189 babies born at Baystate Medical Center, Springfield, MA during 1986. The quantitative response is `bwt` (birth weight in grams), and this is also recorded as `low`, a binary variable corresponding to `bwt` < 2500 (2.5 Kg). The goal is to study how this varies with the available predictor variables. The variables are all recorded as numeric, so in R it may be helpful to convert some of these into factors and possibly collapse some low frequency categories. The code below is just an example of how you might do this for some variables.

```
> data("birthwt", package="MASS")
> birthwt <- within(birthwt, {
+    race <- factor(race, labels = c("white", "black", "other"))
+        ptd <- factor(ptl > 0)    # premature labors
+        ftv <- factor(ftv)        # physician visits
+        levels(ftv)[-(1:2)] <- "2+"
+        smoke <- factor(smoke>0)
+        ht <- factor(ht>0)
+        ui <- factor(ui>0)
+    })
```

(a) Make some exploratory plots showing how low birth weight varies with each of the available predictors. In some cases, it will probably be helpful to add some sort of smoothed summary curves or lines.

(b) Fit several logistic regression models predicting low birth weight from these predictors, with the goal of explaining this phenomenon adequately, yet simply.
(c) Use some graphical displays to convey your findings.

Exercise 7.7 Refer to Exercise 5.9 for a description of the `Accident` data. The interest here is to model the probability that an accident resulted in death rather than injury from the predictors `age`, `mode`, and `gender`. With `glm()`, and the data in the form of a frequency table, you can use the argument `weight=Freq` to take cell frequency into account.

(a) Fit the main effects model, `result=="Died" ~ age + mode + gender`. Use `car::Anova()` to assess the model terms.
(b) Fit the model that allows all two-way interactions. Use `anova()` to test whether this model is significantly better than the main effects model.
(c) Fit the model that also allows the three-way interaction of all factors. Does this offer any improvement over the two-way model?
(d) Interpret the results of the analysis using effect plots for the two-way model, separately for each of the model terms. Describe verbally the nature of the `age*gender` effect. Which mode of transportation leads to greatest risk of death?

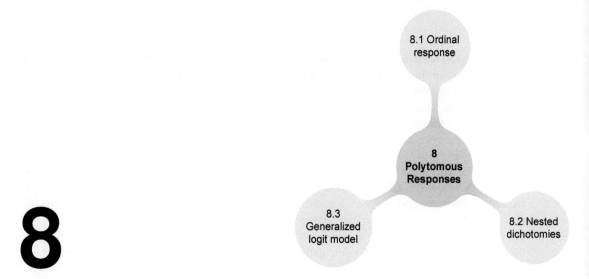

8

Models for Polytomous Responses

This chapter generalizes logistic regression models for a binary response to handle a multi-category (polytomous) response. Different models are available depending on whether the response categories are nominal or ordinal. Visualization methods for such models are mostly straightforward extensions of those used for binary responses.

Ballerinas are often divided into three categories: jumpers, turners and balancers

---

Robert Gottlieb

Polytomous response data arise when the outcome variable, Y, takes on $m > 2$ discrete values. For example, (a) patients may record that their improvement after treatment is "none," "some" or "marked;" (b) high school students may choose a general, vocational, or academic program; (c) women's labor force participation may be recorded in a survey as not working outside the home, working part-time, or working full-time; (d) Canadian voters may express a preference for the Conservative, Liberal, NDP, or Green party. These response categories may be considered *ordered*, as in case (a), or simply *nominal*, as in case (d), and sometimes the response can arguably be treated in either way, as in cases (b) and (c).

In this situation, there are several different ways to model the response probabilities. Let $\pi_{ij} \equiv \pi_j(\boldsymbol{x}_i)$ be the probability of response j for case or group i, given the predictors $\boldsymbol{x}_i$. Because $\sum_j \pi_{ij} = 1$, only $m - 1$ of these probabilities are independent. The essential idea here is to construct a model for the polytomous (or multinomial) response composed of $m - 1$ logit comparisons among the response categories in a similar way to how factors are treated in the predictor variables.

The simplest approach uses the ***proportional odds model***, described in Section 8.1. This model applies *only* when the response is ordinal (as in improvement after therapy) *and* an additional assumption (the proportional odds assumption) holds. This model can be fit using `polr()` in the MASS package, `lrm()` in the rms package, and `vglm()` in VGAM (Yee, 2015).

However, if the response is purely nominal (e.g., vote Conservative, Liberal, NDP, Green), or if the proportional odds assumption is untenable, another particularly simple strategy is to fit separate models to a set of $m - 1$ ***nested dichotomies*** derived from the polytomous response (described in Section 8.2). This method allows you to resolve the differences among the m response categories into *independent* statistical questions (similar to orthogonal contrasts in ANOVA). For example, for women's labor force participation, it might be substantively interesting to contrast not working vs. part-time and full-time and then part-time vs. full-time for women who are working. You fit such nested dichotomies by running the $m - 1$ binary logit models and combining the statistical results.

The most general approach is the ***generalized logit model***, also called the ***multinomial logit model***, described in Section 8.3. This model fits *simultaneously* the $m - 1$ simple logit models against a baseline or reference category, for example, the last category, m. With a 3-category response, there are two generalized logits, $L_{i1} = \log(\pi_{i1}/\pi_{i3})$ and $L_{i2} = \log(\pi_{i2}/\pi_{i3})$, contrasting response categories 1 and 2 against category 3. In this approach, it doesn't matter which response category is chosen as the baseline, because all pairwise comparisons can be recovered from whatever is estimated. This model is conveniently fitted using `multinom()` in nnet.

8.1   Ordinal response: Proportional odds model

For an ordered response Y, with categories $j = 1, 2, \ldots, m$, the ordinal nature of the response can be taken into account by forming logits based on the $m - 1$ adjacent category cutpoints between successive categories. That is, if the cumulative probabilities are

$$\Pr(Y \le j \mid \boldsymbol{x}) = \pi_1(\boldsymbol{x}) + \pi_2(\boldsymbol{x}) + \cdots \pi_j(\boldsymbol{x}) \,,$$

then the ***cumulative logit*** for category j is defined as

$$L_j \equiv \text{logit}[\Pr(Y \le j \mid \boldsymbol{x})] = \log \frac{\Pr(Y \le j \mid \boldsymbol{x})}{\Pr(Y > j \mid \boldsymbol{x})} = \log \frac{\Pr(Y \le j \mid \boldsymbol{x})}{1 - \Pr(Y \le j \mid \boldsymbol{x})} \tag{8.1}$$

for $j = 1, 2, \ldots m - 1$.

In our running example of responses to arthritis treatment, the actual response variable is `Improved`, with ordered levels `"None"` < `"Some"` < `"Marked"`. In this case, the cumulative logits would be defined as

$$L_1 = \log \frac{\pi_1(\boldsymbol{x})}{\pi_2(\boldsymbol{x}) + \pi_3(\boldsymbol{x})} = \text{logit (None vs. [Some or Marked])}$$

$$L_2 = \log \frac{\pi_1(\boldsymbol{x}) + \pi_2(\boldsymbol{x})}{\pi_3(\boldsymbol{x})} = \text{logit ([None or Some] vs. Marked) }\,,$$

where $\boldsymbol{x}$ represents the predictors (sex, treatment and age).

The ***proportional odds model*** (PO) (McCullagh, 1980) proposes a simple and parsimonious account of these effects, where the predictors in (x) are constrained to have the same slopes for all cumulative logits,

$$L_j = \alpha_j + \boldsymbol{x}^\mathsf{T} \boldsymbol{\beta} \qquad j = 1, \ldots, m - 1 \,. \tag{8.2}$$

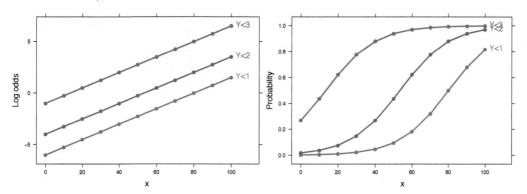

Figure 8.1: Proportional odds model for an ordinal response. The model assumes equal slopes for the cumulative response logits. Left: logit scale; right: probability scale.

That is, the effect of the predictor x_i is the same, β_i, for all cumulative logits. The cumulative logits differ only in their intercepts. In this formulation, the $\{\alpha_j\}$ increase with j, because $\Pr(Y \le j \mid x)$ increases with j for fixed x.[1] Figure 8.1 portrays the PO model for a single quantitative predictor x with $m = 4$ response categories.

The name "proportional odds" stems from the fact that under Eqn. (8.2), for fixed x, the cumulative log odds (logits) for categories j and $j\prime$ are constant and their difference is $(\alpha_j - \alpha_{j\prime})$, so the odds have a constant ratio $\exp(\alpha_j - \alpha_{j\prime}) = \exp(\alpha_j)/\exp(\alpha_{j\prime})$, or are proportional. Similarly, the ratio of the cumulative odds of making a response $Y \le j$ at values of the predictors $x = x_1$ are $\exp((x_1 - x_2)^\mathsf{T}\beta)$ times the odds of this response at $x = x_2$, so the log cumulative odds ratio is proportional to the difference between x_1 and x_2.

8.1.1 Latent variable interpretation

For a binary response, an alternative motivation for logistic regression regards the relation of the observed Y as arising from a continuous, unobserved, (latent) response variable, ξ, representing the propensity for a "success" (1) rather than "failure" (0). The latent response is assumed to be linearly related to the predictors x according to

$$\xi_i = \alpha + x_i^\mathsf{T}\beta + \epsilon_i = \alpha + \beta_1 x_{i1} + \cdots + \beta_p x_{ip} + \epsilon_i. \tag{8.3}$$

However, we can only observe $Y_i = 1$ when ξ_i passes some threshold, that with some convenient scaling can be taken as $\xi_i > 0 \implies Y_i = 1$.[2]

The latent variable motivation extends directly to an ordinal response under the PO model. We now assume that there is a set of $m - 1$ thresholds, $\alpha_1 < \alpha_2 < \cdots < \alpha_{m-1}$ for the latent variable ξ_i in Eqn. (8.3), and we observe

$$Y_i = j \quad \text{if} \quad \alpha_{j-1} < \xi_i \le \alpha_j,$$

---

[1] Some authors and some software describe the PO model in terms of $\mathrm{logit}[\Pr(Y > j \mid x)]$, so the signs and order of the intercepts, α_j, are reversed.

[2] The latent variable derivation of logistic regression (and the related probit model) was fundamental in the history of statistical methods for discrete response outcomes. An early example was Thurstone's (1927) *Law of comparative judgment* designed to account for psychological preference by assuming an underlying latent continuum of "hedonic values." Similarly, the probit model arose from dose-response studies in toxicology (Bliss, 1934, Finney, 1947) where the number killed by some chemical agent was related to its type, dose, or concentration. The idea of a latent variable was also at the heart of the development of factor analysis (Bentler, 1980), and latent class analysis (Lazarsfeld, 1950, 1954) was developed to treat the problem of classifying individuals into discrete latent classes from fallible measurements. See Bollen (2002) for a useful overview of latent variable models in the social sciences.

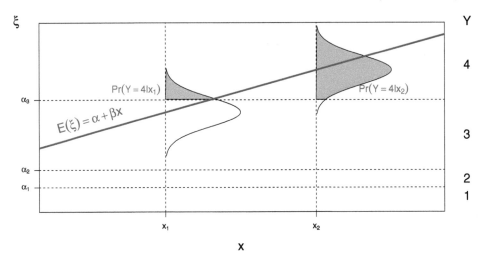

Figure 8.2: Latent variable representation of the proportional odds model for $m = 4$ response categories and a single quantitative predictor, x. *Source*: Adapted from Fox (2008, Fig 14.10), using code provided by John Fox.

with appropriate modifications to the inequalities at the end points.

This is illustrated in Figure 8.2 for a response with $m = 4$ ordered categories and a single quantitative predictor, x. The observable response Y categories are shown on the right vertical axis, and the corresponding latent continuous variable ξ on the left axis together with the thresholds $\alpha_1, \alpha_2, \alpha_3$. The (conditional) logistic distribution of ξ is shown at two values of x, and the shaded areas under the curve give the conditional probabilities $\Pr(Y = 4 \mid x_i)$ for the two values x_1 and x_2.

8.1.2   Fitting the proportional odds model

As mentioned earlier, there are a number of different R packages that provide facilities for fitting the PO model. These have somewhat different capabilities for reporting results, testing hypotheses, and plotting, so we generally use polr() in the MASS package, except where other packages offer greater convenience.

Unless the response variable has numeric values, it is important to ensure that it has been defined as an *ordered* factor (using ordered()). In the *Arthritis* data, the response Improved was set up this way, as we can check by printing some of the values.[3]

```
> data("Arthritis", package = "vcd")
> head(Arthritis$Improved, 8)

[1] Some    None    None    Marked Marked Marked None    Marked
Levels: None < Some < Marked
```

We fit the main effects model for the ordinal response using polr() as shown below. We also specify Hess=TRUE to have the function return the observed information matrix (called the Hessian), that is used in other operations to calculate standard errors.

```
> library(MASS)
> arth.polr <- polr(Improved ~ Sex + Treatment + Age,
+                   data = Arthritis, Hess = TRUE)
```

---

[3] As an unordered factor, the levels would be treated as ordered alphabetically, i.e., Marked, None, Some.

```
> summary(arth.polr)

Call:
polr(formula = Improved ~ Sex + Treatment + Age, data = Arthritis,
    Hess = TRUE)

Coefficients:
                  Value Std. Error t value
SexMale          -1.2517     0.5464   -2.29
TreatmentTreated  1.7453     0.4759    3.67
Age               0.0382     0.0184    2.07

Intercepts:
             Value  Std. Error t value
None|Some    2.532   1.057       2.395
Some|Marked  3.431   1.091       3.144

Residual Deviance: 145.46
AIC: 155.46
```

The output from the `summary()` method, shown above, gives the estimated coefficients (β) and intercepts (α_j) labeled by the cutpoint on the ordinal response. It provides standard errors and t-values ($\beta_i/SE(\beta_i)$), but no significance tests or p-values. The `car::Anova()` method gives the appropriate tests.

```
> library(car)
> Anova(arth.polr)

Analysis of Deviance Table (Type II tests)

Response: Improved
          LR Chisq Df Pr(>Chisq)
Sex           5.69  1    0.01708 *
Treatment    14.71  1    0.00013 ***
Age           4.57  1    0.03251 *
---
Signif. codes:  0 '***' 0.001 '**' 0.01 '*' 0.05 '.' 0.1 ' ' 1
```

8.1.3 Testing the proportional odds assumption

The simplicity of the PO model is achieved only when the proportional odds model holds for a given data set. In essence, a test of this assumption involves a contrast between the PO model and a generalized logit NPO model that allows different effects (slopes) of the predictors across the response categories:

$$\text{PO}: \quad L_j \;=\; \alpha_j + x^\mathsf{T}\beta \qquad j = 1, \ldots, m-1 \tag{8.4}$$

$$\text{NPO}: \quad L_j \;=\; \alpha_j + x^\mathsf{T}\beta_j \qquad j = 1, \ldots, m-1 \tag{8.5}$$

The most general test involves fitting both models and testing the difference in the residual deviance by a likelihood ratio test or using some other measure (such as AIC) for model comparison. The PO model (Eqn. (8.4)) has $(m-1) + p$ parameters, while the NPO model (Eqn. (8.5)) has $(m-1)(1+p) = m(1+p)$ parameters, which may be difficult to fit if this is large relative to the number of observations. An intermediate model, the ***partial proportional odds model*** (Peterson and Harrell, 1990), allows one subset of predictors, x_{po}, to satisfy the proportional odds assumption (equal slopes), while the remaining predictors x_{npo} have slopes varying with the response level:

$$\text{PPO}: \quad L_j = \alpha_j + x_{po}^\mathsf{T}\beta + x_{npo}^\mathsf{T}\beta_j \qquad j = 1, \ldots, m-1 . \tag{8.6}$$

In R, the PO and NPO models can be readily contrasted by fitting them both using `vglm()` in the **VGAM** package. This defines the `cumulative` family of models and allows a `parallel` option. With `parallel=TRUE`, this is equivalent to the `polr()` model, except that the signs of the coefficients are reversed.

```
> library(VGAM)
> arth.po <- vglm(Improved ~ Sex + Treatment + Age, data = Arthritis,
+                 family = cumulative(parallel = TRUE))
> arth.po

Call:
vglm(formula = Improved ~ Sex + Treatment + Age, family = cumulative(parallel = TRUE),
    data = Arthritis)

Coefficients:
   (Intercept):1        (Intercept):2              SexMale
        2.531990             3.430988             1.251671
TreatmentTreated                  Age
       -1.745304            -0.038163

Degrees of Freedom: 168 Total; 163 Residual
Residual deviance: 145.46
Log-likelihood: -72.729
```

The more general NPO model can be fit using `parallel=FALSE`.

```
> arth.npo <- vglm(Improved ~ Sex + Treatment + Age, data = Arthritis,
+                  family = cumulative(parallel = FALSE))
> arth.npo

Call:
vglm(formula = Improved ~ Sex + Treatment + Age, family = cumulative(parallel = FALSE),
    data = Arthritis)

Coefficients:
   (Intercept):1        (Intercept):2              SexMale:1
        2.618539             3.431175             1.509827
        SexMale:2 TreatmentTreated:1 TreatmentTreated:2
        0.866434            -1.836929            -1.704011
            Age:1                Age:2
        -0.040866            -0.037294

Degrees of Freedom: 168 Total; 160 Residual
Residual deviance: 143.57
Log-likelihood: -71.787
```

The **VGAM** package defines a `coef()` method that can print the coefficients in a more readable matrix form giving the category cutpoints:

```
> coef(arth.po, matrix = TRUE)

                 logit(P[Y<=1])  logit(P[Y<=2])
(Intercept)            2.531990        3.430988
SexMale                1.251671        1.251671
TreatmentTreated      -1.745304       -1.745304
Age                   -0.038163       -0.038163

> coef(arth.npo, matrix = TRUE)

                 logit(P[Y<=1])  logit(P[Y<=2])
(Intercept)            2.618539        3.431175
SexMale                1.509827        0.866434
TreatmentTreated      -1.836929       -1.704011
Age                   -0.040866       -0.037294
```

In most cases, nested models can be tested using an `anova()` method, but the **VGAM** package has not implemented this for "vglm" objects. Instead, it provides an analogous function, `lrtest()`:

```
> VGAM::lrtest(arth.npo, arth.po)

Likelihood ratio test

Model 1: Improved ~ Sex + Treatment + Age
Model 2: Improved ~ Sex + Treatment + Age
  #Df LogLik Df Chisq Pr(>Chisq)
1 160  -71.8
2 163  -72.7  3  1.88       0.6
```

The LR test can be also calculated "manually" as shown below using the difference in residual deviance for the two models.

```
> tab <- cbind(
+    Deviance = c(deviance(arth.npo), deviance(arth.po)),
+          df = c(df.residual(arth.npo), df.residual(arth.po))
+          )
> tab <- rbind(tab, diff(tab))
> rownames(tab) <- c("GenLogit", "PropOdds", "LR test")
> tab <- cbind(tab, pvalue=1-pchisq(tab[,1], tab[,2]))
> tab

             Deviance  df  pvalue
GenLogit    143.5741  160 0.81966
PropOdds    145.4579  163 0.83435
LR test       1.8838    3 0.59686
```

The `vglm()` can also fit partial proportional odds models, by specifying a formula giving the terms for which the PO assumption should be taken as TRUE or FALSE. Here we illustrate this using `parallel=FALSE ~ Sex`, to fit separate slopes for males and females, but parallel lines for the other predictors. The same model would be fit using `parallel=TRUE ~ Treatment + Age`.

```
> arth.ppo <- vglm(Improved ~ Sex + Treatment + Age, data = Arthritis,
+    family = cumulative(parallel = FALSE ~ Sex))
> coef(arth.ppo, matrix = TRUE)

                 logit(P[Y<=1]) logit(P[Y<=2])
(Intercept)            2.542452       3.615561
SexMale                1.483336       0.867362
TreatmentTreated      -1.775742      -1.775742
Age                   -0.039622      -0.039622
```

8.1.4   Graphical assessment of proportional odds

There are several graphical methods for visual assessment of the proportional odds assumption. These are all *marginal* methods, in that they treat the predictors one at a time. However, that provides one means to determine if a partial proportional odds model might be more appropriate. Harrell's work *Regression Modeling Strategies* (2001, Chapters 13–14) and the corresponding rms package provide an authoritative treatment and methods in R.

One simple idea is to plot the conditional mean or expected value $E(X \mid Y)$ of a given predictor, X, at each level of the ordered response Y. If the response behaves ordinally in relation to X, these means should be strictly increasing or decreasing with Y. For comparison, one can also plot the estimated conditional means $\widehat{E}(X \mid Y = j)$ under the fitted PO model X as the only predictor. If the PO assumption holds for this X, the model-mean curve should be close to the data mean curve.

```
> library(rms)
> arth.po2 <- lrm(Improved ~ Sex + Treatment + Age, data = Arthritis)
> arth.po2


Logistic Regression Model

lrm(formula = Improved ~ Sex + Treatment + Age, data = Arthritis)
                        Model Likelihood      Discrimination      Rank Discrim.
                          Ratio Test              Indexes            Indexes
Obs              84     LR chi2       24.46   R2       0.291    C       0.750
 None            42     d.f.              3   g        1.335    Dxy     0.500
 Some            14     Pr(> chi2) <0.0001   gr       3.801    gamma   0.503
 Marked          28                           gp       0.280    tau-a   0.309
max |deriv| 1e-07                             Brier    0.187

                      Coef     S.E.     Wald Z  Pr(>|Z|)
y>=Some             -2.5320   1.0570   -2.40    0.0166
y>=Marked           -3.4310   1.0911   -3.14    0.0017
Sex=Male            -1.2517   0.5464   -2.29    0.0220
Treatment=Treated    1.7453   0.4759    3.67    0.0002
Age                  0.0382   0.0184    2.07    0.0382
```

The plot of conditional X means is produced using the plot.xmean.ordinaly() as shown below. It produces one marginal panel for each predictor in the model. For categorical predictors, it plots only the overall most frequent category. The resulting plot is shown in Figure 8.3.

```
> op <- par(mfrow=c(1,3))
> plot.xmean.ordinaly(Improved ~ Sex + Treatment + Age, data=Arthritis,
+                     lwd=2, pch=16, subn=FALSE)
> par(op)
```

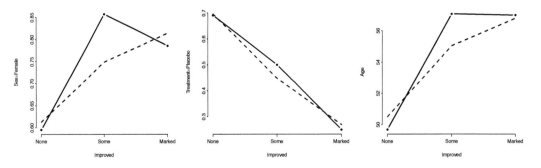

Figure 8.3: Visual assessment of ordinality and the proportional odds assumption for predictors in the Arthritis data. Solid lines connect the stratified means of X given Y. Dashed lines show the estimated expected value of X given Y=j if the proportional odds model holds for X.

In Figure 8.3, there is some evidence that the effect of Sex is non-monotonic and the means differ from their model-implied values under the PO assumption. The effect of Treatment looks good by this method, and the effect of Age hints that the upper two categories may not be well-distinguished as an ordinal response.

Of course, this example has only a modest total sample size, and this method only examines the marginal effects of the predictors. Nevertheless, it is a useful supplement to the statistical methods described earlier.

8.1.5 Visualizing results for the proportional odds model

Results from the PO model (and other models for polytomous responses) can be graphed using the same ideas and methods shown earlier for a binary or binomial response. In particular, full-model plots (described earlier in Section 7.3.2) and effect plots (Section 7.3.3) are still very helpful.

But now there is the additional complication that the response variable has $m > 2$ levels and so needs to be represented by $m - 1$ curves or panels in addition to those related to the predictor variables.

8.1.5.1 Full-model plots

For full-model plots, we continue the idea of appending the fitted response probabilities (or logits) to the data frame and plotting these in relation to the predictors. The predict() method returns the highest probability category label by default (with type="class"), so to get the fitted probabilities you have to ask for type="probs", as shown below.

```
> arth.fitp <- cbind(Arthritis,
+                    predict(arth.polr, type = "probs"))
> head(arth.fitp)

  ID Treatment  Sex Age Improved    None    Some  Marked
1 57   Treated Male  27     Some 0.73262 0.13806 0.12932
2 46   Treated Male  29     None 0.71740 0.14443 0.13816
3 77   Treated Male  30     None 0.70960 0.14763 0.14277
4 17   Treated Male  32   Marked 0.69363 0.15400 0.15237
5 36   Treated Male  46   Marked 0.57025 0.19504 0.23471
6 23   Treated Male  58   Marked 0.45634 0.21713 0.32653
```

For plotting, it is most convenient to reshape these from wide to long format using melt() in the **reshape2** (Wickham, 2014b) package. The response category is named Level.

```
> library(reshape2)
> plotdat <- melt(arth.fitp,
+                 id.vars = c("Sex", "Treatment", "Age", "Improved"),
+                 measure.vars = c("None", "Some", "Marked"),
+                 variable.name = "Level",
+                 value.name = "Probability")
> ## view first few rows
> head(plotdat)

   Sex Treatment Age Improved Level Probability
1 Male   Treated  27     Some  None     0.73262
2 Male   Treated  29     None  None     0.71740
3 Male   Treated  30     None  None     0.70960
4 Male   Treated  32   Marked  None     0.69363
5 Male   Treated  46   Marked  None     0.57025
6 Male   Treated  58   Marked  None     0.45634
```

We can now plot Probability against Age, using Level to assign different colors to the lines for the response categories, as seen in Figure 8.4. Here, facet_grid() is used to split the plot into separate panels by Sex and Treatment. In this example, the **directlabels** package is also used replace the default legend created by ggplot() with category labels on the curves themselves, which is easier to read.

```
> library(ggplot2)
> library(directlabels)
> gg <- ggplot(plotdat, aes(x = Age, y = Probability, colour = Level)) +
+     geom_line(size = 2.5) + theme_bw() + xlim(10, 80) +
+     geom_point(color = "black", size = 1.5) +
```

```
+        facet_grid(Sex ~ Treatment,
+                   labeller = function(x, y) sprintf("%s = %s", x, y)
+                   )
> direct.label(gg)
```

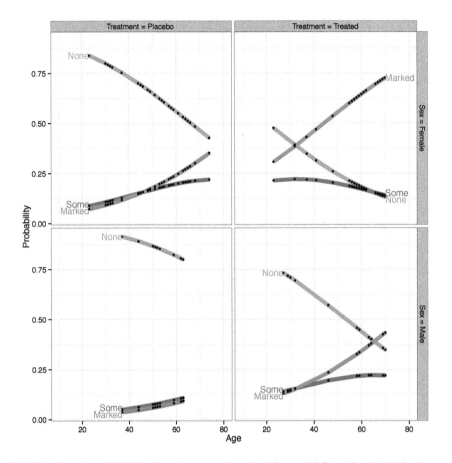

Figure 8.4: Predicted probabilities for the proportional odds model fit to the Arthritis data.

Although we now have three response curves in each panel in Figure 8.4, this plot is relatively easy to understand: (a) In each panel, the probability of no improvement decreases with age, while that for marked improvement increases. (b) It is easy to compare the placebo and treated groups in each row, showing that no improvement decreases, while marked improvement increases with the active treatment. (On the other hand, this layout makes it harder to compare panels vertically for males and females in each condition.) (c) The points show where the observations are located in each panel; so, we can see that the data is quite thin for males given the placebo.[4]

8.1.5.2 Effect plots

For PO models fit using `polr()`, the **effects** package provides two different styles for plotting a given effect. By default, curves are plotted in separate panels for the different response levels of a given effect, together with confidence bands for predicted probabilities. This form provides confidence bands and rug plots for the observations, but the default vertical arrangement of the

---

[4]One way to improve (pun intended) this graph would be to show the points on the lines only for the actual level of `Improve` for each observation.

panels makes it harder to compare the trends for the different response levels. The alternative *stacked* format shows the changes in response level more directly, but doesn't provide confidence bands.

Figure 8.5 shows these two styles for the main effect of Age in the proportional odds model, arth.polr fit earlier.

```
> library(effects)
> plot(Effect("Age", arth.polr))
> plot(Effect("Age", arth.polr), style = "stacked",
+       key.args = list(x = .55, y = .9))
```

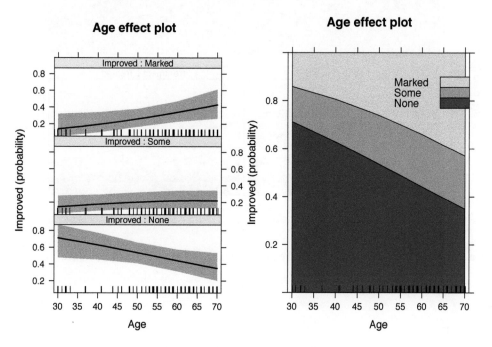

Figure 8.5: Effect plots for the effect of Age in the proportional odds model for the Arthritis data. Left: responses shown in separate panels. Right: responses shown in stacked format.

Even though this model includes only main effects, you can still plot the higher-order effects for more focal predictors in a coherent display. Figure 8.6 shows the predicted probabilities for all three predictors together. Again, visual comparison is easier horizontally for placebo versus treated groups, but you can also see that the prevalence of marked improvement is greater for females than for males.

```
> plot(Effect(c("Treatment", "Sex", "Age"), arth.polr),
+       style = "stacked", key.arg = list(x = .8, y = .9))
```

Finally, the latent variable interpretation of the PO model provides for simpler plots on the logit scale. Figure 8.7 shows this plot for the effects of Treatment and Age (collapsed over Sex) produced with the argument latent=TRUE to Effect(). In this plot, there is a single line in each panel for the effect (slope) of Age on the log odds. The dashed horizontal lines give the thresholds between the adjacent response categories corresponding to the intercepts.

```
> plot(Effect(c("Treatment", "Age"), arth.polr, latent = TRUE), lwd = 3)
```

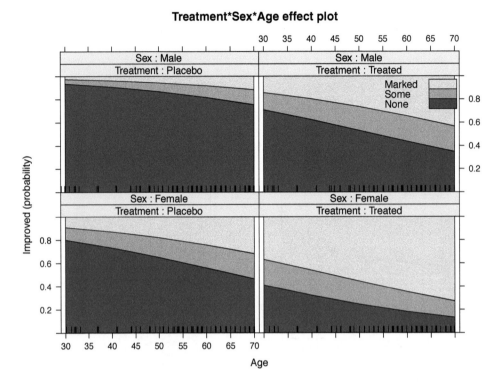

Figure 8.6: Effect plot for the effects of Treatment, Sex, and Age in the Arthritis data.

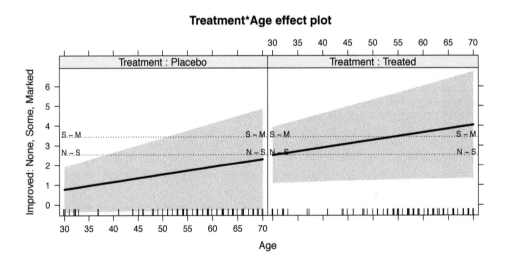

Figure 8.7: Latent variable effect plot for the effects of Treatment and Age in the Arthritis data.

8.2 Nested dichotomies

The method of ***nested dichotomies*** provides another simple way to analyze a polytomous response in the framework of logistic regression (or other generalized linear models). This method does not require an ordinal response or special software. Instead, it uses the familiar binary logistic model and fits $m - 1$ separate models for each of a hierarchically nested set of comparisons among the response categories.

Taken together, this set of models for the dichotomies comprises a complete model for the polytomous response. As well, these models are statistically independent, so test statistics such as G^2 or Wald tests can be added to give overall tests for the full polytomy.

For example, the response categories $Y = \{1,2,3,4\}$ could be divided first as $\{1,2\}$ vs. $\{3,4\}$, as shown in the left side of Figure 8.8. Then these two dichotomies could be divided as $\{1\}$ vs. $\{2\}$, and $\{3\}$ vs. $\{4\}$. Alternatively, these response categories could be divided as shown in the right side of Figure 8.8: first, $\{1\}$ vs. $\{2,3,4\}$, then $\{2\}$ vs $\{3,4\}$, and finally $\{3\}$ vs. $\{4\}$.

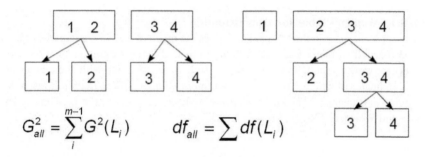

$$G^2_{all} = \sum_i^{m-1} G^2(L_i) \qquad df_{all} = \sum df(L_i)$$

Figure 8.8: Nested dichotomies. The boxes show two different ways a four-category response can be represented as three nested dichotomies. Adapted from Fox (2008).

Such models make the most sense when there are substantive reasons for considering the response categories in terms of such dichotomies. Two examples are shown in Figure 8.9.

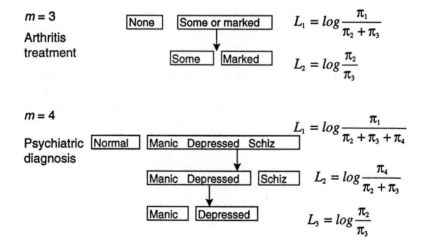

Figure 8.9: Examples of nested dichotomies and the corresponding logits.

- For the `Arthritis` data, it is sensible to consider one dichotomy ("better"), with logit L_1, between the categories of "None" compared to "Some" or "Marked". A second dichotomy, with logit L_2, would then distinguish between the some and marked response categories.
- For a second case where patients are classified into $m = 4$ psychiatric diagnostic categories, the first dichotomy, with logit L_1, distinguishes those considered normal from all others given a clinical diagnosis. Two other dichotomies are defined to further divide the non-normal categories.

Then, consider the separate logit models for these $m - 1$ dichotomies, with different intercepts α_j and slopes β_j for each dichotomy,

$$
\begin{aligned}
L_1 &= \alpha_1 + x^\mathsf{T}\beta_1 \\
L_2 &= \alpha_2 + x^\mathsf{T}\beta_2 \\
\vdots &= \vdots \\
L_{m-1} &= \alpha_{m-1} + x^\mathsf{T}\beta_{m-1} .
\end{aligned}
$$

EXAMPLE 8.1: Women's labor force participation

The data set `Womenlf` in the car package gives the result of a 1977 Canadian survey. It contains data for 263 married women of age 21–30 who indicated their working status (outside the home) as not working, working part time, or working full time, together with their husband's income and a binary indicator of whether they had one or more young children in their household. (Another variable, region of Canada, had no effects in these analyses, and is not examined here.) This example follows Fox and Weisberg (2011a, Section 5.8).

```
> library(car)    # for data and Anova()
> data("Womenlf", package = "car")
> some(Womenlf)

      partic hincome children   region
7    not.work      15  present  Ontario
29   not.work      17  present  Prairie
45   parttime       5  present  Ontario
82   parttime      15  present  Ontario
91   not.work      35   absent  Ontario
97   not.work      17  present  Ontario
129  parttime      13  present  Prairie
138  not.work      13  present  Ontario
175  fulltime       9   absent  Ontario
200  fulltime      11   absent   Quebec
```

In this example, it makes sense to consider a first dichotomy (`working`) between women who are not working vs. those who are (full time or part time). A second dichotomy (`fulltime`) contrasts full time work vs. part time work, among those women who are working at least part time. These two binary variables are created in the data frame using the `recode()` function from the car package.

```
> # create dichotomies
> Womenlf <- within(Womenlf, {
+   working <- recode(partic, " 'not.work' = 'no'; else = 'yes' ")
+   fulltime <- recode(partic,
+     " 'fulltime' = 'yes'; 'parttime' = 'no'; 'not.work' = NA")})
> some(Womenlf)

      partic hincome children   region fulltime working
81   fulltime      13   absent  Ontario      yes     yes
```

```
96  not.work  17  present Ontario  <NA>  no
97  not.work  17  present Ontario  <NA>  no
115 parttime  13  present Prairie   no   yes
123 fulltime   9   absent Ontario   yes  yes
131 parttime  19  present Ontario   no   yes
153 not.work   5   absent     BC   <NA>  no
190 not.work  23  present     BC   <NA>  no
248 not.work  23   absent Quebec   <NA>  no
255 fulltime  11   absent Quebec    yes  yes
```

The tables below show how the response `partic` relates to the recoded binary variables, `working` and `fulltime`. Note that the `fulltime` variable is recoded to NA for women who are not working.

```
> with(Womenlf, table(partic, working))

          working
partic       no yes
   fulltime   0  66
   not.work 155   0
   parttime   0  42

> with(Womenlf, table(partic, fulltime, useNA = "ifany"))

          fulltime
partic       no yes <NA>
   fulltime   0  66    0
   not.work   0   0  155
   parttime  42   0    0
```

We proceed to fit two separate binary logistic regression models for the derived dichotomous variables. For the `working` dichotomy, we get the following results:

```
> mod.working <- glm(working ~ hincome + children, family = binomial,
+                   data = Womenlf)
> summary(mod.working)


Call:
glm(formula = working ~ hincome + children, family = binomial,
    data = Womenlf)

Deviance Residuals:
   Min     1Q  Median     3Q     Max
-1.677  -0.865  -0.777  0.929   1.997

Coefficients:
                Estimate Std. Error z value Pr(>|z|)
(Intercept)       1.3358     0.3838    3.48   0.0005 ***
hincome          -0.0423     0.0198   -2.14   0.0324 *
childrenpresent  -1.5756     0.2923   -5.39    7e-08 ***
---
Signif. codes:  0 '***' 0.001 '**' 0.01 '*' 0.05 '.' 0.1 ' ' 1

(Dispersion parameter for binomial family taken to be 1)

    Null deviance: 356.15  on 262  degrees of freedom
Residual deviance: 319.73  on 260  degrees of freedom
AIC: 325.7

Number of Fisher Scoring iterations: 4
```

And, similarly for the `fulltime` dichotomy:

```
> mod.fulltime <- glm(fulltime ~ hincome + children, family = binomial,
+                     data = Womenlf)
> summary(mod.fulltime)


Call:
glm(formula = fulltime ~ hincome + children, family = binomial,
    data = Womenlf)

Deviance Residuals:
   Min      1Q  Median      3Q     Max
-2.405  -0.868   0.395   0.621   1.764

Coefficients:
                Estimate Std. Error z value Pr(>|z|)
(Intercept)       3.4778     0.7671    4.53  5.8e-06 ***
hincome          -0.1073     0.0392   -2.74   0.0061 **
childrenpresent  -2.6515     0.5411   -4.90  9.6e-07 ***
---
Signif. codes:  0 '***' 0.001 '**' 0.01 '*' 0.05 '.' 0.1 ' ' 1

(Dispersion parameter for binomial family taken to be 1)

    Null deviance: 144.34  on 107  degrees of freedom
Residual deviance: 104.49  on 105  degrees of freedom
  (155 observations deleted due to missingness)
AIC: 110.5

Number of Fisher Scoring iterations: 5
```

Although these were fit separately, we can view this as a combined model for the three-level response, with the following coefficients:

```
> cbind(working = coef(mod.working), fulltime = coef(mod.fulltime))

                   working fulltime
(Intercept)       1.335830  3.47777
hincome          -0.042308 -0.10727
childrenpresent  -1.575648 -2.65146
```

Writing these out as equations for the logits, we have:

$$L_1 = \log \frac{\Pr(\text{working})}{\Pr(\text{notworking})} = 1.336 - 0.042 \, \text{hincome} - 1.576 \, \text{children} \qquad (8.7)$$

$$L_2 = \log \frac{\Pr(\text{fulltime})}{\Pr(\text{parttime})} = 3.478 - 0.1072 \, \text{hincome} - 2.652 \, \text{children} \qquad (8.8)$$

For both dichotomies, increasing income of the husband and the presence of young children decrease the log odds of a greater level of work. However, for those women who are working, the effects of husband's income and and children are greater on the choice between full time and part time work than they are for all women on the choice between working and not working.

As we mentioned above, the use of nested dichotomies implies that the models fit to the separate dichotomies are statistically independent. Thus, we can additively combine χ^2 statistics and degrees of freedom to give overall tests for the polytomous response.

For example, here we define a function, LRtest(), to calculate the likelihood ratio test of the hypothesis $H_0 : \beta = \mathbf{0}$ for all predictors simultaneously. We then use this to display these tests for each sub-model, as well as the combined model based on the sums of the test statistic and degrees of freedom.

```
> LRtest <- function(model)
+    c(LRchisq = model$null.deviance - model$deviance,
+        df = model$df.null - model$df.residual)
>
> tab <- rbind(working = LRtest(mod.working),
+                fulltime = LRtest(mod.fulltime))
> tab <- rbind(tab, All = colSums(tab))
> tab <- cbind(tab, pvalue = 1- pchisq(tab[,1], tab[,2]))
> tab

           LRchisq df      pvalue
working     36.418  2 1.2355e-08
fulltime    39.847  2 2.2252e-09
All         76.265  4 1.1102e-15
```

Similarly, you can carry out tests of individual predictors, $H_0 : \beta_i = 0$, for the polytomy by adding the separate χ^2s from Anova().

```
> Anova(mod.working)

Analysis of Deviance Table (Type II tests)

Response: working
           LR Chisq Df Pr(>Chisq)
hincome     4.82637  1   0.028028 *
children   31.32288  1 2.1849e-08 ***
---
Signif. codes:  0 '***' 0.001 '**' 0.01 '*' 0.05 '.' 0.1 ' ' 1

> Anova(mod.fulltime)

Analysis of Deviance Table (Type II tests)

Response: fulltime
           LR Chisq Df Pr(>Chisq)
hincome      8.9813  1  0.0027275 **
children    32.1363  1 1.4373e-08 ***
---
Signif. codes:  0 '***' 0.001 '**' 0.01 '*' 0.05 '.' 0.1 ' ' 1
```

For example, the test for husband's income gives $\chi^2 = 4.826 + 8.981 = 13.807$ with 2 df.

As before, you can plot the fitted values from such models, either on the logit scale (for the separate logit equations) or in terms of probabilities for the various responses. The general idea is the same: obtain the fitted values from predict() using data frame containing the values of the predictors. However, now we have to combine these for each of the sub-models.

We calculate these values below, on both the logit scale and the response scale of probabilities. The newdata argument to predict() is constructed as the combinations of values for hincome and children.[5]

```
> predictors <- expand.grid(hincome = 1 : 50,
+                            children =c('absent', 'present'))
> fit <- data.frame(predictors,
+      p.working = predict(mod.working, predictors, type = "response"),
+      p.fulltime = predict(mod.fulltime, predictors, type = "response"),
+      l.working = predict(mod.working, predictors, type = "link"),
+      l.fulltime = predict(mod.fulltime, predictors, type = "link")
+ )
> print(some(fit, 5), digits = 3)
```

---

[5]Alternatively, using the predictor values in the *Womenlf* data would give the fitted values for the cases in the data, and allow a more data-centric plot as shown in Figure 8.4.

	hincome	children	p.working	p.fulltime	l.working	l.fulltime
6	6	absent	0.747	0.9445	1.082	2.834
8	8	absent	0.731	0.9321	0.997	2.620
39	39	absent	0.422	0.3306	-0.314	-0.706
68	18	present	0.269	0.2489	-1.001	-1.105
88	38	present	0.136	0.0373	-1.848	-3.250

One wrinkle here is that the probabilities for working full time and part time are conditional on working. We calculate the unconditional probabilities as shown below and choose to display the probability of *not* working as the complement of working.

```
> fit <- within(fit, {
+     `full-time` <- p.working * p.fulltime
+     `part-time` <- p.working * (1 - p.fulltime)
+     `not working` <- 1 - p.working
+ })
```

To plot these fitted values, we will again create a conditional plot using **ggplot2**. Since this requires having all probabilities in one column, together with an additional grouping variable identifying the working status, we need to reshape `fit` from "wide" to "long" format, yet again using the `melt()` from the **reshape2** package:

```
> fit2 <- melt(fit,
+              measure.vars = c("full-time", "part-time", "not working"),
+              variable.name = "Participation",
+              value.name = "Probability")
```

The lines below give the plot shown in Figure 8.10:

```
> gg <- ggplot(fit2,
+              aes(x = hincome, y = Probability, colour= Participation)) +
+         facet_grid(~ children,
+                    labeller = function(x, y) sprintf("%s = %s", x, y)) +
+         geom_line(size = 2) + theme_bw() +
+         scale_x_continuous(limits = c(-3, 55)) +
+         scale_y_continuous(limits = c(0, 1))
>
> direct.label(gg, list("top.bumptwice", dl.trans(y = y + 0.2)))
```

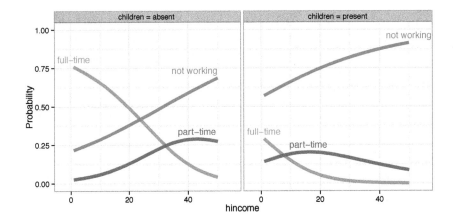

Figure 8.10: Fitted probabilities from the models for nested dichotomies fit to the data on women's labor force participation.

(Note the extension of the axes to avoid label clipping.)

We can see that the decision not to work outside the home increases strongly with husband's income, and is higher when there are children present. As well, among working women, the decision to work full time as opposed to part time decreases strongly with husband's income, and is less likely with young children.

Similarly, we plot the fitted logits for the two dichotomies in `l.working` and `l.fulltime` as shown below, giving Figure 8.11.

```
> fit3 <- melt(fit,
+               measure.vars = c("l.working", "l.fulltime"),
+               variable.name = "Participation",
+               value.name = "LogOdds")
> levels(fit3$Participation) <- c("working", "full-time")
>
> gg <- ggplot(fit3,
+               aes(x = hincome, y = LogOdds, colour = Participation)) +
+          facet_grid(~ children,
+                      labeller = function(x, y) sprintf("%s = %s", x, y)) +
+          geom_line(size = 2) + theme_bw() +
+          scale_x_continuous(limits = c(-3, 50)) +
+          scale_y_continuous(limits = c(-5, 4))
>
> direct.label(gg, list("top.bumptwice", dl.trans(y = y + 0.2)))
```

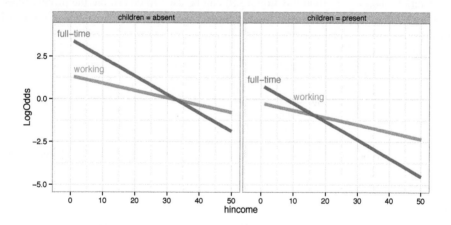

Figure 8.11: Fitted log odds from the models for nested dichotomies fit to the data on women's labor force participation.

This is essentially a graph of the fitted equations for L_1 and L_2 shown in Eqn. (8.7). It shows how the choice of full time work as opposed to part time depends more strongly on husband's income among women who are working than does the choice of working at all among all women. It also illustrates why the proportional odds assumption would not be reasonable for this data: that would require equal slopes for the two lines within each panel. △

8.3 Generalized logit model

The generalized logit (or multinomial logit) approach models the probabilities of the m response categories directly as a set of $m - 1$ logits. These compare each of the first $m - 1$ categories to the last category, which serves as the baseline.[6] The logits for any other pair of categories can be retrieved from the $m - 1$ fitted ones.

When there are p predictors, $x_1, x_2, \ldots, x_p$, which may be quantitative or categorical, the generalized logit model expresses the logits as

$$
\begin{aligned}
L_{jm} \equiv \log \frac{\pi_{ij}}{\pi_{im}} &= \beta_{0j} + \beta_{1j} x_{i1} + \beta_{2j} x_{i2} + \cdots + \beta_{kj} x_{ip} \quad j = 1, \ldots, m - 1 \\
&= \boldsymbol{x}_i^{\mathsf{T}} \boldsymbol{\beta}_j \,.
\end{aligned}
\tag{8.9}
$$

Thus, there is one set of fitted coefficients, $\boldsymbol{\beta}_j$, for each response category except the last. Each coefficient, β_{hj}, gives the effect, for a unit change in the predictor x_h, on the log odds that an observation had a response in category $Y = j$, as opposed to category $Y = m$.

The probabilities themselves can be expressed as

$$
\pi_{ij} = \frac{\exp(\boldsymbol{x}_i^{\mathsf{T}} \boldsymbol{\beta}_j)}{1 + \sum_{\ell=1}^{m-1} \exp(\boldsymbol{x}_i^{\mathsf{T}} \boldsymbol{\beta}_j)} \quad j = 1, 2, \ldots m - 1 \,,
$$

$$
\pi_{im} = 1 - \sum_{i=1}^{m-1} \pi_{ij} \quad \text{for } Y = m \,.
$$

Parameters in the $m - 1$ equations Eqn. (8.9) can be used to determine the probabilities or the predicted log odds for any pair of response categories by subtraction. For instance, for an arbitrary pair of categories, a and b, and two predictors, x_1 and x_2,

$$
\begin{aligned}
L_{ab} &= \log \frac{\pi_{ia}/\pi_{im}}{\pi_{ib}/\pi_{im}} \\
&= \log \frac{\pi_{ia}}{\pi_{im}} - \log \frac{\pi_{ib}}{\pi_{im}} \\
&= (\beta_{0a} - \beta_{0b}) + (\beta_{1a} - \beta_{1b})x_{i1} + (\beta_{2a} - \beta_{2b})x_{i2} \,.
\end{aligned}
$$

For example, the coefficient for x_{i1} in L_{ab} is just $(\beta_{1a} - \beta_{1b})$. Similarly, the predicted logit for any pair of categories can be calculated as

$$
\hat{L}_{ab} = \hat{L}_{am} - \hat{L}_{bm} \,.
$$

The generalized logit model can be fit most conveniently in R using the function `multinom()` in the nnet package, and the effects package has a set of methods for "multinom" models. These models can also be fit using VGAM and the mlogit (Croissant, 2013) package.

EXAMPLE 8.2: Women's labor force participation

To illustrate this method, we fit the generalized logit model to the women's labor force participation data as explained below. The response, `partic`, is a character factor, and, by default `multinom()` treats these in alphabetical order and uses the *first* level as the baseline category.

```
> levels(Womenlf$partic)

[1] "fulltime" "not.work" "parttime"
```

Although the multinomial model does not depend on the baseline category, it makes interpretation easier to choose `"not.work"` as the reference level, which we do with `relevel()`.[7]

---

[6]When the response is a factor, any category can be selected as the baseline level using `relevel()`.

[7]Alternatively, we could declare `partic` an *ordered* factor, using `ordered()`.

```
> # choose not working as baseline category
> Womenlf$partic <- relevel(Womenlf$partic, ref = "not.work")
```

We fit the main effects model for husband's income and children as follows. As we did with
polr() (Section 8.1), specifying Hess=TRUE saves the Hessian and facilitates calculation of
standard errors and hypothesis tests.

```
> library(nnet)
> wlf.multinom <- multinom(partic ~ hincome + children,
+                          data = Womenlf, Hess = TRUE)

# weights:  12 (6 variable)
initial   value 288.935032
iter  10 value 211.454772
final   value 211.440963
converged
```

The summary() method for "multinom" objects doesn't calculate test statistics for the estimated coefficients by default. The option Wald=TRUE produces Wald z-test statistics, calculated
as $z = \beta/SE(\beta)$.

```
> summary(wlf.multinom, Wald = TRUE)

Call:
multinom(formula = partic ~ hincome + children, data = Womenlf,
    Hess = TRUE)

Coefficients:
          (Intercept)      hincome childrenpresent
fulltime      1.9828   -0.0972321       -2.558605
parttime     -1.4323    0.0068938        0.021456

Std. Errors:
          (Intercept)   hincome childrenpresent
fulltime      0.48418  0.028096         0.36220
parttime      0.59246  0.023455         0.46904

Value/SE (Wald statistics):
          (Intercept)   hincome childrenpresent
fulltime      4.0953  -3.46071       -7.064070
parttime     -2.4176   0.29392        0.045744

Residual Deviance: 422.88
AIC: 434.88
```

Notice that the coefficients, their standard errors and the Wald test z values are printed in separate
tables. The first line in each table pertains to the logit comparing full time work with the not working
reference level; the second line compares part time work against not working.

For those who like p-values for significance tests, you can calculate these from the results returned by the summary() method in the Wald.ratios component, using the standard normal
asymptotic approximation:

```
> stats <- summary(wlf.multinom, Wald = TRUE)
> z <- stats$Wald.ratios
> p <- 2 * (1 - pnorm(abs(z)))
> zapsmall(p)

          (Intercept) hincome childrenpresent
fulltime      0.00004 0.00054         0.00000
parttime      0.01562 0.76882         0.96351
```

The interpretation of these tests is that both husband's income and presence of children have highly significant effects on the comparison of working full time as opposed to not working, while neither of these predictors are significant for the comparison of working part time vs. not working.

So far, we have assumed that the effects of husband's income and presence of young children are additive on the log odds scale. We can test this assumption by allowing an interaction of those effects and testing it for significance.

```
> wlf.multinom2 <- multinom(partic ~ hincome * children,
+                           data = Womenlf, Hess = TRUE)

# weights:  15 (8 variable)
initial  value 288.935032
iter  10 value 210.797079
final  value 210.714841
converged

> Anova(wlf.multinom2)

Analysis of Deviance Table (Type II tests)

Response: partic
                LR Chisq Df Pr(>Chisq)
hincome             15.2  2    0.00051 ***
children            63.6  2    1.6e-14 ***
hincome:children     1.5  2    0.48378
---
Signif. codes:  0 '***' 0.001 '**' 0.01 '*' 0.05 '.' 0.1 ' ' 1
```

The test for the interaction term, `hincome:children`, is not significant, so we can abandon this model.

Full model plots of the fitted values can be plotted as shown earlier in Example 8.1: obtain the fitted values over a grid of the predictors and plot these.

```
> predictors <- expand.grid(hincome = 1 : 50,
+                           children = c("absent", "present"))
> fit <- data.frame(predictors,
+                    predict(wlf.multinom, predictors, type = "probs")
+                    )
```

Plotting these fitted values gives the plot shown in Figure 8.12.

```
> fit2 <- melt(fit,
+              measure.vars = c("not.work", "fulltime", "parttime"),
+              variable.name = "Participation",
+              value.name = "Probability")
> levels(fit2$Participation) <- c("not working", "full-time", "part-time")
>
> gg <- ggplot(fit2,
+              aes(x = hincome, y = Probability, colour = Participation)) +
+         facet_grid(~ children,
+                    labeller = function(x, y) sprintf("%s = %s", x, y)) +
+         geom_line(size = 2) + theme_bw() +
+         scale_x_continuous(limits = c(-3, 50)) +
+         scale_y_continuous(limits = c(0, 0.9))
>
> direct.label(gg, list("top.bumptwice", dl.trans(y = y + 0.2)))
```

The results shown in this plot are roughly similar to those obtained from the nested dichotomy models, graphed in Figure 8.10. However, the predicted probabilities of not working under the generalized logit model rise more steeply with husband's income for women with no children and level off sooner for women with young children.

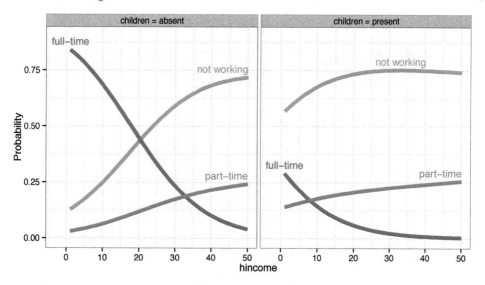

Figure 8.12: Fitted probabilities from the generalized logit model fit to the data on women's labor force participation.

The **effects** package has special methods for "multinom" models. It treats the response levels in the order given by levels(), so before plotting we use ordered() to arrange levels in their natural order. The update() method provides a simple way to get a new fitted model; in the call, the model formula . ~ . means to fit the same model as before, i.e., partic ~ hincome + children.

```
> levels(Womenlf$partic)

[1] "not.work" "fulltime" "parttime"

> Womenlf$partic <- ordered(Womenlf$partic,
+                           levels=c("not.work", "parttime", "fulltime"))
> wlf.multinom <- update(wlf.multinom, . ~ .)

# weights:  12 (6 variable)
initial   value 288.935032
iter   10 value 211.454772
final    value 211.440963
converged
```

As illustrated earlier, you can use plot(allEffects(model), ...) to plot all the high-order terms in the model, either with separate curves for each response level (style="lines") or as cumulative filled polygons (style="stacked"). Here, we simply plot the effects for the combinations of husband's income and children in stacked style, giving a plot (Figure 8.13) that is analogous to the full-model plot shown in Figure 8.12.

```
> plot(Effect(c("hincome", "children"), wlf.multinom),
+      style = "stacked", key.args = list(x = .05, y = .9))
```

△

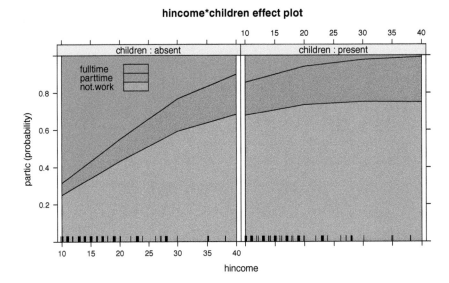

Figure 8.13: Effect plot for the probabilities of not working and working part time and full time from the generalized logit model fit to the women's labor force data.

8.4 Chapter summary

- Polytomous responses may be handled in several ways as extensions of binary logistic regression. These methods require different fitting functions in R; however, the graphical methods for plotting results are relatively straightforward extensions of those used for binary responses.

- The *proportional odds model* (Section 8.1) is simple and convenient, but its validity depends on an assumption of equal slopes for adjacent-category logits.

- *Nested dichotomies* (Section 8.2) among the response categories give a set of statistically independent binary logistic submodels. These may be regarded as a single combined model for the polytomous response.

- *Generalized logit models* (Section 8.3) provide the most general approach. These may be used to construct submodels comparing any pair of categories.

8.5 Lab exercises

Exercise 8.1 For the women's labor force participation data (*Womenlf*), the response variable, partic, can be treated as ordinal by using

```
> Womenlf$partic <- ordered(Womenlf$partic,
+                           levels=c('not.work', 'parttime', 'fulltime'))
```

Use the methods in Section 8.1 to test whether the proportional odds model holds for these data.

Exercise 8.2 The data set *housing* in the **MASS** package gives a $3 \times 3 \times 4 \times 2$ table in frequency form relating (a) satisfaction (Sat) of residents with their housing (High, Medium, Low), (b) perceived degree of influence (Infl) they have on the management of the property (High, Medium, Low), (c) Type of rental (Tower, Atrium, Apartment, Terrace), and (d) contact (Cont) residents have with other residents (Low, High). Consider satisfaction as the ordinal response variable.

(a) Fit the proportional odds model with additive (main) effects of housing type, influence in management, and contact with neighbors to this data. (Hint: Using `polr()`, with the data in frequency form, you need to use the `weights` argument to supply the `Freq` variable.)

(b) Investigate whether any of the two-factor interactions among `Infl`, `Type`, and `Cont` add substantially to goodness of fit of this model. (Hint: use `stepAIC()`, with the scope formula `~ .^2` and `direction="forward"`.)

(c) For your chosen model from the previous step, use the methods of Section 8.1.5 to plot the probabilities of the categories of satisfaction.

(d) Write a brief summary of these analyses, interpreting *how* satisfaction with housing depends on the predictor variables.

Exercise 8.3 The data `TV` on television viewing was analyzed using correspondence analysis in Example 6.4, ignoring the variable `Time`, and extended in Exercise 6.9. Treating `Network` as a three-level response variable, fit a generalized logit model (Section 8.3) to explain the variation in viewing in relation to `Day` and `Time`. The `TV` data is a three-way table, so you will need to convert it to a frequency data frame first.

```
> data("TV", package="vcdExtra")
> TV.df <- as.data.frame.table(TV)
```

(a) Fit the main-effects model, `Network ~ Day + Time`, with `multinom()`. Note that you will have to supply the `weights` argument because each row of `TV.df` represents the number of viewers in the `Freq` variable.

(b) Prepare an effects plot for the fitted probabilities in this model.

(c) Interpret these results in comparison to the correspondence analysis in Example 6.4.

Exercise 8.4 * Refer to Exercise 5.10 for a description of the `Vietnam` data set in vcdExtra. The goal here is to fit models for the polytomous `response` varialble in relation to `year` and `sex`.

(a) Fit the proportional odds model to these data, allowing an interaction of `year` and `sex`.

(b) Is there evidence that the proportional odds assumption does not hold for this data set? Use the methods described in Section 8.1 to assess this.

(c) Fit the multinomial logistic model, also allowing an interaction. Use car::`Anova()` to assess the model terms.

(d) Produce an effect plot for this model and describe the nature of the interaction.

(e) Fit the simpler multinomial model in which there is no effect of year for females and the effect of year is linear for males (on the logit scale). Test whether this model is significantly worse than the general multinomial model with interaction.

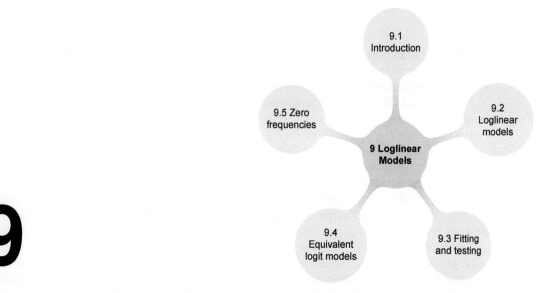

9

Loglinear and Logit Models for Contingency Tables

This chapter extends the model-building approach to loglinear and logit models. These comprise another special case of generalized linear models designed for contingency tables of frequencies. They are most easily interpreted through visualizations, including mosaic displays and effect plots of associated logit models.

Numbers have an important story to tell. They rely on you to give them a clear and convincing voice

Stephen Few

9.1 Introduction

The chapter continues the modeling framework begun in Chapter 7, and takes up the case of log-linear models for contingency tables of frequencies, when all variables are discrete, another special case of generalized linear models. These models provide a comprehensive scheme to describe and understand the associations among two or more categorical variables. Whereas logistic regression models focus on the prediction of one response factor, loglinear models treat all variables symmetrically, and attempt to model all important associations among them.

In this sense, loglinear models are analogous to a correlation analysis of continuous variables, where the goal is to determine the patterns of dependence and independence among a set of variables. When one variable is a response and the others are explanatory, certain loglinear models are equivalent to logistic models for that response. Such models are also particularly useful when there are two or more response variables, a case that would require a multivariate version of the generalized linear model, for which the current theory and implementations are thin at best.

Chapter 5 and Chapter 6 introduced some basic aspects of loglinear models in connection with mosaic displays and correspondence analysis. In this chapter, the focus is on fitting and interpreting loglinear models. The usual analyses with `loglm()` and `glm()` present the results in terms of tables of parameter estimates. Particularly for larger tables, it becomes difficult to understand the nature of these associations from tables of parameter estimates. Instead, we emphasize plots of observed and predicted frequencies, probabilities or log odds (when there are one or more response variables), as well as mosaic and other displays for interpreting a given model. We also illustrate how mosaic displays and correspondence analysis plots may be used in a complementary way to the usual numerical summaries, to provide additional insights into the data.

Section 9.2 gives a brief overview of loglinear models in relation to the more familiar ANOVA and regression models for quantitative data. Methods and software for fitting these models are discussed in Section 9.3. When one variable is a response, logit models for that response provide a simpler, but equivalent means for interpreting and graphing results of loglinear models, as we describe in Section 9.4. In Section 9.5 we consider problems that arise in sparce contingency tables containing cells with frequencies of zero.

9.2 Loglinear models for frequencies

Loglinear models have been developed from two formally distinct, but related perspectives. The first is a discrete analog of familiar ANOVA models for quantitative data, where the multiplicative relations among joint and marginal probabilities are transformed into an additive one by transforming the counts to logarithms. The second is an analog of regression models, where the log of the cell frequency is modeled as a linear function of discrete predictors, with a random component often taken as the Poisson distribution and called ***Poisson regression***; this approach is treated in more detail as generalized linear models for count data in Chapter 11.

9.2.1 Loglinear models as ANOVA models for frequencies

For two discrete variables, A and B, suppose we have a multinomial sample of n_{ij} observations in each cell i, j of an $I \times J$ contingency table. To ease notation, we replace a subscript by $+$ to represent summation over that dimension, so that $n_{i+} = \Sigma_j n_{ij}$, $n_{+j} = \Sigma_i n_{ij}$, and $n_{++} = \Sigma_{ij} n_{ij}$.

Let π_{ij} be the joint probabilities in the table, and let $m_{ij} = n_{++}\pi_{ij}$ be the expected cell frequencies under any model. Conditional on the observed total count, n_{++}, each count has a Poisson distribution, with mean m_{ij}. Any loglinear model may be expressed as a linear model for the $\log m_{ij}$. For example, the hypothesis of independence means that the expected frequencies, m_{ij}, obey

$$m_{ij} = \frac{m_{i+} \, m_{+j}}{m_{++}} \, .$$

This multiplicative model can be transformed to an additive (linear) model by taking logarithms of both sides:

$$\log(m_{ij}) = \log(m_{i+}) + \log(m_{+j}) - \log(m_{++}) \, ,$$

which is usually expressed in an equivalent form in terms of model parameters,

$$\log(m_{ij}) = \mu + \lambda_i^A + \lambda_j^B \tag{9.1}$$

where μ is a function of the total sample size, λ_i^A is the "main effect" for variable A, $\lambda_i^A = \log \pi_{i+} - \sum_k (\log \pi_{k+})/I$, and λ_j^B is the "main effect" for variable B, $\lambda_j^B = \log \pi_{+j} - \sum_k (\log \pi_{+k})/J$. Model Eqn. (9.1) is called the ***loglinear independence model*** for a two-way table.

In this model, there are $1 + I + J$ parameters, but only $(I-1) + (J-1)$ are separately estimable. Hence, the typical ANOVA sum-to-zero restrictions are usually applied to the parameters:

$$\sum_i^I \lambda_i^A = \sum_j^J \lambda_j^B = 0 .$$

These "main effects" in loglinear models pertain to differences among the marginal probabilities of a variable (which are usually not of direct interest).

Other restrictions to make the parameters identifiable are also used. Setting the first values, λ_1^A and λ_1^B, to zero (the default in `glm()`), defines $\lambda_i^A = \log \pi_{i+} - \log \pi_{1+}$, and $\lambda_j^B = \log \pi_{+j} - \log \pi_{+1}$, as deviations from the first, reference category, but these parameterizations are otherwise identical. For modeling functions in R (`lm()`, `glm()`, etc.) the reference category parameterization is obtained using `contr.treatment()`, while the sum-to-zero constraints are obtained with `contr.sum()`.

Model Eqn. (9.1) asserts that the row and column variables are independent. For a two-way table, a model that allows an arbitrary association between the variables is the ***saturated model***, including an additional term, λ_{ij}^{AB}:

$$\log(m_{ij}) = \mu + \lambda_i^A + \lambda_j^B + \lambda_{ij}^{AB} , \qquad (9.2)$$

where again, restrictions must be imposed for estimation:

$$\sum_i^I \lambda_i^A = 0, \quad \sum_j^J \lambda_j^B = 0, \quad \sum_i^I \lambda_{ij}^{AB} = \sum_j^J \lambda_{ij}^{AB} = 0 . \qquad (9.3)$$

There are thus $I-1$ linearly independent λ_i^A row parameters, $J-1$ linearly independent λ_j^B column parameters, and $(I-1)(J-1)$ linearly independent λ_{ij}^{AB} association parameters. This model is called the *saturated model* because the number of parameters in μ, λ_i^A, λ_j^B, and λ_{ij}^{AB} is equal to the number of frequencies in the two-way table,

$$\underset{(\mu)}{1} + \underset{(\lambda_i^A)}{(I-1)} + \underset{(\lambda_j^B)}{(J-1)} + \underset{(\lambda_{ij}^{AB})}{(I-1)(J-1)} = \underset{(n_{ij})}{IJ} .$$

The association parameters λ_{ij}^{AB} express the departures from independence, so large absolute values pertain to cells that differ from the independence model.

Except for the difference in notation, model Eqn. (9.2) is formally the same as a two-factor ANOVA model with an interaction, typically expressed as $E(y_{ij}) = \mu + \alpha_i + \beta_j + (\alpha\beta)_{ij}$. Hence, associations between variables in loglinear models are analogous to interactions in ANOVA models. The use of superscripted symbols, $\lambda_i^A, \lambda_j^B, \lambda_{ij}^{AB}$, rather than separate Greek letters is a convention in loglinear models, and useful mainly for multiway tables.

Models such as Eqn. (9.1) and Eqn. (9.2) are examples of ***hierarchical models***. This means that the model must contain all lower-order terms contained within any high-order term in the model. Thus, the saturated model in Eqn. (9.2) contains λ_{ij}^{AB}, and therefore *must* contain λ_i^A and λ_j^B. As a result, hierarchical models may be identified by the shorthand notation that lists only the high-order terms: model Eqn. (9.2) is denoted $[AB]$, while model Eqn. (9.1) is $[A][B]$.

9.2.2   Loglinear models for three-way tables

Loglinear models for three-way contingency tables were described briefly in Section 5.4.2. Each type of model allows associations among different sets of variables and each has a different independence interpretation, as illustrated in Table 5.2.

For a three-way table, the saturated model, denoted $[ABC]$ is

$$\log m_{ijk} = \mu + \lambda_i^A + \lambda_j^B + \lambda_k^C + \lambda_{ij}^{AB} + \lambda_{ik}^{AC} + \lambda_{jk}^{BC} + \lambda_{ijk}^{ABC} . \tag{9.4}$$

This model allows all variables to be associated; Eqn. (9.4) fits the data perfectly because the number of independent parameters equals the number of table cells. Two-way terms, such as λ_{ij}^{AB}, pertain to the *conditional association* between pairs of factors, controlling for the remaining variable. The presence of the three-way term, λ_{ijk}^{ABC}, means that the partial association (conditional odds ratio) between any pair varies over the levels of the third variable.

Omitting the three-way term in Model Eqn. (9.4) gives the model $[AB][AC][BC]$,

$$\log m_{ijk} = \mu + \lambda_i^A + \lambda_j^B + \lambda_k^C + \lambda_{ij}^{AB} + \lambda_{ik}^{AC} + \lambda_{jk}^{BC} , \tag{9.5}$$

in which all pairs are conditionally dependent given the remaining one. For any pair, the conditional odds ratios are the *same* at all levels of the remaining variable, so this model is often called the **homogeneous association model**.

The interpretation of terms in this model may be illustrated using the Berkeley admissions data (Example 4.11 and Example 4.15), for which the factors are Admit, Gender, and Department, in a $2 \times 2 \times 6$ table. In the homogeneous association model,

$$\log m_{ijk} = \mu + \lambda_i^A + \lambda_j^D + \lambda_k^G + \lambda_{ij}^{AD} + \lambda_{ik}^{AG} + \lambda_{jk}^{DG} , \tag{9.6}$$

the λ-parameters have the following interpretations:

- The main effects, λ_i^A, λ_j^D, and λ_k^G pertain to differences in the one-way marginal probabilities. Thus λ_j^D relates to differences in the total number of applicants to these departments, while λ_k^G relates to the differences in the overall numbers of men and women applicants.

- λ_{ij}^{AD} describes the conditional association between admission and department, that is, different admission rates across departments (controlling for gender).

- λ_{ik}^{AG} relates to the conditional association between admission and gender, controlling for department. This term, if significant, might be interpreted as indicating gender-bias in admissions.

- λ_{jk}^{DG}, the association between department and gender, indicates whether males and females apply differentially across departments.

As we discussed earlier (Section 5.4), loglinear models for three-way (and larger) tables often have an interpretation in terms of various types of independence relations, as illustrated in Table 5.2. The model Eqn. (9.5) has no such interpretation. However the smaller model $[AC][BC]$ can be interpreted as asserting that A and B are (conditionally) independent controlling for C; this independence interpretation is symbolized as $A \perp B \mid C$. Similarly, the model $[AB][C]$ asserts that A and B are jointly independent of C: $(A, B) \perp C$, while the model $[A][B][C]$ is the model of mutual (complete) independence, $A \perp B \perp C$.

9.2.3   Loglinear models as GLMs for frequencies

In the GLM approach, a loglinear model may be cast in the form of a regression model for $\log m$, where the table cells are reshaped to a column vector. One advantage is that models for tables of any size and structure may be expressed in a compact form.

For a contingency table of variables $A, B, C, \cdots$, with $N = I \times J \times K \times \cdots$ cells, let n denote a column vector of the observed counts arranged in standard order, and let m denote a similar vector of the expected frequencies under some model. Then *any* loglinear model may be expressed in the form

$$\log m = X\beta \,,$$

where X is a known design or ***model matrix*** and β is a column vector containing the unknown λ parameters.

For example, for a 2×2 table, the saturated model Eqn. (9.2) with the usual zero-sum constraints Eqn. (9.3) can be represented as

$$\log \begin{pmatrix} m_{11} \\ m_{12} \\ m_{21} \\ m_{22} \end{pmatrix} = \begin{bmatrix} 1 & 1 & 1 & 1 \\ 1 & 1 & -1 & -1 \\ 1 & -1 & 1 & -1 \\ 1 & -1 & -1 & 1 \end{bmatrix} \begin{pmatrix} \mu \\ \lambda_1^A \\ \lambda_1^B \\ \lambda_{11}^{AB} \end{pmatrix} \,.$$

Note that only the linearly independent parameters are represented here. $\lambda_2^A = -\lambda_1^A$, because $\lambda_1^A + \lambda_2^A = 0$, and $\lambda_2^B = -\lambda_1^B$, because $\lambda_1^B + \lambda_2^B = 0$, and so forth.

An additional substantial advantage of the GLM formulation is that it makes it easier to express models with ordinal or quantitative variables. `glm()`, with a model formula of the form `Freq ~ .` involving factors $A, B, \ldots$ and quantitative variables $x_1, x_2, \ldots$, constructs the model matrix X from the terms given in the formula. A factor with K levels gives rise to $K - 1$ columns for its main effect and sets of $K - 1$ columns in each interaction effect. A quantitative predictor, say, x_1 (with a linear effect) creates a single column with its values, and interactions with other terms are calculated at the products of the columns for the main effects.

The parameterization for factors is controlled by the contrasts assigned to a given factor (if any), or by the general `contrasts` option, that gives the contrast functions used for unordered and ordered factors:

```
> options("contrasts")

$contrasts
        unordered           ordered
"contr.treatment"       "contr.poly"
```

This says that, by default, unordered factors use the baseline (first) reference-level parameterization, while ordered factors are given a parameterization based on orthogonal polynomials, allowing linear, quadratic, ... effects, assuming integer-spacing of the factor levels.

9.3 Fitting and testing loglinear models

For a given table, possible loglinear models range from the baseline model of mutual independence, $[A][B][C][\ldots]$ to the saturated model, $[ABC\ldots]$ that fits the observed frequencies perfectly, but offers no simpler description or interpretation than the data itself.

Fitting a loglinear model is usually a process of deciding which association terms are large enough ("significantly different from zero") to warrant inclusion in a model to explain the observed frequencies. Terms that are excluded from the model go into the residual or error term, which reflects the overall badness-of-fit of the model. The usual goal of loglinear modeling is to find a small model (few association terms), which nonetheless achieves a reasonable fit (small residuals).

9.3.1 Model fitting functions

In R, the most basic function for fitting loglinear models is `loglin()` in the **stats** package. This uses the classical iterative proportional fitting (IPF) algorithm described in Haberman (1972) and

Fienberg (1980, Section 3.4). It is designed to work with the frequency data in table form, and a model specified in terms of the (high-order) table margins to be fitted. For example, the model Eqn. (9.5) of homogenous association for a three-way table is specified as

```
> loglin(mytable, margin = list(c(1, 2), c(1, 3), c(2, 3)))
```

The variables are represented by their margin index; margins combined in the same vector represent an interaction term. The function `loglm()` in MASS provides a more convenient front-end to `loglin()` to allow loglinear models to be specified using a model formula. With table variables A, B, and C, the same model can be fit using `loglm()` as

```
> loglm(~ (A + B + C)^2, data = mytable)
```

(Note that the formula expression expands to A*A + A*B + A*C + B*A + B*B + B*C + C*A + C*B + C*C. Since terms like A*A become A and duplicate terms are ignored, this eventually yields A + B + C + A*B + A*C + B*C—all second-order terms and the corresponding main effects.)

When the data is a frequency data frame with frequencies in `Freq`, for example, the result of `mydf <- as.data.frame(mytable)`, you can also use a two-sided formula:

```
> loglm(Freq ~ (A + B + C)^2, data = mydf)
```

As implied in Section 9.2.3, loglinear models can also be fit using `glm()`, using `family=poisson`, which constructs the model for `log(Freq)`. The same model is fit with `glm()` as:

```
> glm(Freq ~ (A + B + C)^2, data = mydf, family = poisson)
```

While all of these fit equivalent models, the details of the printed output, model objects, and available methods differ, as indicated in some of the examples that follow.

It should be noted that both the `loglin()`/`loglm()` methods based on iterative proportional fitting, and the `glm()` approach using the Poisson model for log frequency, give maximum likelihood estimates, $\widehat{m}$, of the expected frequencies, as long as all observed frequencies n are *all* positive. Some special considerations when there are cells with zero frequencies are described in Section 9.5.

9.3.2 Goodness-of-fit tests

For an n-way table, global goodness-of-fit tests for a loglinear model attempt to answer the question, "How well does the model reproduce the observed frequencies?" That is, how close are the fitted frequencies estimated under the model to those of the saturated model or the data?

To avoid multiple subscripts for an n-way table, let $n = (n_1, n_2, \ldots, n_N)$ denote the observed frequencies in a table with N cells, and corresponding fitted frequencies $\widehat{m} = (\widehat{m}_1, \widehat{m}_2, \ldots, \widehat{m}_N)$ according to a particular loglinear model. The standard goodness-of-fit statistics are sums over the cells of measures of the difference between the n and $\widehat{m}$.

The most commonly used are the familiar Pearson chi-square,

$$X^2 = \sum_i^N \frac{(n_i - \widehat{m}_i)^2}{\widehat{m}_i} \,, \tag{9.7}$$

and the likelihood-ratio G^2 or ***deviance*** statistic,

$$G^2 = 2 \sum_i^N n_i \log\left(\frac{n_i}{\widehat{m}_i}\right) \,. \tag{9.8}$$

Both of these statistics have asymptotic χ^2 distributions (as $\Sigma n \rightarrow \infty$), reasonably well-approximated when all expected frequencies are large.[1] The (residual) degrees of freedom are the number of cells (N) minus the number of estimated parameters. The likelihood-ratio test can also be expressed as twice the difference in log-likelihoods under saturated and fitted models,

$$G^2 = 2 \log \left[\frac{\mathcal{L}(n; n)}{\mathcal{L}(\widehat{m}; n)} \right] = 2[\log \mathcal{L}(n; n) - \log \mathcal{L}(\widehat{m}; n)] ,$$

where $\mathcal{L}(n; n)$ is the likelihood for the saturated model and $\mathcal{L}(\widehat{m}; n)$ is the corresponding maximized likelihood for the fitted model.

In practice such global tests are less useful for comparing competing models. You may find that several different models have an acceptable fit or, sadly, that none do (usually because you are "blessed" with a large sample size). It is then helpful to compare competing models *directly*, and two strategies are particularly useful in these cases.

First, the likelihood-ratio G^2 statistic has the property in that one can compare two **nested models** by their difference in G^2 statistics, which has a χ^2 distribution on the difference in degrees of freedom. Two models, M_1 and M_2, are nested when one, say, M_2, is a special case of the other. That is, model M_2 (with ν_2 residual df) contains a subset of the parameters of M_1 (with ν_1 residual df), the remaining ones being effectively set to zero. Model M_2 is therefore more restrictive and cannot fit the data better than the more general model M_1, i.e., $G^2(M_2) \geq G^2(M_1)$. The least restrictive of all models, with $G^2 = 0$ and $\nu = 0$ df, is the saturated model for which $\widehat{m} = n$.

Assuming that the less restrictive model M_1 fits, the difference in G^2,

$$\Delta G^2 \equiv G^2(M_2 \,|\, M_1) \quad = \quad G^2(M_2) - G^2(M_1) \qquad (9.9)$$

$$= \quad 2 \sum_i n_i \, \log(\widehat{m}_{i1}/\widehat{m}_{i2}) \qquad (9.10)$$

has a chi-squared distribution with df $= \nu_2 - \nu_1$. The last equality, Eqn. (9.10), follows from substituting in Eqn. (9.8).

Rearranging terms in Eqn. (9.9), we see that we can partition the $G^2(M_2)$ into two terms,

$$G^2(M_2) = G^2(M_1) + G^2(M_2 \,|\, M_1) .$$

The first term measures the difference between the data and the more general model M_1. If this model fits, the second term measures the additional lack of fit imposed by the more restrictive model. In addition to providing a more focused test, $G^2(M_2 \,|\, M_1)$ also follows the chi-squared distribution more closely when some $\{m_i\}$ are small (Agresti, 2013, Section 10.6.3).

Alternatively, a second strategy uses other measures that combine goodness-of-fit with model parsimony and may also be used to compare non-nested models. The statistics described below are all cast in the form of badness-of-fit relative to degrees of freedom, so that smaller values reflect "better" models.

The simplest idea (Goodman, 1971) is to use G^2/df (or χ^2/df), which has an asymptotic expected value of 1 for a good-fitting model. This type of measure is not routinely reported by R software, but is easy to calculate from output.

The ***Akaike Information Criterion*** (AIC) statistic (Akaike, 1973) is a very general criterion for model selection with maximum likelihood estimation, based on the idea of maximizing the information provided by a fitted model. AIC is defined generally as

$$\text{AIC} = -2 \log \mathcal{L} + 2k ,$$

---

[1]Except in bizarre or borderline cases, these tests provide the same conclusions when expected frequencies are at least moderate (all $\widehat{m} > 5$). However, G^2 approaches the theoretical chi-squared distribution more slowly than does χ^2, and the approximation may be poor when the average cell frequency is less than 5.

where $\log \mathcal{L}$ is the maximized log likelihood and k is the number of parameters estimated in the model. Better models correspond to *smaller* AIC. For loglinear models, minimizing AIC is equivalent to minimizing

$$\text{AIC}^\star = G^2 - 2\nu \, ,$$

where ν is the residual df, but the values of AIC and AIC* differ by an arbitrary constant. This form is easier to calculate by hand from the output of any modeling function if AIC is not reported, or an `AIC()` method is not available.

A third statistic of this type is the ***Bayesian Information Criterion*** (BIC) due to Schwartz (1978) and Raftery (1986),

$$\text{BIC} = G^2 - \log(n)\,\nu \, ,$$

where n is the total sample size. Both AIC and BIC penalize the fit statistic for increasing number of parameters. BIC also penalizes the fit directly with (log) sample size, and so expresses a preference for less complex models than AIC as the sample size increases.

9.3.3 Residuals for loglinear models

Test statistics such as G^2 can determine whether a model has significant lack of fit, and model comparison tests using $\Delta G^2 = G^2(M_2 \mid M_1)$ can assess whether the extra term(s) in model M_1 significantly improves the model fit. Beyond these tests, the pattern of residuals for individual cells offers important clues regarding the nature of lack of fit and can suggest associations that could be accounted for better.

As with logistic regression models (Section 7.5.1), several types of residuals are available for loglinear models. For cell i in the vector form of the contingency table, the ***raw residual*** is simply the difference between the observed and fitted frequencies, $e_i = n_i - \widehat{m}_i$.

The ***Pearson residual*** is the square root of the contribution of the cell to the Pearson χ^2,

$$r_i = \frac{n_i - \widehat{m}_i}{\sqrt{\widehat{m}_i}} \, . \tag{9.11}$$

Similarly, the ***deviance residual*** can be defined as

$$g_i = \text{sign}(n_i - \widehat{m}_i)\sqrt{2n_i \log(n_i/\widehat{m}_i) - 2(n_i - \widehat{m}_i)} \, . \tag{9.12}$$

Both of these attempt to standardize the distribution of the residuals to a standard normal, $N(0, 1)$ form. However, as pointed out by Haberman (1973), the asymptotic variance of these is less than one (with average value df/N) but, worse—the variance decreases with $\widehat{m}_i$. That is, residuals for cells with small expected frequencies have larger sampling variance as might be expected.

Consequently, Haberman suggested dividing the Pearson residual by its estimated standard error, giving what are often called ***adjusted residuals***. When loglinear models are fit using the GLM approach, the adjustment may be calculated using the leverage ("hat value"), h_i to give appropriately standardized residuals,

$$
\begin{aligned}
r_i^\star &= r_i/\sqrt{1 - h_i} \, , \\
g_i^\star &= g_i/\sqrt{1 - h_i} \, .
\end{aligned}
$$

These standardized versions are generally preferable, particularly for visualizing model lack of fit using mosaic displays. The reason for preferring adjusted residuals is illustrated in Figure 9.1, a plot of the factors, $\sqrt{1 - h_i}$, determining the standard errors of the residuals against the fitted values, $\widehat{m}_i$, in the model for the `UCBAdmissions` data described in Example 9.2 below. The values shown in this plot are calculated as:

```
> berkeley <- as.data.frame(UCBAdmissions)
> berk.glm1 <- glm(Freq ~ Dept * (Gender + Admit), data = berkeley,
+                   family = "poisson")
> fit <- fitted(berk.glm1)
> hat <- hatvalues(berk.glm1)
> stderr <- sqrt(1 - hat)
```

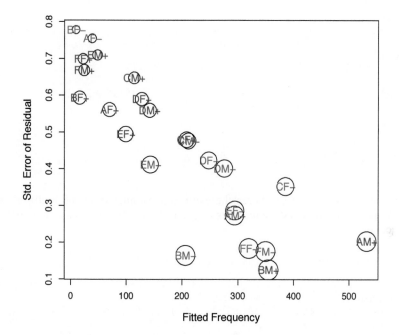

Figure 9.1: Standard errors of residuals, $\sqrt{1 - h_i}$ decrease with expected frequencies. This plot shows why ordinary Pearson and deviance residuals may be misleading. The symbol size in the plot is proportional to leverage, h_i. Labels abbreviate Department, Gender, and Admit, colored by Admit.

In R, raw, Pearson and deviance residuals may be obtained using `residuals(model, type=)`, where `type` is one of `"raw"`, `"pearson"`, and `"deviance"`. Standardized (adjusted) residuals can be calculated using `rstandard(model, type=)`, for `type="pearson"` and `type="deviance"` versions.

9.3.4 Using `loglm()`

Here we illustrate the basics of fitting loglinear models using `loglm()`. As indicated in Section 9.3.1, the model to be fitted is specified by a model formula involving the table variables. The **MASS** package provides a `coef()` method for "loglm" objects that extracts the estimated parameters and a `residuals()` method that calculates various types of residuals according to a `type` argument, one of `"deviance"`, `"pearson"`, `"response"`. **vcd** and **vcdExtra** provide a variety of plotting methods, including `assoc()`, `sieve()`, `mosaic()`, and `mosaic3d()` for "loglm" objects.

EXAMPLE 9.1: Berkeley admissions

The *UCBAdmissions* on admissions to the six largest graduate departments at U.C. Berkeley was examined using graphical methods in Chapter 4 (Example 4.15) and in Chapter 5 (Example 5.14). We can fit and compare several loglinear models as shown below.

The model of mutual independence, $[A][D][G]$, is not substantively reasonable here, because the association of Dept and Gender should be taken into account to control for these variables, but we show it here to illustrate the form of the printed output, giving the Pearson χ^2 and likelihood-ratio G^2 tests of goodness of fit, as well as some optional arguments for saving additional components in the result.

```
> data("UCBAdmissions")
> library(MASS)
> berk.loglm0 <- loglm(~ Dept + Gender + Admit, data = UCBAdmissions,
+                       param = TRUE, fitted = TRUE)
> berk.loglm0

Call:
loglm(formula = ~Dept + Gender + Admit, data = UCBAdmissions,
    param = TRUE, fitted = TRUE)

Statistics:
                  X^2 df P(> X^2)
Likelihood Ratio 2097.7 16        0
Pearson          2000.3 16        0
```

The argument param = TRUE stores the estimated parameters in the loglinear model and fitted = TRUE stores the fitted frequencies $\hat{m}_{ijk}$. The fitted frequencies can be extracted from the model object using fitted().

```
> structable(Dept ~ Admit + Gender, fitted(berk.loglm0))

                  Dept      A       B       C       D       E       F
Admit     Gender
Admitted  Male          215.10 134.87 211.64 182.59 134.64 164.61
          Female        146.68  91.97 144.32 124.51  91.81 112.25
Rejected  Male          339.63 212.95 334.17 288.30 212.59 259.91
          Female        231.59 145.21 227.87 196.59 144.96 177.23
```

Similarly, you can extract the estimated parameters with coef(berk.loglm0), and the Pearson residuals with residuals(berk.loglm0, type = "pearson").

Next, consider the model of conditional independence of gender and admission given department, $[AD][GD]$, which allows associations of admission with department and gender with department.

```
> # conditional independence in UCB admissions data
> berk.loglm1 <- loglm(~ Dept * (Gender + Admit), data = UCBAdmissions)
> berk.loglm1

Call:
loglm(formula = ~Dept * (Gender + Admit), data = UCBAdmissions)

Statistics:
                  X^2 df   P(> X^2)
Likelihood Ratio 21.736  6 0.0013520
Pearson          19.938  6 0.0028402
```

Finally, for this example, the model of homogeneous association, $[AD][AG][GD]$, can be fit as follows.[2]

---

[2]It is useful to note here that the added term $[AG]$ allows a general association of admission with gender (controlling for department). A significance test for this term, or for model berk.loglm2 against berk.loglm1, is a proper test for the assertion of gender bias in admissions.

```
> berk.loglm2 <-loglm(~ (Admit + Dept + Gender)^2, data = UCBAdmissions)
> berk.loglm2

Call:
loglm(formula = ~(Admit + Dept + Gender)^2, data = UCBAdmissions)

Statistics:
                         X^2 df  P(> X^2)
Likelihood Ratio 20.204   5 0.0011441
Pearson          18.823   5 0.0020740
```

Neither of these models fits particularly well, as judged by the goodness-of-fit Pearson χ^2 and likelihood-ratio G^2 test against the saturated model. The anova() method for a nested collection of "loglm" models gives a series of likelihood-ratio tests of the difference, ΔG^2, between each sequential pair of models, according to Eqn. (9.9).

```
> anova(berk.loglm0, berk.loglm1, berk.loglm2, test = "Chisq")

LR tests for hierarchical log-linear models

Model 1:
 ~Dept + Gender + Admit
Model 2:
 ~Dept * (Gender + Admit)
Model 3:
 ~ (Admit + Dept + Gender)^2

          Deviance df Delta(Dev) Delta(df) P(> Delta(Dev)
Model 1   2097.671 16
Model 2     21.736  6  2075.9357        10        0.00000
Model 3     20.204  5     1.5312         1        0.21593
Saturated    0.000  0    20.2043         5        0.00114
```

The conclusion from these results is that the model berk.loglm1 is not much worse than model berk.loglm2, but there is still significant lack of fit. The next example, using glm(), shows how to visualize the lack of fit and account for it.

△

9.3.5 Using glm()

Loglinear models fit with glm() require the data in a data frame in frequency form, for example as produced by as.data.frame() from a table. The model formula expresses the model for the frequency variable, and uses family = poisson to specify the error distribution. More general distributions for frequency data are discussed in Chapter 11.

EXAMPLE 9.2: Berkeley admissions
For the $2 \times 2 \times 6$ *UCBAdmissions* table, first transform this to a frequency data frame:

```
> berkeley <- as.data.frame(UCBAdmissions)
> head(berkeley)

     Admit Gender Dept Freq
1 Admitted   Male    A  512
2 Rejected   Male    A  313
3 Admitted Female    A   89
4 Rejected Female    A   19
5 Admitted   Male    B  353
6 Rejected   Male    B  207
```

Then, the model of conditional independence corresponding to `berk.loglm1` can be fit using `glm()` as shown below.

```
> berk.glm1 <- glm(Freq ~ Dept * (Gender + Admit),
+                  data = berkeley, family = "poisson")
```

Similarly, the all two-way model of homogeneous association is fit using

```
> berk.glm2 <- glm(Freq ~ (Dept + Gender + Admit)^2,
+                  data = berkeley, family = "poisson")
```

These models are equivalent to those fit using `loglm()` in Example 9.1. We get the same residual G^2 as before, and the likelihood-ratio test of ΔG^2 given by `anova()` gives the same result, that the model `berk.glm2` offers no significant improvement over model `berk.glm1`.

```
> anova(berk.glm1, berk.glm2, test = "Chisq")

Analysis of Deviance Table

Model 1: Freq ~ Dept * (Gender + Admit)
Model 2: Freq ~ (Dept + Gender + Admit)^2
  Resid. Df Resid. Dev Df Deviance Pr(>Chi)
1         6       21.7
2         5       20.2  1     1.53     0.22
```

Among other advantages of using `glm()` as opposed to `loglm()` is that an `anova()` method is available for *individual* "**glm**" models, giving significance tests of the contributions of each *term* in the model, as opposed to the tests for individual coefficients provided by `summary()`.[3]

```
> anova(berk.glm1, test = "Chisq")

Analysis of Deviance Table

Model: poisson, link: log

Response: Freq

Terms added sequentially (first to last)


            Df Deviance Resid. Df Resid. Dev Pr(>Chi)
NULL                          23       2650
Dept         5      160        18       2491   <2e-16 ***
Gender       1      163        17       2328   <2e-16 ***
Admit        1      230        16       2098   <2e-16 ***
Dept:Gender  5     1221        11        877   <2e-16 ***
Dept:Admit   5      855         6         22   <2e-16 ***
---
Signif. codes:  0 '***' 0.001 '**' 0.01 '*' 0.05 '.' 0.1 ' ' 1
```

We proceed to consider what is wrong with these models and how they can be improved. A mosaic display can help diagnose the reason(s) for lack of fit of these models. We focus here on the model $[AD][GD]$ that allows an association between gender and department (i.e., men and women apply at different rates to departments).

---

[3]Unfortunately, in the historical development of R, the `anova()` methods for linear and generalized linear models provide only *sequential* ("Type I") tests that are computationally easy, but useful only under special circumstances. The car package provides an analogous `Anova()` method that gives more generally useful *partial* ("Type II") tests for the additional contribution of each term beyond the others, taking marginal relations into account.

The `mosaic()` method for `"glm"` objects in **vcdExtra** provides a `residuals_type` argument, allowing `residuals_type = "rstandard"` for standardized residuals. The `formula` argument here pertains to the order of the variables in the mosaic, not a model formula.

```
> library(vcdExtra)
> mosaic(berk.glm1, shade = TRUE,
+        formula = ~ Dept + Admit + Gender, split = TRUE,
+        residuals_type = "rstandard",
+        main = "Model: [AdmitDept][GenderDept]",
+        labeling = labeling_residuals,
+        abbreviate_labs = c(Gender = TRUE),
+        keep_aspect_ratio = FALSE)
```

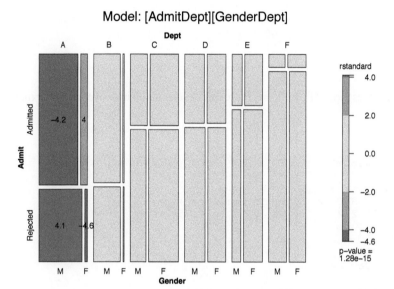

Figure 9.2: Mosaic display for the model [AD][GD], showing standardized residuals for the cell contributions to G^2.

The mosaic display, shown in Figure 9.2, indicates that this model fits well (residuals are small) except in Department A. This suggests a model that allows an association between Admission and Gender in Department A only,

$$\log m_{ijk} = \mu + \lambda_i^A + \lambda_j^D + \lambda_k^G + \lambda_{ij}^{AD} + \lambda_{jk}^{DG} + I(j = 1)\lambda_{ik}^{AG}, \tag{9.13}$$

where the indicator function $I(j = 1)$ equals 1 for Department A ($j = 1$) and is zero otherwise. This model asserts that Admission and Gender are conditionally independent, given Department, except in Department A. It has one more parameter than the conditional independence model, $[AD][GD]$, and forces perfect fit in the four cells for Department A.

Model Eqn. (9.13) may be fit with `glm()` by constructing a variable equal to the interaction of `gender` and `admit` with a dummy variable having the value 1 for Department A and 0 for other departments.

```
> berkeley <- within(berkeley,
+                    dept1AG <- (Dept == "A") *
+                        (Gender == "Female") *
+                        (Admit == "Admitted"))
> head(berkeley)
```

```
      Admit Gender Dept Freq dept1AG
1 Admitted   Male    A  512        0
2 Rejected   Male    A  313        0
3 Admitted Female    A   89        1
4 Rejected Female    A   19        0
5 Admitted   Male    B  353        0
6 Rejected   Male    B  207        0
```

Fitting this model with the extra term dept1AG gives berk.glm3

```
> berk.glm3 <- glm(Freq ~ Dept * (Gender + Admit) + dept1AG,
+                    data = berkeley, family = "poisson")
```

This model does indeed fit well, and represents a substantial improvement over model berk.glm1:

```
> LRstats(berk.glm3)

Likelihood summary table:
          AIC BIC LR Chisq Df Pr(>Chisq)
berk.glm3 200 222   2.68  5        0.75

> anova(berk.glm1, berk.glm3, test = "Chisq")

Analysis of Deviance Table

Model 1: Freq ~ Dept * (Gender + Admit)
Model 2: Freq ~ Dept * (Gender + Admit) + dept1AG
  Resid. Df Resid. Dev Df Deviance Pr(>Chi)
1         6      21.74
2         5       2.68  1    19.1  1.3e-05 ***
---
Signif. codes:  0 '***' 0.001 '**' 0.01 '*' 0.05 '.' 0.1 ' ' 1
```

The parameter estimate for the dept1AG term, $\widehat{\lambda}_{ik}^{AG} = 1.052$, may be interpreted as the log odds ratio of admission for females as compared to males in Dept. A. The odds ratio is $\exp(1.052) = 2.86$, the same as the value calculated from the raw data (see Section 4.4.2).

```
> coef(berk.glm3)[["dept1AG"]]

[1] 1.0521

> exp(coef(berk.glm3)[["dept1AG"]])

[1] 2.8636
```

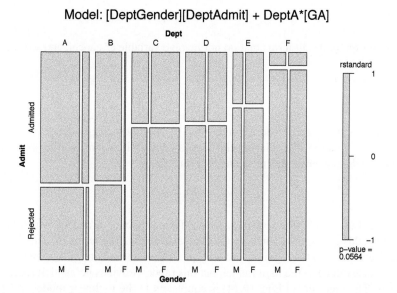

Figure 9.3: Mosaic display for the model `berk.glm3`, allowing an association of gender and admission in Department A. This model now fits the data well.

Finally, Figure 9.3 shows the mosaic for this revised model. The absence of shading indicates a well-fitting model.

```
> mosaic(berk.glm3, shade = TRUE,
+        formula = ~ Dept + Admit + Gender, split = TRUE,
+        residuals_type = "rstandard",
+        main = "Model: [DeptGender][DeptAdmit] + DeptA*[GA]",
+        labeling = labeling_residuals,
+        abbreviate_labs = c(Gender = TRUE),
+        keep_aspect_ratio = FALSE)
```

△

9.4 Equivalent logit models

Because loglinear models are formulated as models for the log (expected) frequency, they make no distinction between response and explanatory variables. In effect, they treat all variables as responses and describe their associations.

Logit (logistic regression) models, on the other hand, describe how the log odds for one variable depends on other, explanatory variables. There is a close connection between the two: When there is a response variable, each logit model for that response is equivalent to a loglinear model.

This relationship often provides a simpler way to formulate and test the model, and to plot and interpret the fitted results. Even when there is no response variable, the logit representation for one variable helps to interpret a loglinear model in terms of odds ratios. The price paid for this simplicity is that associations among the explanatory variables are not expressed in the model.

Consider, for example, the model of homogeneous association, $[AB][AC][BC]$, Eqn. (9.5), for a three-way table, and let variable C be a binary response. Under this model, the logit for variable C is

$$L_{ij} = \log\left(\frac{\pi_{ij|1}}{\pi_{ij|2}}\right) = \log\left(\frac{m_{ij1}}{m_{ij2}}\right)$$

$$= \log(m_{ij1}) - \log(m_{ij2}) \, .$$

Substituting from Eqn. (9.5), all terms that do not involve variable C cancel, and we are left with

$$
\begin{aligned}
L_{ij} = \log(m_{ij1}/m_{ij2}) &= (\lambda_1^C - \lambda_2^C) + (\lambda_{i1}^{AC} - \lambda_{i2}^{AC}) + (\lambda_{j1}^{BC} - \lambda_{j2}^{BC}) \\
&= 2\lambda_1^C + 2\lambda_{i1}^{AC} + 2\lambda_{j1}^{BC} \, ,
\end{aligned}
\tag{9.14}
$$

because all λ terms sum to zero. We are interested in how these logits depend on A and B, so we can simplify the notation by replacing the λ parameters with more familiar ones, $\alpha = 2\lambda_1^C$, $\beta_i^A = 2\lambda_{i1}^{AC}$, etc., which express this relation more directly,

$$L_{ij} = \alpha + \beta_i^A + \beta_j^B \, . \tag{9.15}$$

In the logit model Eqn. (9.15), the response, C, is affected by both A and B, which have additive effects on the log odds of response category C_1 compared to C_2. The terms β_i^A and β_j^B correspond directly to $[AC]$ and $[BC]$ in the loglinear model Eqn. (9.5). The association among the explanatory variables, $[AB]$, is assumed in the logit model, but this model provides no explicit representation of that association. The logit model Eqn. (9.14) is equivalent to the loglinear model $[AB][AC][BC]$ in goodness-of-fit and fitted values, and parameters in the two models correspond directly.

Table 9.1: Equivalent loglinear and logit models for a three-way table, with C as a binary response variable

Loglinear model	Logit model	Logit formula
$[AB][C]$	α	C ~ 1
$[AB][AC]$	$\alpha + \beta_i^A$	C ~ A
$[AB][BC]$	$\alpha + \beta_j^B$	C ~ B
$[AB][AC][BC]$	$\alpha + \beta_i^A + \beta_j^B$	C ~ A + B
$[ABC]$	$\alpha + \beta_i^A + \beta_j^B + \beta_{ij}^{AB}$	C ~ A * B

Table 9.1 shows the equivalent relationships between all loglinear and logit models for a three-way table when variable C is a binary response. Each model necessarily includes the $[AB]$ association involving the predictor variables. The most basic model, $[AB][C]$, is the intercept-only model, asserting constant odds for variable C. The saturated loglinear model, $[ABC]$, allows an interaction in the effects of A and B on C, meaning that the AC association or odds ratio varies with B.

More generally, when there is a binary response variable, say R, and one or more explanatory variables, $A, B, C, \ldots$, any logit model for R has an equivalent loglinear form. Every term in the logit model, such as β_{ik}^{AC}, corresponds to an association of those factors with R, that is, $[ACR]$ in the equivalent loglinear model.

The equivalent loglinear model must also include all associations among the explanatory factors, the term $[ABC \ldots]$. Conversely, any loglinear model that includes all associations among the explanatory variables has an equivalent logit form. When the response factor has more than two categories, models for generalized logits (Section 8.3) also have an equivalent loglinear form.

EXAMPLE 9.3: Berkeley admissions

The homogeneous association model, $[AD][AG][DG]$, did not fit the `UCBAdmissions` data very well, and we saw that the term $[AG]$ was unnecessary. Nevertheless, it is instructive to consider the equivalent logit model. We illustrate the features of the logit model that lead to the same conclusions and simplified interpretation from graphical displays.

Because Admission is a binary response variable, model Eqn. (9.6) is equivalent to the logit model,

$$L_{ij} = \log\left(\frac{m_{\text{Admit}(ij)}}{m_{\text{Reject}(ij)}}\right) = \alpha + \beta_i^{\text{Dept}} + \beta_j^{\text{Gender}}. \tag{9.16}$$

That is, the logit model Eqn. (9.16) asserts that department and gender have additive effects on the log odds of admission. A significance test for the term β_j^{Gender} here is equivalent to the test of the [AG] term for gender bias in the loglinear model. The observed log odds of admission here can be calculated as:

```
> (obs <- log(UCBAdmissions[1,,] / UCBAdmissions[2,,]))

        Dept
Gender       A       B        C       D       E       F
  Male    0.4921  0.5337  -0.5355  -0.704  -0.957  -2.770
  Female  1.5442  0.7538  -0.6604  -0.622  -1.157  -2.581
```

With the data in the form of the frequency data frame `berkeley` we used in Example 9.2, the logit model Eqn. (9.16) can be fit using `glm()` as shown below. In the model formula, the binary response is `Admit == "Admitted"`. The `weights` argument gives the frequency, `Freq` in each table cell.[4]

```
> berk.logit2 <- glm(Admit == "Admitted" ~ Dept + Gender,
+                data = berkeley, weights = Freq, family = "binomial")
> summary(berk.logit2)


Call:
glm(formula = Admit == "Admitted" ~ Dept + Gender, family = "binomial",
    data = berkeley, weights = Freq)

Deviance Residuals:
    Min      1Q   Median      3Q     Max
-25.342  -13.058  -0.163  16.017  21.320

Coefficients:
              Estimate Std. Error z value Pr(>|z|)
(Intercept)     0.5821     0.0690    8.44   <2e-16 ***
DeptB          -0.0434     0.1098   -0.40     0.69
DeptC          -1.2626     0.1066  -11.84   <2e-16 ***
DeptD          -1.2946     0.1058  -12.23   <2e-16 ***
DeptE          -1.7393     0.1261  -13.79   <2e-16 ***
DeptF          -3.3065     0.1700  -19.45   <2e-16 ***
GenderFemale    0.0999     0.0808    1.24     0.22
---
Signif. codes:  0 '***' 0.001 '**' 0.01 '*' 0.05 '.' 0.1 ' ' 1

(Dispersion parameter for binomial family taken to be 1)

    Null deviance: 6044.3  on 23  degrees of freedom
Residual deviance: 5187.5  on 17  degrees of freedom
AIC: 5201

Number of Fisher Scoring iterations: 6
```

As in logistic regression models, parameter estimates may be interpreted as increments in the log odds, or $\exp(\beta)$ may be interpreted as the multiple of the odds associated with the explanatory categories. Because `glm()` uses a baseline category parameterization (by default), the coefficients of the first category of `Dept` and `Gender` are set to zero. You can see from the `summary()`

---

[4]Using weights gives the same fitted values, but not the same LR tests for model fit.

output that the coefficients for the departments decline steadily from A–F.[5] The coefficient $\beta_F^{\text{Gender}} = 0.0999$ for females indicates that, overall, women were $\exp(0.0999) = 1.105$ times as likely as male applicants to be admitted to graduate school at U.C. Berkeley, a 10% advantage.

Similarly, the logit model equivalent of the loglinear model Eqn. (9.13) `berk.glm3` containing the extra 1 df term for an effect of gender in Department A is

$$L_{ij} = \alpha + \beta_i^{\text{Dept}} + I(j = 1)\beta^{\text{Gender}} . \tag{9.17}$$

This model can be fit as follows:

```
> berkeley <- within(berkeley,
+                    dept1AG <- (Dept == "A") * (Gender == "Female"))
> berk.logit3 <- glm(Admit == "Admitted" ~ Dept + Gender + dept1AG,
+                    data = berkeley, weights = Freq, family = "binomial")
```

In contrast to the tests for individual coefficients, the `Anova()` method in the **car** package gives likelihood-ratio tests of the terms in a model. As mentioned earlier, this provides *partial* ("Type II") tests for the additional contribution of each term beyond all others.

```
> library(car)
> Anova(berk.logit2)

Analysis of Deviance Table (Type II tests)

Response: Admit == "Admitted"
       LR Chisq Df Pr(>Chisq)
Dept      763.4  5     <2e-16 ***
Gender      1.5  1      0.216
---
Signif. codes:  0 '***' 0.001 '**' 0.01 '*' 0.05 '.' 0.1 ' ' 1

> Anova(berk.logit3)

Analysis of Deviance Table (Type II tests)

Response: Admit == "Admitted"
       LR Chisq Df Pr(>Chisq)
Dept      646.7  5    < 2e-16 ***
Gender      0.1  1      0.724
dept1AG    17.6  1   2.66e-05 ***
---
Signif. codes:  0 '***' 0.001 '**' 0.01 '*' 0.05 '.' 0.1 ' ' 1
```

Plotting logit models

Logit models are easier to interpret than the corresponding loglinear models because there are fewer parameters, and because these parameters pertain to the odds of a response category rather than to cell frequency. Nevertheless, interpretation is often easier still from a graph than from the parameter values.

The simple interpretation of these logit models can be seen by plotting the logits for a given model. To do that, it is necessary to construct a data frame containing the observed (`obs`) and fitted (`fit`) for the combinations of gender and department.

```
> pred2 <- cbind(berkeley[,1:3], fit = predict(berk.logit2))
> pred2 <- cbind(subset(pred2, Admit == "Admitted"), obs = as.vector(obs))
> head(pred2)
```

---

[5] In fact, the departments were labeled A–F in decreasing order of rate of admission.

	Admit	Gender	Dept	fit	obs
1	Admitted	Male	A	0.58205	0.49212
3	Admitted	Female	A	0.68192	1.54420
5	Admitted	Male	B	0.53865	0.53375
7	Admitted	Female	B	0.63852	0.75377
9	Admitted	Male	C	-0.68055	-0.53552
11	Admitted	Female	C	-0.58068	-0.66044

In this form, these results can be plotted as a line plot of the fitted logits vs. department, with separate curves for males and females, and adding points to show the observed values. Here, we use ggplot2 as shown below, with the aes() arguments group = Gender, color = Gender. This produces the left panel in Figure 9.4. The same steps for the model berk.logit3 gives the right panel in this figure. The observed logits, of course, are the same in both plots.

```
> library(ggplot2)
> ggplot(pred2, aes(x = Dept, y = fit, group = Gender, color = Gender)) +
+    geom_line(size = 1.2) +
+    geom_point(aes(x = Dept, y = obs, group = Gender, color = Gender),
+               size = 4) +
+    ylab("Log odds (Admitted)") + theme_bw() +
+    theme(legend.position = c(.8, .9),
+          legend.title = element_text(size = 14),
+          legend.text = element_text(size = 14))
```

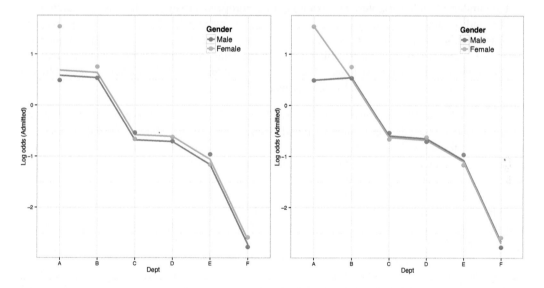

Figure 9.4: Observed (points) and fitted (lines) log odds of admissions in the logit models for the *UCBAdmissions* data. Left: the logit model Eqn. (9.16) corresponding to the loglinear model [AD] [AG] [DG]. Right: the logit model Eqn. (9.17), allowing only a 1 df term for Department A.

The effects seen in our earlier analyses (Examples 5.14, 5.15, and 9.2) may all be observed in these plots. In the left panel of Figure 9.4, corresponding to the loglinear model $[AD][AG][DG]$, the effect of gender, β_j^{Gender}, in the equivalent logit model is shown by the constant separation between the two curves. From the plot we see that this effect is very small (and nonsignificant). In the right panel, corresponding to the logit model Eqn. (9.17), there is no effect of gender on admission, except in department A, where the extra parameter allows perfect fit.

$\triangle$

9.5   Zero frequencies

Cells with frequencies of zero create problems for loglinear and logit models. For loglinear models, most of the derivations of expected frequencies by maximum likelihood and other quantities that depend on these (e.g., G^2 tests) assume that all $n_{ijk\cdots} > 0$. In analogous logit models, the observed log odds (e.g., for a three-way table), $\log(n_{ij1}/n_{ij2})$, will be undefined if either frequency is zero.

Zero frequencies may occur in contingency tables for two different reasons:

- *structural zeros* (also called *fixed zeros*) will occur when it is impossible to observe values for some combinations of the variables. For these cases we should have $\widehat{m}_i = 0$ wherever $n_i = 0$. For example, suppose we have three different methods of contacting people at risk for some obscure genetically inherited disease: newspaper advertisement, telephone campaign, and radio appeal. If each person contacted in any way is classified dichotomously by the three methods of contact, there can never be a non-zero frequency in the 'No-No-No' cell.[6] Similarly, in a tabulation of seniors by gender and health concerns, there can never be males citing menopause or females citing prostate cancer. Square tables, such as wins and losses for sporting teams often have structural zeros in the main diagonal.

- *sampling zeros* (also called *random zeros*) occur when the total size of the sample is not large enough in relation to the probabilities in each of the cells to assure that someone will be observed in every cell. Here, it is permissible to have $\widehat{m}_i > 0$ when $n_i = 0$. This problem increases with the number of table variables. For example, in a European survey of religious affiliation, gender, and occupation, we may not happen to observe any female Muslim vineyard-workers in France, although such individuals surely exist in the population. Even when zero frequencies do not occur, tables with many cells relative to the total frequency tend to produce small expected frequencies in at least some cells, which tends to make the G^2 statistics for model fit and likelihood-ratio statistics for individual terms unreliable.

Following Birch (1963b), Haberman (1974) and many others (e.g., Bishop et al., 1975) identified conditions under which the maximum likelihood estimate for a given loglinear model does not exist, meaning that the algorithms used in `loglin()` and `glm()` do not converge to a solution. The problem depends on the number and locations of the zero cells, but not on the size of the frequencies in the remaining cells. Fienberg and Rinaldo (2007) give a historical overview of the problem and current approaches, and Agresti (2013, Section 10.6) gives a compact summary.

In R, the mechanism to handle structural zeros in the IPF approach of `loglin()` and `loglm()` is to supply the argument `start`, giving a table conforming to the data, containing values of 0 in the locations of the zero cells, and non-zero elsewhere.[7] In the `glm()` approach, the argument `subset=Freq > 0` can be used to remove the cells with zero frequencies from the data; alternatively, zero frequencies can be set to `NA`. This usually provides the correct degrees of freedom; however, some estimated coefficients may be infinite.

For a complete table, the residual degrees of freedom are determined as

$$df = \#\text{ of cells} - \#\text{ of fitted parameters} .$$

For tables with structural zeros, an analogous general formula is

$$df = (\#\text{ cells} - \#\text{ of parameters}) - (\#\text{ zero cells} - \#\text{ of NA parameters}) , \qquad (9.18)$$

---

[6]Yet, if we fit an unsaturated model, expected frequencies may be estimated for all cells, and provide a means to estimate the total number at risk in the population. See Lindsey (1995, Section 5.4).

[7]If structural zeros are present, the calculation of degrees of freedom may not be correct. `loglm()` deducts one degree of freedom for each structural zero, but cannot make allowance for patterns of zeros based on the fitted margins that lead to gains in degrees of freedom due to smaller dimension in the parameter space. `loglin()` makes no such correction.

where NA parameters refers to parameters that cannot be estimated due to zero marginal totals in the model formula.

In contrast, sampling zeros are often handled by some modification of the data frequencies to ensure all non-zero cells. Some suggestions are:

- Add a small positive quantity (0.5 is often recommended) to *every* cell in the contingency table (Goodman, 1970), as is often done in calculating empirical log odds (Example 10.10); this simple approach over-smooths the data for unsaturated models, and should be deprecated, although it is widely used in practice.
- Replace sampling zeros by some small number, typically 10^{-10} or smaller (Agresti, 1990).
- Add a small quantity, like 0.1, to *all* zero cells, sampling or structural (Evers and Namboordiri, 1977).

In complex, sparse tables, a sensitivity analysis, comparing different approaches, can help determine if the substantive conclusions vary with the approach to zero cells.

EXAMPLE 9.4: Health concerns of teenagers

Fienberg (1980, Table 8-3) presented a classic example of structural zeros in the analysis of the $4 \times 2 \times 2$ table shown in Table 9.2. The data come from a survey of health concerns among teenagers, originally from Brunswick (1971). Among the health concerns, the two zero entries for menstrual problems among males are clearly structural zeros and therefore one structural zero in the concern-by-gender marginal table. As usual, we abbreviate the table variables concern, age, and gender by their initial letters, C, A, G below.

Table 9.2: Results from a survey of teenagers, regarding their health concerns. *Note*: Two cells with structural zeros are highlighted. *Source:* Fienberg (1980, Table 8-3)

| Health | Gender: | Male | | Female | |
Concerns	Age:	12–15	16–17	12–15	16–17
sex, reproduction		4	2	9	7
menstrual problems		**0**	**0**	4	8
how healthy I am		42	7	19	10
nothing		57	20	71	21

Note: Two cells with structural zeros are highlighted. *Source:* Fienberg (1980, Table 8-3)

The `Health` data is created as a frequency data frame as follows.

```
> Health <- expand.grid(concerns = c("sex", "menstrual",
+                                     "healthy", "nothing"),
+                       age      = c("12-15", "16-17"),
+                       gender   = c("M", "F"))
> Health$Freq <- c(4, 0, 42, 57, 2, 0, 7, 20,
+                  9, 4, 19, 71, 7, 8, 10, 21)
```

In this form, we first use `glm()` to fit two small models, neither of which involves the $\{CG\}$ margin. Model `health.glm0` is the model of mutual independence, $[C][A][G]$. Model `health.glm1` is the model of joint independence, $[C][AG]$, allowing an association between age and gender, but neither with concern. As noted above, the argument `subset = (Freq>0)` is used to eliminate the structural zero cells.

```
> health.glm0 <- glm(Freq ~ concerns + age + gender, data = Health,
+                    subset = (Freq > 0), family = poisson)
> health.glm1 <- glm(Freq ~ concerns + age * gender, data = Health,
+                    subset = (Freq > 0), family = poisson)
```

Neither of these fits the data well. To conserve space, we show only the results of the G^2 tests for model fit.

```
> LRstats(health.glm0, health.glm1)

Likelihood summary table:
             AIC BIC LR Chisq Df Pr(>Chisq)
health.glm0 100.7 105     27.7  8    0.00053 ***
health.glm1  99.9 104     24.9  7    0.00080 ***
---
Signif. codes:  0 '***' 0.001 '**' 0.01 '*' 0.05 '.' 0.1 ' ' 1
```

To see why, Figure 9.5 shows the mosaic display for model `health.glm1`, $[C][AG]$. Note that `mosaic()` takes care to make cells of zero frequency more visible by marking them with a small "0," as these have an area of zero.

```
> mosaic(health.glm1, ~ concerns + age + gender,
+        residuals_type = "rstandard",
+        rot_labels = 0,
+        just_labels = c(left = "right"),
+        margin = c(left = 5))
```

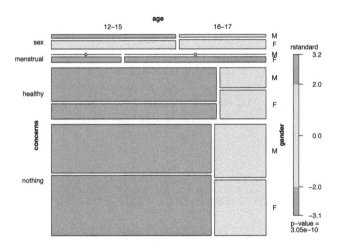

Figure 9.5: Mosaic display for the Health data, model `health.glm1`.

This suggests that there are important associations at least between concern and gender ($[CG]$) and between concern and age ($[CA]$). These are incorporated into the next model:

```
> health.glm2 <- glm(Freq ~ concerns*gender + concerns*age, data = Health,
+                    subset = (Freq > 0), family = poisson)
> LRstats(health.glm2)

Likelihood summary table:
             AIC   BIC LR Chisq Df Pr(>Chisq)
health.glm2 87.7 94.7    4.66  3        0.2
```

The degrees of freedom are correct here. Eqn. (9.18), with 2 zero cells and 1 NA parameter due to the zero in the $\{CG\}$ margin, gives $df = (16 - 12) - (2 - 1) = 3$. The loss of one estimable parameter can be seen in the output from `summary`.

```
> summary(health.glm2)


Call:
glm(formula = Freq ~ concerns * gender + concerns * age, family = poisson,
    data = Health, subset = (Freq > 0))

Deviance Residuals:
     1        3        4        5        7        8        9       10       11       12
 0.236    0.585   -0.173   -0.300   -1.202    0.302   -0.149    0.000   -0.795    0.158
    13       14       15       16
 0.176    0.000    1.348   -0.282

Coefficients: (1 not defined because of singularities)
                             Estimate Std. Error z value Pr(>|z|)
(Intercept)                     1.266      0.445    2.84   0.0045 **
concernsmenstrual              -0.860      0.586   -1.47   0.1425
concernshealthy                 2.380      0.471    5.05  4.4e-07 ***
concernsnothing                 2.800      0.462    6.07  1.3e-09 ***
genderF                         0.981      0.479    2.05   0.0405 *
age16-17                       -0.368      0.434   -0.85   0.3964
concernsmenstrual:genderF          NA         NA      NA       NA
concernshealthy:genderF        -1.505      0.533   -2.82   0.0047 **
concernsnothing:genderF        -0.803      0.503   -1.60   0.1105
concernsmenstrual:age16-17      1.061      0.750    1.41   0.1574
concernshealthy:age16-17       -0.910      0.513   -1.77   0.0761 .
concernsnothing:age16-17       -0.771      0.469   -1.64   0.1005
---
Signif. codes:  0 '***' 0.001 '**' 0.01 '*' 0.05 '.' 0.1 ' ' 1

(Dispersion parameter for poisson family taken to be 1)

    Null deviance: 252.4670  on 13  degrees of freedom
Residual deviance:    4.6611  on  3  degrees of freedom
AIC: 87.66

Number of Fisher Scoring iterations: 4
```

In contrast, loglm() reports the degrees of freedom incorrectly for models containing zeros in any fitted margin. For use with loglm(), we convert it to a $4 \times 2 \times$ table.

```
> health.tab <- xtabs(Freq ~ concerns + age + gender, data = Health)
```

The same three models are fitted with loglm() as shown below. The locations of the positive frequencies are marked in the array nonzeros and supplied as the value of the start argument.

```
> nonzeros <- ifelse(health.tab>0, 1, 0)
> health.loglm0 <- loglm(~ concerns + age + gender,
+                          data = health.tab, start = nonzeros)
> health.loglm1 <- loglm(~ concerns + age * gender,
+                          data = health.tab, start = nonzeros)
> # df is wrong
> health.loglm2 <- loglm(~ concerns*gender + concerns*age,
+                          data = health.tab, start = nonzeros)
> LRstats(health.loglm0, health.loglm1, health.loglm2)

Likelihood summary table:
                 AIC  BIC LR Chisq Df Pr(>Chisq)
health.loglm0  104.7  111    27.74  8    0.00053 ***
health.loglm1  103.9  111    24.89  7    0.00080 ***
health.loglm2   93.7  104     4.66  2    0.09724 .
---
Signif. codes:  0 '***' 0.001 '**' 0.01 '*' 0.05 '.' 0.1 ' ' 1
```

The results agree with those of glm(), except for the degrees of freedom for the last model.

$\triangle$

9.6   Chapter summary

- Loglinear models provide a comprehensive scheme to describe and understand the associations among two or more categorical variables. It is helpful to think of these as discrete analogs of ANOVA models, or of regression models, where the log of cell frequency is modelled as a linear function of predictors.

- Loglinear models typically make no distinction between response and explanatory variables. When one variable *is* a response, however, any logit model for that response has an equivalent loglinear model. The logit form is usually simpler to formulate and test, and plots of the observed and fitted logits are easier to interpret.

- In all these cases, the interplay between graphing and fitting is important in arriving at an understanding of the relationships among variables, and an adequate descriptive model that is faithful to the details of the data.

- Cells with zero frequencies create problems for estimation and testing hypotheses in loglinear models. Different methods are available to handle *structural zeros* and *sampling zeros*.

9.7   Lab exercises

Exercise 9.1   Consider the data set `DaytonSurvey` (described in Example 2.6), giving results of a survey of use of alcohol (A), cigarettes (C), and marijuana (M) among high school seniors. For this exercise, ignore the variables `sex` and `race`, by working with the marginal table `Dayton.ACM`, a $2 \times 2 \times 2$ table in frequency data frame form.

```
> Dayton.ACM <- aggregate(Freq ~ cigarette + alcohol + marijuana,
+                         data=DaytonSurvey, FUN=sum)
```

(a) Use `loglm()` to fit the model of mutual independence, [A][C][M].
(b) Prepare mosaic display(s) for associations among these variables. Give a verbal description of the association between cigarette and alcohol use.
(c) Use `fourfold()` to produce fourfold plots for each pair of variables, AC, AM, and CM, stratified by the remaining one. Describe these associations verbally.

Exercise 9.2   Continue the analysis of the `DaytonSurvey` data by fitting the following models:

(a) Joint independence, [AC][M]
(b) Conditional independence, [AM][CM]
(c) Homogeneous association, [AC][AM][CM]
(d) Prepare a table giving the goodness-of-fit tests for these models, as well as the model of mutual independence, [A][C][M], and the saturated model, [ACM]. *Hint*: `anova()` and `LRstats()` are useful here. Which model appears to give the most reasonable fit?

Exercise 9.3   The data set `Caesar` in **vcdExtra** gives a 3×2^3 frequency table classifying 251 women who gave birth by Caesarian section by `Infection` (three levels: none, Type 1, Type2) and `Risk`, whether `Antibiotics` were used, and whether the Caesarian section was `Planned` or not. `Infection` is a natural response variable, but the table has quite a few zeros.

(a) Use `structable()` and `mosaic()` to see the locations of the zero cells in this table.
(b) Use `loglm()` to fit the baseline model [I][RAP]. Is there any problem due to zero cells indicated in the output?

(c) For the purpose of this excercise, treat all the zero cells as *sampling zeros* by adding 0.5 to all cells, e.g., `Caesar1 <- Caesar + 0.5`. Refit the baseline model.

(d) Now fit a "main effects" model [IR][IA][IP][RAP] that allows associations of `Infection` with each of the predictors.

Exercise 9.4 The `Detergent` in **vcdExtra** gives a $2^3 \times 3$ table classifying a sample of 1,008 consumers according to their preference for (a) expressed `Preference` for Brand "X" or Brand "M" in a blind trial, (b) `Temperature` of laundry water used, (c) previous use (`M_user`) of detergent Brand "M," and (d) the softness (`Water_softness`) of the laundry water used.

(a) Make some mosaic displays to visualize the associations among the table variables. Try using different orderings of the table variables to make associations related to `Preference` more apparent.

(b) Use a `doubledecker()` plot to visualize how `Preference` relates to the other factors.

(c) Use `loglm()` to fit the baseline model [P][TMW] for `Preference` as the response variable. Use a mosaic display to visualize the lack of fit for this model.

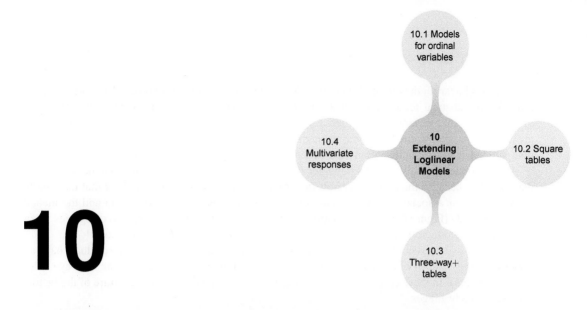

10

Extending Loglinear Models

Loglinear models have special forms to represent additional structure in the variables in contingency tables. Models for ordinal factors allow a more parsimonious description of associations. Models for square tables allow a wide range of specific models for the relationship between variables with the same categories. Another extended class of models arise when there are two or more response variables.

> The universe is built on a plan the profound symmetry of which is somehow present in the inner structure of our intellect.

Paul Valery, 1871–1945

This chapter extends the analysis of loglinear models to some important special cases allowing us to represent additional structure in the variables in contingency tables in a way that provides a more parsimonious description of associations than is available from models for general association. One class of such simplified models (Section 10.1) occurs when one or more of the explanatory variables are ordinal, and discrete levels might be replaced by numerical values.

Models for square tables (Section 10.2) with the same row and column categories comprise another special case giving simpler descriptions than the saturated model of general association. These important special cases are extended to three-way and higher-dimensional tables in Section 10.3.

Finally, Section 10.4 describes some methods for dealing with situations where there are several response variables, and it is useful to understand both the marginal relations of the responses with the predictors as well as how their association varies with the predictors.

10.1   Models for ordinal variables

Standard loglinear models treat all classification variables as nominal, unordered factors. In these models, all statistical tests are identical and parameter estimates are equivalent if the categories of any of the table variable are reordered. Yet we have seen that the ordering of categories often provides important information about the nature of associations, and we showed (Section 4.2.4) that non-parametric tests which take into account the ordered nature of a factor are more powerful.

Correspondence analysis plots (Chapter 6) make it easy to see the relationships between ordinal variables, because the method assigns quantitative scores to the table variables that maximally account for their association. As we saw for the hair–eye color data (Figure 6.1) and the mental impairment data (Figure 6.2), an association can be interpreted in terms of ordered categories when the points for two factors are ordered similarly, usually along the first CA dimension.

Similarly, in a mosaic display, an ordered associative effect is seen when the residuals have an opposite-corner pattern of positive and negative signs and magnitudes (e.g., for the hair–eye color data, Figure 5.4). In these cases loglinear and logit models that use the ordered nature of the factors offer several advantages:

- Because they are more focused, tests that use the ordinal structure of the table variables are more powerful when the association varies systematically with the ordered values of a factor.

- Because they consume fewer degrees of freedom, we can fit unsaturated models where the corresponding model for nominal factors would be saturated. In a two-way table, for example, a variety of models for ordinal factors may be proposed that are intermediate between the independence model and the saturated model.

- Parameter estimates from these models are fewer in number, are easier to interpret, and quantify the nature of effects better than corresponding quantities in models for nominal factors. Estimating fewer parameters typically gives smaller standard errors.

These advantages are analogous to the use of tests for trends or polynomial contrasts in ANOVA models. More importantly, in some research areas in the social sciences (where categorical data is commonplace), models for ordinal variables have proved crucial in theory construction and debates, giving more precise tests of hypotheses than are available from less focused or descriptive methods (Agresti, 1984).

10.1.1   Loglinear models for ordinal variables

For a two-way table, when either the row variable or the column variable, or both, are ordinal, one simplification comes from assigning ordered scores, $a = (a_i), a_1 \leq a_2 \leq \cdots a_I$, and/or $b = (b_j), b_1 \leq b_2 \leq \cdots b_J$, to the categories so that the ordinal relations are necessarily included in the model. Typically, equally spaced scores are used; for example, integer scores, $a_i = i$, or the zero-sum equivalent, $a_i = i - (I + 1)/2$ (e.g., $(a_i) = (-1, 0, 1)$ for $I = 3$).

Using such scores gives simple interpretations of the association parameters in terms of *local odds ratios* for adjacent 2×2 subtables,

$$\theta_{ij} = \frac{m_{ij}\, m_{i+1,j+1}}{m_{i,j+1}\, m_{i+1,j}} \,, \tag{10.1}$$

which is the odds ratio for pairs of adjacent rows and adjacent columns.

When both variables are assigned scores, this gives the ***linear-by-linear model*** $(L \times L)$

$$\log(m_{ij}) = \mu + \lambda_i^A + \lambda_j^B + \gamma\, a_i b_j \,. \tag{10.2}$$

Because the scores a and b are fixed, this model has only one extra parameter, γ, compared to the independence model, which is the special case, $\gamma = 0$. In contrast, the saturated model, allowing general association λ_{ij}^{AB}, uses $(I - 1)(J - 1)$ additional parameters.

The terms $\gamma a_i b_j$ in Eqn. (10.2) describe a pattern of association where deviations from independence increase linearly with a_i and b_j in opposite directions towards the opposite corners of the table, as we have often observed in mosaic displays.

In the linear-by-linear association model, the local log odds ratios are

$$\log(\theta_{ij}) = \gamma(a_{i+1} - a_i)(b_{j+1} - b_j) \,,$$

which reduces to

$$\log(\theta_{ij}) = \gamma$$

for integer-spaced scores, so γ is the common local log odds ratio. As a result, the linear-by-linear model is sometimes called the ***uniform association model*** (Goodman, 1979).

Generalizations of the linear-by-linear model result when only one variable is assigned scores. In the ***row effects model*** (R), the row variable, A, is treated as nominal, while the column variable, B, is assigned ordered scores (b_j). The loglinear model is then

$$\log(m_{ij}) = \mu + \lambda_i^A + \lambda_j^B + \alpha_i b_j \,, \tag{10.3}$$

where the α_i parameters are the *row effects*. An additional constraint, $\sum_i \alpha_i = 0$ or $\alpha_1 = 0$, is imposed, so that model Eqn. (10.3) has only $(I - 1)$ more parameters than the independence model. The linear-by-linear model is the special case where the row effects are equally spaced, and the independence model is the special case where all $\alpha_i = 0$.

The row-effects model Eqn. (10.3) also has a simple odds ratio interpretation. The local log odds ratio for adjacent pairs of rows and columns is

$$\log(\theta_{ij}) = \alpha_{i+1} - \alpha_i \,,$$

which is constant for all pairs of adjacent columns. Plots of the local log odds ratio against i would appear as a set of coincident curves.

In the analogous ***column effects model*** (C), $(J - 1)$ linearly independent column effect parameters β_j are estimated for the column variable, while fixed scores $\{a_i\}$ are assigned to the row variable. It is also possible to fit a ***row plus column effects model*** (R+C), which assigns specified scores to both the rows and column variables. Plots of the local odds ratios for the R+C model appear as parallel curves.

Nesting relationships among these models and others (RC(1) and RC(2)) described in Section 10.1.3 are shown in Figure 10.1. Any set of models connected by a path can be directly compared with likelihood-ratio tests of the form $G^2(M_2|M_1)$.

In **R**, the $L \times L$, row effects, and column effects models can all be fit using `glm()` simply by replacing the appropriate table factor variable(s) with their `as.numeric()` equivalents.

EXAMPLE 10.1: Mental impairment and parents' SES

The `Mental` data on the mental health status of young New York residents in relation to their parents' socioeconomic status was examined in Example 4.7 using CMH tests for ordinal association and in Example 6.2 using correspondence analysis. Figure 6.2 showed that nearly all of the association in the table was accounted for by a single dimension along which both factors were ordered, consistent with the view that mental health increased in relation to parents' SES.

Because these models provide their interpretations in terms of local odds ratios, Eqn. (10.1), it is helpful to see these values for the observed data, corresponding to the saturated model. The values $\log(\theta_{ij})$ are calculated by `loddsratio()` in **vcd**, with the data in table form.

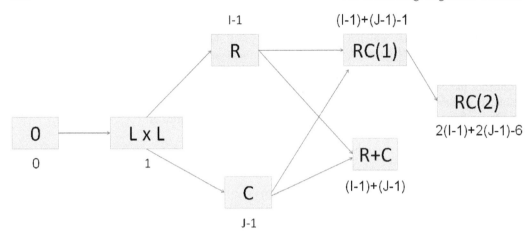

Figure 10.1: Nesting relationships among some association models for an $I \times J$ table specifying the association parameters, λ_{ij}^{AB}. Model **0** is the independence model. Formulas near the boxes give the number of identifiable association parameters. Arrows point from one nested model to another that is a more general version.

```
> library(vcd)
> data("Mental", package = "vcdExtra")
> (mental.tab <- xtabs(Freq ~ mental + ses, data=Mental))

          ses
mental       1   2   3   4   5   6
  Well      64  57  57  72  36  21
  Mild      94  94 105 141  97  71
  Moderate  58  54  65  77  54  54
  Impaired  46  40  60  94  78  71

> (LMT <- loddsratio(mental.tab))

log odds ratios for mental and ses

                   ses
mental                1:2     2:3      3:4     4:5     5:6
  Well:Mild        0.1158  0.1107   0.0612  0.3191  0.227
  Mild:Moderate   -0.0715  0.0747  -0.1254  0.0192  0.312
  Moderate:Impaired -0.0683  0.2201  0.2795  0.1682 -0.094
```

A simple plot of these values, using area- and color-proportional shaded squares, is shown in Figure 10.2. This plot is drawn using the corrplot (Wei, 2013) package. It is easy to see that most of the local odds ratios are mildly positive.[1]

```
> library(corrplot)
> corrplot(as.matrix(LMT), method = "square", is.corr = FALSE,
+          tl.col = "black", tl.srt = 0, tl.offset = 1)
```

For comparison with the $L \times L$ model fitted below, the mean local log odds ratio is 0.103.

---

[1]Using plot(loddsratio(mental.tab)) would give a line plot of the odds ratios as illustrated in Section 5.9.2 (e.g., Figure 5.37).

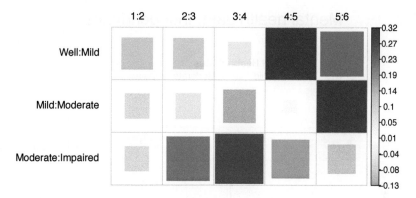

Figure 10.2: Shaded-square plot of the local log odds ratios in the `Mental` data.

```
> mean(LMT$coefficients)

[1] 0.10323
```

As a baseline, we first fit the independence model (testing $H_0 : \log(\theta_{ij}) = 0$) with `glm()`. As expected, this model fits quite badly, with $G^2 (15) = 47.418$.

```
> indep <- glm(Freq ~ mental + ses, data = Mental, family = poisson)
> LRstats(indep)

Likelihood summary table:
        AIC BIC LR Chisq Df Pr(>Chisq)
indep 210 220     47.4 15    3.2e-05 ***
---
Signif. codes:  0 '***' 0.001 '**' 0.01 '*' 0.05 '.' 0.1 ' ' 1
```

The mosaic display of standardized residuals from this model is shown in Figure 10.3. The argument `labeling=labeling_residuals` is used to show the numerical values in the cells with absolute values greater than `suppress=1`.

```
> long.labels <- list(set_varnames = c(mental="Mental Health Status",
+                                       ses="Parent SES"))
> mosaic(indep,
+        gp=shading_Friendly,
+        residuals_type="rstandard",
+        labeling_args = long.labels,
+        labeling=labeling_residuals, suppress=1,
+        main="Mental health data: Independence")
```

This figure shows the classic opposite-corner pattern of the signs and magnitudes of the residuals that would arise if the association between mental health and SES could be explained by the ordinal relation of these factors using one of the $L \times L$, R, or C models.

To fit such ordinal models, you can use `as.numeric()` on a factor variable to assign integer scores, or assign other values if integer spacing is not appropriate.

```
> Cscore <- as.numeric(Mental$ses)
> Rscore <- as.numeric(Mental$mental)
```

Then, the $L \times L$, R, C, and $R + C$ models can be fit as follows, using `update()`, where beyond the main effects of `mental` and `ses`, their association is represented as the interaction of the numeric score(s) or factor(s), as appropriate in each case.

Mental health data: Independence

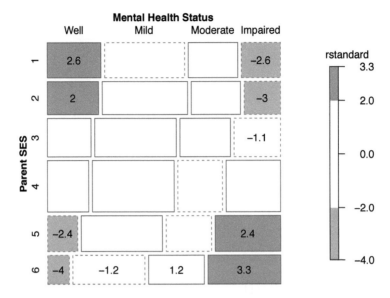

Figure 10.3: Mosaic display of the independence model for the mental health data.

```
> linlin <- update(indep, . ~ . + Rscore:Cscore)
> roweff <- update(indep, . ~ . + mental:Cscore)
> coleff <- update(indep, . ~ . + Rscore:ses)
> rowcol <- update(indep, . ~ . + Rscore:ses + mental:Cscore)
```

Goodness-of-fit tests for these models are shown below. They show that all of the $L \times L$, R, and C models are acceptable in terms of the likelihood-ratio G^2. The $L \times L$ model, with only one more parameter than the independence model, is judged the best by both AIC and BIC.

```
> LRstats(indep, linlin, roweff, coleff, rowcol)

Likelihood summary table:
         AIC    BIC LR Chisq Df Pr(>Chisq)
indep  209.6 220.2    47.42 15   3.16e-05 ***
linlin 174.1 185.8     9.90 14      0.770
roweff 174.4 188.6     6.28 12      0.901
coleff 179.0 195.5     6.83 10      0.741
rowcol 179.2 198.1     3.05  8      0.931
---
Signif. codes:  0 '***' 0.001 '**' 0.01 '*' 0.05 '.' 0.1 ' ' 1
```

In cases where such overall tests are unclear, you can carry out tests of nested sets of models using `anova()`, giving tests of ΔG^2.

```
> anova(indep, linlin, roweff, test = "Chisq")

Analysis of Deviance Table

Model 1: Freq ~ mental + ses
Model 2: Freq ~ mental + ses + Rscore:Cscore
Model 3: Freq ~ mental + ses + mental:Cscore
  Resid. Df Resid. Dev Df Deviance Pr(>Chi)
```

```
1          15        47.4
2          14         9.9  1      37.5      9e-10 ***
3          12         6.3  2       3.6       0.16
---
Signif. codes:  0 '***' 0.001 '**' 0.01 '*' 0.05 '.' 0.1 ' ' 1

> anova(indep, linlin, coleff, test = "Chisq")

Analysis of Deviance Table

Model 1: Freq ~ mental + ses
Model 2: Freq ~ mental + ses + Rscore:Cscore
Model 3: Freq ~ mental + ses + ses:Rscore
  Resid. Df Resid. Dev Df Deviance Pr(>Chi)
1         15       47.4
2         14        9.9  1     37.5     9e-10 ***
3         10        6.8  4      3.1      0.55
---
Signif. codes:  0 '***' 0.001 '**' 0.01 '*' 0.05 '.' 0.1 ' ' 1
```

Under the $L \times L$ model, the estimate of the coefficient of Rscore:Cscore is $\hat{\gamma} = 0.0907$ (s.e.=0.015) with unit-spaced scores, as shown below.

```
> # interpret linlin association parameter
> coef(linlin)[["Rscore:Cscore"]]

[1] 0.090687

> exp(coef(linlin)[["Rscore:Cscore"]])

[1] 1.0949
```

This corresponds to a local odds ratio, $\hat{\theta}_{ij} = \exp(0.0907) = 1.095$. This single number describes the association succinctly: each step down the socioeconomic scale increases the odds of being classified one step poorer in mental health by 9.5%.

$\triangle$

10.1.2 Visualizing model structure

In Section 5.8 we illustrated how to use mosaic displays to visualize the *structure* of loglinear models. The basic idea was just to use mosaic plots or mosaic matrices to show the *fitted* values implied by a given model. As just described, loglinear models for ordinal variables have very simple structures in terms of log odds ratios, and you can similarly understand their structure by calculating or plotting the local odds ratios from the fitted frequencies for a given model.

EXAMPLE 10.2: Mental impairment and parents' SES
We illustrate this idea numerically here, for the row effects (R) model, roweff, fit to the *Mental* data. fitted() gets the fitted frequencies for this model, and using loddsratio() on the result show that these are constant in each row.

```
> roweff.fit <- matrix(fitted(roweff), 4, 6,
+                      dimnames=dimnames(mental.tab))
> round(as.matrix(loddsratio(roweff.fit)), 3)

                  ses
mental              1:2   2:3   3:4   4:5   5:6
  Well:Mild        0.145 0.145 0.145 0.145 0.145
  Mild:Moderate    0.018 0.018 0.018 0.018 0.018
  Moderate:Impaired 0.143 0.143 0.143 0.143 0.143
```

Similarly, the column effects (C) model, `coleff`, shows these values to be constant in each column.

```
> coleff.fit <- matrix(fitted(coleff), 4, 6,
+                        dimnames = dimnames(mental.tab))
> round(as.matrix(loddsratio(coleff.fit)), 3)

                   ses
mental              1:2    2:3    3:4    4:5    5:6
  Well:Mild        -0.013 0.125 0.053 0.142 0.139
  Mild:Moderate    -0.013 0.125 0.053 0.142 0.139
  Moderate:Impaired -0.013 0.125 0.053 0.142 0.139
```

Using `plot(loddsratio(model))` for these cases and the R+C model gives the plots in Figure 10.4.

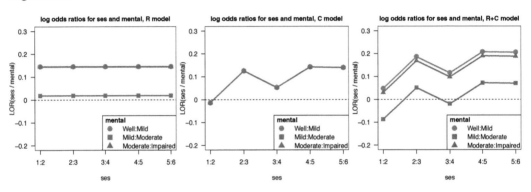

Figure 10.4: Log odds ratio plots for the R (left), C (middle), and R+C (right) models fit to the mental health data.

In contrast, the (more parsimonious) $L \times L$ model has a constant log odds ratio, $\widehat{\gamma} = 0.0907$. △

10.1.3 Log-multiplicative (RC) models

The association models described above are all more parsimonious and easier to interpret than the saturated model. However, they depend on assigning fixed and possibly arbitrary scores to the variable categories. A generalization of the $L \times L$ model that treats *both* row and column scores as parameters is the ***row-and-column effects model*** (RC(1)) suggested by Goodman (1979),

$$\log(m_{ij}) = \mu + \lambda_i^A + \lambda_j^B + \gamma \, \alpha_i \beta_j \,, \tag{10.4}$$

where γ, α, and β comprise additional parameters to be estimated beyond the independence model.[2] This model has a close connection with correspondence analysis (Goodman, 1985), where the estimated scores α and β are analogous to correspondence analysis scores on a first dimension.[3] γ, called the *intrinsic association coefficient*, is analogous to the same parameter in the $L \times L$ model.

For identifiability and interpretation it is necessary to impose some normalization constraints on the α and β. An *unweighted, unit standardized* solution forces $\sum_i \alpha_i = \sum_j \beta_j = 0$ and

---

[2]In contrast to the R, C, and R+C models, RC models do not assume that the categories are appropriately ordered because the category scores are estimated from the data.

[3]However, when estimated by maximum likelihood, the RC(1) model allows likelihood-ratio tests of parameters and model fit, AIC and BIC statistics, and methods for estimating standard errors of the parameters. Such model-based methods are not available for correspondence analysis.

$\sum_i \alpha_i^2 = \sum_j \beta_j^2 = 1$. Alternatively, and more akin to correspondence analysis solutions, the *marginally weighted* solution uses the marginal probabilities π_{i+} of the row variable and π_{+j} of the columns as weights:

$$\sum_i \alpha_i \pi_{i+} = \sum_j \beta_j \pi_{+j} = 0 , \tag{10.5}$$

$$\sum_i \alpha_i^2 \pi_{i+} = \sum_j \beta_j^2 \pi_{+j} = 1 .$$

Goodman (1986) generalized this to multiple bilinear terms of the form $\gamma_k \alpha_{ik} \beta_{jk}$, with M terms (the RC(M) model) and showed that *all* associations in the saturated model could be expressed exactly as

$$\lambda_{ij}^{AB} = \sum_{k=1}^{M} \gamma_k \alpha_{ik} \beta_{jk} \qquad M = \min(I - 1, J - 1) . \tag{10.6}$$

In practice, models with fewer terms usually suffice. For example, an RC(2) model with two multiplicative terms is analogous to a two-dimensional correspondence analysis solution. In addition to the normalization constraints for the RC(1) model, parameters in an RC(M) model must satisfy the additional constraints that the (possibly weighted) scores for distinct dimensions are orthogonal (uncorrelated), similar to correspondence analysis solutions.

The RC model is *not* a loglinear model because it contains a multiplicative term in the parameters. This model and a wide variety of other nonlinear models for categorical data can be fit using gnm() in the gnm package. This provides the basic machinery for extending glm() models to nonlinear terms, quite generally. The function rc() in the logmult package uses gnm() for fitting, and offers greater convenience in normalizing the category scores, calculating standard errors, and plotting.

EXAMPLE 10.3: Mental impairment and parents' SES

The gnm package provides a number of functions that can be used in model formulas for non-linear association terms. Among these, Mult() expresses a multiplicative association in terms of two (or more) factors. The RC(1) model for factors A, B uses Mult(A,B) for the association term in Eqn. (10.4). Multiple multiplicative RC terms, as in Eqn. (10.6), can be expressed using instances(Mult(A,B), m).

To illustrate, we fit the RC(1) and RC(2) models to the *Mental* data using gnm(). In this table, both factors are ordered, but we don't want to use the default polynomial contrasts, so we set their contrast attributes to treatment.

```
> library(gnm)
> contrasts(Mental$mental) <- contr.treatment
> contrasts(Mental$ses) <- contr.treatment
> indep <- gnm(Freq ~ mental + ses, data = Mental, family = poisson)
> RC1 <- update(indep, . ~ . + Mult(mental, ses), verbose = FALSE)
> RC2 <- update(indep, . ~ . + instances(Mult(mental, ses), 2),
+               verbose = FALSE)
```

For comparison with the loglinear association models fit in Example 10.1 we show the G^2 goodness-of-fit tests for all these models. The ordinal loglinear models and the RC models all fit well, with the $L \times L$ model preferred on the basis of parsimony by AIC and BIC.

```
> LRstats(indep, linlin, roweff, coleff, RC1, RC2)

Likelihood summary table:
         AIC    BIC LR Chisq Df Pr(>Chisq)
indep  209.6 220.2    47.42 15   3.16e-05 ***
```

```
linlin 174.1 185.8      9.90 14      0.770
roweff 174.4 188.6      6.28 12      0.901
coleff 179.0 195.5      6.83 10      0.741
RC1     179.7 198.6     3.57  8      0.894
RC2     186.7 211.4     0.52  3      0.914
---
Signif. codes:  0 '***' 0.001 '**' 0.01 '*' 0.05 '.' 0.1 ' ' 1
```

The substantive difference between the $L \times L$ model and the RC(1) model is whether the categories of mental health status and SES can be interpreted as equally spaced along some latent continua, versus the alternative that category spacing is unequal. We can test this directly using the likelihood-ratio test, $G^2(L \times L \,|\, RC(1))$. Similarly, model RC1 is nested within model RC2, so $G^2(RC(1) \,|\, RC(2))$ gives a direct test of the need for a second dimension.

```
> anova(linlin, RC1, RC2, test = "Chisq")

Analysis of Deviance Table

Model 1: Freq ~ mental + ses + Rscore:Cscore
Model 2: Freq ~ mental + ses + Mult(mental, ses)
Model 3: Freq ~ mental + ses + Mult(mental, ses, inst = 1) + Mult(mental,
    ses, inst = 2)
  Resid. Df Resid. Dev Df Deviance Pr(>Chi)
1        14       9.90
2         8       3.57  6     6.32     0.39
3         3       0.52  5     3.05     0.69
```

We see that estimated scores for the categories in the model RC1 do not provide a significantly better fit, and there is even less evidence for a second dimension of category parameters in the RC2 model.

Nevertheless, for cases where RC models *do* provide some advantage, it is useful to know how to visualize the estimated category parameters. The key to this is the function getContrasts(), which computes contrasts or scaled contrasts for a set of (non-eliminated) parameters from a "gnm" model, together with standard errors for the estimated contrasts following the methods of Firth (2003) and also Firth and Menezes (2004). The details are explained in help(getContrasts) and in vignette("gnmOverview"), which comes with the gnm package.

The coefficients in the marginally weighted solution Eqn. (10.5) can be obtained as follows.

```
> rowProbs <- with(Mental, tapply(Freq, mental, sum) / sum(Freq))
> colProbs <- with(Mental, tapply(Freq, ses, sum) / sum(Freq))
> mu <- getContrasts(RC1, pickCoef(RC1, "[.]mental"),
+                 ref = rowProbs, scaleWeights = rowProbs)
> nu <- getContrasts(RC1, pickCoef(RC1, "[.]ses"),
+                 ref = colProbs, scaleWeights = colProbs)
```

In our notation, the coefficients α and β can be extracted as the qvframe component of the "qv" object returned by getContrasts().

```
> (alpha <- mu$qvframe)

                            Estimate Std. Error
Mult(., ses).mentalWell      1.67378    0.19043
Mult(., ses).mentalMild      0.14009    0.20018
Mult(., ses).mentalModerate -0.13669    0.27948
Mult(., ses).mentalImpaired -1.41055    0.17418

> (beta  <- nu$qvframe)
```

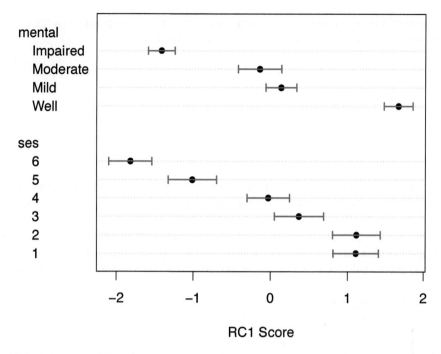

Figure 10.5: Dotchart of the scaled category scores for the RC(1) model fit to the mental health data. Error bars show ±1 standard error.

```
                       Estimate Std. Error
Mult(mental, .).ses1  1.111360    0.29921
Mult(mental, .).ses2  1.120459    0.31422
Mult(mental, .).ses3  0.370752    0.31915
Mult(mental, .).ses4 -0.027006    0.27328
Mult(mental, .).ses5 -1.009480    0.31470
Mult(mental, .).ses6 -1.816647    0.28095
```

For plotting this RC(1) solution for the scaled category scores together with their estimated standard errors, a `dotchart()`, shown in Figure 10.5, provides a reasonable visualization.

To create this plot, first combine the row and column scores in a data frame, and add columns `lower`, `upper` corresponding to ±1 standard error (or some other multiple).

```
> scores <- rbind(alpha, beta)
> scores <- cbind(scores,
+                 factor = c(rep("mental", 4), rep("ses", 6)) )
> rownames(scores) <- c(levels(Mental$mental), levels(Mental$ses))
> scores$lower <- scores[,1] - scores[,2]
> scores$upper <- scores[,1] + scores[,2]
> scores
```

```
          Estimate Std. Error factor   lower    upper
Well         1.674      0.190 mental  1.4834   1.864
Mild         0.140      0.200 mental -0.0601   0.340
Moderate    -0.137      0.279 mental -0.4162   0.143
Impaired    -1.411      0.174 mental -1.5847  -1.236
1            1.111      0.299    ses  0.8121   1.411
2            1.120      0.314    ses  0.8062   1.435
3            0.371      0.319    ses  0.0516   0.690
4           -0.027      0.273    ses -0.3003   0.246
5           -1.009      0.315    ses -1.3242  -0.695
6           -1.817      0.281    ses -2.0976  -1.536
```

The dotchart shown in Figure 10.5 is then a plot of `Estimate`, grouped by `factor`, with arrows showing the range of `lower` to `upper` for each parameter.

```
> with(scores, {
+    dotchart(Estimate, groups = factor, labels = rownames(scores),
+            cex = 1.2, pch = 16, xlab = "RC1 Score",
+            xlim = c(min(lower), max(upper)))
+    arrows(lower, c(8 + (1 : 4), 1 : 6), upper, c(8 + (1 : 4), 1 : 6),
+            col = "red", angle = 90, length = .05, code = 3, lwd = 2)
+    })
```

In this plot, the main substantive difference from the $L \times L$ model is in the spacing of the lowest two categories of `ses` and the middle two categories of `mental`, which are not seen to differ in the RC1 model.

The coefficients in the `RC2` model can also be plotted (in a 2D plot) by extracting the coefficients from the "gnm" object and reshaping them to 2-column matrices. The function `pickCoef()` is handy here to get the indices of a subset of parameters by matching a pattern in their names.

```
> alpha <- coef(RC2)[pickCoef(RC2, "[.]mental")]
> alpha <- matrix(alpha, ncol=2)
> rownames(alpha) <- levels(Mental$mental)
> colnames(alpha) <- c("Dim1", "Dim2")
> alpha
```

```
             Dim1      Dim2
Well      0.497610 -0.192275
Mild      0.042652 -0.039651
Moderate  0.127071  0.208333
Impaired -0.505584 -0.012349
```

```
> beta <- coef(RC2)[pickCoef(RC2, "[.]ses")]
> beta <- matrix(beta, ncol=2)
> rownames(beta) <- levels(Mental$ses)
> colnames(beta) <- c("Dim1", "Dim2")
> beta
```

```
       Dim1      Dim2
1   0.547607 -0.167775
2   0.574184 -0.082320
3   0.205737  0.078252
4  -0.087662 -0.349440
5  -0.502335  0.059357
6  -0.758785  1.233043
```

For plotting and interpretation, these dimension scores need to be standardized as described at the start of this section (e.g., Eqn. (10.5)). We don't show the steps for doing this or producing a plot, because it is much simpler to use the logmult package, as described next. A basic plot using the marginal-weighted scaling is shown in Figure 10.6.

The patterns of the row and column category scores in Figure 10.6 are quite similar to the 2D correspondence analysis solution shown in Figure 6.2. The main difference is in the relative scaling of the axes. In Figure 10.6, the variances of the two dimensions are equated; in the correspondence analysis plot, the axes are scaled in relation to their contributions to Pearson χ^2, allowing an interpretation of distance between points in terms of χ^2-distance.

$\triangle$

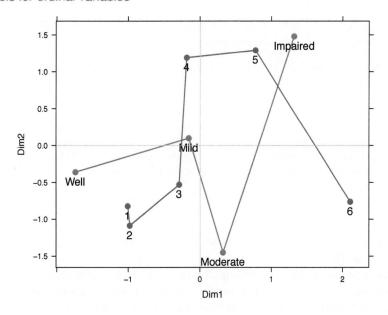

Figure 10.6: Scaled category scores for the RC(2) model fit the mental health data.

10.1.3.1 Using logmult

It takes a fair bit of work to extract the coefficients from "gnm" objects and carry out the scaling necessary for informative plots. Much of this effort is now performed by the logmult package with several convenience functions that do the heavy lifting.

rc() fits the class of RC(M) models, allowing an argument nd to specify the number of dimensions, and also providing for standard errors estimated using jackknife and bootstrap methods (Milan and Whittaker, 1995), which are computationally intensive. For square tables, a symmetric argument constrains the row and column scores to be equal, and a diagonal option fits parameters for each diagonal cell, providing for models of quasi-independence and quasi-symmetry (see Section 10.2).

It returns an object of class "rc" with the components of the "gnm" object. An assoc component is also returned, containing the normalized association parameters for the categories.

rcL() fits extensions of RC models to tables with multiple layers, called RC(M)-L models by Wong (2010).

plot.rc() is a plot method for visualizing scores for RC(M) models in two selected dimensions. Among other options, it can plot confidence ellipses for the category scores, using the estimated covariance matrix (assuming a normal distribution of the category scores). The plot method returns (invisibly) the coordinates of the scores as plotted, facilitating additional plot annotation.

EXAMPLE 10.4: Mental impairment and parents' SES
Here we use rc() to estimate the RC(1) and RC(2) models for the *Mental* data. In contrast to gnm(), which has a formula interface for a data argument, rc() requires the input in the form of a two-way table, given here as mental.tab.

```
> library(logmult)
> rc1 <- rc(mental.tab, verbose = FALSE, weighting = "marginal",
+              se = "jackknife")
> rc2 <- rc(mental.tab, verbose = FALSE, weighting = "marginal", nd = 2,
+              se = "jackknife")
```

The option `weighting="marginal"` gives the marginally weighted solution and `se = "jackknife"` estimates the covariance matrix using the leave-one-out jackknife.[4]

A plot of the scaled category scores similar to Figure 10.6, with 1 standard error confidence ellipses (making them comparable to the 1D solution shown in Figure 10.5) but no connecting lines can then be easily produced with the `plot()` method for "rc" objects.

```
> coords   <- plot(rc2, conf.ellipses = 0.68, cex = 1.5,
+                  rev.axes = c(TRUE, FALSE))
```

The orientation of the axes is arbitrary in RC(M) models, so the horizontal axis is reversed here to conform with Figure 10.6.

This produces (in Figure 10.7) a symmetric biplot in which the scaled coordinates of points for rows (α_{ik}) and columns (β_{jk}) on both axes are the product of normalized scores and the square root of the intrinsic association coefficient (γ_k) corresponding to each dimension.

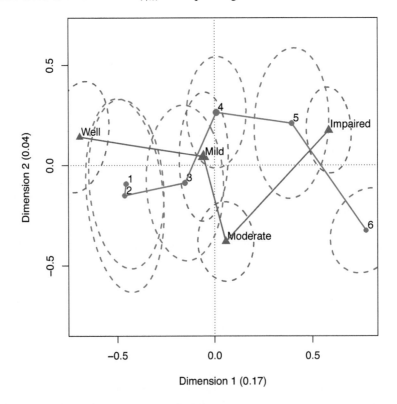

Figure 10.7: Scaled category scores for the RC(2) model fit and plotted using the logmult package. The 68% confidence ellipses correspond to bivariate ±1 confidence intervals for the category parameters.

---

[4]Becker and Clogg (1989) recommend using unweighted solutions, `weighting="none"` (they call them "uniformly weighted") to preserve independence of inferences about association and marginal effects and estimates of the intrinsic association parameters, γ_k. That choice makes very little difference in the plots for this example, but the γ_k parameters are affected considerably.

Such plots can be customized using the category coordinates (`coords`) returned by the `plot()` method. As in other biplots, joining the row and column points by lines (sorted by the first dimension) makes it easier to see their relationships across the two dimensions. The following code draws the lines shown in Figure 10.7.

```
> scores <- rbind(coords$row, coords$col)
> lines(scores[1 : 4,], col = "blue", lwd = 2)
> lines(scores[-(1 : 4),], col = "red", lwd = 2)
```

We saw earlier that there was not strong evidence supporting the need for a second RC dimension to describe the relationship between mental health and SES. This is apparent in the sizes of the confidence ellipses, which overlap much more along Dimension 2 than Dimension 1. △

10.2 Square tables

Square tables, where the row and column variables have the same categories, comprise an important special case for loglinear models that can account for associations more parsimoniously than the saturated model. Some examples are the data on visual acuity in Example 4.14, categorical ratings of therapy clients by two observers, and mobility tables, tracking the occupational categories between generations in the same families, or migration tables, giving movement of people between regions. The latter topic has been important in sociological and geographic research and has spurred the development of a wide range of specialized loglinear models for this purpose.

10.2.1 Quasi-independence, symmetry, quasi-symmetry, and topological models

In many square tables, such as the *Vision* data, independence is not a credible hypothesis because the diagonal cells, representing equal values of the row and column variables, tend to be very large and often contribute most of the lack of fit. A substantively more interesting hypothesis is whether the table exhibits independence, ignoring the diagonal cells. This leads to what is called the **quasi-independence model**, that specifies independence only in the off-diagonal cells.

For a two-way table, quasi-independence can be expressed as

$$\pi_{ij} = \pi_{i+}\pi_{+j} \qquad \text{for } i \neq j$$

or in loglinear form as

$$\log m_{ij} = \mu + \lambda_i^A + \lambda_j^B + \delta_i I(i = j).$$

This model effectively adds one parameter, δ_i, for each main diagonal cell that fits those frequencies perfectly.

Another hypothesis of substantive interest for square tables, particularly those concerning occupational and geographical mobility, is that the joint distribution of row and column variables is symmetric, that is, $\pi_{ij} = \pi_{ji}$ for all $i \neq j$. For example, this **symmetry model** (S) asserts that sons are as likely to move from their father's occupation i to another, j, as the reverse. This form of symmetry is quite strong, because it also implies **marginal homogeneity** (MH), that the marginal probabilities of the row and column variables are equal, $\pi_{i+} = \sum_j \pi_{ij} = \sum_j \pi_{ji} = \pi_{+i}$ for all i.

To separate marginal homogeneity from symmetry of the association terms per se, the model of **quasi-symmetry** (QS) uses the standard main-effect terms in the loglinear model,

$$\log m_{ij} = \mu + \lambda_i^A + \lambda_j^B + \lambda_{ij}, \tag{10.7}$$

where $\lambda_{ij} = \lambda_{ji}$. It can be shown (Caussinus, 1966) that

$$\begin{aligned} \text{symmetry} &= \text{quasi-symmetry} + \text{marginal homogeneity} \\ G^2(S) &= G^2(QS) + G^2(MH) \end{aligned}$$

where $G^2(MH)$ is defined by the likelihood-ratio test of the difference between the S and QS models,

$$G^2(MH) \equiv G^2(S \,|\, QS) = G^2(S) - G^2(QS). \tag{10.8}$$

The gnm package provides several model building convenience functions that facilitate fitting these and related models:

- `Diag(row, col, ...)` constructs a diagonals association factor for two (or more) factors with integer levels where the original factors are equal, and `"."` otherwise.
- `Symm(row, col, ...)` constructs an association factor giving equal levels to sets of symmetric cells. The QS model is specified using `Diag() + Symm()`.
- `Topo(row, col, ..., spec)` creates an association factor for two or more factors, as specified by an array of levels, which may be arbitrarily structured. Both `Diag()` and `Symm()` factors are special cases of `Topo()`.

The factor levels representing these association effects for a 4×4 table are shown below by their unique values in each array.

$$\text{Diag}_{4\times4} = \begin{bmatrix} 1 & . & . & . \\ . & 2 & . & . \\ . & . & 3 & . \\ . & . & . & 4 \end{bmatrix} \quad \text{Symm}_{4\times4} = \begin{bmatrix} 11 & 12 & 13 & 14 \\ 12 & 22 & 23 & 24 \\ 13 & 23 & 33 & 34 \\ 14 & 24 & 34 & 44 \end{bmatrix} \quad \text{Topo}_{4\times4} = \begin{bmatrix} 2 & 3 & 4 & 4 \\ 3 & 3 & 4 & 4 \\ 4 & 4 & 5 & 5 \\ 4 & 4 & 5 & 1 \end{bmatrix}.$$

EXAMPLE 10.5: Visual acuity

Example 4.14 presented the data on tests of visual acuity in the left and right eyes of a large sample of women working in the Royal Ordnance factories in World War II. A sieve diagram (Figure 4.10) showed that, as expected, most women had the same acuity in both eyes, but the off-diagonal cells had a pattern suggesting some form of symmetry.

The data set `VisualAcuity` contains data for both men and women in frequency form and for this example we subset this to include only the 4×4 table for women.

```
> data("VisualAcuity", package="vcd")
> women <- subset(VisualAcuity, gender=="female", select=-gender)
```

The four basic models of independence, quasi-independence, symmetry, and quasi-symmetry for square tables are fit as shown below. We use `update()` to highlight the relations among these models in two pairs.

```
> #library(vcdExtra)
> indep <- glm(Freq ~ right + left,  data = women, family = poisson)
> quasi <- update(indep, . ~ . + Diag(right, left))
>
> symm <- glm(Freq ~ Symm(right, left), data = women, family = poisson)
> qsymm <- update(symm, . ~ right + left + .)
```

The brief summary of goodness of fit of these models below shows that the QS model fits reasonably well, but none of the others do by likelihood-ratio tests or AIC or BIC.

```
> LRstats(indep, quasi, symm, qsymm)

Likelihood summary table:
      AIC  BIC LR Chisq Df Pr(>Chisq)
indep 6803 6808     6672  9     <2e-16 ***
quasi  338  347      199  5     <2e-16 ***
symm   157  164       19  6     0.0038 **
qsymm  151  161        7  3     0.0638 .
---
Signif. codes:  0 '***' 0.001 '**' 0.01 '*' 0.05 '.' 0.1 ' ' 1
```

Beyond just saying that the QS model fits best, the reasons *why* it does can be seen in mosaic displays. Figure 10.8 compares the mosaics for the models of quasi-independence (accounting only for the diagonal cells) and quasi-symmetry (also accounting for symmetry). It can be seen in the left panel that the non-diagonal associations are largely symmetric, and also that when they differ, visual acuity in the two eyes is most likely to differ by only one eye grade.

```
> labs <- c("High", "2", "3", "Low")
> largs <- list(set_varnames = c(right = "Right eye grade",
+                                 left = "Left eye grade"),
+              set_labels=list(right = labs, left = labs))
> mosaic(quasi, ~ right + left, residuals_type = "rstandard",
+        gp = shading_Friendly,
+        labeling_args = largs,
+        main = "Quasi-Independence (women)")
> mosaic(qsymm, ~ right + left, residuals_type = "rstandard",
+        gp = shading_Friendly,
+        labeling_args = largs,
+        main = "Quasi-Symmetry (women)")
```

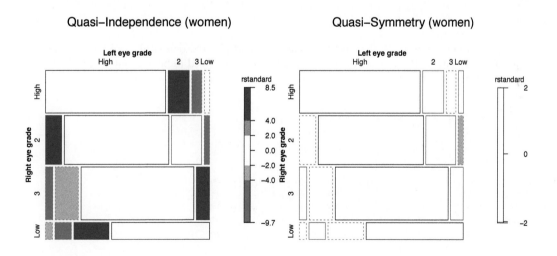

Figure 10.8: Mosaic displays comparing the models of quasi-independence and quasi-symmetry for visual acuity in women.

Finally, as usual, `anova()` can be used to carry out specific tests of nested models. For example, the test of marginal homogeneity, Eqn. (10.8), compares models S and QS and shows here that the marginal probabilities for the left and right eyes differ.

```
> anova(symm, qsymm, test = "Chisq")

Analysis of Deviance Table

Model 1: Freq ~ Symm(right, left)
Model 2: Freq ~ right + left + Symm(right, left)
  Resid. Df Resid. Dev Df Deviance Pr(>Chi)
1         6      19.25
2         3       7.27  3       12   0.0075 **
---
Signif. codes:  0 '***' 0.001 '**' 0.01 '*' 0.05 '.' 0.1 ' ' 1
```

                                                                                    △

EXAMPLE 10.6: Hauser's occupational mobility table

The data *Hauser79* in **vcdExtra**, from Hauser (1979), gives a 5×5 table in frequency form cross-classifying 19,912 individuals in the United States by father's occupation and son's first occupation. The occupational categories are represented by abbreviations, of Upper Non-Manual (UpNM), Lower Non-Manual (LoNM), Upper Manual (UpM), Lower Manual (LoM), and Farm. These data were also analyzed by Powers and Xie (2008, Chapter 4).

```
> data("Hauser79", package = "vcdExtra")
> structable(~ Father + Son, data = Hauser79)

        Son UpNM LoNM  UpM  LoM Farm
Father
UpNM        1414  521  302  643   40
LoNM         724  524  254  703   48
UpM          798  648  856 1676  108
LoM          756  914  771 3325  237
Farm         409  357  441 1611 1832
```

Before fitting any models, it is useful to calculate and plot the observed local log odds ratios, as we did in Example 10.1, to see the patterns in the data that need to be accounted for. These are calculated using `loddsratio()`.

```
> hauser.tab <- xtabs(Freq ~ Father + Son, data = Hauser79)
> (lor.hauser <- loddsratio(hauser.tab))

log odds ratios for Father and Son

            Son
Father      UpNM:LoNM LoNM:UpM UpM:LoM LoM:Farm
  UpNM:LoNM     0.675   -0.179   0.262   0.0931
  LoNM:UpM      0.115    1.003  -0.346  -0.0579
  UpM:LoM       0.398   -0.449   0.790   0.1009
  LoM:Farm     -0.326    0.381  -0.166   2.7697
```

This 4×4 table is graphed using `plot(lor.hauser)`, giving Figure 10.9.

```
> plot(lor.hauser, confidence = FALSE, legend_pos = "topleft",
+       xlab = "Father's status comparisons")
> m <- mean(lor.hauser$coefficients)         # mean LOR
> grid.lines(x = unit(c(0, 1), "npc"),
+            y = unit(c(m, m), "native"))
```

Among the features here, you can see that there is a tendency for the odds ratio contrasting sons in the non-manual categories (UpNM:LoNM) to decline with the adjacent comparisons of their

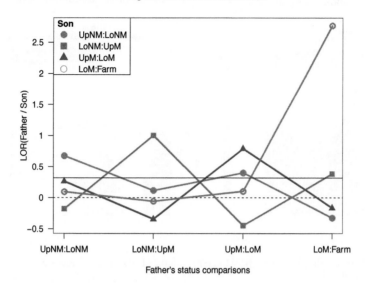

Figure 10.9: Plot of observed local log odds ratios in the Hauser79 data. The dotted horizontal line at zero shows local independence; the solid black horizontal line shows the mean.

fathers' occupations. As well, the 2 × 2 table for fathers and sons in the `LoM:Farm` stands out as deserving some attention. These observed features will be smoothed by fitting models, as described below. For additional interpretation, you can always construct similar plots of the log odds ratios using the `fitted()` values (see: Section 10.1.2) from any of the models described below.

We begin by fitting the independence model and the quasi-independence model, where the diagonal parameters in the latter are specified as `Diag(Father, Son)`. As expected, given the large frequencies in the diagonal cells, the quasi-independence model is a considerable improvement, but the fit is still very poor.

```
> hauser.indep <- gnm(Freq ~ Father + Son, data = Hauser79,
+                     family = poisson)
> hauser.quasi <-  update(hauser.indep, ~ . + Diag(Father, Son))
> LRstats(hauser.indep, hauser.quasi)

Likelihood summary table:
              AIC  BIC   LR Chisq Df Pr(>Chisq)
hauser.indep 6391 6402      6170 16      <2e-16 ***
hauser.quasi  914  931       683 11      <2e-16 ***
---
Signif. codes:  0 '***' 0.001 '**' 0.01 '*' 0.05 '.' 0.1 ' ' 1
```

The pattern of associations can be seen in the mosaic displays for both models, shown in Figure 10.10.

```
> mosaic(hauser.indep, ~ Father + Son, main = "Independence model",
+        gp = shading_Friendly)
> mosaic(hauser.quasi, ~ Father + Son, main = "Quasi-independence model",
+        gp = shading_Friendly)
```

The mosaic for quasi-independence shows an approximately symmetric pattern of residuals, so we proceed to add `Symm(Father, Son)` to the model to specify quasi-symmetry.

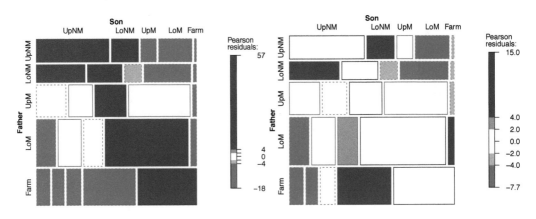

Figure 10.10: Mosaic displays for the Hauser79 data. Left: independence model; right:quasi-independence model.

```
> hauser.qsymm <- update(hauser.indep,
+                        ~ . + Diag(Father, Son) + Symm(Father, Son))
> LRstats(hauser.qsymm)

Likelihood summary table:
             AIC BIC LR Chisq Df Pr(>Chisq)
hauser.qsymm 268 291    27.4   6    0.00012 ***
---
Signif. codes:  0 '***' 0.001 '**' 0.01 '*' 0.05 '.' 0.1 ' ' 1
```

This model represents a huge improvement in goodness of fit. With such a large sample size, it might be considered an acceptable fit. The remaining lack of fit is shown in the mosaic for this model, Figure 10.11.

```
> mosaic(hauser.qsymm, ~ Father + Son, main = "Quasi-symmetry model",
+        gp = shading_Friendly, residuals_type = "rstandard")
```

The cells with the largest lack of symmetry (using standardized residuals) are those for the upper and lower non-manual occupations, where the son of an upper manual worker is less likely to move to lower non-manual work than the reverse.

For cases like this involving structured associations in square tables, Hauser (1979) developed the more general idea of grouping the row and column categories into levels of an association factor based on similar values of residuals or local odds ratios observed from the independence model. Such models are called *topological models* or *levels models*, which are implemented in the Topo() function.

To illustrate, Hauser suggested the following matrix of levels to account for the pattern of associations seen in Figure 10.10. The coding here takes the diagonal cell for the Farm category as the reference cell. Four other parameters are assigned by the numbers 2–5 to account for lack of independence.

```
> levels <- matrix(c(
+    2,   4,   5,   5,   5,
+    3,   4,   5,   5,   5,
```

Quasi–symmetry model

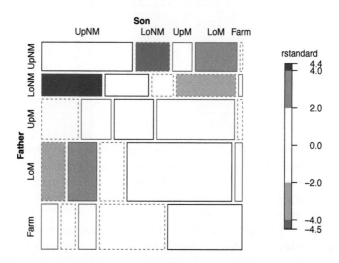

Figure 10.11: Mosaic display for the model of quasi-symmetry fit to the Hauser79 data.

```
+    5,    5,    5,    5,    5,
+    5,    5,    5,    4,    4,
+    5,    5,    5,    4,    1
+    ),
+    5, 5, byrow = TRUE)
```

This model is fit using `Topo()` as shown below. It also provides a huge improvement over the independence model, with 4 additional parameters.

```
> hauser.topo <- update(hauser.indep,
+                       ~ . + Topo(Father, Son, spec = levels))
> LRstats(hauser.topo)

Likelihood summary table:
            AIC BIC LR Chisq Df Pr(>Chisq)
hauser.topo 295 311     66.6 12    1.4e-09 ***
---
Signif. codes:  0 '***' 0.001 '**' 0.01 '*' 0.05 '.' 0.1 ' ' 1
```

As with other models fit using `gnm()`, you can extract the coefficients for particular terms using `pickCoef()`.

```
> as.vector((coef(hauser.topo)[pickCoef(hauser.topo, "Topo")]))

[1] -1.8128 -2.4973 -2.8035 -3.4026
```

The models fit in this example are summarized below. Both AIC and BIC prefer the quasi-symmetry model, `hauser.quasi`.

```
> LRstats(hauser.indep, hauser.quasi, hauser.qsymm, hauser.topo)

Likelihood summary table:
            AIC   BIC LR Chisq Df Pr(>Chisq)
```

```
hauser.indep 6391 6402      6170 16    < 2e-16 ***
hauser.quasi  914  931       683 11    < 2e-16 ***
hauser.qsymm  268  291        27  6    0.00012 ***
hauser.topo   295  311        67 12    1.4e-09 ***
---
Signif. codes:  0 '***' 0.001 '**' 0.01 '*' 0.05 '.' 0.1 ' ' 1
```

△

10.2.2 Ordinal square tables

The theory presented in Section 10.2.1 treats the row and column variables as nominal. In many instances, such as Example 10.6, the variable categories are also ordered, yet these models do not exploit their ordinal nature. In such cases, the models such as uniform association ($L \times L$), row effects, RC, and others discussed in Section 10.1 can be combined with terms for quasi-independence and symmetry of the remaining associations.

For example, the $L \times L$ model Eqn. (10.2) of uniform association applies directly to square tables, and, for square tables, can also be amended to include a diagonals term, Diag(), giving a model of *quasi-uniform association*. In this model, all adjacent 2×2 sub-tables not involving diagonal cells have a common local odds ratio.

A related model is the ***crossings model*** (Goodman, 1972). This hypothesizes that there are different difficulty parameters for crossing from one category to the next, and that the associations between categories decreases with their separation. In the crossings model for an $I \times I$ table, there are $I - 1$ crossings parameters, $\nu_1, \nu_2, \ldots, \nu_{I-1}$. The association parameters, λ_{ij}^{AB} have the form of the product of the intervening ν parameters,

$$\lambda_{ij}^{AB} = \begin{cases} \prod_{k=j}^{k=i-1} \nu_k & : \quad i > j \\ \prod_{k=i}^{k=j-1} \nu_k & : \quad i < j \end{cases}.$$

This model can also be cast in *quasi* form, by addition of a Diag term to fit the main diagonal cells. See Powers and Xie (2008, Section 4.4.7) for further details of this model. The Crossings() function in vcdExtra implements such crossings terms.

EXAMPLE 10.7: Hauser's occupational mobility table

Without much comment or detail, for reference we first fit some of the ordinal models to the Hauser79 data: Uniform association ($L \times L$), row effects, and the RC(1) model.

```
> Fscore <- as.numeric(Hauser79$Father)    # numeric scores
> Sscore <- as.numeric(Hauser79$Son)       # numeric scores
>
> # uniform association
> hauser.UA <- update(hauser.indep, ~ . + Fscore * Sscore)
> # row effects model
> hauser.roweff <- update(hauser.indep, ~ . + Father * Sscore)
> # RC model
> hauser.RC <- update(hauser.indep,
+                     ~ . + Mult(Father, Son), verbose = FALSE)
```

All of these fit very poorly, yet they are all substantial improvements over the independence model.

```
> LRstats(hauser.indep, hauser.UA, hauser.roweff, hauser.RC)

Likelihood summary table:
                AIC  BIC LR Chisq Df Pr(>Chisq)
hauser.indep   6391 6402    6170 16     <2e-16 ***
hauser.UA      2503 2516    2281 15     <2e-16 ***
hauser.roweff  2309 2325    2080 12     <2e-16 ***
hauser.RC       920  940     685  9     <2e-16 ***
---
Signif. codes:  0 '***' 0.001 '**' 0.01 '*' 0.05 '.' 0.1 ' ' 1
```

The $L \times L$ model, `hauser.UA` might be improved by ignoring the diagonals, and, indeed it is.

```
> hauser.UAdiag <- update(hauser.UA, ~ . + Diag(Father, Son))
> anova(hauser.UA, hauser.UAdiag, test = "Chisq")

Analysis of Deviance Table

Model 1: Freq ~ Father + Son + Fscore + Sscore + Fscore:Sscore
Model 2: Freq ~ Father + Son + Fscore + Sscore + Fscore:Sscore + Diag(Father,
    Son)
  Resid. Df Resid. Dev Df Deviance Pr(>Chi)
1        15       2281
2        10         73  5     2208   <2e-16 ***
---
Signif. codes:  0 '***' 0.001 '**' 0.01 '*' 0.05 '.' 0.1 ' ' 1
```

In this model, the estimated common local log odds ratio—the coefficient for the linear-by-linear term `Fscore:Sscore`, is

```
> coef(hauser.UAdiag)[["Fscore:Sscore"]]

[1] 0.1584
```

For comparisons not involving the diagonal cells, each step down the scale of occupational categories for the father multiplies the odds that the son will also be in one lower category by $\exp(0.158) = 1.172$, an increase of 17%.

The crossings model, with and without the diagonal cells, can be fit as follows:

```
> hauser.CR <- update(hauser.indep, ~ . + Crossings(Father, Son))
> hauser.CRdiag <- update(hauser.CR, ~ . + Diag(Father, Son))
> LRstats(hauser.CR, hauser.CRdiag)

Likelihood summary table:
               AIC BIC LR Chisq Df Pr(>Chisq)
hauser.CR      319 334    89.9 12    5.1e-14 ***
hauser.CRdiag  299 318    64.2  9    2.0e-10 ***
---
Signif. codes:  0 '***' 0.001 '**' 0.01 '*' 0.05 '.' 0.1 ' ' 1
```

The quasi-crossings model `hauser.CRdiag` has a reasonable G^2 fit statistic, and its interpretation and lack of fit is worth exploring further. The crossings coefficients ν can be extracted as follows.

```
> nu <- coef(hauser.CRdiag)[pickCoef(hauser.CRdiag, "Crossings")]
> names(nu) <- gsub("Crossings(Father, Son)C", "nu", names(nu),
+                    fixed = TRUE)
> nu

     nu1      nu2      nu3      nu4
-0.42275 -0.38768 -0.27500 -1.40244
```

They indicate the steps between adjacent categories in terms of the barriers for a son moving to a lower occupational category. The numerically largest gap separates the lower non-manual category from farming.

In contrast to the `UAdiag` model, the quasi-crossing model with diagonal terms implies that all 2×2 off-diagonal sub-tables are independent, i.e., the local odds ratios are all equal to 1.0. The reasons for lack of fit of this model can be seen in the corresponding mosaic display, shown in Figure 10.12.

```
> mosaic(hauser.CRdiag, ~ Father + Son,
+          gp = shading_Friendly, residuals_type = "rstandard",
+          main = "Crossings() + Diag()")
```

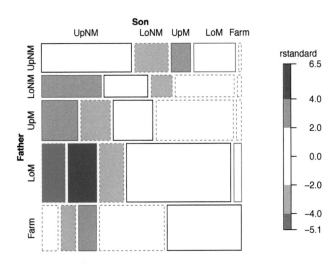

Figure 10.12: Mosaic display for the quasi-crossings model fit to the Hauser79 data.

It can be seen that lack of fit for this model is largely concentrated in the lower triangle, where the father's occupation is lower than that of his son.

In this example and the last, we have fit quite a few different models to the Hauser (1979) data. In presentations, articles and books, it is common to summarize such a collection in a table, sorted by G^2, degrees of freedom, AIC or BIC, to show their ordering along some metric. For instance, here we collect all the models fit in Example 10.6 and this example in a `glmlist()` and sort in decreasing order of BIC to show model fit by this measure.

```
> modlist <- glmlist(hauser.indep, hauser.roweff, hauser.UA,
+                    hauser.UAdiag, hauser.quasi, hauser.qsymm,
+                    hauser.topo, hauser.RC, hauser.CR, hauser.CRdiag)
> LRstats(modlist, sortby = "BIC")

Likelihood summary table:
                AIC  BIC LR Chisq Df Pr(>Chisq)
hauser.indep   6391 6402    6170 16   < 2e-16 ***
hauser.UA      2503 2516    2281 15   < 2e-16 ***
hauser.roweff  2309 2325    2080 12   < 2e-16 ***
```

```
hauser.RC        920    940      685   9     < 2e-16 ***
hauser.quasi     914    931      683  11     < 2e-16 ***
hauser.CR        319    334       90  12     5.1e-14 ***
hauser.UAdiag    306    324       73  10     1.2e-11 ***
hauser.CRdiag    299    318       64   9     2.0e-10 ***
hauser.topo      295    311       67  12     1.4e-09 ***
hauser.qsymm     268    291       27   6     0.00012 ***
---
Signif. codes:   0 '***' 0.001 '**' 0.01 '*' 0.05 '.' 0.1 ' ' 1
```

When there are more than just a few models, a more useful display is a ***model comparison plot*** of measures like G^2/df, AIC, or BIC against degrees of freedom. For example, Figure 10.13 plots BIC against Df from the result of LRstats(). Because interest is focused on the smallest values of BIC and these values span a large range, BIC is shown on the log scale using log="y".

```
> sumry <- LRstats(modlist)
> mods <- substring(rownames(sumry), 8)
> with(sumry, {
+    plot(Df, BIC, cex = 1.3, pch = 19,
+         xlab = "Degrees of freedom", ylab = "BIC (log scale)",
+         log = "y", cex.lab = 1.2)
+    pos <- ifelse(mods == "UAdiag", 1, 3)
+    text(Df, BIC + 55, mods, pos = pos, col = "red", xpd = TRUE, cex = 1.2)
+    })
```

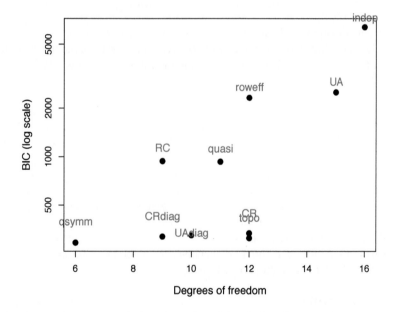

Figure 10.13: Model comparison plot for the models fit to the Hauser79 data.

Compared with the sorted tabular display shown above, such a plot sorts the models *both* by a measure of fit and by model complexity (degrees of freedom). Figure 10.13 shows that the quasi-symmetry model is best by BIC, but also shows that the next four best models by this measure are quite similar in terms of BIC. Similar plots for AIC and G^2/df show that the model of quasi-symmetry is favored by these measures.

△

10.3   Three-way and higher-dimensional tables

The models and methods for ordinal factors and square tables described in Section 10.1 and Section 10.2 extend readily to multidimensional tables with these properties for some of the factors. In three-way tables, these models provide a more parsimonious account than the saturated model, $[ABC]$, and also allow simpler models than the general model of homogeneous association, $[AB][AC][BC]$, using scores for ordinal factors or terms for symmetry and diagonal factors in square layers.

For example, consider the case where all three factors are ordinal and the model of homogeneous association $[AB][AC][BC]$ fits poorly. In this case we can generalize the model of uniform association by assigning scores a, b, and c and model the three-way association, λ_{ijk}^{ABC} as

$$\lambda_{ijk}^{ABC} = \gamma a_i b_j c_k$$

with only one more parameter. This gives the model of **uniform interaction** (or *homogeneous uniform association*)

$$\log(m_{ijk}) = \mu + \lambda_i^A + \lambda_j^B + \lambda_k^C + \lambda_{ij}^{AB} + \lambda_{ik}^{AC} + \lambda_{jk}^{BC} + \gamma a_i b_j c_k \,. \tag{10.9}$$

This model posits that (with equally spaced scores) all local odds ratios θ_{ijk} in adjacent rows, columns, and layers are constant,

$$\log(\theta_{ijk}) = \gamma \qquad \forall \quad i, j, k \,.$$

The homogeneous association model is the special case of $\log \theta_{ijk} = \gamma = 0$.

A less restricted model of **heterogeneous uniform association** retains the linear-by-linear form of association for factors A and B, but allows the strength of this association to vary over layers, C, representing λ_{ijk}^{ABC} as

$$\lambda_{ijk}^{ABC} = (\gamma + \gamma_k) a_i b_j$$

with the constraint $\sum_k \gamma_k = 0$. This model is equivalent to fitting separate models of uniform association at each level k of factor C and gives estimates of the conditional local log odds ratios, $\log \theta_{ij(k)} = \gamma + \gamma_k$.

Following the development in Section 10.1 there is a large class of other models for ordinal factors (see Figure 10.1), where not all factors are assigned scores. For three-way tables, these can be represented in homogeneous form when the two-way association of A and B is the same for all levels of C, or in a heterogeneous form, when it varies over C.

Similarly, the models for square tables described in Section 10.2 extend to three-way tables with several layers (strata), allowing both homogeneous and heterogeneous terms for diagonals and symmetry describing the AB association over levels of C.

EXAMPLE 10.8: Visual acuity

We continue the analysis of the `VisualAcuity` data, but now consider the three-way, $4 \times 4 \times 2$ table comprising both men and women. The main questions here are whether the pattern of quasi-symmetry observed in the analysis for women also pertains to men and whether there is heterogeneity of the association between right and left acuity across gender.

A useful first step for n-dimensional tables is to consider the models composed of all 1-way, 2-way, ... n-way terms as a quick overview. The function `Kway()` in the **vcdExtra** package does this automatically, returning a "glmlist" object containing the fitted models. That is, for this problem, `Kway()` generates (and fits) the following model formulae, also including the 0-way model, corresponding to $\log m_{ijk...} = \mu$.

```
> Freq ~ 1
> Freq ~ right + left + gender
> Freq ~ (right + left + gender)^2
> Freq ~ (right + left + gender)^3
```

We use Kway() as follows:

```
> vis.kway <- Kway(Freq ~ right + left + gender, data = VisualAcuity)
> LRstats(vis.kway)

Likelihood summary table:
         AIC   BIC LR Chisq Df Pr(>Chisq)
kway.0 13857 13858    13631 31    < 2e-16 ***
kway.1  9925  9937     9686 24    < 2e-16 ***
kway.2   298   332       28  9    0.00079 ***
kway.3   287   334        0  0    < 2e-16 ***
---
Signif. codes:  0 '***' 0.001 '**' 0.01 '*' 0.05 '.' 0.1 ' ' 1
```

This shows that the model of homogeneous association kway.2 ($[RL][RG][LG]$) does not fit well, but it doesn't account for diagonal agreement or symmetry to simplify the associations.

As a basis for comparison, we first fit the simple models of quasi-independence and quasi-symmetry that do not involve gender, asserting the same pattern of diagonal and off-diagonal cells for males and females.

```
> vis.indep <- glm(Freq ~ right + left + gender,  data = VisualAcuity,
+                  family = poisson)
> vis.quasi <- update(vis.indep, . ~ . + Diag(right, left))
> vis.qsymm <- update(vis.indep, . ~ . + Diag(right, left)
+                                      + Symm(right, left))
>
> LRstats(vis.indep, vis.quasi, vis.qsymm)

Likelihood summary table:
            AIC  BIC LR Chisq Df Pr(>Chisq)
vis.indep  9925 9937     9686 24     <2e-16 ***
vis.quasi   696  714      449 20     <2e-16 ***
vis.qsymm   435  456      184 18     <2e-16 ***
---
Signif. codes:  0 '***' 0.001 '**' 0.01 '*' 0.05 '.' 0.1 ' ' 1
```

The model of homogeneous quasi-symmetry fits quite badly, even worse than the all two-way association model. We can see why in the mosaic for this model, shown in Figure 10.14.

```
> mosaic(vis.qsymm, ~ gender + right + left, condvars = "gender",
+        residuals_type = "rstandard", gp = shading_Friendly,
+        labeling_args = largs, rep = FALSE,
+        main = "Homogeneous quasi-symmetry")
```

It can be seen in Figure 10.14 that the pattern of residuals for men and women are nearly completely opposite in the upper and lower portions of the plot: men have positive residuals in the same right, left cells where women have negative residuals, and vice-versa. In particular, the diagonal cells of both tables have large absolute residuals, because the term Diag(right, left) fits a common set of diagonals for both men and women.

We can correct for this by allowing separate diagonal and symmetry terms, given as interactions of gender with Diag() and Symm().

Homogeneous quasi–symmetry

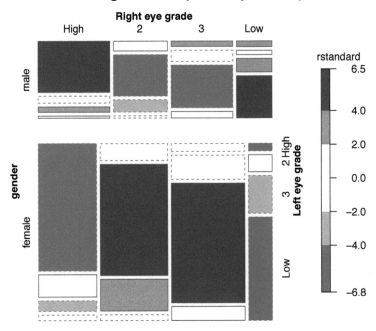

Figure 10.14: Mosaic display for the model of homogeneous quasi-symmetry fit to the VisualAcuity data.

```
> vis.hetdiag <- update(vis.indep, . ~ . + gender * Diag(right, left) +
+                       Symm(right, left))
> vis.hetqsymm <- update(vis.indep, . ~ . + gender * Diag(right, left) +
+                       gender * Symm(right, left))
> LRstats(vis.qsymm, vis.hetdiag, vis.hetqsymm)

Likelihood summary table:
              AIC BIC LR Chisq Df Pr(>Chisq)
vis.qsymm     435 456    183.7 18    < 2e-16 ***
vis.hetdiag   312 338     52.3 14    2.5e-06 ***
vis.hetqsymm  287 321     17.7  9      0.038 *
---
Signif. codes:  0 '***' 0.001 '**' 0.01 '*' 0.05 '.' 0.1 ' ' 1
```

Note that the model `vis.hetqsymm` fits better than the model `vis.hetdiag` in absolute terms, but the latter, with fewer parameters, fits better by AIC and BIC. The mosaic for the model `vis.hetqsymm` is shown in Figure 10.15.

```
> mosaic(vis.hetqsymm, ~ gender + right + left, condvars="gender",
+        residuals_type = "rstandard", gp = shading_Friendly,
+        labeling_args = largs, rep = FALSE,
+        main="Heterogeneous quasi-symmetry")
```

As in the two-way case, this model now fits the diagonal cells in each table exactly, effectively ignoring this part of the association between right and left eye acuity. All remaining residuals are relatively small in magnitude, except for the two opposite off-diagonal cells (`Low`, `High`) and (`High`, `Low`) in the table for women.

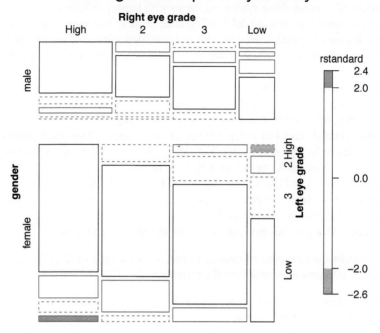

Figure 10.15: Mosaic display for the model of heterogeneous quasi-symmetry fit to the VisualAcuity data.

The substantive interpretation of this example is that visual acuity is largely the same (diagonal cells) in the right and left eyes of both men and women. Ignoring the diagonal cells, when visual acuity differs, both men and women exhibit approximately symmetric associations. However, deviations from symmetry (Figure 10.14) are such that men are slightly more likely to have a lower grade in the right eye, while women are slightly more likely to have a higher grade in the right eye.

△

10.4 Multivariate responses*

In many studies, there may be *several* categorical responses observed along with one or more explanatory variables. In a clinical trial, for example, the efficacy of a drug might be the primary response, but the occurrence of side-effects might give rise to additional response variables of substantive interest. Or, in a study of occupational health, the occurrence of two or more distinct symptoms might be treated as response variables.

If there are *no* explanatory variables, then the problem is simply to understand the joint distribution of the response categories, and the loglinear models and graphical displays described earlier are sufficient. Otherwise, in these cases we usually wish to understand how the various responses are affected by the explanatory variables. Moreover, it may also be important to understand how the association between the categorical responses depends on the explanatory variables. That is, we would like to study how *both* the marginal distributions of the responses, and their joint distribution depends on the predictors. In the occupational health example, the goal might be to understand both how the prevalence of several symptoms varies with one or more predictors, and how the association (loosely, "correlation") among those symptoms varies with those predictors.

Although the general loglinear model is often used in these situations, there are special reparameterizations that may be used to separate the *marginal* dependence of each response on the explanatory variables from the relationship of the *association* among the responses on the explanatory variables.

Let us say that categorical responses, Y_1, Y_2, ... have been observed, together with possible explanatory variables, X_1, X_2, ..., and let $\pi_{ij...}$ be the joint probability of all the responses and explanatory variables; we also use x to refer to the values of X_1, X_2, ...

Note that the minimal model of independence of all responses from each other and from the explanatory variables is the loglinear model $[Y_1][Y_2] \cdots [X_1 X_2 \cdots]$ (i.e., all associations among the X_i must be included). A no-effect model, in which the responses do not depend on the explanatory variables, but may be associated among themselves, is $[Y_1 Y_2 \cdots][X_1 X_2 \cdots]$. However, these models do not separate the individual (marginal) effects of $X_1, X_2 \ldots$ on each Y_i from their associative effects on the joint relationships among the Y_i.

There are three useful general approaches that *do* separate these effects:

1. Model the marginal dependence of each response, Y_i, separately on X_1, X_2, ..., and, in addition, model the interdependence among the responses, $Y_1, Y_2, \ldots$.[5]

2. Model the joint dependence of all responses on X_1, X_2, ..., but parameterized so that marginal and associative effects are delineated.

3. Construct simultaneous models, estimated together, for the marginal and joint dependence of the responses on the explanatory variables.

The first approach is the simplest, an informative starting place, and is satisfactory in the (often unlikely) case that the responses are not associated, or if the associations among responses do not vary much over the explanatory variables (i.e., no terms like $[Y_1 Y_2 X_j]$ are required). In the clinical trial example, we would construct separate loglinear or logit models for efficacy of the drug, and for occurrence of side-effects, and supplement these analyses with mosaic or other displays showing the relations between efficacy and side-effects and a model for their joint association. If those who improve with the drug also show more serious side effects, the worth of the treatment would be questioned. A limitation of this method is that it does not provide an overall model comprising these effects.

In the second approach, the joint probabilities, $\pi_{ij...}$, are recast to give separate information regarding the dependence of the univariate marginal probabilities $\pi_{i\bullet}, \pi_{\bullet j}, \ldots$, on the explanatory variables and the dependence of the intra-response associations on the explanatory variables. The **VGAM** package provides several versions of this approach with the function vglm() (for *vector generalized linear model*).

The third approach, developed, for example, by Lang and Agresti (1994), is the most general, and provides a scheme to represent a model $\mathcal{J}(\bullet)$ for the joint distributions of the X, Y variables together with a model $\mathcal{M}(\bullet)$ for their first-order marginal distributions: The joint models are typically loglinear models, ranging from the mutual independence model, $\mathcal{J}(I) = [Y_1][Y_2][\cdots][X_1][X_2][\cdots]$, to the saturated model, $\mathcal{J}(S) = [Y_1 Y_2 \cdots X_1 X_2 \cdots]$, while the marginal models are logit models for the response variables. The combined model, denoted $\mathcal{J}(\bullet) \cap \mathcal{M}(\bullet)$, is estimated simultaneously by maximum likelihood. This approach is implemented in R in the **hmmm** (Roberto et al., 2014) package (hierarchical multinomial marginal models). However, model specification in this implementation is complicated, and it will not be considered further here.

---

[5]For quantitative responses, this is roughly analogous to fitting univariate response models for each Y_i, followed by something like a principal component analysis of the relationships among the Y_i. But in this case, the multivariate linear model, $Y = XB + E$ provides a general solution.

10.4.1 Bivariate, binary response models

We focus here on two related models reflecting the second approach, as discussed by McCullagh and Nelder (1989, Section 6.5). We consider here only the case of two binary responses, though the general approach can be applied to $R > 2$ responses $Y_1, Y_2, \ldots, Y_R$, and these may be polytomous or ordinal.

Let x refer to the values of all the explanatory variables and let $\pi_{ij}(x)$ be the joint probabilities in cell $Y_1 = i$, $Y_2 = j$. The essential idea of the *bivariate logistic model* arises from a linear transformation of the cell probabilities π to interpretable functions of the marginal probabilities (logits) and their association (odds ratio), a mapping of $\pi \to \eta$,

$$
\begin{aligned}
\eta_1 &= \operatorname{logit}(\pi_{1\bullet}) \\
\eta_2 &= \operatorname{logit}(\pi_{\bullet 1}) \\
\eta_{12} &= \log\left(\frac{\pi_{11}\,\pi_{22}}{\pi_{12}\,\pi_{21}} \right).
\end{aligned}
\tag{10.10}
$$

The predictors in x are then taken into account by considering models that relate π to x through η,

$$
\begin{aligned}
\eta_1 &= x_1^\mathsf{T} \beta_1 \\
\eta_2 &= x_2^\mathsf{T} \beta_2 \\
\eta_{12} &= x_{12}^\mathsf{T} \beta_{12},
\end{aligned}
\tag{10.11}
$$

where x_1, x_2, and x_{12} are subsets of the predictors in x for each sub-model, and β_1, β_2, and β_{12} are the corresponding parameters to be estimated.

McCullagh and Nelder (1989) arrive at this joint bivariate model in two steps. First, transform the cell probabilities π to a vector of probabilities γ, which also includes the univariate margins, given by

$$
\gamma = L\pi,
\tag{10.12}
$$

where L is a matrix of 0s and 1s of the form of a factorial design matrix. In the 2×2 case,

$$
\gamma =
\begin{pmatrix}
\pi_{1\bullet} \\
\pi_{2\bullet} \\
\pi_{\bullet 1} \\
\pi_{\bullet 2} \\
\pi_{11} \\
\pi_{12} \\
\pi_{21} \\
\pi_{22}
\end{pmatrix}
=
\begin{bmatrix}
1 & 1 & 0 & 0 \\
0 & 0 & 1 & 1 \\
1 & 0 & 1 & 0 \\
0 & 1 & 0 & 1 \\
1 & 0 & 0 & 0 \\
0 & 1 & 0 & 0 \\
0 & 0 & 1 & 0 \\
0 & 0 & 0 & 1
\end{bmatrix}
\begin{pmatrix}
\pi_{11} \\
\pi_{12} \\
\pi_{21} \\
\pi_{22}
\end{pmatrix}.
\tag{10.13}
$$

There are of course only three linearly independent probabilities, because $\sum\sum \pi_{ij} = 1$. In the second step, the bivariate logistic model is formulated in terms of factorial contrasts on the elements of γ, which express separate models for the two logits and the log odds. The model is expressed as

$$
\eta = C \log \gamma = C \log(L\pi),
\tag{10.14}
$$

where C is a matrix of contrasts. In the 2×2 case, the usual contrasts may be defined by

$$
\boldsymbol{\eta} = \begin{pmatrix} \eta_1 \\ \eta_2 \\ \eta_{12} \end{pmatrix} = \begin{pmatrix} \text{logit } \pi_{1\bullet} \\ \text{logit } \pi_{\bullet 1} \\ \log(\theta_{12}) \end{pmatrix} = \begin{bmatrix} 1 & -1 & 0 & 0 & 0 & 0 & 0 & 0 \\ 0 & 0 & 1 & -1 & 0 & 0 & 0 & 0 \\ 0 & 0 & 0 & 0 & 1 & -1 & -1 & 1 \end{bmatrix} \begin{pmatrix} \log \pi_{1\bullet} \\ \log \pi_{2\bullet} \\ \log \pi_{\bullet 1} \\ \log \pi_{\bullet 2} \\ \log \pi_{11} \\ \log \pi_{12} \\ \log \pi_{21} \\ \log \pi_{22} \end{pmatrix}.
$$

$$(10.15)$$

Thus, we are modeling the marginal log odds of each response, together with the log odds ratio $\log(\theta_{12})$ simultaneously.

Specific models are then formulated for the dependence of $\eta_1(\boldsymbol{x}), \eta_2(\boldsymbol{x})$, and $\eta_{12}(\boldsymbol{x})$ on some or all of the explanatory variables. For example, with one quantitative explanatory variable, x, the model

$$
\begin{pmatrix} \eta_1 \\ \eta_2 \\ \eta_{12} \end{pmatrix} = \begin{pmatrix} \alpha_1 + \beta_1 x \\ \alpha_2 + \beta_2 x \\ \log(\theta) \end{pmatrix} \tag{10.16}
$$

asserts that the log odds of each response changes linearly with x, while the odds ratio between the responses remains constant. In the general form given by McCullagh and Nelder (1989) the submodels in Eqn. (10.16) may each depend on the explanatory variables in different ways. For example, the logits could both depend quadratically on x, while an intercept-only model could be posited for the log odds ratio.

The second model is the **bivariate loglinear model**, the special case obtained by taking $\boldsymbol{L} = \boldsymbol{I}$ in Eqn. (10.12) and Eqn. (10.14) so that $\boldsymbol{\gamma} = \boldsymbol{\pi}$. Then a loglinear model of the form

$$
\boldsymbol{\eta}(\boldsymbol{x}) = \boldsymbol{C} \log \boldsymbol{\pi}
$$

expresses contrasts among log probabilities as linear functions of the explanatory variables. For the 2×2 case, we take the contrasts C as shown below

$$
\boldsymbol{\eta} = \begin{pmatrix} l_1 \\ l_2 \\ \eta_{12} \end{pmatrix} = \begin{bmatrix} 1 & 1 & -1 & -1 \\ 1 & -1 & 1 & -1 \\ 1 & -1 & 1 & -1 \end{bmatrix} \begin{pmatrix} \log \pi_{11} \\ \log \pi_{12} \\ \log \pi_{21} \\ \log \pi_{22} \end{pmatrix} \tag{10.17}
$$

and models for the dependence of $l_1(\boldsymbol{x})$, $l_2(\boldsymbol{x})$, and $\eta_{12}(\boldsymbol{x})$ are expressed in the same way as in Eqn. (10.16). The estimates of the odds ratio, η_{12}, are the same under both models. The marginal functions are parameterized differently, however, but lead to similar predicted probabilities.

In R, bivariate logistic models of the form Eqn. (10.10) and Eqn. (10.11) can be fit using `vglm()` with the `binom2.or()` family in the **VGAM** package.[6] The fitting and graphing of these models is illustrated in the next example.

EXAMPLE 10.9: Breathlessness and wheeze in coal miners

In Example 4.12 we examined the association between the occurrence of two pulmonary conditions, breathlessness and wheeze, among coal miners classified by age (Ashford and Sowden, 1970). Figure 4.7 showed fourfold displays focused on the odds ratio for the co-occurrence of these symptoms, and Figure 4.8 plotted these odds ratios against age directly. Here, we consider models that examine the changes in prevalence of the two symptoms over age, together with the changes in their association.

---

[6]This package also provides for bivariate and trivariate loglinear models with `loglinb2()` and `loglinb2`.

10.4.1.1   Plotting bivariate response data

As a starting point and overview of what is necessary for bivariate response models, we calculate
the empirical log odds for breathlessness and for wheeze, and the log odds ratio for their association
in each 2×2 table. The log odds ratios are the same values plotted in Figure 4.8 (but the youngest
age group was not included in the earlier analysis).

The *CoalMiners* data is a $2 \times 2 \times 9$ table. For convenience in this analysis (and for use with
VGAM) we convert it to a 4×9 data frame, and relabel the columns to use the combinations of
(`"B"`, `"b"`) and (`"W"`, `"w"`) to represent the conditions of breathlessness and wheeze, where
the upper case letter indicates presence of the condition. A variable `age` is also created, using the
midpoints of the age categories.

```
> data("CoalMiners", package = "vcd")
> coalminers <- data.frame(t(matrix(aperm(CoalMiners, c(2, 1, 3)),
+                          4, 9)))
> colnames(coalminers) <- c("BW", "Bw", "bW", "bw")
> coalminers$age <- c(22, 27, 32, 37, 42, 47, 52, 57, 62)
> coalminers

   BW  Bw  bW   bw age
1   9   7  95 1841  22
2  23   9 105 1654  27
3  54  19 177 1863  32
4 121  48 257 2357  37
5 169  54 273 1778  42
6 269  88 324 1712  47
7 404 117 245 1324  52
8 406 152 225  967  57
9 372 106 132  526  62
```

With the data in this form, a simple function `blogits()` in **vcdExtra** calculates the logits and
log odds ratios corresponding to Eqn. (10.10). The `add` argument accommodates cases where there
were very small, or 0 frequencies in some cells, and it is common to add a small constant, such as 0.5,
to each cell in calculating *empirical logits*. This function is used to calculate the empirical logits
and log odds as follows:

```
> logitsCM <- blogits(coalminers[, 1 : 4], add = 0.5)
> colnames(logitsCM)[1:2] <- c("logitB", "logitW")
> logitsCM

       logitB   logitW  logOR
[1,] -4.73568 -2.86844 3.1956
[2,] -3.97656 -2.55717 3.6583
[3,] -3.31713 -2.09388 3.3790
[4,] -2.73322 -1.84818 3.1327
[5,] -2.21492 -1.42014 3.0069
[6,] -1.73870 -1.10922 2.7770
[7,] -1.10116 -0.79681 2.9217
[8,] -0.75808 -0.57219 2.4368
[9,] -0.31902 -0.22591 2.6318
```

We plot these as shown below, using `matplot()`, which is convenient for plotting multiple
columns against a given horizontal variable, `age` here.[7] For ease of interpretation of the log odds,
we also use right vertical axis showing the equivalent probabilities for breathlessness and wheeze.

---

[7]It is actually a small graphical misdemeanor to plot logits and odds ratios on the same vertical axis because they are not
strictly commensurable. We plead guilty with the explanation that this graph shows what we want to see here and does not
distort the data.

```
> col <- c("blue", "red", "black")
> pch <- c(15, 17, 16)
> age <- coalminers$age
>
> op <- par(mar = c(4, 4, 1, 4)+.2)
> matplot(age, logitsCM, type = "p",
+   col = col, pch = pch, cex = 1.2, cex.lab = 1.25,
+   xlab = "Age", ylab = "Log Odds or Odds Ratio")
> abline(lm(logitsCM[,1] ~ age), col = col[1], lwd = 2)
> abline(lm(logitsCM[,2] ~ age), col = col[2], lwd = 2)
> abline(lm(logitsCM[,3] ~ age), col = col[3], lwd = 2)
>
> # right probability axis
> probs <- c(.01, .05, .10, .25, .5)
> axis(4, at = qlogis(probs), labels = probs)
> mtext("Probability", side = 4, cex = 1.2, at = -2, line = 2.5)
> # curve labels
> text(age[2], logitsCM[2, 1] + .5, "Breathlessness",
+     col = col[1], pos = NULL, cex = 1.2)
> text(age[2], logitsCM[2, 2] + .5, "Wheeze",
+     col = col[2], pos = NULL, cex = 1.2)
> text(age[2], logitsCM[2, 3] - .5, "log OR\n(B|W)/(B|w)",
+     col = col[3], pos = 1, cex = 1.2)
> par(op)
```

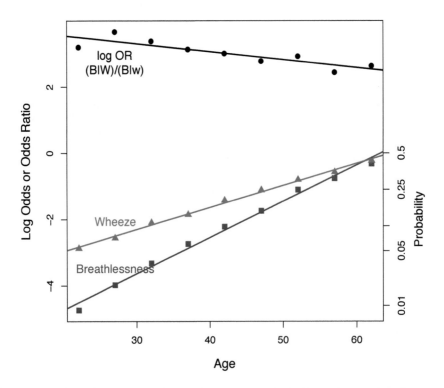

Figure 10.16: Empirical logits and log odds ratio for breathlessness and wheeze in the CoalMiners data. The lines show separate linear regressions for each function. The right vertical axis shows equivalent probabilities for the logits.

In Figure 10.16 we see that both symptoms, while quite rare among young miners, increase steadily with age (or years working in the mine). By age 60, the probability is nearly 0.5 of having either condition. There is a hint of curvilinearity, particularly in the logit for breathlessness. The decline in the odds ratio with age may reflect selection, as miners who had retired for health or other reasons were excluded from the study.

10.4.1.2 Fitting `glm` models

Next, we illustrate what can easily be achieved using the standard `glm()` approach for loglinear models and why the bivariate models we described are more useful in this situation. `glm()` requires a data frame as input, so first reshape *CoalMiners* to a frequency data frame. For convenience, we simplify the variable names to B and W.

```
> CM <- as.data.frame(CoalMiners)
> colnames(CM)[1:2] <- c("B", "W")
> head(CM)

    B    W   Age Freq
1   B    W 20-24    9
2 NoB    W 20-24   95
3   B  NoW 20-24    7
4 NoB  NoW 20-24 1841
5   B    W 25-29   23
6 NoB    W 25-29  105
```

As a point of comparison, we fit the mutual independence model, [B][W][Age], and the baseline model for associated responses, [BW][Age], which asserts that the association between B and W is independent of Age.

```
> cm.glm0 <- glm(Freq ~ B + W + Age, data = CM, family = poisson)
> cm.glm1 <- glm(Freq ~ B * W + Age, data = CM, family = poisson)
> LRstats(cm.glm0, cm.glm1)


Likelihood summary table:
           AIC  BIC LR Chisq Df Pr(>Chisq)
cm.glm0 7217 7234     6939 25      <2e-16 ***
cm.glm1 2981 3000     2702 24      <2e-16 ***
---
Signif. codes:  0 '***' 0.001 '**' 0.01 '*' 0.05 '.' 0.1 ' ' 1
```

The baseline model `cm.glm1` fits very badly. We can see the pattern of the residual association in a mosaic display for this model shown in Figure 10.17. The formula argument here specifies the order of the variables in the mosaic.

```
> vnames <- list(set_varnames = c(B = "Breathlessness", W = "Wheeze"))
> lnames <- list(B=c("B", "b"), W = c("W", "w"))
> mosaic(cm.glm1, ~ Age + B + W,
+        labeling_args = vnames, set_labels = lnames)
```

As structured here, it is easy to see the increase in the prevalence of breathlessness and wheeze with age and the changing pattern of their association with age.

From Figure 10.16 and Figure 10.17, it is apparent that both breathlessness and wheeze increase with age, so we can model this by adding terms [B Age][W Age] to the baseline model. This is the no-three-way interaction model, which could also be specified as `Freq ~ (B + W + Age)^2`.

```
> cm.glm2 <- glm(Freq ~ B * W + (B + W) * Age, data = CM, family = poisson)
> LRstats(cm.glm1, cm.glm2)


Likelihood summary table:
           AIC  BIC LR Chisq Df Pr(>Chisq)
```

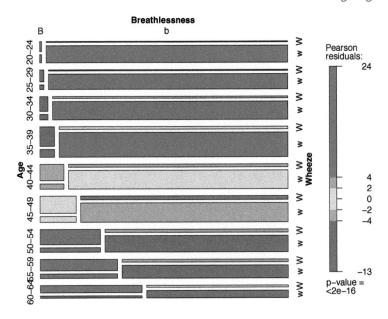

Figure 10.17: Mosaic display for the baseline model, [BW][Age], fit to the CoalMiners data.

```
cm.glm1 2981 3000        2702 24       <2e-16 ***
cm.glm2  338  383          27  8       8e-04 ***
---
Signif. codes:  0 '***' 0.001 '**' 0.01 '*' 0.05 '.' 0.1 ' ' 1
```

The improvement in fit is substantial, and all terms are highly significant, yet, the residual $G^2(8)$ indicates there is still lack of fit.

```
> library(car)
> Anova(cm.glm2)

Analysis of Deviance Table (Type II tests)

Response: Freq
      LR Chisq Df Pr(>Chisq)
B        11026  1     <2e-16 ***
W         7038  1     <2e-16 ***
Age        887  8     <2e-16 ***
B:W       3025  1     <2e-16 ***
B:Age     1130  8     <2e-16 ***
W:Age      333  8     <2e-16 ***
---
Signif. codes:  0 '***' 0.001 '**' 0.01 '*' 0.05 '.' 0.1 ' ' 1
```

One way to improve the model using the `glm()` framework is to make use of `Age` as a quantitative variable and add a term to allow the odds ratio for the [BW] association to vary linearly with age. Here, we construct the variable `age` using the midpoints of the Age intervals.

```
> CM$age <- rep(seq(22, 62, 5), each = 4)
```

In the `glm()` approach, the odds ratio cannot be modeled directly, but we can use the following trick: For each 2×2 subtable, the odds ratio can be parameterized in terms of the frequency in any one cell, say, n_{11k}, given that the marginal total n_{++k} is included in the model. We do this by adding a new interaction variable, ageOR, having the value of age for the $(1, 1, k)$ cells and 0 otherwise.

```
> CM$ageOR <- (CM$B == "B") * (CM$W == "W") * CM$age
> cm.glm3 <- update(cm.glm2, . ~ . + ageOR)
> LRstats(cm.glm0, cm.glm1, cm.glm2, cm.glm3)

Likelihood summary table:
          AIC  BIC LR Chisq Df Pr(>Chisq)
cm.glm0 7217 7234     6939 25     <2e-16 ***
cm.glm1 2981 3000     2702 24     <2e-16 ***
cm.glm2  338  383       27  8     0.0008 ***
cm.glm3  320  366        7  7     0.4498
---
Signif. codes:  0 '***' 0.001 '**' 0.01 '*' 0.05 '.' 0.1 ' ' 1
```

The model cm.glm3, with one more parameter, now fits reasonably well, having residual $G^2(7) = 6.80$. The likelihood ratio test of model cm.glm3 against cm.glm2, which assumes equal odds ratios over age, can be regarded as a test of the hypothesis of homogeneity of odds ratios, against the alternative that the [BW] association changes linearly with age. The glm() models fit in this example are summarized above. As usual, anova() can be used to compare competing nested models.

```
> anova(cm.glm2, cm.glm3, test = "Chisq")

Analysis of Deviance Table

Model 1: Freq ~ B * W + (B + W) * Age
Model 2: Freq ~ B + W + Age + ageOR + B:W + B:Age + W:Age
  Resid. Df Resid. Dev Df Deviance Pr(>Chi)
1         8       26.7
2         7        6.8  1     19.9  8.2e-06 ***
---
Signif. codes:  0 '***' 0.001 '**' 0.01 '*' 0.05 '.' 0.1 ' ' 1
```

This analysis, while useful, also shows the limitations of the glm() approach: (a) It doesn't easily allow us to represent and test the substantively interesting hypotheses regarding *how* the prevalence of the binary responses, B and W, vary with Age, such as seen in Figure 10.16. (b) It doesn't represent the odds ratio for the [BW] association directly, but only through the coding trick we used here. Thus, it is difficult to interpret the coefficient for ageOR = -0.02613 in a substantively meaningful way, except that is shows that the odds ratio is decreasing.[8]

10.4.1.3 Fitting vglm models

The vglm() function in the **VGAM** package provides a very general implementation of these and other models for discrete multivariate responses. The family function, binom2.or() for binary logistic models, allows some or all of the logits or odds ratio submodels to be constrained to be intercept-only (e.g., as in Eqn. (10.16)) and the two marginal distributions can be constrained to be equal.

Quantitative predictors (such as age, here), can be modeled linearly or nonlinearly, using poly() for a parametric fit, or smooth regression splines, as provided by the functions ns(), bs(), and others in model formulas. In this illustration, we fit bivariate linear and quadratic models in age.

vglm() takes its input data in the wide form we called coalminers at the beginning of this

---

[8]Actually, the interpretability of the coefficient for the log odds ratio can be enhanced here by centering age, and representing its units in steps of 5 years, as we do below.

example. We could use the 9-level factor, Age, as we did with glm(), but we plan to use age as a numeric variable in all three submodels. The coefficients in these models will be more easily interpreted if we center age and express it as agec in units of five years, as shown below.

```
> coalminers <- transform(coalminers, agec = (age - 42) / 5)
> coalminers$Age <- dimnames(CoalMiners)[[3]]
> coalminers

   BW  Bw  bW   bw age agec   Age
1   9   7  95 1841  22   -4 20-24
2  23   9 105 1654  27   -3 25-29
3  54  19 177 1863  32   -2 30-34
4 121  48 257 2357  37   -1 35-39
5 169  54 273 1778  42    0 40-44
6 269  88 324 1712  47    1 45-49
7 404 117 245 1324  52    2 50-54
8 406 152 225  967  57    3 55-59
9 372 106 132  526  62    4 60-64
```

vglm() takes the 2×2 response frequencies as a 4-column matrix on the left-hand side of the model formula. However, denoting the responses of failure and success by 0 and 1, respectively, it takes these in the order $y_{00}, y_{01}, y_{10}, y_{11}$. We specify the order below so that the logits are calculated for the occurrence of breathlessness or wheeze, rather than their absence.

```
> library(VGAM)
> #                     00   01   10   11
> cm.vglm1 <- vglm(cbind(bw, bW, Bw, BW) ~ agec,
+                  binom2.or(zero = NULL), data = coalminers)
> cm.vglm1

Call:
vglm(formula = cbind(bw, bW, Bw, BW) ~ agec, family = binom2.or(zero = NULL),
    data = coalminers)

Coefficients:
(Intercept):1 (Intercept):2 (Intercept):3      agec:1
     -2.26247      -1.48776       3.02191     0.51451
         agec:2        agec:3
        0.32545      -0.13136

Degrees of Freedom: 27 Total; 21 Residual
Residual deviance: 30.394
Log-likelihood: -100.53
```

In this call, the argument zero = NULL indicates that none of the linear predictors, $\eta_1, \eta_2, \eta_{12}$, are modeled as constants.[9]

At this writing, there is no anova() method for the "vgam" objects produced by vglm(), but we can test the residual deviance of the model (against the saturated model) as follows, showing that this model has an acceptable fit.

```
> (G2 <- deviance(cm.vglm1))

[1] 30.394

> # test residual deviance
> 1-pchisq(deviance(cm.vglm1), cm.vglm1@df.residual)

[1] 0.084355
```

---

[9]The default, zero=3 gives the model shown in Eqn. (10.16), with the odds ratio constant.

The estimated coefficients in this model are usefully shown as below, using the argument `matrix=TRUE` in `coef()`.

Using `exp()` on the result gives values of odds that can be easily interpreted:

```
> coef(cm.vglm1, matrix = TRUE)

            logit(mu1) logit(mu2) loge(oratio)
(Intercept)   -2.26247   -1.48776      3.02191
agec           0.51451    0.32545     -0.13136

> exp(coef(cm.vglm1, matrix = TRUE))

            logit(mu1) logit(mu2) loge(oratio)
(Intercept)    0.10409    0.22588      20.5304
agec           1.67282    1.38465       0.8769
```

Thus, the odds of a miner showing breathlessness are multiplied by 1.67, a 67% increase, for each 5 years' increase in age; similarly, the odds of wheeze are multiplied by 1.38, a 38% increase. The odds ratio for the association between the two symptoms are multiplied by 0.88, a 12% decrease over each 5-year interval.

The **VGAM** package has no special plot methods for **"vglm"** objects, but it is not hard to construct these using the methods we showed earlier in this example. First, we can obtain the fitted probabilities for the 4 response combinations using `fitted()`, and the corresponding observed probabilities using `depvar()`.

```
> age <- coalminers$age
> P <- fitted(cm.vglm1)
> colnames(P) <- c("bw", "bW", "Bw", "BW")
> head(P)

       bw       bW        Bw        BW
1 0.93747 0.049409 0.0046356 0.0084831
2 0.91461 0.063636 0.0069757 0.0147776
3 0.88411 0.080029 0.0104965 0.0253679
4 0.84394 0.097484 0.0158138 0.0427671
5 0.79188 0.113839 0.0238598 0.0704196
6 0.72578 0.125910 0.0359684 0.1123366

> Y <- depvar(cm.vglm1)
```

In the left panel of Figure 10.18, we plot the fitted probabilities in the matrix `P` using `matplot()` and the observed probabilities in `Y` using `matpoints()`.

```
> col <- c("red", "blue", "red", "blue")
> pch <- c(1, 2, 16, 17)
>
> op <- par(mar = c(5, 4, 1, 1) + .1)
> matplot(age, P, type = "l",
+    col = col,
+    lwd = 2, cex = 1.2, cex.lab = 1.2,
+    xlab = "Age", ylab = "Probability",
+    xlim = c(20,65))
> matpoints(age, Y, pch = pch, cex = 1.2, col = col)
> # legend
> text(64, P[9,]+ c(0,.01, -.01, 0), labels = colnames(P), col = col, cex = 1.2)
> text(20, P[1,]+ c(0,.01, -.01, .01), labels = colnames(P), col = col, cex = 1.2)
> par(op)
```

The right panel of Figure 10.18 shows these on the log odds scale, produced using the same code as above, applied to the probabilities transformed using `qlogis()`, the quantile function for the logistic distribution.

```
> lP <- qlogis(P)
> lY <- qlogis(Y)
```

In Figure 10.16 we plotted the empirical logits and log odds using the function `blogits()` to transform frequencies to these values. An essentially identical plot can be produced by transforming the fitted and observed probabilities, as calculated below.

```
> # blogits, but for B and W
> logitsP <- blogits(P[, 4 : 1])
> logitsY <- blogits(Y[, 4 : 1])
```

To test for nonlinearity in the prevalence of the symptoms or their odds ratio with age, we can fit a similar model using `poly()` or a smoothing spline, such as `ns()`. We illustrate this here using a bivariate model allowing quadratic effects of age on all three components.

```
> cm.vglm2 <- vglm(cbind(bw, bW, Bw, BW) ~ poly(agec, 2),
+                  binom2.or(zero = NULL), data = coalminers)
```

This model has a residual $G^2 = 16.963$ with 18 df. Compared to the linear model `cm.vglm1`, this represents a significant improvement in goodness of fit.

```
> (LR <- deviance(cm.vglm1) - deviance(cm.vglm2))

[1] 13.43

> 1 - pchisq(LR, cm.vglm1@df.residual - cm.vglm2@df.residual)

[1] 0.0037925
```

A plot of the fitted logits and log odds ratios under this model is shown in Figure 10.19. You can interpret this plot as showing that the statistical evidence for the quadratic model indicates some slight tendency for the prevalence of breathlessness and wheeze to level off slightly with age, particularly the former.

$\triangle$

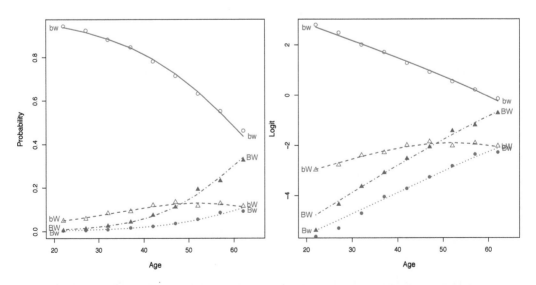

Figure 10.18: Observed and fitted values for the combinations of breathlessness and wheeze in the binary logistic regression model `cm.vglm1`. Left: probabilities; right: on the log odds scale.

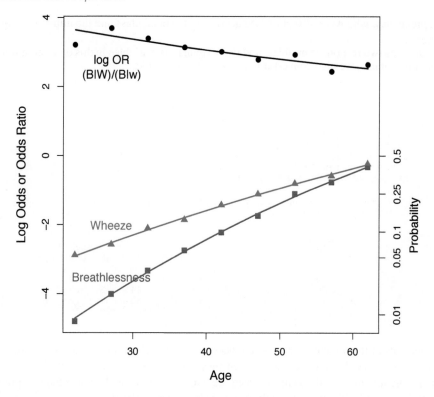

Figure 10.19: Observed (points) and fitted (lines) logits and log odds ratios for the quadratic binary logistic regression model `cm.vglm2`.

10.4.2 More complex models

When there is more than one explanatory variable and several responses, the methods described above using `glm()` and `vglm()` still apply. However, it is useful to begin with a more thorough visual examination of the relations within and between these sets. Some useful graphical displays include:

- mosaic displays showing the marginal relations among the response variables and of the explanatory variables, each collapsed over the other set;

- conditional mosaics or fourfold displays of the associations among the responses, stratified by one or more of the explanatory variables;

- plots of empirical logits and log odds ratios, as in Figure 10.16 or model-based plots, such as Figure 10.19, showing a model-smoothed summary.

These displays can, and should, inform our search for an adequate descriptive or explanatory model. Some of these ideas are illustrated in the following example.

EXAMPLE 10.10: Toxaemic symptoms in pregnancy

Brown et al. (1983) gave the data used here on two signs of *toxaemia*, an abnormal condition during pregnancy characterized by high blood pressure (hypertension) and high levels of protein in the urine. If untreated, both the mother and baby are at risk of complications or death. The data frame *Toxaemia* in **vcdExtra** represents 13,384 expectant mothers in Bradford, England in their first pregnancy, who were also classified according to social class and the number of cigarettes smoked per day.

There are thus two response variables, and two explanatory variables in this data set in frequency form. For convenience, we also convert it to a $2 \times 2 \times 5 \times 3$ table.

```
> data("Toxaemia", package = "vcdExtra")
> str(Toxaemia)

'data.frame': 60 obs. of  5 variables:
 $ class: Factor w/ 5 levels "1","2","3","4",..: 1 1 1 1 1 1 1 1 1 1 ...
 $ smoke: Factor w/ 3 levels "0","1-19","20+": 1 1 1 2 2 2 2 3 3 ...
 $ hyper: Factor w/ 2 levels "Low","High": 2 2 1 1 2 2 1 1 2 2 ...
 $ urea : Factor w/ 2 levels "Low","High": 2 1 2 1 2 1 2 1 2 1 ...
 $ Freq : int  28 82 21 286 5 24 5 71 1 3 ...

> tox.tab <- xtabs(Freq ~ class + smoke + hyper + urea, Toxaemia)
> ftable(tox.tab, row.vars = 1)
```

	smoke	0				1-19				20+			
	hyper	Low		High		Low		High		Low		High	
	urea	Low	High	Low	High	Low	High	Low	High	Low	High	Low	High
class													
1		286	21	82	28	71	5	24	5	13	0	3	1
2		785	34	266	50	284	17	92	13	34	3	15	0
3		3160	164	1101	278	2300	142	492	120	383	32	92	16
4		656	52	213	63	649	46	129	35	163	12	40	7
5		245	23	78	20	321	34	74	22	65	4	14	7

The questions of main interest are how the occurrence of each symptom varies with social class and smoking, and how the association between these symptoms varies. It is useful, however, to examine first the marginal relationship between the two responses, and between the two predictors. The calls to mosaic() below produce the two panels in Figure 10.20.

```
> mosaic(~ smoke + class, data = tox.tab, shade = TRUE,
+          main = "Predictors", legend = FALSE)
> mosaic(~ hyper + urea, data = tox.tab, shade = TRUE,
+          main = "Responses", legend = FALSE)
```

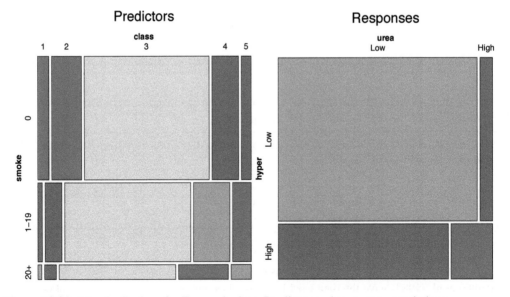

Figure 10.20: Mosaic displays for Toxaemia data: Predictor and response associations.

We see in Figure 10.20 that the majority of the mothers are in the third social class, and that smoking is negatively related to social class, with the highest levels of smoking in classes 4 and 5. (Social class 1 is the highest in status here.) More than 50% are non-smokers. Within the responses, the great majority of women exhibit neither symptom, but showing one symptom makes it much more likely to show the other. Marginally, hypertension is somewhat more prevalent than protein urea.

We next examine how the association between responses varies with social class and with smoking. Figure 10.21 shows a collection of conditional mosaic plots using `cotabplot()` of the association between hypertension and urea, for each level of smoking, collapsed over social class.

```
> cotabplot(~ hyper + urea | smoke, tox.tab, shade = TRUE,
+           legend = FALSE, layout = c(1, 3))
```

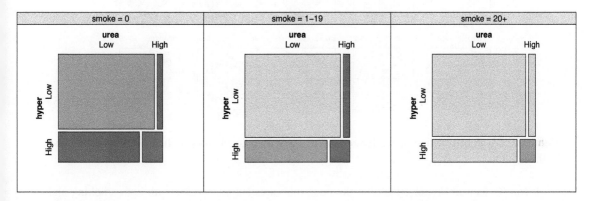

Figure 10.21: Toxaemia data: Response association conditioned on smoking level.

Figure 10.22 is similar, but stratified by social class. The marginal frequencies of the conditioning variable is not represented in these plots. (For example, as can be seen in Figure 10.20, the greatest number of women are in class 3.)

```
> cotabplot(~ hyper + urea | class, tox.tab, shade = TRUE,
+           legend = FALSE, layout = c(1, 5))
```

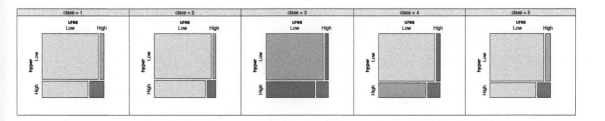

Figure 10.22: Toxaemia data: Response association conditioned on social class.

Ignoring social class, the association between hypertension and protein urea decreases with smoking. Ignoring smoking, the association is greatest in social class 3. However, these displays don't show directly how the two symptoms are associated in the combinations of social class and smoking. The fourfold display in Figure 10.23 does that.

```
> fourfold(aperm(tox.tab), fontsize = 16)
```

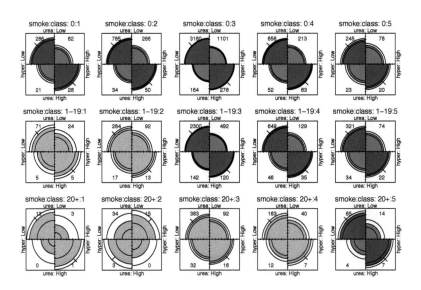

Figure 10.23: Fourfold display for Toxaemia data. Smoking levels vary in the rows and social class in the columns.

It can be seen in Figure 10.23 that the odds ratio appears to increase with both smoking and social class number and these two symptoms are positively associated in nearly all cases. In only two cases the odds ratio is not significantly different from 1: mothers in classes 1 and 2, who smoke more than 20 cigarettes a day, but the frequency in this cell is quite small.

```
> margin.table(tox.tab, 2 : 1)

       class
smoke    1     2     3    4    5
  0     417  1135  4703  984  366
  1-19  105   406  3054  859  451
  20+    17    52   523  222   90
```

From these plots, it is useful to examine the association between hypertension and urea more directly, by calculating and plotting the odds ratios. For a $2 \times 2 \times K \times L \times \cdots$ table, the function loddsratio() in **vcd** calculates these for each 2×2 subtable, and returns an array of dimension $K \times L \times \cdots$, together with similar array of standard errors.

```
> (LOR <- loddsratio(urea ~ hyper | smoke + class, data = tox.tab))

log odds ratios for urea and hyper by smoke, class

       class
smoke    1         2        3        4        5
  0     1.5268   1.46196  1.58056  1.31351  1.0036
  1-19  1.0710   0.86401  1.37370  1.34260  1.0348
  20+   2.4485  -1.14579  0.74425  0.88469  2.0187
```

The plot() method for the resulting **"logoddsratio"** object treats the conditioning variables in the formula argument as strata, and plots the log odds ratios for the first such variable on the horizontal axis with curves for the subsequent strata variables. The lines below produce Figure 10.24.

```
> plot(t(LOR), confidence = FALSE, legend_pos = "bottomright",
+      xlab = "Social class of mother")
```

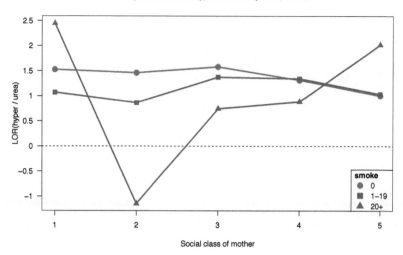

Figure 10.24: Log odds ratios for protein urea given hypertension, by social class and level of maternal smoking.

The association between the response symptoms, shown in Figure 10.24, is clearer, once we take the variation in sample sizes into account. Except for the heavy smokers, particularly in social classes 1 and 2, the log odds ratio appears to range only between 1–1.5, meaning that, given one symptom, the odds of also having the other range between $\exp(1) = 2.72$ and $\exp(1.5) = 4.48$.

This initial overview of the data is completed by calculating and plotting the log odds for each symptom within each class–smoke population. This could be done in the same way as in Example 10.9 (except that there are now two explanatory factors). The steps used there were: (a) Reshape the $2 \times 2 \times K \cdots$ table to a matrix with four columns corresponding to the binary response combinations. (b) Calculate the logits (and log odds ratio) using `blogits()`.

Here, it is more useful to make separate plots for each of the logits, and we illustrate a more general approach that applies to two or more binary responses, with two or more predictor variables. The essential idea is to fit a separate logit model for each response separately, using the *highest-order interaction* of all predictors (the saturated model). The fitted logits in these models then match those in the data.

```
> tox.hyper <- glm(hyper == "High" ~ class * smoke, weights = Freq,
+                  data = Toxaemia, family = binomial)
> tox.urea <- glm(urea == "High" ~ class * smoke, weights = Freq,
+                 data = Toxaemia, family = binomial)
```

It is then simple to plot these results using the **effects** package as shown in Figure 10.25. Each plot shows the logit for the response measure against class, with separate curves for the levels of smoking.[10]

---

[10]As is usual for effect plots of binary response `glm()` models, the vertical axis is plotted on the scale of log odds, but labeled in terms of probabilities.

```
> library(effects)
>
> plot(allEffects(tox.hyper),
+    ylab = "Probability (hypertension)",
+    xlab = "Social class of mother",
+    main = "Hypertension: class*smoke effect plot",
+    colors = c("blue", "black", "red"),
+    lwd=3,   multiline = TRUE,
+    key.args = list(x = 0.05, y = 0.2, cex = 1.2, columns = 1)
+    )
>
> plot(allEffects(tox.urea),
+    ylab = "Probability (Urea)",
+    xlab = "Social class of mother",
+    main = "Urea: class*smoke effect plot",
+    colors = c("blue", "black", "red"),
+    lwd=3,   multiline = TRUE,
+    key.args = list(x = 0.65, y = 0.2, cex = 1.2, columns = 1)
+    )
```

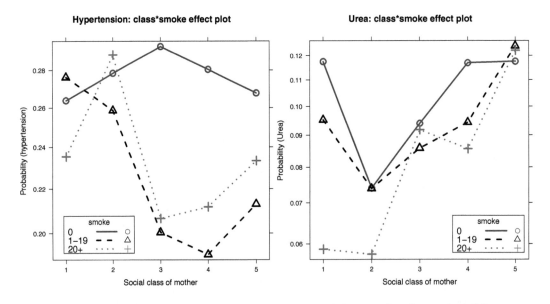

Figure 10.25: Effect plots for hypertension and urea, by social class of mother and smoking.

From Figure 10.25, it can be seen that the prevalence of these symptoms has a possibly complex relation to social class and smoking. However, the mosaic for these predictors in Figure 10.20 has shown us that several of the class–smoking categories are quite small (particularly heavy smokers in Classes 1 and 2), so the response effects for these classes will be poorly estimated. Taking this into account, we suspect that protein urea varies with social class, but not with smoking, while the prevalence of hypertension may truly vary with neither, just one, or both of these predictors.

10.4.2.1 Fitting models

The plots shown so far in this example are all essentially *data-based*, in that they use the observed frequencies or transformations of them and don't allow for a simpler view, based on a reasonable model. That is, abbreviating the table variables by their initial letters, the plots in Figure 10.24 and Figure 10.25 are plots of the saturated model, [CSHU], which fits perfectly, but with the data transformed for each 2×2 subtable to the log odds ratio and the two log odds for `hyper` and `urea`.

The bivariate logistic model fit by `vglm()` still applies when there are two or more predictors; however, like other multivariate response models, it doesn't easily allow the logits to depend on *different* predictor terms. To illustrate this, we first transform the `Toxaemia` to a 15×4 data frame in the form required by `vglm()`.

```
> tox.tab <- xtabs(Freq~class + smoke + hyper + urea, Toxaemia)
> toxaemia <- t(matrix(aperm(tox.tab), 4, 15))
> colnames(toxaemia) <- c("hu", "hU", "Hu", "HU")
> rowlabs <- expand.grid(smoke = c("0", "1-19", "20+"),
+                            class = factor(1:5))
> toxaemia <- cbind(toxaemia, rowlabs)
> head(toxaemia)

  hu hU  Hu HU smoke class
1 286 21  82 28     0     1
2  71  5  24  5  1-19     1
3  13  0   3  1   20+     1
4 785 34 266 50     0     2
5 284 17  92 13  1-19     2
6  34  3  15  0   20+     2
```

In the model specification for `vglm()`, the `zero` argument in `binom.or()` allows any one or more of the two log odds and log odds ratio to be fit as a constant (intercept-only) in Eqn. (10.11). However, in that equation, the predictors x_1, x_2, x_{12}, must be the *same* in all three submodels. For example, the model `tox.vglm1` below uses main effects of `class` and `smoke` in both models for the logits, and `zero=3` for a constant log odds ratio.

```
> tox.vglm1 <- vglm(cbind(hu, hU, Hu, Hu) ~ class + smoke,
+                     binom2.or(zero = 3), data = toxaemia)
> coef(tox.vglm1, matrix=TRUE)

               logit(mu1)  logit(mu2)  loge(oratio)
(Intercept)   -0.50853648  -1.2214518       2.7808
class2         0.18156457   0.0382046       0.0000
class3         0.06332765  -0.0087552       0.0000
class4        -0.02227055  -0.0031541       0.0000
class5        -0.00077172   0.0821863       0.0000
smoke1-19     -0.41298650  -0.2198673       0.0000
smoke20+      -0.30562472  -0.1245019       0.0000
```

Instead, when there are no quantitative predictors, and when the odds ratio is relatively constant (as here) it is easier to fit ordinary loglinear models than to use the bivariate logit formulation of the previous example. These allow the responses H and U to depend on the class-smoking combinations separately, by including the terms $[CSH]$ or $[CSU]$, respectively.

The minimal, null model, $[CS][H][U]$, fits the marginal association of the numbers in each class–smoking category, but asserts that the responses, H and U, are independent, which we have already seen is contradicted by the data. We take $[CS][HU]$ as the baseline model (Model 1), asserting no relation between response and predictor variables, but associations within each set are allowed. These models are fit as shown below.

```
> # null model
> tox.glm0 <- glm(Freq ~ class*smoke + hyper + urea,
+                   data = Toxaemia, family = poisson)
> # baseline model: no association between predictors and responses
> tox.glm1 <- glm(Freq ~ class*smoke + hyper*urea,
+                   data = Toxaemia, family = poisson)
```

We proceed to fit a collection of other models, adding terms to allow more associations between the responses and predictors. Summary measures of goodness of fit and parsimony are shown in Table 10.1.

Table 10.1: Loglinear models, `tox.glm*`, fit to the Toxaemia data

Model	Terms	df	G^2	p-value	G^2/df	AIC	BIC	R^2
0	CS H U	43	672.85	0.0000	15.65	586.85	264.27	.
1	CS HU	42	179.03	0.0000	4.26	95.03	-220.04	0.000
2	CS HU SH CU	36	46.12	0.1203	1.28	-25.88	-295.94	0.742
3	CS CH CU HU SH CU	30	40.47	0.0960	1.35	-19.53	-244.58	0.774
4	CSH CU HU	24	26.00	0.3529	1.08	-22.00	-202.04	0.855
5	CSH CU SU HU	22	25.84	0.2588	1.17	-18.16	-183.20	0.856
6	CSH CSU HU	14	22.29	0.0729	1.59	-5.71	-110.74	0.875
7	CSH CSU SHU	12	15.65	0.2079	1.30	-8.35	-98.37	0.913
8	CSH CSU CHU SHU	8	12.68	0.1233	1.59	-3.32	-63.33	0.929
9	CSHU	0	0.00	0	0	0.00	0.00	1.000

```
> tox.glm2 <- update(tox.glm1, . ~ . + smoke*hyper + class*urea)
>
> tox.glm3 <- glm(Freq ~ (class + smoke + hyper + urea)^2,
+               data=Toxaemia, family=poisson)
>
> tox.glm4 <- glm(Freq ~ class*smoke*hyper + hyper*urea + class*urea,
+               data=Toxaemia, family=poisson)
>
> tox.glm5 <- update(tox.glm4, . ~ . + smoke*urea)
>
> tox.glm6 <- update(tox.glm4, . ~ . + class*smoke*urea)
>
> tox.glm7 <- update(tox.glm6, . ~ . + smoke*hyper*urea)
>
> tox.glm8 <- glm(Freq ~ (class + smoke + hyper + urea)^3,
+               data = Toxaemia, family = poisson)
>
> tox.glm9 <- glm(Freq ~ (class + smoke + hyper + urea)^4,
+               data = Toxaemia, family = poisson)
```

Model 2 adds the simple dependence of hypertension on smoking ($[SH]$) and that of urea on class ($[CU]$). Model 3 includes all two-way terms. In Model 4, hypertension is allowed to depend on both class and smoking jointly ($[CSH]$). In Model 5 an additional dependence of urea on smoking ($[SU]$) is included, while in Model 6 urea depends on class and smoking jointly ($[CSU]$).

None of these models contain three-way terms involving both H and U, so these models assume that the log odds ratio for hypertension given urea is constant over the explanatory variables. Recalling the conditional mosaics (Figure 10.21 and Figure 10.22), Models 7 and 8 add terms that allow the odds ratio to vary, first with smoking ($[SHU]$), then with class ($[CHU]$) as well. Finally, Model 9 is the saturated model, that fits perfectly.

How do we choose among these models? Model 2 is the smallest model whose deviance is non-significant. Models 4 and 5 both have a smaller ratio of G^2/df. For comparing nested models, we can also examine the change in deviance as terms are added (or dropped). Thus, going from Model 2 to Model 3 decreases the deviance by 5.65 on 6 df, while the step from Model 3 to Model 4 gives a decrease of 14.47, also on 6 df. These tests can be performed using `lrtest()` in the lmtest package, shown below for models `tox.glm1–tox.glm5`.

```
> library(lmtest)
> lmtest::lrtest(tox.glm1, tox.glm2, tox.glm3, tox.glm4, tox.glm5)

Likelihood ratio test

Model 1: Freq ~ class * smoke + hyper * urea
Model 2: Freq ~ class + smoke + hyper + urea + class:smoke + hyper:urea +
    smoke:hyper + class:urea
```

```
Model 3: Freq ~ (class + smoke + hyper + urea)^2
Model 4: Freq ~ class * smoke * hyper + hyper * urea + class * urea
Model 5: Freq ~ class + smoke + hyper + urea + class:smoke + class:hyper +
    smoke:hyper + hyper:urea + class:urea + smoke:urea + class:smoke:hyper
  #Df LogLik Df  Chisq Pr(>Chisq)
1  18   -260
2  24   -194  6 132.91      <2e-16 ***
3  30   -191  6   5.65       0.464
4  36   -184  6  14.47       0.025 *
5  38   -184  2   0.17       0.920
---
Signif. codes:  0 '***' 0.001 '**' 0.01 '*' 0.05 '.' 0.1 ' ' 1
```

The AIC and BIC statistics, balancing parsimony and goodness-of-fit, have their minimum value for Model 2, which we adopt here for this example.

10.4.2.2 Plotting model results

Whatever model is chosen, as a final step, it is important to determine what that model implies about the original research questions. Because our focus here is on the prevalence of each symptom, and their association, it is helpful to graph the fitted logits and log odds ratios implied by the model, as was done in Figure 10.18 and Figure 10.19.

The presentation goal here is to produce plots showing the observed logits and log odds ratios as in Figure 10.25 and Figure 10.24, supplemented by lines showing these values according to the fitted model. In Example 10.9 we fit the bivariate logit model, for which the response functions were the desired logits and log odds. Here, where we have fit ordinary loglinear models, the observed and fitted logits can be calculated from the observed and fitted frequencies. The calculations require a bit of R calisthenics to arrange these into forms suitable for plotting.

As we did earlier, we first reshape the `Toxaemia` to wide format, as a 15×4 table of observed frequencies. Because there are now two predictor variables, we take care to include the levels of `smoke` and `class` as additional columns.

```
> # reshape to 15 x 4 table of frequencies
> tox.tab <- xtabs(Freq ~ class + smoke + hyper + urea, Toxaemia)
> toxaemia <- t(matrix(aperm(tox.tab), 4, 15))
> colnames(toxaemia) <- c("hu", "hU", "Hu", "HU")
> rowlabs <- expand.grid(smoke = c("0", "1-19", "20+"),
+                        class = factor(1:5))
> toxaemia <- cbind(toxaemia, rowlabs)
```

Applying `blogits()`, we get the observed logits and log odds ratios in `logitsTox`.

```
> # observed logits and log odds ratios
> logitsTox <- blogits(toxaemia[,4:1], add=0.5)
> colnames(logitsTox)[1:2] <- c("logitH", "logitU")
> logitsTox <- cbind(logitsTox, rowlabs)
> head(logitsTox)

     logitH  logitU    logOR  smoke class
1 -1.02057 -1.9988  1.52679      0     1
2 -0.94261 -2.1665  1.07102   1-19     1
3 -1.02962 -2.1401  2.44854    20+     1
4 -0.95040 -2.5158  1.46196      0     2
5 -1.04699 -2.4983  0.86401   1-19     2
6 -0.86500 -2.5257 -1.14579    20+     2
```

The fitted frequencies are extracted using `predict(tox.glm2, type="response")`, and then manipulated in a similar way to give `logitsFit`.

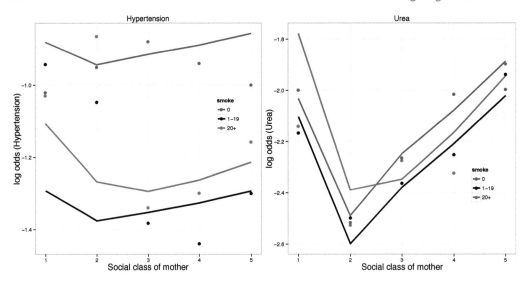

Figure 10.26: Observed (points) and fitted (lines) logits for the `Toxaemia` data under Model 2.

```
> # fitted frequencies, as a 15 x 4 table
> Fit <- t(matrix(predict(tox.glm2, type = "response"), 4, 15))
> colnames(Fit) <- c("HU", "Hu", "hU", "hu")
> Fit <- cbind(Fit, rowlabs)
> logitsFit <- blogits(Fit[, 1 : 4], add=0.5)
> colnames(logitsFit)[1 : 2] <- c("logitH", "logitU")
> logitsFit <- cbind(logitsFit, rowlabs)
```

In tabular form, you can examine any of these components; for example, the log odds ratios from the fitted values shown below.

```
> matrix(logitsFit$logOR, 3, 5,
+          dimnames = list(smoke = c("0", "1-19", "20+"), class = 1 : 5))

      class
smoke        1       2       3       4       5
   0    1.3588  1.3638  1.3675  1.3643  1.3582
 1-19   1.3582  1.3678  1.3683  1.3674  1.3658
  20+   1.2799  1.3471  1.3662  1.3622  1.3511
```

Finally, we can plot the observed values in `logitsTox` (as points) and the fitted values under Model 2 in `logitsFit` (as lines), separately for the `logitH`, `logitU`, and `logOR` components. The code below uses **ggplot2** for the log odds of hypertension, and is repeated for urea and the log odds ratio. These graphs are shown in Figure 10.26 and Figure 10.27.

```
> ggplot(logitsFit, aes(x = as.numeric(class), y = logitH,
+                          color = smoke)) +
+    theme_bw() +
+    geom_line(size = 1.2) +
+    scale_color_manual(values = c("blue", "black", "red")) +
+    ylab("log odds (Hypertension)") +
+    xlab("Social class of mother") +
+    ggtitle("Hypertension") +
+    theme(axis.title = element_text(size = 16)) +
+    geom_point(data = logitsTox,
+                 aes(x = as.numeric(class), y = logitH, color = smoke),
+                 size = 3) +
+    theme(legend.position = c(0.85, .6))
```

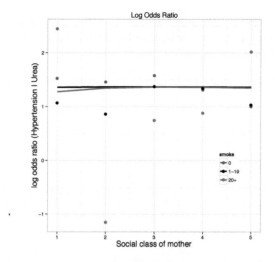

Figure 10.27: Observed (points) and fitted (lines) log odds ratios for the `Toxaemia` data under Model 2.

According to this model, Figure 10.27 shows that the fitted log odds ratio is in fact nearly constant, while Figure 10.26 shows that the log odds for hypertension depend mainly on smoking (with a large difference of the non-smoking mothers from the rest), and that for protein urea depends mainly on social class.[11]

Yet the great variability of the observed points around the fitted curves indicates that these relationships are not well-determined. Adding error bars showing the standard error around each fitted point would indicate that the data conforms as closely to the model as can be expected, given the widely different sample sizes. However, this would make the plots more complex, and so was omitted here. In addition to showing the pattern of the results according to the fitted model, such graphs also help us to appreciate the model's limitations.

$\triangle$

10.5 Chapter summary

- Standard loglinear models treat all variables as unordered factors. When one or more factors are ordinal, however, loglinear and logit models may be simplified by assigning quantitative scores to the levels of an ordered factor. Such models are often more sensitive and have greater power because they are more focused.

- Models for square tables, with the same row and column categories, are an important special case. For these and other structured tables, a variety of techniques provide the opportunity to fit models more descriptive than the independence model and more parsimonious than the saturated model.

- When there are several categorical responses, along with one or more explanatory variables, some special forms of loglinear and logit models may be used to separate the marginal dependence of each response on the explanatory variables from the interdependence among the responses.

---

[11]Some possible enhancements to these graphs include (a) plotting on the scale of probabilities or including a right vertical axis showing corresponding probabilities; (b) using the same vertical axis limits for the two graphs for direct comparison.

- In all these cases, the interplay between graphing and fitting is important in arriving at an understanding of the relationships among variables and an adequate descriptive model that is faithful to the details of the data.

- In particular, mosaic-like displays show all the data by areas, and indicate goodness of fit of a model by shading. In contrast, for more complex models, plots of derived quantities like log odds and log odds ratios can be more effective.

10.6   Lab exercises

Exercise 10.1   Example 10.5 presented an analysis of the data on visual acuity for the subset of women in the `VisualAcuity` data. Carry out a parallel analysis of the models fit there for the men in this data set, given by:

```
> data("VisualAcuity", package="vcd")
> men <- subset(VisualAcuity, gender=="male", select=-gender)
```

Exercise 10.2   Table 10.2 gives a 4×4 table of opinions about premarital sex and whether methods of birth control should be made available to teenagers aged 14–16, from the 1991 General Social Survey (Agresti, 2013, Table 10.3). Both variables are ordinal, and their grades are represented by the case of the row and column labels.

Table 10.2: Opinions about premarital sex and availability of teenage birth control. *Source*: Agresti (2013, Table 10.3).

Premarital sex	Birth control			
	DISAGREE	disagree	agree	AGREE
WRONG	81	68	60	38
Wrong	24	26	29	14
wrong	18	41	74	42
OK	36	57	161	157

(a) Fit the independence model to these data using `loglm()` or `glm()`.
(b) Make a mosaic display showing departure from independence and describe verbally the pattern of association.
(c) Treating the categories as equally spaced, fit the $L \times L$ model of uniform association, as in Section 10.1. Test the difference against the independence model with a likelihood-ratio test.
(d) Fit the RC(1) model with `gnm()`, and test the difference of this against the model of uniform association.
(e) Write a brief summary of these results, including plots useful for explaining the relationships in this data set.

Exercise 10.3   For the data on attitudes toward birth control in Table 10.2,

(a) Calculate and plot the observed local log odds ratios.
(b) Also fit the R, C, and R+C models.
(c) Use the method described in Section 10.1.2 to visualize the structure of fitted local log odds ratios implied by each of these models, together with the RC(1) model.

Exercise 10.4   The data set `gss8590` in logmult gives a $4 \times 5 \times 4$ table of education levels and occupational categories for the four combinations of gender and race from the General Social Surveys, 1985–1990, as reported by Wong (2001, Table 2). Wong (2010, Table 2.3B) later used the subset pertaining to women to illustrate RC(2) models. This data is created below as `Women.tab`, correcting an inconsistency to conform with the 2010 table.

```
> data("gss8590", package="logmult")
> Women.tab <- margin.table(gss8590[,,c("White Women", "Black Women")], 1:2)
> Women.tab[2,4] <- 49
> colnames(Women.tab)[5] <- "Farm"
```

(a) Fit the independence model, and also the RC(1) and RC(2) models using `rc()` with marginal weights, as illustrated in Example 10.4. Summarize these statistical tests in a table.

(b) Plot the solution for the RC(2) model with 68% confidence ellipses. What verbal labels would you use for the two dimensions?

(c) Is there any indication that a simpler model, using integer scores for the row (Education) or column (Occupation) categories, or both, might suffice? If so, fit the analogous column effects, row effects, or $L \times L$ model, and compare with the models fit in part (a).

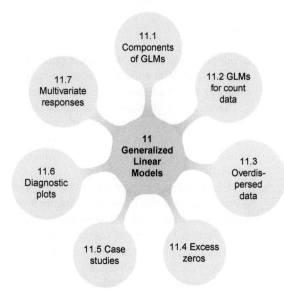

11

Generalized Linear Models for Count Data

Generalized linear models extend the familiar linear models of regression and ANOVA to include counted data, frequencies, and other data for which the assumptions of independent normal errors are not reasonable. We rely on the analogies between ordinary and generalized linear models (GLMs) to develop visualization methods to explore the data, display the fitted relationships, and check model assumptions. The main focus of this chapter is on models for count data.

> In one word, to draw the rule from experience, one must generalize; this is a necessity that imposes itself on the most circumspect observer.

Henri Poincaré, *The Value of Science: Essential Writings of Henri Poincaré*

In the modern history of statistics, most developments occur incrementally, with small additions to existing models and theory that extend their range and applicability to new problems and data. Occasionally, there is a major synthesis that unites a wide class of existing methods in a general framework and provides opportunities for far greater growth.

A prime example is the theory of generalized linear models, introduced originally by Nelder and Wedderburn (1972), that extended the familiar (classical) linear models for regression and ANOVA to include related models, such as logistic regression and logit models (described in Chapter 7) and loglinear models (described in Chapter 9), and other variations, as "families" within a single general system.

This approach has proved attractive because it: (a) integrates many familiar statistical models in a general theory where they are just special cases; (b) provides the basis for extending these and developing new models within the same or similar framework; (c) simplifies the implementation of these models in software, since the same algorithm can be used for estimation, inference, and assessing model adequacy for all generalized linear models.

Section 11.1 gives a brief sketch of the GLM framework. The focus of this book is on visualization methods for categorical data, and the two important topics concern models and methods for binomial response data and for count data. The first of these was described extensively in Chapter 7, with extensions to multinomial data (Chapter 8), and there is little to add here, except for changes in notation.

GLM models for count data, however, provide the opportunity to extend the scope of these methods beyond what was covered in Chapter 9, and this topic is introduced in Section 11.2. Extensions to the GLM framework also provide the opportunity to deal with common problems of overdispersion (Section 11.3) and an overabundance of zero counts (Section 11.4), giving some new models and visualization methods that help to understand such data in greater detail. These are illustrated with two case studies in Section 11.5. Section 11.6 illustrates other graphical methods for diagnostic model checking, some of which were introduced in earlier chapters. Finally, Section 11.7 outlines some simple extensions of these models to handle multivariate responses.

11.1 Components of generalized linear models

The motivation for the **generalized linear model** (GLM) and its structure are most easily seen by considering the classical linear model,

$$y_i = \boldsymbol{x}_i^\mathsf{T}\boldsymbol{\beta} + \epsilon_i \, ,$$

where y_i is the response variable for case $i, i = 1, \ldots n$, $\boldsymbol{x}_i$ is the vector of explanatory variables or regressors, $\boldsymbol{\beta}$ is the vector of model parameters, and the ϵ_i are random errors. In the classical linear model, the ϵ_i are assumed to (a) have constant variance, σ^2_ϵ, (b) follow a normal (Gaussian) distribution (conditional on $\boldsymbol{x}_i$), and (c) be independent across observations.

Thus, Nelder and Wedderburn (1972) generalized this Gaussian linear model to consist of the following three components, by relaxing assumptions (a) and (b) above:[1]

random component: The conditional distribution of the $y_i \,|\, \boldsymbol{x}_i$, with mean $\mathcal{E}(y_i) = \mu_i$. Under classical assumptions, this is independent, normal with constant variance σ^2, i.e., $y_i \overset{\text{iid}}{\sim} N(\mu_i, \sigma^2)$. In the GLM, the probability distribution of the y_i can be any member of the **exponential family**, including the normal, Poisson, binomial, gamma, and others. Subsequent work has extended this framework to include multinomial distributions and some non-exponential families such as the negative binomial distribution.

systematic component: The idea that the predicted value of y_i itself is a linear combination of the regressors is replaced by that of a **linear predictor**, η, that captures this aspect of linear models,

$$\eta_i = \boldsymbol{x}_i^\mathsf{T}\boldsymbol{\beta} \, .$$

link function: The connection between the mean of the response, μ_i, and the linear predictor, η_i, is specified by the **link function**, $g(\bullet)$, giving

$$g(\mu_i) = \eta_i = \boldsymbol{x}_i^\mathsf{T}\boldsymbol{\beta} \, .$$

---

[1]The remaining assumption of independent observations is relaxed in **generalized linear mixed models** (GLMMs), in which random effects to account for non-independence are added to the linear predictor. This allows the modeling of correlated (responses of family members), clustered (residents in different communities), or hierarchical data (patients within hospitals within regions). See: McCulloch and Neuhaus (2005) and Hedeker (2005)

Table 11.1: Common link functions and their inverses used in generalized linear models

Link name	Function: $\eta_i = g(\mu_i)$	Inverse: $\mu_i = g^{-1}(\eta_i)$
identity	μ_i	η_i
square-root	$\sqrt{\mu_i}$	η_i^2
log	$\log_e(\mu_i)$	$\exp(\eta_i)$
inverse	μ_i^{-1}	η_i^{-1}
inverse-square	μ_i^{-2}	$\eta_i^{-1/2}$
logit	$\log_e \frac{\mu_i}{1-\mu_i}$	$\frac{1}{1+\exp(-\eta_i)}$
probit	$\Phi^{-1}(\mu_i)$	$\Phi(\eta_i)$
log-log	$-\log_e[-\log_e(\mu_i)]$	$\exp[-\exp(-\eta_i)]$
comp. log-log	$\log_e[-\log_e(1-\mu_i)]$	$1-\exp[-\exp(\eta_i)]$

The link function $g(\bullet)$ must be both *smooth* and *monotonic*, meaning that it is one-to-one, so an inverse transformation, $g^{-1}(\bullet)$ exists,

$$\mu_i = g^{-1}(\eta_i) = g^{-1}(\boldsymbol{x}_i^\mathsf{T}\boldsymbol{\beta}) \,,$$

which allows us to obtain and plot the predicted values on their original scale. The link function captures the familiar idea that linear models are often estimated with a transformation of the response, such as $\log(y_i)$ for a frequency variable or $\text{logit}(y_i)$ for a binomial variable. The inverse function $g^{-1}(\bullet)$ is also called the ***mean function***.

Some commonly used link functions are shown in Table 11.1. Some of these link functions have restrictions on the range of y_i to which they can be applied. For example, the square-root and log links apply only to non-negative and positive values, respectively. The last four link functions in this table are for binomial data, where y_i represents the observed proportion of successes in n_i independent trials, and thus the mean μ_i represents the probability of success (symbolized by π_i in Chapter 7). Binary data are the special case where $n_i = 1$.

11.1.1 Variance functions

The GLM has the additional property that, for distributions in the exponential family, the conditional variance of $y_i \mid \eta_i$ is a known function, $\mathcal{V}(\mu_i)$, of the mean and possibly one other parameter called the ***scale parameter*** or ***dispersion parameter***, ϕ. Some commonly used distributions in the exponential family and their variance functions are shown in Table 11.2.

- In the classical Gaussian linear model, the conditional variance is constant, $\phi = \sigma_\epsilon^2$.

- In the Poisson family, $\mathcal{V}(\mu_i) = \mu_i$ and the dispersion parameter is fixed at $\phi = 1$. In practice, it is common for count data to exhibit ***overdispersion***, meaning that $\mathcal{V}(\mu_i) > \mu_i$. One way to correct for this is to extend the GLM to allow the dispersion parameter to be estimated from the data, giving what is called the ***quasi-Poisson*** family, with $\mathcal{V}(\mu_i) = \widehat{\phi}\mu_i$.

- Similarly, for binomial data, the variance function is $\mathcal{V}(\mu_i) = \mu_i(1 - \mu_i)/n_i$, with ϕ fixed at 1. Overdispersion often results from failures of the assumptions of the binomial model: supposedly independent observations may be correlated or clustered and the probability of success may not be constant, or vary with unmeasured or unmodeled variables.

Table 11.2: Common distributions in the exponential family used with generalized linear models and their canonical link and variance functions

Family	Notation	Canonical link	Range of y	Variance function, $\mathcal{V}(\mu \mid \eta)$
Gaussian	$N(\mu, \sigma^2)$	identity: μ	$(-\infty, +\infty)$	ϕ
Poisson	$\text{Pois}(\mu)$	$\log_e(\mu)$	$0, 1, \ldots, \infty$	μ
Negative-Binomial	$\text{NBin}(\mu, \theta)$	$\log_e(\mu)$	$0, 1, \ldots, \infty$	$\mu + \mu^2/\theta$
Binomial	$\text{Bin}(n, \mu)/n$	$\text{logit}(\mu)$	$\{0, 1, \ldots, n\}/n$	$\mu(1-\mu)/n$
Gamma	$G(\mu, \nu)$	μ^{-1}	$(0, +\infty)$	$\phi\mu^2$
Inverse-Gaussian	$IG(\mu, \nu)$	μ^2	$(0, +\infty)$	$\phi\mu^3$

- The gamma and inverse-Gaussian families are distributions useful for modeling a continuous and positive response variable with no upper bound (e.g., reaction time). They both have the property that conditional variance increases with the mean, and for the inverse-Gaussian, variance increases at a faster rate. Their dispersion parameters ϕ are simple functions of their intrinsic "shape" parameters, indicated as ν in the table.

The important points from this discussion are that the GLM together with the exponential family of distributions:

- provide for simple linear relations between the response and the predictors via the link function and the linear predictor.

- allow a very flexible relationship between the mean and conditional variance to be specified in terms of a set of known families.

- incorporate a dispersion parameter ϕ that in some cases can be estimated or tested for departure from that entailed in a given family.

- have allowed further extensions of this framework outside the exponential family, ranging from simple adjustments for statistical inference ("quasi" families, adjusted "sandwich" covariances) to separate modeling of the variance relation to the predictors.

Further details of generalized linear models are beyond the scope of this book, but the interested reader should consult Fox (2008, Section 15.3) and Agresti (2013, Ch. 4) for a comprehensive treatment.

11.1.2 Hypothesis tests for coefficients

GLMs are fit using maximum likelihood estimation, and implemented in software using an iterative algorithm known as *iteratively weighted least squares* that generalizes the least squares method for classical linear models. This provides estimates $\widehat{\boldsymbol{\beta}}$ of the model coefficients for the predictors in $\boldsymbol{x}$, as well as an estimated asymptotic (large sample) variance matrix of $\widehat{\boldsymbol{\beta}}$, given by

$$\mathcal{V}(\widehat{\boldsymbol{\beta}}) = \phi(\boldsymbol{X}^{\mathsf{T}}\boldsymbol{W}\boldsymbol{X}),\tag{11.1}$$

where $\boldsymbol{W}$ is a diagonal matrix of weights computed in the final iteration. In the standard Poisson GLM, the weight matrix is $\boldsymbol{W} = \text{diag}\,(\widehat{\boldsymbol{\mu}})$ and $\phi = 1$ is assumed.

Asymptotic standard errors, $\text{se}(\widehat{\beta}_j)$, for the coefficients are then the square roots of the diagonal elements of $\mathcal{V}(\widehat{\boldsymbol{\beta}})$, and tests of hypotheses regarding an individual coefficient, e.g., $H_0 : \beta_j = 0$, can be carried out using the Wald test statistic, $z_j = \widehat{\beta}_j/\text{se}(\widehat{\beta}_j)$. When the null hypothesis is true, z_j has a standard normal $\mathcal{N}(0,1)$ distribution, providing p-values for significance tests.[2]

More generally, we can test any **linear hypothesis**, of the form $H_0 : \boldsymbol{L\beta} = \boldsymbol{c}$, where $\boldsymbol{L}$ is a constant hypothesis matrix of size $h \times p$ giving h linear combinations of the coefficients, to be tested for equality with the constants in $\boldsymbol{c}$, typically taken as $\boldsymbol{c} = \boldsymbol{0}$. The test statistic is the Wald chi-square,

$$Z^2 = (\boldsymbol{L\widehat{\beta}} - \boldsymbol{c})^{\mathsf{T}}\,[\boldsymbol{L}\mathcal{V}(\widehat{\boldsymbol{\beta}})\boldsymbol{L}^{\mathsf{T}}]^{-1}\,(\boldsymbol{L\widehat{\beta}} - \boldsymbol{c})\,, \tag{11.2}$$

which has a χ^2 distribution on h degrees of freedom.[3]

For example, to test the hypothesis that all of $\beta_1 = \beta_2 = \beta_3 = 0$ in a model with three predictors, you can use

$$\boldsymbol{L} = \begin{bmatrix} 0 & 1 & 0 & 0 \\ 0 & 0 & 1 & 0 \\ 0 & 0 & 0 & 1 \end{bmatrix} = \begin{bmatrix} \boldsymbol{0} & \boldsymbol{I} \end{bmatrix}, \qquad \boldsymbol{c} = \begin{pmatrix} 0 \\ 0 \\ 0 \end{pmatrix}.$$

Similarly, to test the hypothesis that $\beta_1 = \beta_2$ in the same model, you can use $\boldsymbol{L} = [0, 1, -1, 0]$ and $\boldsymbol{c} = [0]$.[4]

In R, such tests are most conveniently carried out using `linearHypothesis()` in the car package, supporting Fox and Weisberg (2011b). The hypothesis matrix $\boldsymbol{L}$ can be supplied as a numeric matrix, or more conveniently, the hypothesis can be specified symbolically as a character vector of the names of the coefficients involved in each row of $\boldsymbol{L}$. For example, the first hypothesis test above could be specified using the vector `c("x1=0", "x2=0", "x3=0")`, and the test of equality as `"x1-x2=0"`.

11.1.3 Goodness-of-fit tests

The basic ideas for testing goodness-of-fit were discussed in Section 9.3.2 in connection with log-linear models for contingency tables. As before, these assess the overall performance of a model in reproducing the data. The commonly used measures include the Pearson chi-square and likelihood-ratio deviance statistics, which can be seen as weighted sums of residuals. We re-state these test statistics here in the wider context of the GLM.

Let $y_i, i = 1, 2, \ldots, n$ be the response and $\widehat{\mu}_i = g^{-1}(\boldsymbol{x}_i^{\mathsf{T}}\widehat{\boldsymbol{\beta}})$ the fitted mean using the estimated coefficients, having estimated variance $\widehat{\omega}_i = \mathcal{V}(\widehat{\mu}_i \,|\, \eta_i)$ as in Table 11.2. Then the normalized squared residual for observation i is $(y_i - \widehat{\mu}_i)^2/\widehat{\omega}_i$, and the Pearson statistic is

$$X_P^2 = \sum_{i=1}^{n} \frac{(y_i - \widehat{\mu}_i)^2}{\widehat{\omega}_i}\,. \tag{11.3}$$

In the GLM for count data, the main focus of this chapter, the Poisson family sets $\omega = \mu$ with the dispersion parameter fixed at $\phi = 1$.

The **residual deviance** statistic, as in logistic regression and loglinear models, is defined as twice the difference between the maximum possible log-likelihood for the *saturated model* that fits perfectly and maximized log-likelihood for the fitted model. The deviance can be defined as

$$D(\boldsymbol{y}, \widehat{\boldsymbol{\mu}}) \equiv 2[\log_e \mathcal{L}(\boldsymbol{y}; \boldsymbol{y}) - \log_e \mathcal{L}(\boldsymbol{y}; \widehat{\boldsymbol{\mu}})]\,.$$

For classical linear models under normality, the deviance is simply the residual sum of squares,

---

[2] Wald tests are sometimes carried out using z^2, which has an equivalent χ_1^2 distribution with 1 degree of freedom.

[3] When a dispersion parameter ϕ has been estimated from the data, it is common to use an F-test, using the statistic $F = Z^2/h$, with h and $n - p$ degrees of freedom.

[4] Such a test is only sensible if the predictors $\boldsymbol{x_1}$ and $\boldsymbol{x_2}$ are on the same scale, so their coefficients are commensurable.

$\sum_i^n (y_i - \widehat{\mu}_i)$. This has led to the deviance being taken in the GLM framework as a generalization of the sum of squares used in ANOVA, and hence, an analogous *analysis of deviance* to carry out tests for individual terms in GLMs, or to compare nested models.

In R, `anova(mod)` for the `"glm"` object `mod` gives *sequential* ("Type I") tests of successive terms in a model, while `Anova()` in the **car** package gives the more generally useful "Type II" (and "Type III") *partial* tests, that assess the additional contribution of each term above all others, taking marginality into account.

For Poisson models with a log link giving $\boldsymbol{\mu} = \exp(\boldsymbol{x}^\mathsf{T}\boldsymbol{\beta})$, the deviance takes the form[5]

$$D(\boldsymbol{y}, \widehat{\boldsymbol{\mu}}) = 2 \sum_{i=1}^{n} \left[y_i \log_e \left(\frac{y_i}{\widehat{\mu}_i} \right) - (y_i - \widehat{\mu}_i) \right] . \tag{11.4}$$

For a GLM with p parameters, both the Pearson and residual deviance statistics follow approximate χ^2_{n-p} distributions with $n - p$ degrees of freedom.

11.1.4 Comparing non-nested models

The flexibility of the GLM and its extensions allows us to fit models to the same data using different families and different link functions, and to fit models that allow for overdispersion (Section 11.3) or that make special provisions for zero counts (Section 11.4). One price paid for this additional versatility is that standard LR tests and F tests (such as provided by `anova()` and `linearHypothesis()` in the **car** package) do not apply to models that are not nested; that is, where one model cannot be represented as a restricted, special case of another.

For models estimated by maximum likelihood, one general route to comparing non-nested models is through the AIC information criterion proposed initially by Akaike (1973) and the related BIC criterion (Schwartz, 1978), based on the fitted log-likelihood function:

$$\text{AIC} = -2 \log_e \mathcal{L} + 2k . \tag{11.5}$$

$$\text{BIC} = -2 \log_e \mathcal{L} + \log_e(n)k . \tag{11.6}$$

As noted in Section 9.3.2, these both penalize models with larger k, the number of parameters in the model, with BIC adding a greater penalty with larger sample size. However, because they are based only on the maximized log-likelihood, they are agnostic as to whether models are nested or not, and give comparable results (lower is better) provided the same observations have been used in all models.

In R, these results are given for a collection of models by the generic functions `AIC()` and `BIC()`; these can be calculated for any model for which `logLik()` and (for BIC) `nobs()` methods exist. The **vcdExtra** function `LRstats()` is a convenient wrapper for these methods.

AIC and BIC do not give significance tests for assessing whether one model is significantly "better" than another. A series of tests that *do* this was proposed by Vuong (1989): they are based on comparing the predicted probabilities or the pointwise log-likelihoods of the two models, and test the null hypothesis that each is equally close to the saturated model, against the alternative that one model is closer. The different tests handle nested, partially nested and non-nested cases. However, whenever *Vuong's test* is mentioned in literature, this typically refers to the test assuming that both models are *strictly* non-nested, which may not be obvious to see in all cases.[6] For example, some models may neither be nested or non-nested, but *overlapping*, that is, yield the same moments and fit statistics only for some, not all data. However, our use of Vuong's test will be confined to count data models, precluding most of these issues.

---

[5] In the context of the loglinear models discussed in Section 9.3.2, this is also referred to as the likelihood-ratio G^2 statistic.

[6] The test versions for (partially) nested models are difficult to compute in practice, and at the time of writing, no implementation for them is available for R.

For two such models, let $f_1(y_i \mid x_i, \theta_1)$ be the density function under model 1, with parameters θ_1 and similarly $f_2(y_i \mid x_i, \theta_2)$ under model 2 with parameters θ_2, where $f_1(\bullet)$ and $f_2(\bullet)$ need not be the same. Vuong's test compares these based on the observation-wise log-likelihood ratios,

$$\ell_i = \log_e \left(\frac{f_1(y_i \mid x_i, \widehat{\theta}_1)}{f_2(y_i \mid x_i, \widehat{\theta}_2)} \right) .$$

The test statistic is

$$V = \frac{\bar{\ell} - \text{penalty}}{\sqrt{n} s_\ell} ,$$

where $\bar{\ell}$ is the mean of the ℓ_i, s_ℓ is their variance, and penalty is an adjustment for model parsimony, typically taken as $\log(n)(k_1 - k_2)/2$ when model 1 has k_1 parameters in θ_1 and model 2 has k_2 parameters in θ_2.

The test statistic V has an asymptotic normal $N(0, 1)$ distribution, and is directional, with large positive values favoring model 1, and large negative values favoring model 2. This test is implemented as the `vuong()` function in the pscl (Jackman et al., 2015) package, and a more flexible version is provided by `vuongtest()` in nonnest2 (Merkle and You, 2014) package[7].

11.2 GLMs for count data

The prototypical GLM for count data, where the response y_i takes on non-negative values $0, 1, 2, \ldots$, uses the Poisson family with the log link. We used this model extensively throughout all of Chapter 9. There the focus was on the special case of the loglinear model applied largely to contingency tables, where the loglinear model could be seen as a fairly direct extension of ANOVA models for a quantitative response applied to the log of cell frequency.

The advantage there was that models for two-way, three-way, and by implication, n-way tables could be discussed and illustrated using notation and graphs that separated the parameters and effects for one-way terms ("main effects"), two-way terms ("simple associations"), and higher-way terms ("conditional associations").

The disadvantage is that these models as formulated there do not easily accommodate general quantitative predictors and were limited to the log link and the Poisson family. For example, the models discussed in Section 10.1 for ordinal variables allow one or more table factors to be assigned quantitative scores or have such scores estimated from the data, as in RC() models (Section 10.1.3). Yet the contingency table approach for loglinear models breaks down if there are continuous predictors, and count data often exhibits features that make the equivalent Poisson regression model unsuitable or incomplete. We consider some extended models here.

EXAMPLE 11.1: Publications of PhD candidates

In Example 3.24 we considered the distribution of the number of publications by PhD candidates in their last three years of study, but without taking any available predictors into account. For these data, a simple calculation shows why the Poisson distribution is unsuitable (for the marginal distribution), because the variance is 2.19 times the mean.

```
> data("PhdPubs", package = "vcdExtra")
> with(PhdPubs, c(mean = mean(articles), var = var(articles),
+                 ratio = var(articles) / mean(articles)))

  mean    var  ratio
1.6929 3.7097 2.1914
```

---

[7]This also allows for testing *nested* models where the full model is not assumed to be correct as required by classical likelihood ratio tests, and also for other models than count regression models, such as Structural Equation Models (SEMs).

The earlier example showed rootograms (in Figure 3.25) of the number of articles, but here it is useful to consider some more basic exploratory displays. A basic barplot of the frequency distribution of number of articles published is shown in the left panel of Figure 11.1. A quick look indicates that the distribution is highly skewed and there is a large number of counts of zero.

Another problem is that the frequencies of 0–2 articles account for over 75% of the total, so that the frequencies of the larger counts get lost in the display. The rootogram corrects for this by plotting frequency on the square-root scale. However, because we are contemplating a model with a log link, the same goal can be achieved by plotting log of frequency, as shown in the right panel of Figure 11.1. To accommodate the zero frequencies, the plot shows log(Frequency+1), avoiding errors from `log(0)`. It can be seen that log frequency decreases steadily up to 7 articles and then levels off approximately.

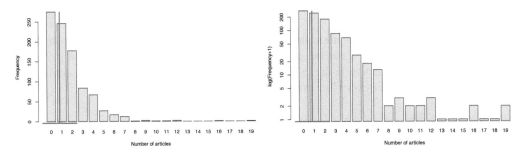

Figure 11.1: Barplots showing the frequency distribution of number of publications by PhD candidates. Left: raw scale; right: a log scale makes the smaller counts more visible. The vertical red lines show the mean and horizontal lines show mean ± 1 standard deviation.

These plots are produced as shown below. The frequency distribution of `articles` can be tabulated by `table()`, but there is a subtle wrinkle here: By default, `table()` excludes the values of `articles` that do not occur in the data (zero frequencies). To include all values in the entire range, it is necessary to treat `articles` as a factor with levels $0:19$.

```
> art.fac <- factor(PhdPubs$articles, levels = 0 : 19)   # include zero frequencies
> art.tab <- table(art.fac)
> art.tab
```

```
art.fac
  0   1   2   3   4   5   6   7   8   9  10  11  12  13  14  15  16  17  18  19
275 246 178  84  67  27  17  12   1   2   1   1   2   0   0   1   0   0   1
```

Then, the basic plot on the frequency scale is created using `barplot()`, to which some annotations can be added using standard plotting tools, such as the mean or an interval showing the variance (e.g., with a range of one standard deviation).

```
> barplot(art.tab, xlab = "Number of articles", ylab = "Frequency",
+         col = "lightblue")
> abline(v = mean(PhdPubs$articles), col = "red", lwd = 3)
> ci <- mean(PhdPubs$articles) + c(-1, 1) * sd(PhdPubs$articles)
> lines(x = ci, y = c(-4, -4), col = "red", lwd = 3, xpd = TRUE)
```

Similarly, the plot on the log scale in the right panel of Figure 11.1 is produced with `barplot()`, but using `art.tab+1` to start frequency at one and `log="y"` to scale the vertical axis to log.

```
> barplot(art.tab + 1, ylab = "log(Frequency+1)",
+         xlab = "Number of articles", col = "lightblue", log = "y")
```

Other useful exploratory plots for count data include boxplots of the response (on a log scale) and scatterplots against continuous predictors, where jittering the response is often necessary to avoid overplotting and a smooth nonparametric curve can show possible nonlinearity. The `log="y"` option is again handy, and the formula method allows adding a start value to the response. Figure 11.2 illustrates these ideas, for the factor `married` and the covariate `mentor`.

```
> boxplot(articles + 1 ~ married, data = PhdPubs, log = "y",
+         varwidth = TRUE, ylab = "log(articles + 1)", xlab = "married",
+         cex.lab = 1.25)
> plot(jitter(articles + 1) ~ mentor, data = PhdPubs, log = "y",
+      ylab="log(articles + 1)", cex.lab = 1.25)
> lines(lowess(PhdPubs$mentor, PhdPubs$articles + 1), col = "blue",
+       lwd = 3)
```

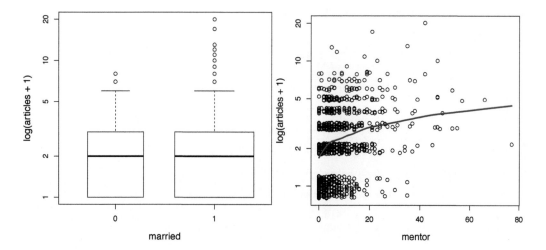

Figure 11.2: Exploratory plots for the number of articles in the PhdPubs data. Left: boxplots for married (1) vs. non-married (0); right: jittered scatterplot vs. mentor publications with a lowess smoothed curve.

It can be seen that the distribution of articles for married and non-married are quite similar, except that for the married students there are quite a few observations with a large number of publications. The relationship between log(articles) and mentor publications seems largely linear except possibly at the very low end. The large number of zero counts at the lower left corner stands out; this would not be seen without jittering.

Plots similar to those in Figure 11.2 can also be produced using **ggplot2** with greater flexibility, but perhaps greater effort to get the details right. One key feature is the use of `scale_y_log10()` to plot the response, and all other features on a log scale. The following code gives a plot similar to the right panel of Figure 11.2, but also plots a confidence band around the smoothed curve, and adds a linear regression line of log(articles) on mentor publications. This plot is not shown here, but it is a good exercise to reproduce it for yourself.

```
> ggplot(PhdPubs, aes(mentor, articles + 1)) +
+   geom_jitter(position = position_jitter(h = 0.05)) +
+   stat_smooth(method = "loess", size = 2, fill = "blue", alpha = 0.25) +
+   stat_smooth(method = "lm", color = "red", size = 1.25, se = FALSE) +
+   scale_y_log10(breaks = c(1, 2, 5, 10, 20)) +
+   labs(y = "log(articles + 1)", x = "Mentor publications")
```

To start analysis, we fit the Poisson model using all predictors—female, married, kid5, phdprestige, and mentor. As recorded in *PhdPubs*, female and married are both dummy (0/1) variables, and it slightly more convenient for plotting purposes to make them factors.

```
> PhdPubs <- within(PhdPubs, {
+    female <- factor(female)
+    married <- factor(married)
+ })
```

The model is fit as shown below and summarized using summary().

```
> phd.pois <- glm(articles ~ ., data = PhdPubs, family = poisson)
> summary(phd.pois)


Call:
glm(formula = articles ~ ., family = poisson, data = PhdPubs)

Deviance Residuals:
    Min      1Q  Median      3Q     Max
 -3.488  -1.538  -0.365   0.577   5.483

Coefficients:
             Estimate Std. Error z value Pr(>|z|)
(Intercept)   0.26562    0.09962    2.67   0.0077 **
female1      -0.22442    0.05458   -4.11  3.9e-05 ***
married1      0.15732    0.06125    2.57   0.0102 *
kid5         -0.18491    0.04012   -4.61  4.0e-06 ***
phdprestige   0.02538    0.02527    1.00   0.3153
mentor        0.02523    0.00203   12.43  < 2e-16 ***
---
Signif. codes:  0 '***' 0.001 '**' 0.01 '*' 0.05 '.' 0.1 ' ' 1

(Dispersion parameter for poisson family taken to be 1)

    Null deviance: 1817.4  on 914  degrees of freedom
Residual deviance: 1633.6  on 909  degrees of freedom
AIC: 3313

Number of Fisher Scoring iterations: 5
```

Significance tests for the individual coefficients show that all are significant, except for phdprestige. We ignore this here, and continue to interpret and extend the full main effects model.[8]

The estimated coefficients β for the predictors are shown below. Recall that using the log link means, for example, that being married increases the log of the expected number of articles published by 0.157, holding all other predictors constant. Each additional child of age 5 or less decreases this by 0.185.

```
> round(cbind(beta = coef(phd.pois),
+             expbeta = exp(coef(phd.pois)),
+             pct = 100 * (exp(coef(phd.pois)) - 1)), 3)

             beta expbeta     pct
(Intercept)  0.266   1.304  30.425
female1     -0.224   0.799 -20.102
married1     0.157   1.170  17.037
kid5        -0.185   0.831 -16.882
phdprestige  0.025   1.026   2.570
mentor       0.025   1.026   2.555
```

---

[8]It is usually less harmful to include a non-significant predictor (which in any case may be a variable useful to control, as phdprestige here), than to omit a potentially important predictor, or worse—to fail to account for an important interaction.

It is somewhat easier to interpret the exponentiated coefficients, $\exp(\beta)$, as multiplicative effects on the expected number of articles and convert these to percentage change, again holding other predictors constant. For example, expected publications by married candidates are 1.17 times that of non-married, a 17% increase, while each additional child multiplies articles by 0.831, a 16.88% decrease. Alternatively, we recommend visual displays for model interpretation, and effect plots do well in most cases, as shown in Figure 11.3. For a Poisson GLM, an important feature is that the response is plotted on the log scale, so that effects in the model appear as linear functions, while the values of the response (number of articles) are labeled on their original scale, facilitating interpretation. The confidence bands and error bars give 95% confidence intervals around the fitted effects.

```
> library(effects)
> plot(allEffects(phd.pois), band.colors = "blue", lwd = 3,
+      ylab = "Number of articles", main = "")
```

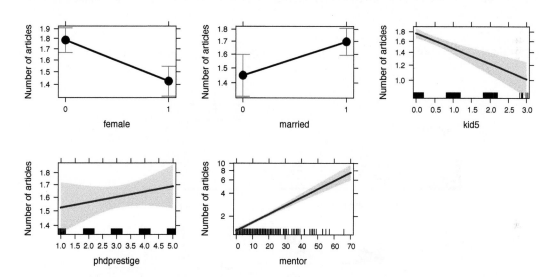

Figure 11.3: Effect plots for the predictors in the Poisson regression model for the PhdPubs data. Jittered values of the continuous predictors are shown at the bottom as rug-plots.

In Figure 11.3 we can see the decrease in published articles with number of young children, but also that the confidence band gets wider with increasing children. The predicted effect here of number of publications by the student's mentor is more dramatic, particularly for those whose mentor was truly prolific. You should note that the panels for the predictors in Figure 11.3 are scaled individually for the range of the fitted main effects. This is often a sensible default and all predictors except mentor give a similar range here. To make all of these plots strictly comparable, provide a ylim argument, giving the range of the response on the log scale, as below (but not shown here).

```
> plot(allEffects(phd.pois), band.colors = "blue", ylim = c(0, log(10)))
```

All of the above is useful, but still leaves aside the question of how well the Poisson model fits the data. The output from `summary(phd.pois)` above showed that the Poisson model fits quite badly. The residual deviance of 1633.6 with 909 degrees of freedom is highly significant.

△

EXAMPLE 11.2: Mating of horseshoe crabs

Brockmann (1996) studied the mating behavior of female horseshoe crabs in the Gulf of Mexico. In the mating season, crabs arrive on the beach in female/male pairs to lay and fertilize eggs. However, unattached males, called "satellites", also come to the beach, crowd around the nesting couples and compete with attached males for fertilizations, contributing to reproductive success. Some females are ignored by satellite males, and some attract more satellites than others, and the question is: what factors contribute to the number of satellites for each female? Or, perhaps better, how do unattached males choose among available females? This is another example in which zero counts may require special treatment.

The data, given in *CrabSatellites* in the **countreg** (Zeileis and Kleiber, 2014) package, give the response variable `satellites` for 173 females. Possible predictors are the female's `color` and `spine` condition, given as ordered factors, as well as her `weight` and carapace (shell) `width`.

```
> data("CrabSatellites", package = "countreg")
> str(CrabSatellites)
```

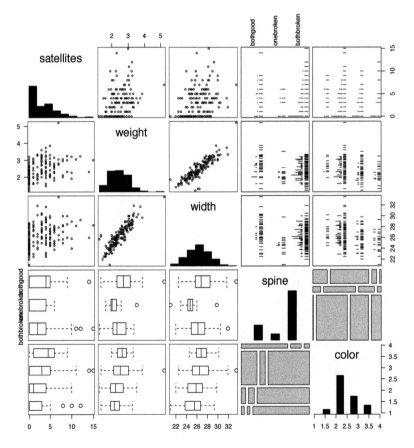

Figure 11.4: Generalized pairs plot for the CrabSatellites data.

```
'data.frame':  173 obs. of  5 variables:
 $ color     : Ord.factor w/ 4 levels "lightmedium"<..: 2 3 3 4 2 1 4 2 2 2 ...
 $ spine     : Ord.factor w/ 3 levels "bothgood"<"onebroken"<..: 3 3 3 2 3 2 3 2 3 3 1 3 ...
 $ width     : num  28.3 26 25.6 21 29 25 26.2 24.9 25.7 27.5 ...
 $ weight    : num  3.05 2.6 2.15 1.85 3 2.3 1.3 2.1 2 3.15 ...
 $ satellites: int  8 4 0 0 1 3 0 0 8 6 ...
```

Agresti (2013, Section 4.3) analyzes the number of satellites using count data GLMs, and in his Chapter 5, describes separate logistic regression models for the binary outcome of one or more satellites vs. none. Later in this chapter (Section 11.4) we consider hurdle and zero-inflated models for count data. These have the advantage of modeling the zero counts together with a model for the positive counts.

A useful overview plot of the data is shown using `gpairs()` in Figure 11.4. You can see that the distribution of `satellites` is quite positively skewed, with many zero counts. `width` and `weight` are highly correlated (0.89), and both relate to the size of the female. Their scatterplots in the first row show that larger females attract more satellites. The categorical ordered factors `spine` condition and `color` are strongly associated, with the lightest colored crabs having the best conditions.

```
> library(vcd)
> library(gpairs)
> gpairs(CrabSatellites[, 5 : 1],
+        diag.pars = list(fontsize = 16))
```

Figure 11.5 shows the scatterplots of `satellites` against `width` and `weight` together with smoothed lowess curves.

```
> plot(jitter(satellites) ~ width, data = CrabSatellites,
+    ylab = "Number of satellites (jittered)", xlab = "Carapace width",
+    cex.lab = 1.25)
> with(CrabSatellites, lines(lowess(width, satellites), col = "red", lwd = 2))
> plot(jitter(satellites) ~ weight, data = CrabSatellites,
+    ylab = "Number of satellites (jittered)", xlab = "Weight",
+    cex.lab = 1.25)
> with(CrabSatellites, lines(lowess(weight, satellites), col = "red", lwd = 2))
```

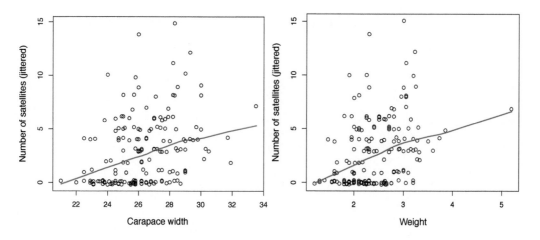

Figure 11.5: Scatterplots of number of satellites vs. width and weight, with lowess smooths.

Both variables show approximately linear relations to the mean number of satellites, so it would not be unreasonable to fit models using the identity link ($\mu \sim x$) rather than the log link ($\mu \sim \log(x)$) with the Poisson family GLM.

In these plots, we reduce the problem of overplotting of the discrete response by jittering, but an alternative technique is to transform a numeric count or continuous predictor to a factor (for visualization purposes only), thereby giving boxplots. A convenience function for this purpose, `cutfac()`, is defined in **vcdExtra**. It acts like `cut()`, but gives nicer labels for the factor levels and by default chooses convenient breaks among the values based on deciles. Using this, the plots in Figure 11.5 can be re-drawn as boxplots, giving Figure 11.6.

```
> plot(satellites ~ cutfac(width), data = CrabSatellites,
+      ylab = "Number of satellites", xlab = "Carapace width (deciles)")
> plot(satellites ~ cutfac(weight), data = CrabSatellites,
+      ylab = "Number of satellites", xlab = "Weight (deciles)")
```

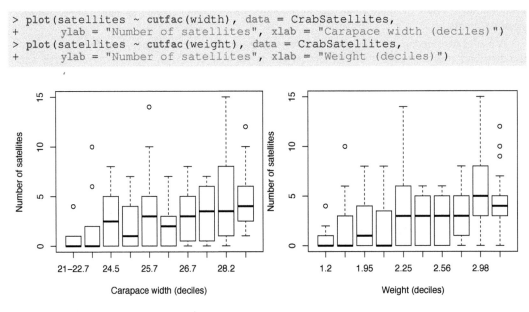

Figure 11.6: Boxplots of number of satellites vs. width and weight.

With this visual overview, we proceed to an initial Poisson GLM model, using all predictors. Note that `color` and `spine` are ordered factors, so `glm()` represents them as polynomial contrasts, as if they were coded numerically.

```
> crabs.pois <- glm(satellites ~ ., data = CrabSatellites,
+                                  family = poisson)
> summary(crabs.pois)


Call:
glm(formula = satellites ~ ., family = poisson, data = CrabSatellites)

Deviance Residuals:
    Min      1Q  Median      3Q     Max
 -3.029  -1.863  -0.599   0.933   4.945

Coefficients:
            Estimate Std. Error z value Pr(>|z|)
(Intercept)  -0.7057     0.9344   -0.76   0.4501
color.L      -0.4120     0.1567   -2.63   0.0085 **
color.Q       0.1237     0.1231    1.00   0.3150
color.C       0.0481     0.0914    0.53   0.5983
spine.L       0.0618     0.0848    0.73   0.4660
spine.Q       0.1585     0.1609    0.99   0.3244
width         0.0165     0.0489    0.34   0.7358
```

```
weight            0.4971        0.1663      2.99     0.0028 **
---
Signif. codes:  0 '***' 0.001 '**' 0.01 '*' 0.05 '.' 0.1 ' ' 1

(Dispersion parameter for poisson family taken to be 1)

    Null deviance: 632.79  on 172  degrees of freedom
Residual deviance: 549.56  on 165  degrees of freedom
AIC: 920.9

Number of Fisher Scoring iterations: 6
```

The Wald tests for the coefficients show that only the linear effect of color and the effect of width are significant. Effect plots, in Figure 11.7, show the nature of these effects—lighter colored females attract more satellites, as do wider and heavier females.

```
> plot(allEffects(crabs.pois), main = "")
```

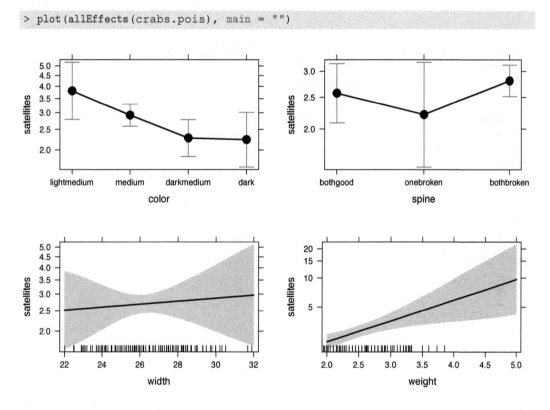

Figure 11.7: Effect plots for the predictors in the Poisson regression model for the CrabSatellites data.

A simpler model can be constructed using `color` as a numeric variable, and either width or weight to represent female size. We choose weight here.[9]

```
> CrabSatellites1 <- transform(CrabSatellites, color = as.numeric(color))
>
> crabs.pois1 <- glm(satellites ~ weight + color, data = CrabSatellites1,
```

---

[9]Agresti (2013, Section 4.3) and others who have analyzed this example uses carapace width as the main quantitative predictor, possibly because width might be more biologically salient to the single males than weight. This is a case where two highly correlated predictors are each strongly related to the outcome, yet partial tests (controlling for all others) may prefer one over the other.

```
+                           family = poisson)
> summary(crabs.pois1)


Call:
glm(formula = satellites ~ weight + color, family = poisson,
    data = CrabSatellites1)

Deviance Residuals:
    Min      1Q  Median      3Q     Max
 -2.978  -1.916  -0.547   0.918   4.834

Coefficients:
             Estimate Std. Error z value Pr(>|z|)
(Intercept)    0.0888     0.2544    0.35    0.727
weight         0.5458     0.0675    8.09    6e-16 ***
color         -0.1728     0.0615   -2.81    0.005 **
---
Signif. codes:  0 '***' 0.001 '**' 0.01 '*' 0.05 '.' 0.1 ' ' 1

(Dispersion parameter for poisson family taken to be 1)

    Null deviance: 632.79  on 172  degrees of freedom
Residual deviance: 552.77  on 170  degrees of freedom
AIC: 914.1

Number of Fisher Scoring iterations: 6
```

From the statistical and graphical analysis so far, the answer to the question posed in this example is clear: unattached male horseshoe crabs prefer light-colored, big fat mamas!

Yet neither of these models fit well, as can be seen from their residual deviances and likelihood-ratio tests.

```
> LRstats(crabs.pois, crabs.pois1)

Likelihood summary table:
            AIC BIC LR Chisq  Df Pr(>Chisq)
crabs.pois  921 946      550 165     <2e-16 ***
crabs.pois1 914 924      553 170     <2e-16 ***
---
Signif. codes:  0 '***' 0.001 '**' 0.01 '*' 0.05 '.' 0.1 ' ' 1
```

Perhaps there is something else to be learned here.

$\triangle$

11.3   Models for overdispersed count data

In practice, the Poisson model is often very useful for describing the relationship between the mean μ_i and the linear predictors, but typically underestimates the variance in the data. The consequence is that the Poisson standard errors are too small, rendering the Wald tests of coefficients, $z_j = \widehat{\beta}_j / \mathrm{se}(\widehat{\beta}_j)$ (and other hypothesis test statistics) too large, and thus overly liberal.

In applications of the GLM, overdispersion is usually assessed by the likelihood-ratio test of the deviance (or the Pearson statistic) given in Section 11.1.3, but there is a subtle problem here. Lack of fit in a GLM for count data can result either from a mis-specified model for the systematic component (omitted or unmeasured predictors, nonlinear relations, etc.) or from failure of the Poisson mean = variance assumption. Thus, use of these methods requires some high degree of confidence that the systematic part of the model has been correctly specified, so that any lack of fit can be attributed to overdispersion.

One way of dealing with this is to base inference on so-called *sandwich* covariance estimators that are robust against some types of model mis-specification. In R, this is provided by the `sandwich()` function in the **sandwich** (Lumley and Zeileis, 2015) package, and can be used with `coeftest(model, vcov = sandwich)` to give overdispersion-corrected hypothesis tests[10] (Zeileis, 2004, 2006). Alternatively, the Poisson model variance assumption can be relaxed in the quasi-Poisson model and the negative-binomial model as discussed below.

11.3.1 The quasi-Poisson model

One obvious solution to the problem of overdispersion for count data is the relaxed assumption that the conditional variance is merely *proportional* to the mean,

$$\mathcal{V}(y_i|\eta_i) = \phi\mu_i .$$

Overdispersion is the common case of $\phi > 1$, implying that the conditional variance increases faster than the mean, but the opposite case of underdispersion, $\phi < 1$, is also possible, though relatively rare in practice. This strategy entails estimating the dispersion parameter ϕ from the data, and gives the *quasi-Poisson model* for count data.

One possible estimate is the residual deviance divided by degrees of freedom. However, it is more common to use the Pearson statistic, giving a method-of-moments estimate with improved statistical properties:

$$\widehat{\phi} = \frac{X_P^2}{n-p} = \sum_{i=1}^{n} \frac{(y_i - \widehat{\mu}_i)^2}{\widehat{\mu}_i} / (n-p) .$$

It turns out that this model gives the same coefficient estimates as the standard Poisson GLM, but inference is adjusted for over/under dispersion. In particular, following Eqn. (11.1), the standard errors of the model coefficients are multiplied by $\widehat{\phi}^{1/2}$ and so are inflated when overdispersion is present. In R, the quasi-Poisson model with this estimated dispersion parameter is fitted with the `glm()` function, by setting `family=quasipoisson`.

EXAMPLE 11.3: Publications of PhD candidates

For the *PhdPubs* data, the deviance and Pearson estimates of dispersion ϕ can be calculated using the results of the Poisson model saved in the `phd.pois` object. The Pearson estimate, 1.83, indicates that standard errors of coefficients in this model should be multiplied by $\sqrt{1.83} = 1.35$, a 35% increase, to correct for overdispersion.

```
> with(phd.pois, deviance / df.residual)

[1] 1.7971

> sum(residuals(phd.pois, type = "pearson")^2) / phd.pois$df.residual

[1] 1.8304
```

The quasi-Poisson model is then fitted using `glm()` as:

```
> phd.qpois <- glm(articles ~ ., data = PhdPubs, family = quasipoisson)
```

For use in other computation, the dispersion parameter estimate $\widehat{\phi}$ can be obtained as the `dispersion` value of the `summary()` method for a quasi-Poisson model.

---

[10]More precisely, given that the mean function of the model is correctly specified, the sandwich standard errors guard against misspecifications of the remaining likelihood, including overdispersion and heteroskedasticity.

```
> (phi <- summary(phd.qpois)$dispersion)

[1] 1.8304
```

Note that this value can be compared to the variance/mean ratio of 2.91 calculated for the marginal distribution in Example 11.1; there is considerable improvement taking the predictors into account.

△

11.3.2   The negative-binomial model

The negative-binomial (NB) model for count data was introduced in Section 3.2.3 as a different generalization of the Poisson model that allows for overdispersion. In the context of the GLM, this can be developed as the extended form where the distribution of $y_i \mid x_i$ where the mean μ_i for fixed x_i can vary across observations i according to a gamma distribution with mean μ_i and a constant shape parameter, θ, reflecting the additional variation due to heterogeneity.

For a fixed value of θ, the negative-binomial is another special case of the GLM. The expected value of the response is again $\mathcal{E}(y_i) = \mu_i$, but the variance function is $\mathcal{V}(y_i) = \mu_i + \mu_i^2/\theta$, so the variance of y increases more rapidly than that of the Poisson distribution. Some authors (e.g., Agresti (2013), Hilbe (2014)) prefer to parameterize the variance function in terms of $\alpha = 1/\theta$, giving

$$\mathcal{V}(y_i) = \mu_i + \mu_i^2/\theta = \mu_i + \alpha\mu_i^2 \,,$$

so that α is a kind of dispersion parameter. Note that as $\alpha \to 0$, $\mathcal{V}(y_i) \to \mu_i$ and the negative-binomial converges to the Poisson.

The **MASS** package provides the family function `negative.binomial(theta)` that can be used directly with `glm()` provided that the argument `theta` is specified. One example would be the related geometric distribution (Section 3.2.4) that is the special case of $\theta = 1$. This can be fitted in R by setting `family=negative.binomial(theta=1)` in the call to `glm()`.

Most often, θ is unknown and must be estimated from the data. In this case, the negative-binomial model is not a special case of the GLM, but it is possible to obtain maximum likelihood estimates of both β and θ, by iteratively estimating β for fixed θ and vice-versa. This method is implemented in the `glm.nb()` in the package **MASS**.

EXAMPLE 11.4:   Mating of horseshoe crabs

For example, for the *CrabSatellites* data, we can fit the general negative-binomial model with θ free.

```
> library(MASS)
> crabs.nbin <- glm.nb(satellites ~ weight + color,
+                      data = CrabSatellites1)
> crabs.nbin$theta

[1] 0.95562
```

The estimated value $\widehat{\theta}$ returned by `glm.nb()` is not very far from 1. Hence, we might also consider fixing $\theta = 1$, as illustrated below.

```
> crabs.nbin1 <- glm(satellites ~ weight + color, data = CrabSatellites1,
+                    family = negative.binomial(1))
```

△

11.3.3 Visualizing the mean–variance relation

The quasi-Poisson and negative-binomial models have different variance functions, and one way to visualize which provides a better fit to the data is to group the data according to the fitted value of the linear predictor, calculate the mean and variance for each group, and then plot the variances against the means. A smoothed curve will then approximate the *empirical* mean–variance relationship. To this, we can add curves showing the mean–variance function implied by various models.[11]

EXAMPLE 11.5: Publications of PhD candidates

For the *PhdPubs* data, the fitted values are obtained with `fitted()` for the Poisson and negative binomial models. Either set can be used to categorize the observations into groups for the purpose of calculating means and variances of the response.

```
> fit.pois <- fitted(phd.pois, type = "response")
> fit.nbin <- fitted(phd.nbin, type = "response")
```

Here we use a simpler version of the `cutfac()` function to group a numeric variable into quantile-based groups. `cutq()` also uses deciles by default, and just uses simple integer values for the factor labels.

```
> cutq <- function(x, q = 10) {
+      quantile <- cut(x, breaks = quantile(x, probs = (0 : q) / q),
+          include.lowest = TRUE, labels = 1 : q)
+      quantile
+ }
```

Using this, we create a variable `group` giving 20 quantile groups of the fitted values, and then use `aggregate()` to find the mean and variance of the number of articles in each group.

```
> group <- cutq(fit.nbin, q = 20)
> qdat <- aggregate(PhdPubs$articles,
+          list(group),
+          FUN = function(x) c(mean = mean(x), var = var(x)))
> qdat <- data.frame(qdat$x)
> qdat <- qdat[order(qdat$mean),]
```

We can then calculate the theoretical variances implied by the quasi-Poisson and negative-binomial models:

```
> phi <- summary(phd.qpois)$dispersion
> qdat$qvar <- phi * qdat$mean
> qdat$nbvar <- qdat$mean + (qdat$mean^2) / phd.nbin$theta
> head(qdat)

    mean     var    qvar   nbvar
1 0.61224 0.78401 1.1206  0.7776
2 1.14894 1.78168 2.1030  1.7312
8 1.24444 2.46162 2.2778  1.9276
4 1.26087 1.70821 2.3079  1.9622
6 1.27273 1.83087 2.3296  1.9873
7 1.29787 4.34413 2.3756  2.0409
```

The plot, shown in Figure 11.8, then simply plots the points and uses `lines()` to plot the model-implied variances.

---

[11]This idea and the example that follows was suggested by Germán Rodrigues in a Stata example given at `http://data.princeton.edu/wws509/stata/overdispersion.html`.

```
> with(qdat, {
+    plot(var ~ mean, xlab = "Mean number of articles", ylab = "Variance",
+         pch = 16, cex = 1.2, cex.lab = 1.2)
+    abline(h = mean(PhdPubs$articles), col = gray(.40), lty = "dotted")
+    lines(mean, qvar, col = "red", lwd = 2)
+    lines(mean, nbvar, col = "blue", lwd = 2)
+    lines(lowess(mean, var), lwd = 2, lty = "dashed")
+    text(3, mean(PhdPubs$articles), "Poisson", col = gray(.40))
+    text(3, 5, "quasi-Poisson", col = "red")
+    text(3, 6.7, "negbin", col = "blue")
+    text(3, 8.5, "lowess")
+ })
```

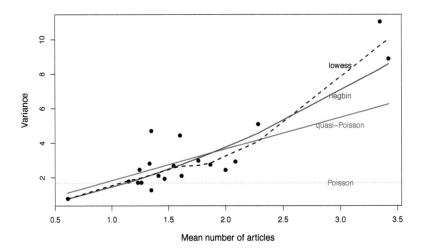

Figure 11.8: Mean–variance functions for the PhdPubs data. Points show the observed means and variances for 20 quantile groups based on the fitted values in the negative-binomial model. The labeled lines and curves show the variance functions implied by various models.

We can see from this plot that the variances implied by the quasi-Poisson and negative-binomial models are in reasonable accord with the data and with each other up to a mean of about 2.5. They diverge substantially at the upper end, for the 20–30% of the most productive candidates, where the quadratic variance function of the negative-binomial provides a better fit.

Finally, we can also compare the standard errors of coefficients for the various methods designed to correct for overdispersion. These are extracted as the diagonal elements of the vcov() and sandwich() methods from the model objects.

```
> library(sandwich)
> phd.SE <- sqrt(cbind(
+    pois = diag(vcov(phd.pois)),
+    sand = diag(sandwich(phd.pois)),
+    qpois = diag(vcov(phd.qpois)),
+    nbin = diag(vcov(phd.nbin))))
> round(phd.SE, 4)

             pois    sand   qpois    nbin
(Intercept) 0.0996 0.1382 0.1348 0.1327
female1     0.0546 0.0714 0.0738 0.0726
married1    0.0613 0.0823 0.0829 0.0819
kid5        0.0401 0.0560 0.0543 0.0528
```

```
phdprestige 0.0253 0.0392 0.0342 0.0343
mentor      0.0020 0.0039 0.0027 0.0032
```

For this example, the sandwich, quasi-Poisson, and negative-binomial methods give similar results, all about 40% larger on average than those from the Poisson model. △

11.3.4 Testing overdispersion

The forms of overdispersion seen in these examples and in Figure 11.8 give rise to a statistical test (Cameron and Trivedi 1990; Cameron and Trivedi 1998, Section 3.4) for the null hypothesis of Poisson variation, $H_0 : \mathcal{V}(y) = \mu$, against an alternative that the variance has a particular form depending on the mean,

$$\mathcal{V}(y) = \mu + \alpha \times f(\mu),$$

where $f(\mu)$ is a given transformation function of the mean.

Overdispersion corresponds to $\alpha > 0$ and underdispersion to $\alpha < 0$. The coefficient α can be estimated by an auxiliary OLS regression (without an intercept), i.e., of the form

```
lm(var ~ -1 + f(mean))
```

and tested with the corresponding t (or z) statistic, which is asymptotically standard normal under the null hypothesis.

Common specifications of the transformation function are $f(\mu) = \mu$ and $f(\mu) = \mu^2$. The first corresponds to an NB model with a linear variance function (called NB1 by various authors) or a quasi-Poisson model with dispersion parameter ϕ, i.e.,

$$\mathcal{V}(y) = (1 + \alpha)\mu = \phi\mu.$$

The second is the more traditional form with quadratic variance function described in Section 11.3.2 (called NB2 by some authors).

These tests are carried out using the `dispersiontest()` function in the **AER** (Kleiber and Zeileis, 2015) package, the companion software of Kleiber and Zeileis (2008). The first argument is a Poisson GLM model; the second specifies the alternative hypothesis, either as an integer power of μ or a function of the mean.

```
> library(AER)
> dispersiontest(phd.pois)


        Overdispersion test

data:  phd.pois
z = 5.73, p-value = 4.9e-09
alternative hypothesis: true dispersion is greater than 1
sample estimates:
dispersion
    1.8259

> dispersiontest(phd.pois, 2)


        Overdispersion test

data:  phd.pois
z = 6.46, p-value = 5.3e-11
alternative hypothesis: true alpha is greater than 0
sample estimates:
  alpha
0.50877
```

These tests use a specified alternative hypothesis, so there is no way to compare directly which of the NB1 or NB2 models is better or worse, except by using methods such as AIC or BIC described in Section 11.1.4.

11.3.5 Visualizing goodness-of-fit

Even with correction for overdispersion, goodness-of-fit tests provide only an overall summary of model fit. Some specialized tests for particular forms of overdispersion are also available (e.g., see Cameron and Trivedi (1998, Chapter 5)), but these only identify general problems and cannot provide detailed indications of the possible source of these problems.

In Chapter 3, we illustrated the use of rootograms for visualizing goodness-of-fit to a wide variety of discrete distributions using the `plot()` method for class **"goodfit"** objects with the **vcd** package. However, those methods were developed for one-way discrete distributions without explanatory variables.

Kleiber and Zeileis (2014) have generalized this idea to the wider class of GLM-related count regression models considered here. The **countreg** package provides a new implementation of `rootogram()` with methods for all of these models (and others not mentioned). We illustrate these plots for the models considered to this point, and then extend this use for models allowing for excess zero counts in Section 11.4.

EXAMPLE 11.6: Publications of PhD candidates

For the *PhdPubs* data, Figure 11.9 shows hanging rootograms for the Poisson and negative-binomial models produced using `countreg::rootogram`[12] on the fitted model objects. We are looking both for general patterns of under/over fit, as well as counts that stand out as poorly fitted against the background.

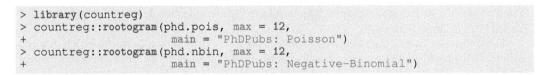

```
> library(countreg)
> countreg::rootogram(phd.pois, max = 12,
+                     main = "PhDPubs: Poisson")
> countreg::rootogram(phd.nbin, max = 12,
+                     main = "PhDPubs: Negative-Binomial")
```

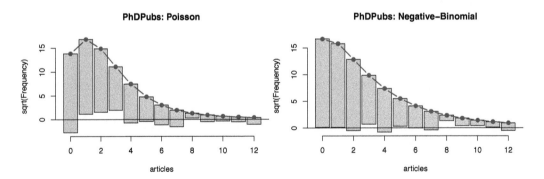

Figure 11.9: Hanging rootograms for the PhdPubs data.

The Poisson model shows a systematic, wave-like pattern with excess zeros, too few observed frequencies for counts of 1–3, but generally greater frequencies for counts of 4 or more. The negative-binomial model clearly fits much better, though there is a peculiar tendency among the smaller frequencies for 8 or more articles.   △

---

[12]At the time of this writing, `rootogram` in **countreg** conflicts with the version in **vcd**, so we qualify the use here with the package name.

EXAMPLE 11.7: Mating of horseshoe crabs

Figure 11.10 shows similar plots for the same two models fit to the number of crab satellites. The fit of the Poisson model clearly reveals the excess of zero male satellites. For the negative-binomial, the rootogram no longer exhibits same wave-like pattern, however, the underfitting of the count for 0 and overfitting for counts 1–2 is characteristic of data with excess zeros.

```
> countreg::rootogram(crabs.pois, max = 15,
+                     main = "CrabSatellites: Poisson")
> countreg::rootogram(crabs.nbin, max = 15,
+                     main = "CrabSatellites: Negative-Binomial")
```

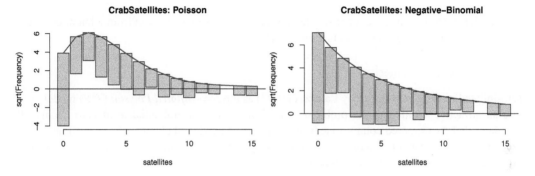

Figure 11.10: Hanging rootograms for the CrabSatellites data.

$\triangle$

11.4   Models for excess zero counts

In addition to overdispersion, many sets of empirical data exhibit a greater prevalence of zero counts than can be accommodated by the Poisson or negative-binomial models. We saw this in the *PhdPubs* data set, where there were many candidates who had not published at all, and in the *CrabSatellites* data where a large number of females attracted no unattached males. Other examples abound in many different fields: studies of the use of health care services often find that many people never visit a hospital in some time frame; similarly, the distribution of insurance claims often shows large numbers who make no claims (Yip and Yau, 2005) because of under-reporting of small claims, policy deductible provisions, and desire to avoid premium increases.

Beyond simply identifying this as a problem of lack-of-fit, understanding the reasons for excess zero counts can make a contribution to a more complete explanation of the phenomenon of interest, and this requires both new statistical models and visualization techniques illustrated in this section.

In the first example, Long (1997) argued that the PhD candidates might fall into two distinct groups: "publishers" (perhaps striving for an academic career) and "non-publishers" (seeking other career paths). Of the 275 observations having `articles==0`, some might not have published due to chance or unmeasured factors. One reasonable form of explanation is that the observed zero counts reflect a mixture of the two latent classes—those who simply have not yet published and those who will likely never publish. A statistical formulation of this idea leads to the class of *zero-inflated* models described below.

A different form of explanation is that there may be some special circumstance or "hurdle" required to achieve a positive count, like publishing the master's thesis (such as being driven internally by a personality trait or externally by pressure from a mentor). This idea leads to the class of *hurdle* models that entertain and fit (simultaneously) two separate models: one for the occurrence of the zero counts, and one for the positive counts. These two approaches are illustrated in Figure 11.11

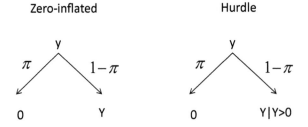

Figure 11.11: Models for excess zeros. The observed response y is derived from a latent or parent distribution for Y yielding zero counts with probability π.

11.4.1 Zero-inflated models

Zero-inflated models, introduced by Lambert (1992) as the ***zero-inflated Poisson*** (ZIP) model, provide an attractive solution to the problem of dealing with an overabundance of zero counts. It postulates that the observed counts arise from a mixture of two latent classes of observations: some structural zeros for whom y_i will always be 0, and the rest, sometimes giving random zeros. The ZIP model is comprised of two components:

- A model for the binary event of membership in the unobserved (latent) class of those for whom the count is necessarily zero (e.g., "non-publishers"). This is typically taken as a logistic regression for the probability π_i that observation i is in this class, with predictors $z_1, z_2, \ldots, z_q$, giving

$$\text{logit}(\pi_i) = z_i^\mathsf{T}\gamma = \gamma_0 + \gamma_1 z_{i1} + \gamma_2 z_{i2} + \cdots + \gamma_q z_{iq} . \tag{11.7}$$

- A Poisson model for the other class (e.g., "publishers"), for whom the observed count may be 0 or positive. This model typically uses the usual log link to predict the mean, using predictors $x_1, x_2, \ldots, x_p$, so

$$\log_e \mu(x_i) = x_i^\mathsf{T}\beta = \beta_0 + \beta_1 x_{i1} + \beta_2 x_{i2} + \cdots + \beta_q x_{ip} . \tag{11.8}$$

In application, it is permissible and not uncommon to use the same set of predictors $x = z$ in both submodels, but the notation indicates that this is not required. Some simple special cases arise when the model for the always-zero latent class is an intercept-only model, $\text{logit}(\pi_i) = \gamma_0$, implying the same probability for all individuals, and (less commonly) when the Poisson mean model is intercept-only with no predictors, but there might be excess zero counts.

With this setup, one can show that the probability of observing counts of $y_i = 0$ and $y_i > 0$ are

$$\Pr(y_i = 0 \mid x, z) = \pi_i + (1 - \pi_i)e^{-\mu_i} \tag{11.9}$$

$$\Pr(y_i \mid x, z) = (1 - \pi_i) \times \left[\frac{\mu_i^{y_i} e^{-\mu_i}}{y_i!} \right], \qquad y_i \geq 0 ,$$

where the term in brackets in the second equation is the Poisson probability $\Pr(y = y_i)$ with rate parameter $\text{Pois}(\mu_i)$. In these equations, $\pi_i = \text{logit}^{-1}(z_i^\mathsf{T}\gamma)$ depends on the z through Eqn. (11.7), and $\mu_i = \exp(x^\mathsf{T}\beta)$ depends on the x through Eqn. (11.8).

The conditional expectation and variance of y_i then have the forms

$$\mathcal{E}(y_i) = (1 - \pi_i)\mu_i$$

$$\mathcal{V}(y_i) = (1 - \pi_i)\mu_i(1 + \mu_i\pi_i) .$$

Thus, when $\pi_i > 0$, the mean of y is always less than μ_i, and the variance of y is greater than its mean by a dispersion factor of $(1 + \mu_i \pi_i)$.

There is nothing special about the use of the Poisson distribution here. The model for the count variable could also be taken as the negative-binomial, giving a *zero-inflated negative-binomial* (ZINB) model using $\text{NBin}(\mu, \theta)$ or a *zero-inflated geometric* model using $\text{NBin}(\mu, \theta = 1)$.

EXAMPLE 11.8: **Simulating zero-inflated data**

A simple way of understanding the effects of zero-inflation on count data is to simulate data from their distribution and plot it. For the standard Poisson and negative-binomial, random values can be generated using rpois() and rnegbin() (in **MASS**), respectively. Their zero-inflated counterparts are implemented in the **VGAM** package as rzipois() and rzinegbin().

To illustrate this use, we generate two random data sets using rzipois() having constant mean $\mu = 3$. The first is a standard Poisson ($\pi = 0$), while the second has a constant probability $\pi = 0.3$ of an excess zero.

```
> library(VGAM)
> set.seed(1234)
> data1 <- rzipois(200, 3, 0)
> data2 <- rzipois(200, 3, .3)
```

Barplots of the frequencies in these data sets are shown in Figure 11.12. The sample mean in data1 is 2.925, quite close to $\mu = 3$. In the zero-inflated data2, the mean is only 2.25 due to the excess zeros.

```
> tdata1 <- table(data1)
> barplot(tdata1, xlab = "Count", ylab = "Frequency",
+          main = "Poisson(3)")
> tdata2 <- table(data2)
> barplot(tdata2, xlab = "Count", ylab = "Frequency",
+          main = expression("ZI Poisson(3, " * pi * "= .3)"))
```

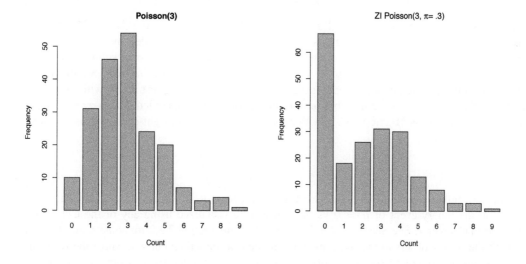

Figure 11.12: Bar plots of simulated data from Poisson and zero-inflated Poisson distributions.

$\triangle$

There are several packages in R capable of fitting zero-inflated models. The most mature and complete of these is `zeroinfl()` in the **countreg** package (a successor to the **pscl** package). The function `zeroinfl()` is modeled after `glm()`, but provides an extended syntax for the model formula.

If the `formula` argument is supplied in the form `y ~ x1 + x2 + ...`, it not only describes the count regression of y on $x_1, x_2, \ldots$, but also implies that the *same* set of regressors, $z_j = x_j$, is used for the zero count binary submodel. The extended syntax uses the notation `y ~ x1 + x2 + ... | z1 + z2 + ...` to specify the x variables separately, conditional on (`|`) the always-zero count model `y ~ z1 + z2 + ...`. The model for the not-always-zero class can be specified using the `dist` argument, with possible values `"poisson"`, `"negbin"`, and `"geometric"`.

11.4.2 Hurdle models

A different class of models capable of accounting for excess zero counts is the ***hurdle model*** (also called the ***zero-altered model***) proposed initially by Cragg (1971) and developed further by Mullahy (1986). This model also uses a separate logistic regression submodel to distinguish counts of $y = 0$ from larger counts, $y > 0$. The submodel for the positive counts is expressed as a (left) *truncated* Poisson or negative-binomial model, excluding the zero counts. As an example, consider a study of behavioral health in which one outcome is the number of cigarettes smoked in one month. All the zero counts will come from non-smokers and smokers will nearly always smoke a positive number.

This differs from the set of ZIP models in that classes of $y = 0$ and $y > 0$ are now considered fully observed, rather than latent. Conceptually, there is one process and submodel accounting for the zero counts and a separate process accounting for the positive counts, once the "hurdle" of $y = 0$ has been passed. In other words, for ZIP models, the first process generates only extra zeros beyond those of the regular Poisson distribution. For hurdle models, the first process generates all of the zeros. The probability equations corresponding to Eqn. (11.9) are:

$$\Pr(y_i = 0 \,|\, \boldsymbol{x}, \boldsymbol{z}) = \pi_i \tag{11.10}$$

$$\Pr(y_i \,|\, \boldsymbol{x}, \boldsymbol{z}) = \frac{(1 - \pi_i)}{1 - e^{-\mu_i}} \times \left[\frac{\mu_i{}^{y_i} e^{-\mu_i}}{y_i!} \right], \qquad y_i \geq 0 \,.$$

The hurdle model can be fitted in **R** using the `hurdle()` function from the **countreg** package. The syntax for the model formula is the same extended form provided by `zeroinfl()`, where `y ~ x1 + x2` uses the same regressors for the zero and positive count submodels, while `y ~ x1 + x2 | z1 + z2` uses `y ~ z1 + z2` for the zero hurdle model. Similarly, the count distribution can be given as `"poisson"`, `"negbin"`, or `"geometric"` with the `dist` argument. For `hurdle()`, the distribution for zero model can be specified with a `zero.dist` argument. The default is `"binomial"` (with a logit `link`), but other right-censored distributions can also be specified.

11.4.3 Visualizing zero counts

Both the zero-inflated and hurdle models treat the zero counts $y = 0$ specially with separate submodels, so the binary event of $y = 0$ vs. $y > 0$ can be visualized using any of the techniques illustrated in Chapter 7. See Section 7.2.3, Section 7.3.1, and Section 7.3.2 for some examples that plot both the binary observations and a model summary or smoothed curve to show the relationships with one or more regressors. To apply these ideas in the current context, simply define or plot a logical variable corresponding to the expression `y==0`, giving values of TRUE or FALSE.

A different, and simpler idea is illustrated here using what is called a ***spineplot*** Hummel (1996) when a predictor x is a discrete factor or ***spinogram*** when x is continuous. Both are forms of mosaic plots with special formatting of spacing and shading, and in this context they plot $\Pr(y = 0 | x)$ against $\Pr(x)$; when x is numerical, it is first made discrete, as in a histogram.

Then, in the spine plot or spinogram, the widths of the bars correspond to the relative frequencies of x and heights of the bars correspond to the conditional relative frequencies of $y = 0$ in every x group. In R, spine plots are implemented in the function `spineplot()`; however, this is what you get by default if you use `plot(y==0 ~ x)` to plot the binary factor against any regressor x.

A related graphical method is the ***conditional density plot*** (Hofmann and Theus, 2005). The conditional probabilities $\Pr(y = 0|x)$ are derived using a smoothing approach (via `density()`) over x rather than by making x discrete. These plots are provided by `cdplot()` in the **graphics** package and a similar `cd_plot()` in **vcd**. The smoothing method for the density estimate is controlled by a `bw` (bandwidth) method and other arguments.

EXAMPLE 11.9: Mating of horseshoe crabs

For the *CrabSatellites* data, we can examine the relationship of the zero counts (females who attract no unattached male satellites) to the predictors using spinograms or conditional density plots. Here, we consider `weight` and `color` (treated numerically) as predictors.

Spinograms for the occurrence of zero satellites against `weight` and `color` are shown in Figure 11.13, where we have used quantiles of those distributions to define the breaks on the horizontal axis. Using `ylevels=2:1` reverses the order of the vertical categories. You can easily see that the zeros decrease steadily with weight and increase with darkness.

```
> plot(factor(satellites == 0) ~ weight, data = CrabSatellites,
+       breaks = quantile(weight, probs = seq(0,1,.2)), ylevels = 2:1,
+       ylab = "No satellites")
> plot(factor(satellites == 0) ~ color, data = CrabSatellites,
+       breaks = quantile(color, probs = seq(0,1,.33)),  ylevels = 2:1,
+       ylab = "No satellites")
```

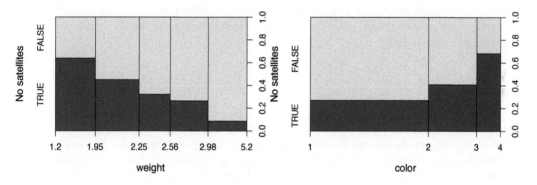

Figure 11.13: Spinograms for the CrabSatellites data. The variables weight (left) and color (right) have been made discrete using quantiles of their distributions.

Similar plots in the form of conditional density plots are shown in Figure 11.14, with a similar interpretation.

```
> cdplot(factor(satellites == 0) ~ weight, data = CrabSatellites,
+         ylevels = 2:1, ylab = "No satellites")
> cdplot(factor(satellites == 0) ~ color, data = CrabSatellites,
+         ylevels = 2:1, , ylab = "No satellites")
```

△

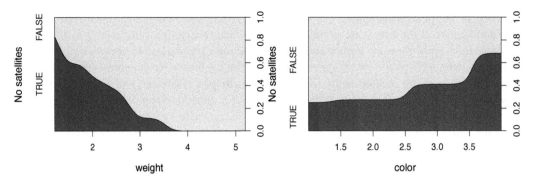

Figure 11.14: Conditional density plots for the CrabSatellites data. The region shaded below shows the conditional probability density estimate for a count of zero.

11.5 Case studies

In this section, we introduce two extended examples, designed to illustrate aspects of exploratory analysis, visualization, model fitting, and interpretation for count data GLMs. The first (Section 11.5.1) concerns another well-known data set from ethology, where (a) excess zeros require special treatment, (b) the occurrence of zero counts has substantive meaning, and (c) an interaction between two factors is important.

The second case study (Section 11.5.2) uses a larger, also well-known data set from health economics, with more predictors and more potential interactions. The emphasis shifts here from fitting and comparing models with different distributional forms and link functions to selecting terms for an adequate descriptive and explanatory model. Another feature of these examples is that the relatively large sample size in this data supports a wider range of model complexity than is available in smaller samples.

11.5.1 Cod parasites

The cod fishery is extremely important to the economy of Norway, so anything that affects the health of the cod population and its ecosystem can have severe consequences. The red king crab *Paralithodes camtschaticus* was deliberately introduced by Russian scientists to the Barents Sea in the 1960s and 1970s from its native area in the North Pacific. The carapace of these crabs is used by the leech *Johanssonia arctica* to deposit its eggs. This leech in turn is a vector for the blood parasite *Trypanosoma murmanensis* that can infect marine fish, including cod.

Hemmingsen et al. (2005) examined cod for trypanosome infections during annual cruises along the coast of Finnmark in North Norway over three successive years and in four different areas (A1: Sørøya; A2: Magerøya; A3: Tanafjord; A4: Varangerfjord). They show that trypanosome infections are strongest in the area Varangerfjord where the density of red king crabs is highest. Thus, there is evidence that the introduction of the foreign red king crabs had an indirect detrimental effect on the health of the native cod population. This situation stands out because it is not an introduced *parasite* that is dangerous for a native host, but rather an introduced *host* that promotes transmission of two endemic parasites. They call the connections among these factors "an unholy trinity."[13]

---

[13] The four areas A1–A4 are arranged from east to west, with Varangerfjord (A4) closest to the Russian Kola Peninsula where the red king crabs initially migrated. A more specific test of the "Russian hypothesis" could be developed by treating area as an ordered factor and testing the linear component. We leave this analysis to an exercise for the reader.

EXAMPLE 11.10: Cod parasites

The data from Hemmingsen et al. (2005) is contained in *CodParasites* in the countreg package. It gives the results for 1, 254 cod caught by one ship in annual autumn cruises from 1999–2001. The main response variable, `intensity`, records the counted number of *Trypanosoma* parasites found in blood samples from these fish. To distinguish between infected vs. non-infected fish, a secondary response, `prevalence`, is also recorded, corresponding to the expression

```
> CodParasites$prevalence <-
+      ifelse(CodParasites$intensity == 0, "no", "yes")
```

Thus, `intensity` is the basic count response variable, and `prevalence` reflects the zero count that would be assessed in zero-inflated and hurdle models. In substantive terms, in a hurdle model, `prevalence` corresponds to whether a fish is infected or not; once infected, `intensity` gives the degree of infection. In a zero-inflated model, infected could be considered a latent variable; there are extra zeros from non-infected fish, but some infected fish are measured as "normal" zeros.

Hemmingsen et al. (2005) consider only three explanatory predictors: `area`, `year` (both factors) and `length` of the fish.[14] A quick numerical summary of the univariate properties of these variables is shown below. The intensity values are indeed extremely skewed, with a median of 0 and a maximum of 257. However, there are some missing values (NAs) among the response variables and a few in the length variable.

```
> data("CodParasites", package = "countreg")
> summary(CodParasites[, c(1 : 4, 7)])

   intensity       prevalence           area           year          length
 Min.   :  0.00   no  :654   soroya        :272   1999:567   Min.   : 17.0
 1st Qu.:  0.00   yes :543   mageroya      :255   2000:230   1st Qu.: 44.0
 Median :  0.00   NA's: 57   tanafjord     :415   2001:457   Median : 54.0
 Mean   :  6.18              varangerfjord:312              Mean   : 53.4
 3rd Qu.:  4.00                                             3rd Qu.: 62.0
 Max.   :257.00                                             Max.   :101.0
 NA's   :57                                                 NA's   : 6
```

Even better, a quick univariate and bivariate summary of these variables can be shown in a generalized pairs plot (Figure 11.15).

```
> library(vcd)
> library(gpairs)
> gpairs(CodParasites[, c(1 :4, 7)],
+        diag.pars = list(fontsize = 16),
+        mosaic.pars = list(gp = shading_Friendly))
```

In this plot, among the categorical variables, prevalence is strongly associated with area, but also with year. As well, there seems to be an association between area and year, meaning the number of cod samples collected in different areas varied over time. In the univariate plots on the diagonal, intensity stands out as extremely skewed, and the distribution of length appears reasonably symmetric.

Before fitting any models, some more detailed exploratory plots are helpful for understanding the relationship of both prevalence and intensity to the predictors. The general idea is to make separate plots of prevalence and intensity and to try to show both the data and some simple summaries. In their Table 1, Hemmingsen et al. (2005) counted the missing observations as infected and we do the same to get a similar contingency table.

---

[14]Other potential predictors include weight, sex, age, and developmental stage, as well as the depth at which the fish were caught.

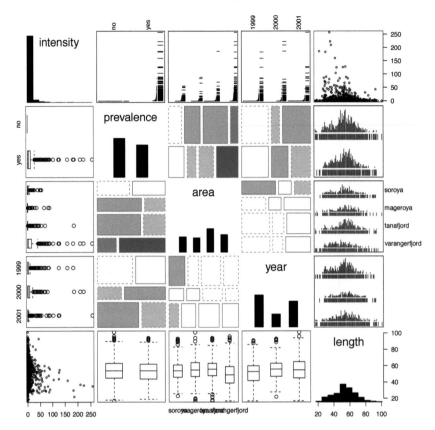

Figure 11.15: Generalized pairs plot for the CodParasites data.

```
> cp.tab <- xtabs(~ area + year + factor(is.na(prevalence) |
+                                      prevalence == "yes"),
+              data = CodParasites)
> dimnames(cp.tab)[3] <- list(c("No", "Yes"))
> names(dimnames(cp.tab))[3] <- "prevalence"
```

For the factors `area` and `year`, we can visualize prevalence as before (Example 11.9) using spineplots, but, for two (or more) factors, doubledecker and mosaic plots are better because they are more flexible and keep the factors distinct. The doubledecker plot (Figure 11.16) highlights the infected fish, and shows that prevalence is indeed highest in all years in Varangerfjord.

```
> doubledecker(prevalence ~ area + year, data = cp.tab,
+            margins = c(1, 5, 3, 1))
```

A similar plot can be drawn shading the tiles according to a model for the expected counts. It makes sense here to consider the null loglinear model for prevalence as a response, independent of the combinations of area and year. This plot (Figure 11.17) shows further that prevalence differs substantially over the area-year combinations, so we should expect an interaction in the model for zero counts. As well, Varangerfjord stands out as having consistently greater prevalence in all years than expected under this model.

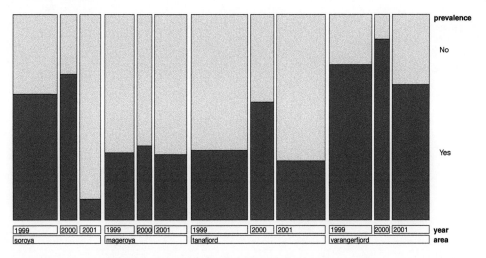

Figure 11.16: Doubledecker plot for prevalence against area and year in the CodParasites data. The cases of infected fish are highlighted.

```
> doubledecker(prevalence ~ area + year, data = cp.tab,
+              gp = shading_hcl, expected = ~ year:area + prevalence,
+              margins = c(1, 5, 3, 1))
```

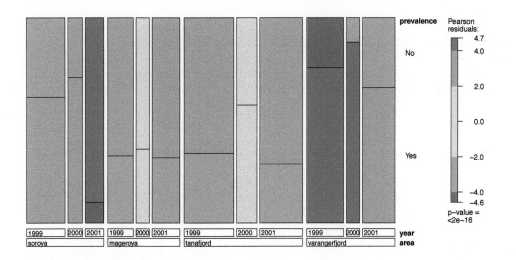

Figure 11.17: Mosaic plot for prevalence against area and year in the CodParasites data, in the doubledecker format. Shading reflects departure from a model in which prevalence is independent of area and year jointly.

The effect of fish `length` on `prevalence` can be most easily seen by treating the factor as a numeric (0/1) variable and smoothing, as shown in Figure 11.18. The loess smoothed curve shows an apparent U-shaped relationship; however, the plotted observations and the confidence bands make clear that there is very little data in the extremes of `length`.

```
> library(ggplot2)
> ggplot(CodParasites, aes(x = length, y = as.numeric(prevalence) - 1)) +
+   geom_jitter(position = position_jitter(height = .05), alpha = 0.25) +
```

```
+     geom_rug(position = "jitter", sides = "b") +
+     stat_smooth(method = "loess", color = "red",
+                 fill = "red", size = 1.5) +
+     labs(y = "prevalence")
```

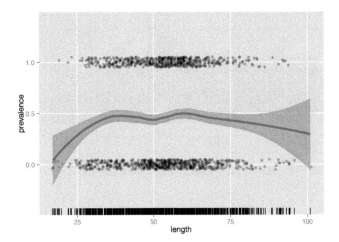

Figure 11.18: Jittered scatterplot of prevalence against length of fish, with loess smooth.

For the positive counts of `intensity`, boxplots by area and year show the distributions of parasites, and it is again useful to display these on a log scale. In Figure 11.19, we have used `ggplot2`, with `geom_boxplot()` and `geom_jitter()` to also plot the individual observations. Note that `facet_grid()` makes it easy to organize the display with separate panels for each area, a technique that could extend to additional factors.

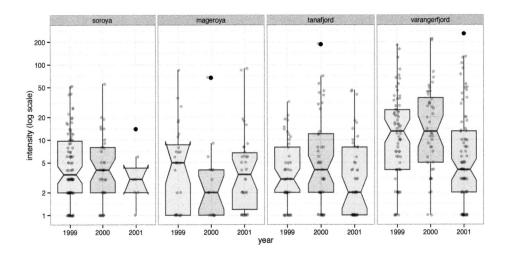

Figure 11.19: Notched boxplots for log (intensity) of parasites by area and year in the CodParasites data. Significant differences in the medians are signaled when the notches of two groups do not overlap.

```
> # plot only positive values of intensity
> CPpos <- subset(CodParasites, intensity > 0)
> ggplot(CPpos, aes(x = year, y = intensity)) +
+    geom_boxplot(outlier.size = 3, notch = TRUE, aes(fill = year),
+                 alpha = 0.2) +
+    geom_jitter(position = position_jitter(width = 0.1), alpha = 0.25) +
+    facet_grid(. ~ area) +
+    scale_y_log10(breaks = c(1, 2, 5, 10, 20, 50, 100, 200)) +
+    theme(legend.position = "none") +
+    labs(y = "intensity (log scale)")
```

Most of these distributions are positively skewed and there are a few high outliers, but probably not more than would be expected in a sample of this size. The positive counts (degree of infection) are also higher in all years in Varangerfjord than other areas. You can also see that the intensity values were generally lower in 2001 than other years.

For the effect of length of fish, we want to know if log (intensity) is reasonably linear on length. A jittered scatterplot produced with **ggplot2** is shown in Figure 11.20. The smoothed loess curve together with the linear regression line show no indication of nonlinearity.

```
> ggplot(CPpos, aes(x = length, y = intensity)) +
+    geom_jitter(position = position_jitter(height = .1), alpha = 0.25) +
+    geom_rug(position = "jitter", sides = "b") +
+    scale_y_log10(breaks = c(1, 2, 5, 10, 20, 50, 100, 200)) +
+    stat_smooth(method = "loess", color = "red", fill = "red", size = 2) +
+    stat_smooth(method = "lm", size = 1.5)
```

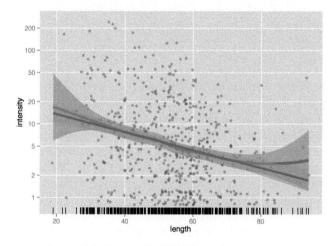

Figure 11.20: Jittered scatterplot of log (intensity) for the positive counts against length of fish, with loess smooth and linear regression line.

△

11.5.1.1 Fitting models

The simple summary of these exploratory analyses is that both the zero component (prevalence) and non-zero component (intensity) involve an interaction of `area` and `year`, and that at least intensity depends on `length`. We proceed to fit some count data models.

EXAMPLE 11.11: Cod parasites

For a baseline reference, we first fit the standard Poisson and negative-binomial models, not allowing for excess zeros.

```
> library(MASS)
> library(countreg)
> cp_p  <- glm(intensity ~ length + area * year,
+              data = CodParasites, family = poisson)
> cp_nb <- glm.nb(intensity ~ length + area * year,
+                   data = CodParasites)
```

Next, we fit analogous hurdle and zero-inflated models, in each case allowing the non-zero count component to be either Poisson or negative-binomial. The zero components are fit as logistic regressions with the same predictors and the logit link.

```
> cp_hp  <- hurdle(intensity ~ length + area * year,
+                  data = CodParasites, dist = "poisson")
> cp_hnb <- hurdle(intensity ~ length + area * year,
+                  data = CodParasites, dist = "negbin")
> cp_zip <- zeroinfl(intensity ~ length + area * year,
+                    data = CodParasites, dist = "poisson")
> cp_znb <- zeroinfl(intensity ~ length + area * year,
+                    data = CodParasites, dist = "negbin")
```

Following Section 11.3.5, we can compare the fit of these models using rootograms. The details of fit of these six models are shown in Figure 11.21.

```
> countreg::rootogram(cp_p, max = 50, main = "Poisson")
> countreg::rootogram(cp_nb, max = 50, main = "Negative Binomial")
> countreg::rootogram(cp_hp, max = 50, main = "Hurdle Poisson")
> countreg::rootogram(cp_hnb, max = 50, main = "Hurdle Negative Binomial")
> countreg::rootogram(cp_zip, max = 50, main = "Zero-inflated Poisson")
> countreg::rootogram(cp_znb, max = 50,
+                              main = "Zero-inflated Negative Binomial")
```

The basic Poisson model of course fits terribly due to the excess zero counts. The hurdle Poisson and zero-inflated Poisson fit the zero counts perfectly, but at the expense of underfitting the counts for low-intensity values. All of the negative binomial models show a reasonable fit (at the scale shown in this plot), and none show a systematic pattern of under/overfitting.

These models are all in different GLM and extended-GLM families, and there are no anova() methods for hurdle and zero-inflated models. Each pair of Poisson and negative-binomial models are a nested set, because the Poisson is a special case of the negative-binomial where $\theta \to \infty$, and so can be compared using likelihood-ratio tests available with lrtest() from lmtest. However, this cannot be used to compare models of different classes, such as a hurdle model vs. a zero-inflated model. (In Figure 11.21, each pair in the same row are nested models, while all other pairs are non-nested.) Yet, they all have logLik() methods to calculate their log likelihood, and so AIC() and BIC() can be used.

```
> LRstats(cp_p, cp_nb, cp_hp, cp_hnb, cp_zip, cp_znb, sortby = "BIC")

Likelihood summary table:
         AIC   BIC LR Chisq   Df Pr(>Chisq)
cp_p   20378 20444    20352 1178    <2e-16 ***
cp_hp  13688 13820    13636 1165    <2e-16 ***
cp_zip 13687 13819    13635 1165    <2e-16 ***
cp_nb   5031  5102     5003 1178    <2e-16 ***
cp_znb  4955  5092     4901 1164    <2e-16 ***
cp_hnb  4937  5074     4883 1164    <2e-16 ***
---
Signif. codes:  0 '***' 0.001 '**' 0.01 '*' 0.05 '.' 0.1 ' ' 1
```

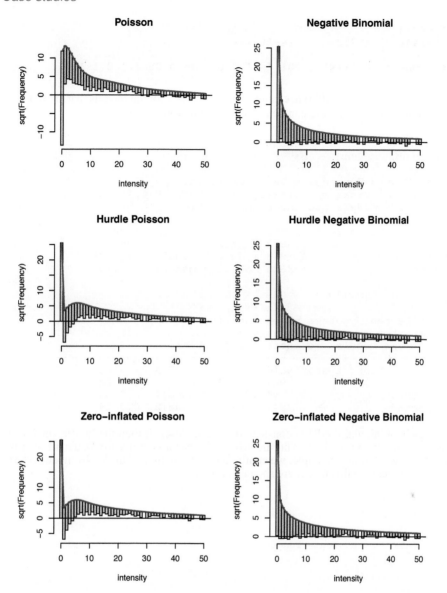

Figure 11.21: Rootograms for six models fit to the CodParasites data.

These show that all the Poisson models fit quite badly, and among the negative-binomial models, the hurdle version, cp_hnb, is preferred by both AIC and BIC. If you want to carry out formal tests, lrtest() can be used to compare a given Poisson model to its negative-binomial counterpart, which are nested. For example, the test below compares the hurdle Poisson to the hurdle negative-binomial and confirms that the latter is a significant improvement.

```
> library(lmtest)
> lrtest(cp_hp, cp_hnb)

Likelihood ratio test

Model 1: intensity ~ length + area * year
Model 2: intensity ~ length + area * year
  #Df LogLik Df Chisq Pr(>Chisq)
```

```
1   26  -6818
2   27  -2442  1  8752      <2e-16 ***
---
Signif. codes:  0 '***' 0.001 '**' 0.01 '*' 0.05 '.' 0.1 ' ' 1
```

Of greater interest is the difference among the negative-binomial models that are not nested. As described in Section 11.1.4, these can be compared using Voung's test[15]

```
> library(pscl)
> vuong(cp_nb, cp_hnb)      # nb vs. hurdle nb

Vuong Non-Nested Hypothesis Test-Statistic:
(test-statistic is asymptotically distributed N(0,1) under the
 null that the models are indistinguishible)
-------------------------------------------------------------
              Vuong z-statistic        H_A  p-value
Raw                     -5.4873 model2 > model1 2.04e-08
AIC-corrected           -4.2943 model2 > model1 8.76e-06
BIC-corrected           -1.2625 model2 > model1    0.103

> vuong(cp_hnb, cp_znb)     # hurdle nb vs znb

Vuong Non-Nested Hypothesis Test-Statistic:
(test-statistic is asymptotically distributed N(0,1) under the
 null that the models are indistinguishible)
-------------------------------------------------------------
              Vuong z-statistic        H_A  p-value
Raw                      1.7941 model1 > model2  0.0364
AIC-corrected            1.7941 model1 > model2  0.0364
BIC-corrected            1.7941 model1 > model2  0.0364
```

The negative-binomial model is considered to be a closer fit than the hurdle version (because it is more parsimonious), while the hurdle NB model has a significantly better fit than the zero-inflated NB model. For this example, we continue to work with the hurdle NB model. The tests for individual coefficients in this model are shown below.

```
> summary(cp_hnb)


Call:
hurdle(formula = intensity ~ length + area * year, data = CodParasites,
    dist = "negbin")

Pearson residuals:
   Min     1Q Median     3Q    Max
-0.696 -0.407 -0.336 -0.108 11.114

Count model coefficients (truncated negbin with log link):
                           Estimate Std. Error z value Pr(>|z|)
(Intercept)                 3.37580    0.39947    8.45  < 2e-16 ***
length                     -0.03748    0.00587   -6.38  1.7e-10 ***
areamageroya                0.37898    0.38105    0.99   0.3199
areatanafjord              -0.50480    0.31238   -1.62   0.1061
areavarangerfjord           0.89159    0.29161    3.06   0.0022 **
year2000                   -0.03957    0.32857   -0.12   0.9041
year2001                   -0.75388    0.68925   -1.09   0.2741
areamageroya:year2000      -0.63981    0.61667   -1.04   0.2995
areatanafjord:year2000      1.19387    0.49479    2.41   0.0158 *
areavarangerfjord:year2000  0.51074    0.47719    1.07   0.2845
areamageroya:year2001       0.70444    0.82036    0.86   0.3905
areatanafjord:year2001      0.90824    0.77685    1.17   0.2424
areavarangerfjord:year2001  0.59838    0.74738    0.80   0.4233
```

---

[15]Note that the Poisson (NB) and ZIP (ZINB) models are, in fact, nested (against popular belief). The Poisson (NB) and HP (HNB) may or may not be nested, depending on which binary zero hurdle is employed. The HNB (HP) and ZINB (ZIP) models may be nested for certain types of covariates.

```
Log(theta)                     -1.49866    0.23904   -6.27  3.6e-10 ***
Zero hurdle model coefficients (binomial with logit link):
                             Estimate Std. Error z value Pr(>|z|)
(Intercept)                   0.08526    0.29505    0.29    0.773
length                        0.00693    0.00465    1.49    0.136
areamageroya                 -1.32137    0.28526   -4.63  3.6e-06 ***
areatanafjord                -1.44918    0.24388   -5.94  2.8e-09 ***
areavarangerfjord             0.30073    0.27111    1.11    0.267
year2000                      0.39507    0.34382    1.15    0.251
year2001                     -2.65201    0.43340   -6.12  9.4e-10 ***
areamageroya:year2000        -0.08034    0.50797   -0.16    0.874
areatanafjord:year2000        0.87058    0.45027    1.93    0.053 .
areavarangerfjord:year2000    0.86462    0.59239    1.46    0.144
areamageroya:year2001         2.73749    0.53291    5.14  2.8e-07 ***
areatanafjord:year2001        2.71899    0.49949    5.44  5.2e-08 ***
areavarangerfjord:year2001    2.54144    0.51825    4.90  9.4e-07 ***
---
Signif. codes:  0 '***' 0.001 '**' 0.01 '*' 0.05 '.' 0.1 ' ' 1

Theta: count = 0.223
Number of iterations in BFGS optimization: 25
Log-likelihood: -2.44e+03 on 27 Df
```

From the above and from Figure 11.18, it appears that `length` is not important as a linear effect in the submodel for prevalence. A revised model excludes this from the zero formula.

```
> cp_hnb1 <- hurdle(intensity ~ length + area * year | area * year,
+                   data = CodParasites, dist = "negbin")
```

A likelihood-ratio test shows no advantage for the smaller model; however, Vuong's test leads to the conclusion that this reduced model is preferable:

```
> lrtest(cp_hnb, cp_hnb1)

Likelihood ratio test

Model 1: intensity ~ length + area * year
Model 2: intensity ~ length + area * year | area * year
  #Df LogLik Df Chisq Pr(>Chisq)
1  27  -2442
2  26  -2443 -1  2.23       0.14

> vuong(cp_hnb, cp_hnb1)

Vuong Non-Nested Hypothesis Test-Statistic:
(test-statistic is asymptotically distributed N(0,1) under the
 null that the models are indistinguishible)
-----------------------------------------------------------------
              Vuong z-statistic             H_A p-value
Raw               0.741801 model1 > model2   0.2291
AIC-corrected     0.076907 model1 > model2   0.4693
BIC-corrected    -1.612770 model2 > model1   0.0534
```

$\triangle$

11.5.1.2 Model interpretation: Effect plots

Interpreting these models from their coefficients is very difficult because an interaction is present and there are separate submodels for the zero and count components. This task is much easier with effects plots. The **effects** package has methods for any GLM, but cannot handle the extended forms of the zero-inflated and hurdle models.

When the same predictors are used in both submodels, and a standard GLM such as the negative-binomial provides a reasonable fit, you can use the standard **effects** functions to visualize the (total) expected count, which for the zeros would include both the extra zeros and those that derive from the count submodel. For visual interpretation, these will be sufficiently similar, even though the hurdle and zero-inflated models differ with respect to explaining overdispersion and/or excess zeros.

Alternatively, if you want to visualize and interpret the zero and nonzero components separately, perhaps with different predictors, you can fit the implied submodels separately, and then use **effects** functions for the effects in each. These ideas are illustrated in the next example.

EXAMPLE 11.12: Cod parasites

The expected counts for `intensity`, including both zero and positive counts, can be plotted using **effects** for the `cp_nb` NB model. Figure 11.21 gives some confidence that the fitted values are similar to those in the hurdle and zero-inflated versions.

We use `allEffects()` to calculate the effects for the high-order terms—the main effect of `length` and the interaction of `area` and `year`. These could be plotted together by plotting the resulting `eff.nb` object, but we plot them separately to control the plot details. In these plots, the argument `type="response"` gives plots on the response scale, and we use `ylim` to equate the ranges to make the plots directly comparable. The code below produces Figure 11.22.

```
> library(effects)
> eff.nb <- allEffects(cp_nb)
> plot(eff.nb[1], type = "response", ylim = c(0,30),
+      main  ="NB model: length effect")
>
> plot(eff.nb[2], type = "response", ylim = c(0,30),
+      multiline = TRUE, ci.style = "bars",
+      key.args = list(x = .05, y = .95, columns = 1),
+      colors = c("black", "red", "blue") ,
+      symbols = 15 : 17, cex = 2,
+      main = "NB model: area*year effect")
```

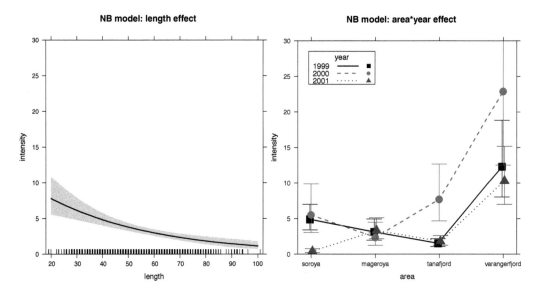

Figure 11.22: Effect plots for total intensity of parasites from the negative-binomial model.

This helps to interpret the nature of the area by year effect. The pattern of mean expected intensity of cod parasites is similar in 1999 and 2001, except for the Sørøya area. The results in year 2000 differ mainly in greater intensity in Tanafjord and Varangerfjord. Varangerfjord shows larger infection counts overall, but particularly in year 2000. The effect plot for length on this scale is roughly comparable to the variation in areas and years.

In this example, the submodels for zero and positive counts have substantively different interpretations. To visualize the fitted effects in these submodels using **effects**, first fit the equivalent submodels separately using GLM methods. The following models for `prevalence`, using the binomial family, and the positive counts for `intensity`, using `glm.nb()`, give similar fitted results to those obtained from the hurdle negative-binomial model, `cp_hnb` discussed earlier.

```
> cp_zero   <- glm(prevalence ~ length + area * year,
+                  data = CodParasites, family = binomial)
> cp_nzero <- glm.nb(intensity ~ length + area * year,
+                    data = CodParasites, subset = intensity > 0)
```

We could construct effect plots for each of these submodels, but interest here is largely on the binomial model for the zero counts, `cp_zero`. Effect plots for the terms in this model are shown in Figure 11.23. Again, we set the `ylim` values to equate the vertical ranges to make the plots comparable.

```
> eff.zero <- allEffects(cp_zero)
> plot(eff.zero[1], ylim=c(-2.5, 2.5),
+      main="Hurdle zero model: length effect")
>
> plot(eff.zero[2],  ylim=c(-2.5, 2.5),
+      multiline=TRUE,
+      key.args=list(x=.05, y=.95, columns=1),
+      colors=c("black", "red", "blue"),
+      symbols=15:17, cex=2,
+      main="Hurdle zero model: area*year effect")
```

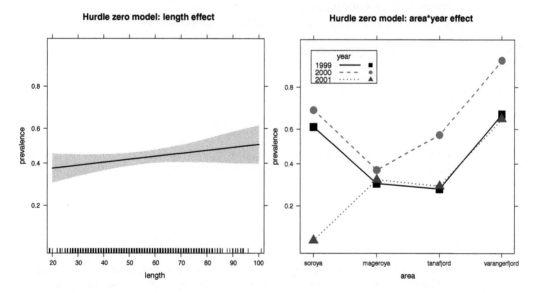

Figure 11.23: Effect plots for prevalence of parasites analogous to the hurdle negative-binomial model, fitted using a binomial GLM model.

The effect of `length` on prevalence is slightly increasing, but we saw earlier that this is not significant. For the area–year interaction, the three curves have similar shapes, except for the aberrant value for Sørøya in 2001 and the closeness of the values at Magerøya in all years. Overall, prevalence was highest in 2000, and also in the Varangerfjord samples.

△

11.5.2   Demand for medical care by the elderly

A large cross-sectional study was carried out by the U.S. National Medical Expenditure Survey (NMES) in 1987–1988 to assess the demand for medical care, as measured by the number of physician/non-physician office visits and the number of hospital outpatient visits to a physician/non-physician. The survey was based upon a representative national probability sample of the civilian non-institutionalized population and individuals admitted to long-term care facilities during 1987. A subsample of 4, 406 individuals aged 66 and over, all of whom are covered by Medicare, is contained in the *NMES1988* data set in the **AER** package. These data were previously analyzed by Deb and Trivedi (1997) and Zeileis et al. (2008), from which this account borrows. The objective of the study and these analyses is to create a descriptive, and hopefully predictive, model for the demand for medical care in this elderly population.

EXAMPLE 11.13: Demand for medical care

The potential response variables in the *NMES1988* data set form a 2×2 set of the combinations of *place of visit* (office vs. hospital) and (physician vs. non-physician) *practitioner*. Here, we focus on the highest total frequency variable `visits`, recording office visits to a physician. There are quite a few potential predictors, but here we consider only the following:

- `hospital`: number of hospital stays[16]
- `health`: a factor indicating self-perceived health status, with categories `"poor"`, `"average"` (reference category), `"excellent"`
- `chronic`: number of chronic conditions
- `gender`
- `school`: number of years of education
- `insurance`: a factor. Is the individual covered by private insurance?

For convenience, these variables are extracted to a reduced data set, `nmes`.

```
> data("NMES1988", package = "AER")
> nmes <- NMES1988[, c(1, 6:8, 13, 15, 18)]
```

A quick overview of the response variable, `visits`, is shown as simple (unbinned) histograms on the frequency and log(frequency) scales in Figure 11.24. The zero counts are not as extreme as we have seen in other examples. On the log scale, there is a small, but noticeable spike at 0, followed by a progressive, nearly linear decline, up to about 30 visits.

```
> plot(table(nmes$visits),
+      xlab = "Physician office visits", ylab = "Frequency")
> plot(log(table(nmes$visits)),
+      xlab = "Physician office visits", ylab = "log(Frequency)")
```

However, as a benchmark, without taking any predictors into account, there is very substantial overdispersion relative to a Poisson distribution, the variance being nearly 8 times the mean.

```
> with(nmes, c(mean = mean(visits),
+              var = var(visits),
+              ratio = var(visits) / mean(visits)))

   mean      var    ratio
 5.7744  45.6871   7.9120
```

---

[16]It is arguable that number of hospitalizations should be regarded as a dependent variable, reflecting another aspect of demand for medical care, rather than as a predictor. We include it here as a predictor to control for its relationship to the outcome `visits`.

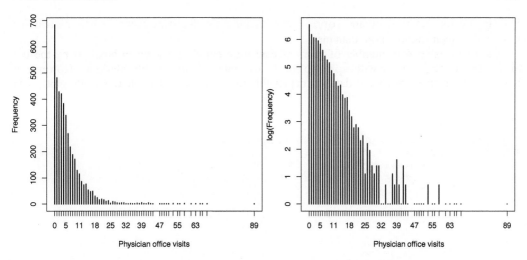

Figure 11.24: Frequency distributions of the number of physician office visits.

As before, it is useful to precede the formal analysis with a variety of exploratory plots. Figure 11.25 shows a few of these as boxplots, using `cutfac()` to make predictors discrete, and plotting `visits` on a log scale, started at 1. All of these show the expected relationships, e.g., number of office visits increases with numbers of chronic conditions and hospital stays, but decreases with better perceived health status.

```
> plot(log(visits + 1) ~ cutfac(chronic), data = nmes,
+      ylab = "Physician office visits (log scale)",
+      xlab = "Number of chronic conditions", main = "chronic")
> plot(log(visits + 1) ~ health, data = nmes, varwidth = TRUE,
+      ylab = "Physician office visits (log scale)",
+      xlab = "Self-perceived health status", main = "health")
> plot(log(visits + 1) ~ cutfac(hospital, c(0:2, 8)), data = nmes,
+      ylab = "Physician office visits (log scale)",
+      xlab = "Number of hospital stays", main = "hospital")
```

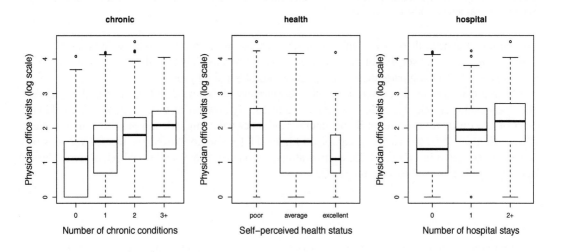

Figure 11.25: Number of physician office visits plotted against some of the predictors.

Similar plots for insurance and gender show that those with private insurance have more office visits and women slightly more than men.

The relationship with number of years of education could be shown in boxplots by the use of cutfac(school), or with spineplot() by making both variables discrete. However, it is more informative (shows the data) to depict this in a smoothed and jittered scatterplot, as in Figure 11.26.

```
> library(ggplot2)
> ggplot(nmes, aes(x = school, y = visits + 1)) +
+    geom_jitter(alpha = 0.25) +
+    stat_smooth(method = "loess", color = "red", fill = "red",
+                size = 1.5, alpha = 0.3) +
+    labs(x = "Number of years of education",
+         y = "log(Physician office visits + 1)") +
+    scale_y_log10(breaks = c(1, 2, 5, 10, 20, 50, 100))
```

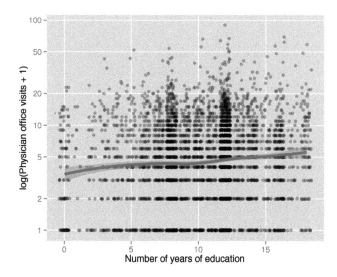

Figure 11.26: Jittered scatterplot of physician office visits against number of years of education, with nonparametric (loess) smooth.

As you might expect, there is a small but steady increase in mean office visits with years of education. It is somewhat surprising that there are quite a few individuals with 0 years of education; jittering also shows the greater density of observations at 8 and 12 years.

As in previous examples, a variety of other exploratory plots would be helpful in understanding the relationships among these variables *jointly*, particularly how office visits depends on combinations of two (or more) predictors. Some natural candidates would include mosaic and doubledecker plots (using cutfac(visits)), e.g., as in Figure 11.17, and conditional or faceted versions of the boxplots shown in Figure 11.25, each stratified by one (or more) additional predictors. These activities are left as exercises for the reader.

△

11.5.2.1   Fitting models

Most previous analyses of these data have focused on exploring and comparing different types of count data regression models. Deb and Trivedi (1997) compared the adequacy of fit of the negative-binomial, a hurdle NB, and models using finite mixtures of NB models. Zeileis et al. (2008) used this data to illustrate hurdle and zero-inflated models using the countreg package, while Cameron and Trivedi (1998, 2013) explored a variety of competing models, including 1- and 2-parameter NB models and *C*-component finite mixture models that can be thought of as generalizations of the 2-component models described in Section 11.4.

In most cases, the full set of available predictors was used, and models were compared using the standard methods for model selection: likelihood-ratio tests for nested models, AIC, BIC, and so forth. An exception is Kleiber and Zeileis (2014), who used a reduced set of predictors similar to those employed here, and illustrated the use of rootograms and plots of predicted values for visualizing and comparing fitted models.

This is where model comparison and selection for count data models (and other GLMs) adds another layer of complexity beyond what needs to be considered for classical (Gaussian) linear models, standard logistic regression models, and the special case of loglinear models treated earlier. Thus, when we consider and compare different distribution types or link functions, we have to be reasonably confident that the systematic part of the model has been correctly specified (as we noted in Section 11.3), and is the *same* in all competing models, so that any differences can be attributed to the distribution type. However, lack-of-fit may still arise because the systematic part of the model is incorrect.

In short, we cannot easily compare apples to oranges (different distributions with different regressors), but we also have to make sure we have a good apple to begin with. The important questions are:

- Have all important predictors and control variables been included in the model?
- Are quantitative predictors represented on the correct scale (via transformations or nonlinear terms) so their effects are reasonably additive for the linear predictor?
- Are there important interactions among the explanatory variables?

EXAMPLE 11.14: Demand for medical care

In this example, we start with the all main-effects model of the predictors in the nmes data, similar to that considered by Zeileis et al. (2008). We first fit the basic Poisson and NB models, as points of reference.

```
> nmes.pois <-    glm(visits ~ ., data = nmes, family = poisson)
> nmes.nbin <- glm.nb(visits ~ ., data = nmes)
```

A quick check with lmtest() shows that the NB model is clearly superior to the standard Poisson regression model as we expect (and also to the quasi-Poisson).

```
> library(lmtest)
> lrtest(nmes.pois, nmes.nbin)

Likelihood ratio test

Model 1: visits ~ hospital + health + chronic + gender + school + insurance
Model 2: visits ~ hospital + health + chronic + gender + school + insurance
  #Df LogLik Df Chisq Pr(>Chisq)
1   8 -17972
2   9 -12171  1 11602      <2e-16 ***
---
Signif. codes:  0 '***' 0.001 '**' 0.01 '*' 0.05 '.' 0.1 ' ' 1
```

The model summary for the NB model below shows the coefficients area all significant. Moreover, the signs of the coefficients are all as we would expect from our exploratory plots. For example, log(visits) increases with number of hospital stays, chronic conditions, and education, and is greater for females and those with private health insurance. So, what's not to like?

```
> summary(nmes.nbin)


Call:
glm.nb(formula = visits ~ ., data = nmes, init.theta = 1.206603534,
    link = log)

Deviance Residuals:
    Min      1Q  Median      3Q      Max
  -3.047  -0.995  -0.295   0.296    5.818

Coefficients:
                 Estimate Std. Error z value Pr(>|z|)
(Intercept)       0.92926    0.05459   17.02  < 2e-16 ***
hospital          0.21777    0.02018   10.79  < 2e-16 ***
healthpoor        0.30501    0.04851    6.29  3.2e-10 ***
healthexcellent  -0.34181    0.06092   -5.61  2.0e-08 ***
chronic           0.17492    0.01209   14.47  < 2e-16 ***
gendermale       -0.12649    0.03122   -4.05  5.1e-05 ***
school            0.02682    0.00439    6.10  1.0e-09 ***
insuranceyes      0.22440    0.03946    5.69  1.3e-08 ***
---
Signif. codes:  0 '***' 0.001 '**' 0.01 '*' 0.05 '.' 0.1 ' ' 1

(Dispersion parameter for Negative Binomial(1.2066) family taken to be 1)

    Null deviance: 5743.7  on 4405  degrees of freedom
Residual deviance: 5044.5  on 4398  degrees of freedom
AIC: 24359

Number of Fisher Scoring iterations: 1


              Theta:  1.2066
          Std. Err.:  0.0336

 2 x log-likelihood:  -24341.1070
```

This all-main-effects model is relatively simple to interpret, but a more important question is whether it adequately explains the relations of the predictors to the outcome, visits.

Significant interactions among the predictors could substantially change the interpretation of the model, and in the end, could affect policy recommendations based on this analysis. This question turns out to be far more interesting and important than the subtle differences among models for handling overdispersion and zero counts.

One simple way to consider whether there are important interactions among the predictors that better explain patient visits is to get simple tests of the additional contribution of each two-way (or higher-way) interaction using the add1() function. The formula argument in the call below specifies to test the addition of all two-way terms.

```
> add1(nmes.nbin, . ~ .^2, test = "Chisq")

Single term additions

Model:
visits ~ hospital + health + chronic + gender + school + insurance
                 Df Deviance    AIC  LRT Pr(>Chi)
<none>                  5045  24357
hospital:health   2     5025  24341 19.9  4.7e-05 ***
hospital:chronic  1     5009  24324 35.2  3.0e-09 ***
hospital:gender   1     5044  24358  0.8   0.3650
hospital:school   1     5041  24355  4.0   0.0453 *
```

```
hospital:insurance   1     5036 24351  8.0    0.0046 **
health:chronic       2     5005 24322 39.5    2.6e-09 ***
health:gender        2     5040 24357  4.3    0.1172
health:school        2     5030 24347 14.3    0.0008 ***
health:insurance     2     5032 24348 12.9    0.0016 **
chronic:gender       1     5045 24359  0.0    0.9008
chronic:school       1     5043 24357  1.9    0.1705
chronic:insurance    1     5039 24354  5.1    0.0246 *
gender:school        1     5040 24354  4.8    0.0290 *
gender:insurance     1     5042 24357  2.5    0.1169
school:insurance     1     5037 24352  7.2    0.0072 **
---
Signif. codes:  0 '***' 0.001 '**' 0.01 '*' 0.05 '.' 0.1 ' ' 1
```

From this, we decide to add all two-way interactions among `health`, `hosp`, and `numchron`, and also the two-way interaction `health:school`. Other significant interactions could also be explored, but we don't do this here.

```
> nmes.nbin2 <- update(nmes.nbin,
+                      . ~ . + (health + chronic + hospital)^2
+                      + health : school)
```

This model clearly fits much better than the main effects model, as shown by a likelihood ratio test. The same conclusion would result from `anova()`.

```
> lrtest(nmes.nbin, nmes.nbin2)

Likelihood ratio test

Model 1: visits ~ hospital + health + chronic + gender + school + insurance
Model 2: visits ~ hospital + health + chronic + gender + school + insurance
    + health:chronic + hospital:health + hospital:chronic + health:school
  #Df LogLik Df Chisq Pr(>Chisq)
1   9 -12171
2  16 -12133  7  74.3      2e-13 ***
---
Signif. codes:  0 '***' 0.001 '**' 0.01 '*' 0.05 '.' 0.1 ' ' 1
```

△

11.5.2.2 Model interpretation: Effect plots

Complex models with more than a few predictors are difficult to understand and explain, even more so when there are interactions among the predictors. As we have noted previously, effect plots (Fox, 1987, Fox and Andersen, 2006) provide a ready solution.

They have the advantage that each plot shows the correct *partial* relation between the response and the variables in the term shown, controlling (adjusting) for all other variables in the model, as opposed to *marginal* plots that ignore all other variables. From these, it is possible to read an interpretation of a given model effect directly from the effect plot graphs, knowing that all variables not shown in a given graph have been controlled (adjusted for) by setting them equal to average or typical values.

A disadvantage is that these plots show only the predicted (fitted) effects under the *given model* (and not the *data*). If relationships of the response to predictors are nonlinear, or important interactions are not included in the model, you won't see this in an effect plot. We illustrate this point using the results of the main effect NB model, `nmes.nbin`, as shown in Figure 11.27.

EXAMPLE 11.15: Demand for medical care

```
> library(effects)
> plot(allEffects(nmes.nbin), ylab = "Office visits")
```

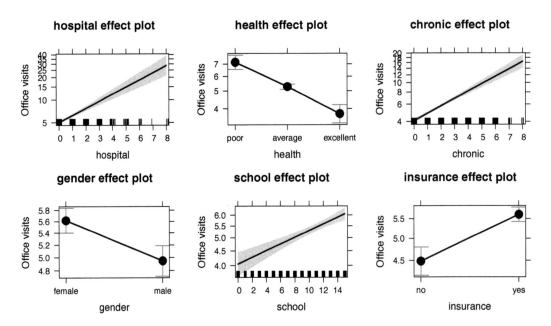

Figure 11.27: Effect plots for the main effects of each predictor in the negative binomial model nmes.nbin.

All of these panels show the expected relations of the predictors to the visits response, and the confidence bands and error bars provide visual tests of the sizes of differences. But they don't tell the full story, because the presence of an important interaction (such as health:chronic) means that the effect of one predictor (health) differs over the values of the other (chronic).

We can see this clearly in effect plots for the model nmes.nbin2 with interactions. For display purposes, it is convenient here to calculate the fitted effects for model terms over a smaller but representative subset of the levels of the integer-valued predictors, using the xlevels= argument to allEffects().

```
> eff_nbin2 <- allEffects(nmes.nbin2,
+     xlevels = list(hospital = c(0 : 3, 6, 8),
+     chronic = c(0:3, 6, 8),
+     school = seq(0, 20, 5)))
```

The result of allEffects(), eff_nbin2, is a **"efflist"** object, a list of effects for each *high-order term* in the model. Note that only the terms gender and insurance, not involved in any interaction, appear as main effects here.

```
> names(eff_nbin2)

[1] "gender"          "insurance"        "health:chronic"
[4] "hospital:health" "hospital:chronic" "health:school"
```

Plotting the entire "efflist" object gives a collection of plots, one for each high-order term, as in Figure 11.27, and is handy for a first look. However, the plot() methods for effects objects offer greater flexibility when you plot terms individually using additional options. For example, Figure 11.28 plots the effect for the interaction of health and number of chronic conditions with a few optional arguments. See help(plot.eff, package="effects") for the available options.

```
> plot(eff_nbin2, "health:chronic", layout = c(3, 1),
+       ylab = "Office visits", colors = "blue")
```

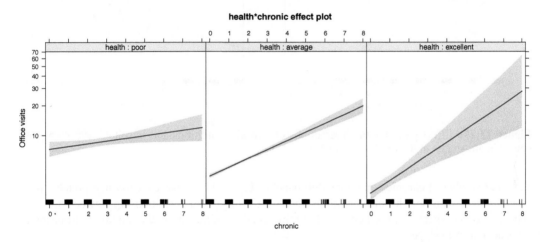

Figure 11.28: Effect plot for the interaction of health and number of chronic conditions in the model nmes.nbin2.

The default style shown in Figure 11.28 is a conditional or faceted plot, graphing the response against the X variable with the greatest number of levels, with separate panels for the levels of the other predictor. Alternatively, the effects for a given term can be shown overlaid in a single plot, using the multiline=TRUE argument, as shown in Figure 11.29 for the two interactions involving health status. Not only is this style more compact, but it also makes direct comparison of the trends for the other variable easier.

```
> plot(eff_nbin2,
+       "health:chronic", multiline = TRUE, ci.style = "bands",
+       ylab = "Office visits", xlab ="# Chronic conditions",
+       key.args = list(x = 0.05, y = .80, corner = c(0, 0), columns = 1))
>
> plot(eff_nbin2,
+       "hospital:health", multiline = TRUE, ci.style = "bands",
+       ylab = "Office visits", xlab = "Hospital stays",
+       key.args = list(x = 0.05, y = .80, corner = c(0, 0), columns = 1))
```

From both Figure 11.28 and the left panel of Figure 11.29, it can be seen that for people with poor health status, the relationship of chronic conditions to office visits is relatively flat. For those who view their health status as excellent, their use of office visits is much more strongly related to their number of chronic conditions.

The interaction of perceived health status with number of hospital stays (right panel of Figure 11.29) shows that the difference in office visits according to health status is mainly important only for those with 0 or 1 hospital stays.

The remaining two interaction effects are plotted in Figure 11.30. The interaction of hospital

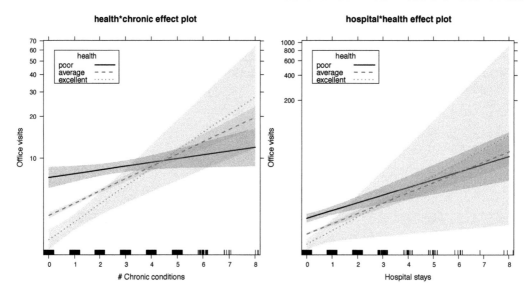

Figure 11.29: Effect plots for the interactions of chronic conditions and hospital stays with perceived health status in the model nmes.nbin2.

stays and number of chronic conditions (left panel of Figure 11.30) has a clearly interpretable pattern: for those with few chronic conditions, there is a strong positive relationship between hospital stays and office visits. As the number of chronic conditions increases, the relation with hospital stays decreases in slope.

```
> plot(eff_nbin2, "hospital:chronic", multiline = TRUE, ci.style = "bands",
+       ylab = "Office visits", xlab = "Hospital stays",
+       key.args = list(x = 0.05, y = .70, corner = c(0, 0), columns = 1))
>
> plot(eff_nbin2, "health:school", multiline = TRUE, ci.style = "bands",
+       ylab = "Office visits", xlab = "Years of education",
+       key.args = list(x = 0.65, y = .1, corner = c(0, 0), columns = 1))
```

Finally, the interaction of health:school is shown in the right panel of Figure 11.30. It can be readily seen that for those of poor health, office visits are uniformly high, and have no relation to years of education. Among those of average or excellent health, office visits increase with years of education in roughly similar ways.   △

11.5.2.3   More model wrinkles: Nonlinear terms

Effect plots such as those above are much easier to interpret than tables of fitted coefficients. However, we emphasize that these only reflect the *fitted model*. It might be that the effects of both hospital and chronic are nonlinear (on the scale of log(visits)). In assessing this question, we increase the complexity of model and try to balance parsimony against goodness-of-fit, but also assure that the model retains a sensible interpretation.

EXAMPLE 11.16: Demand for medical care

The simplest approach is to use poly(hosp,2) and/or poly(numchron,2) to add possible quadratic (or higher power) relations to the model nmes.nbin2 containing interactions studied above. A slightly more complex model could use poly(hosp, numchron, degree=2) for a response-surface model in these variables. A significantly improved fit of such a model is evidence for nonlinearity of the effects of these predictors. This is easily done using update():

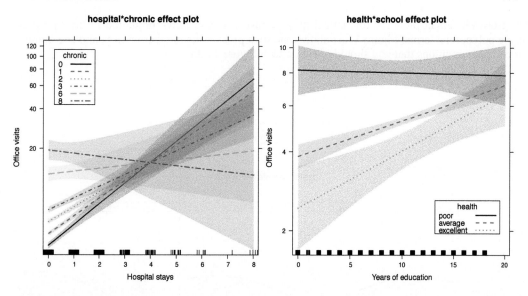

Figure 11.30: Effect plots for the interactions of chronic conditions and hospital stays and for health status with years of education in the model nmes.nbin2.

```
> nmes.nbin3 <- update(nmes.nbin2, . ~ . + I(chronic^2) + I(hospital^2))
```

This model is equivalent to the long-form version below:

```
> nmes.nbin3 <- glm.nb(visits ~ poly(hospital, 2) + poly(chronic, 2) +
+                      insurance + school + gender +
+                      (health + chronic + hospital)^2 + health : school,
+                      data = nmes)
```

Comparing these models using anova(), we see that there is a substantial improvement in the model fit by including these nonlinear terms. The quadratic model also fits best by AIC and BIC.

```
> ret <- anova(nmes.nbin, nmes.nbin2, nmes.nbin3)
> ret$Model <- c("nmes.nbin", "nmes.nbin2", "nmes.nbin3")
> ret

Likelihood ratio tests of Negative Binomial Models

Response: visits
        Model  theta Resid. df  2 x log-lik.   Test  df
1  nmes.nbin 1.2066      4398       -24341
2 nmes.nbin2 1.2354      4391       -24267 1 vs 2    7
3 nmes.nbin3 1.2446      4389       -24245 2 vs 3    2
   LR stat.    Pr(Chi)
1
2   74.307 1.9829e-13
3   22.278 1.4537e-05

> LRstats(nmes.nbin, nmes.nbin2, nmes.nbin3)

Likelihood summary table:
             AIC   BIC LR Chisq   Df Pr(>Chisq)
nmes.nbin  24359 24417     5045 4398    2.2e-11 ***
nmes.nbin2 24299 24401     5047 4391    1.1e-11 ***
nmes.nbin3 24281 24396     5049 4389    8.5e-12 ***
---
Signif. codes:  0 '***' 0.001 '**' 0.01 '*' 0.05 '.' 0.1 ' ' 1
```

However, effect plots for this model quickly reveal a *substantive* limitation of this approach using polynomial terms. Figure 11.31 shows one such plot for the interaction of health and number of chronic conditions that you should compare with Figure 11.28.

```
> eff_nbin3 <- allEffects(nmes.nbin3,
+    xlevels = list(hospital = c(0 : 3, 6, 8),
+                        chronic = c (0 : 3, 6, 8),
+                        school = seq(0, 20, 5)))
> plot(eff_nbin3, "health : chronic", layout = c(3, 1))
```

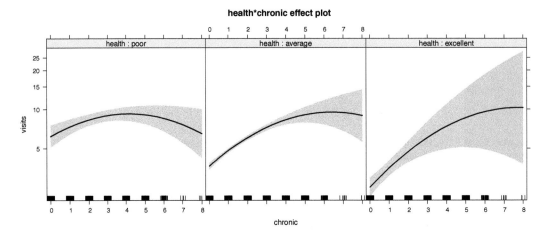

Figure 11.31: Effect plot for the interaction of health and number of chronic conditions in the quadratic model nmes.nbin3.

The quadratic fits for each level of health in Figure 11.31 imply that office visits increase with chronic conditions up to a point and then decrease—with a quadratic, what goes up must come down, the same way it went up! This makes no sense here, particularly for those with poor health status. As well, the confidence bands in this figure are uncomfortably wide, particularly at higher levels of chronic conditions, compared to those in Figure 11.28. The quadratic model is thus preferable statistically and descriptively, but serves less well for explanatory, substantive, and predictive goals.

An alternative approach to handle nonlinearity is to use regression splines (as in Example 7.9) or a ***generalized additive model*** (Hastie and Tibshirani, 1990) for these terms. The latter specifies the linear predictor as a sum of smooth functions,

$$g(\mathcal{E}(y)) = \beta_0 + f_1(x_1) + f_2(x_2) + \cdots + f_m(x_m) .$$

where each $f_j(x_j)$ may be a function with a specified parametric form (for example, a polynomial) or may be specified non-parametrically, simply as "smooth functions," to be estimated by non-parametric means.

In R, a very general implementation of the generalized additive model (GAM) is provided by gam() in the mgcv (Wood, 2015) package and described in detail by Wood (2006). Particular features of the package are facilities for automatic smoothness selection (Wood, 2004), and the provision of a variety of smooths of more than one variable. This example just scratches the surface of GAM methodology.

In the context of the NB model we are considering here, the analog of model nmes.nbin3 fitted using gam() is nmes.gamnb shown below. The negative-binomial distribution can be specified using family=nb() when the parameter θ is also estimated from the data (as with `glm.nb())`, or `family=negbin(theta)` when θ is taken as fixed, for example using the value `theta=1.24` available from models `nmes.nbin2`, and `nmes.nbin3`.

```
> library(mgcv)
> nmes.gamnb <- gam(visits ~ s(hospital, k = 3) + s(chronic, k = 3) +
+                            insurance + school + gender +
+                            (health + chronic + hospital)^2 +
+                            health : school,
+               data = nmes, family = nb())
```

The key feature here is the specification of the smooth terms for `s(hospital, k=3)` and `s(chronic, k=3)`, where `k=3` specifies the dimension of the basis used to represent the smooth term. There are many other possibilities with `gam()`, but these are beyond the scope of this example.

We could again visualize the predicted values from this model using effect plots. However, a different approach is to visualize the *fitted surface* in 3D, using a range of values for two of the predictors, and controlling for the others.

The rsm package provides extensions of the standard `contour()`, `image()` and `persp()` functions for this purpose. The package provides S3 methods (e.g., `persp.lm()`) for `"lm"` objects, or classes (such as `"negbin"` and `"glm"`) that inherit methods from `lm`. The calculation of fitted values in these plots use the applicable `predict()` method for the model object. As in effect plots, the remaining predictors are controlled at their average values (or other values specified in the `at` argument).

Two such plots are shown in Figure 11.32. The left panel shows the interaction of hospital stays and chronic conditions, included in the model with smoothed terms for their main effects. The right panel shows the joint effects of years of education and chronic conditions on office visits, but there is no interaction of these variables in the GAM model `nmes.gamnb`. These plots use `rainbow()` colors (in HCL space) to depict the predicted values of office visits. Contours of these values are projected into the bottom or top plane with corresponding color coding.[17]

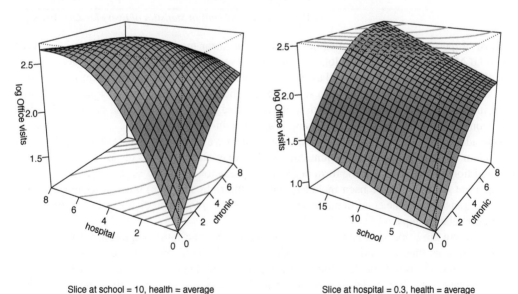

Slice at school = 10, health = average   Slice at hospital = 0.3, health = average

Figure 11.32: Fitted response surfaces for the relationships among chronic conditions, number of hospital stays, and years of education to office visits in the generalized additive model, nmes.gamnb.

---

[17]The vignette `vignette("rsm-plots", package="rsm")` illustrates some of these options.

```
> library(rsm)
> library(colorspace)
> persp(nmes.gamnb, hospital ~ chronic, zlab = "log Office visits",
+    col = rainbow_hcl(30), contour = list(col = "colors", lwd = 2),
+    at = list(school = 10, health = "average"), theta = -60)
>
> persp(nmes.gamnb, school ~ chronic, zlab = "log Office visits",
+    col = rainbow_hcl(30),
+    contour = list(col = "colors", lwd = 2, z = "top"),
+    at = list(hospital = 0.3, health = "average"), theta = -60)
```

A simple, credible interpretation of the plot in the left panel is that office visits rise steeply initially with both hospital stays and number of chronic conditions, and then level off. For those with no chronic conditions, the effect of hospital stays rises to a higher level compared with the effect of chronic conditions among those who have had no hospital stays. However, as we have seen before, the data is quite thin at the upper end of these predictors, and this plot does not show model uncertainty.

The right panel of Figure 11.32 illustrates the form of model predictions for a term where one variable (`chronic`) is treated as possibly nonlinear using a smooth `s()` effect, the other is treated as linear (`school`), and no interaction between these is included in the model. At each fixed value of `chronic`, increasing education results in greater office visits. At each fixed value of `school`, the number of chronic conditions shows a steep increase in office visits initially, leveling off toward higher levels, but these all have the same predicted shape.

$\triangle$

11.6   Diagnostic plots for model checking

> Models, of course, are never true, but fortunately it is only necessary that they be useful.

> G. E. P. Box, *Some Problems of Statistics of Everyday Life*, 1979, p. 2

Most of the model diagnostic methods for classical linear models extend in a relatively direct way to GLMs. These include (a) plots of residuals of various types, (b) diagnostic measures and plots of leverage and influence, as well as some (c) more specialized plots (component-plus-residual plots, added-variable plots) designed to show the specific contribution of a given predictor among others in a linear model. These methods were described in Section 7.5 in the context of logistic regression, and most of that discussion is applicable here in wider GLM class.

One additional complication here is that in any GLM we are specifying: (a) the distribution of the random component, which for count data models may also involve a dispersion parameter or other additional parameters; (b) the form of the linear predictor, $\eta = x^{\mathsf{T}}\beta = \beta_0 + \beta_1 x_1 + \cdots$, where all important regressors have been included, and on the right scale; (c) the correct link function, $g(\mu) = \eta$ transforming the conditional mean of the response y to the predictor variables where they have linear relationships.

Thus, there are a lot of things that could go wrong, but the famous quote from George Box should remind us that all models are approximate, and the goal for model diagnosis should be an adequate model, useful for description, estimation, or prediction as the case may be. What is most important is that our models should not be misleadingly wrong, that is, they should not affect substantive conclusions or interpretation.

11.6.1   Diagnostic measures and residuals for GLMs

Estimation of GLMs by maximum likelihood uses an iterative weighted least squares (IWLS) algorithm, and many of the diagnostic measures for these models are close counterparts of their forms for classical linear models. Roughly speaking, these follow from replacing y and $\widehat{y}$ in least squares diagnostics by a "working response" and $\widehat{\eta}$, replacing the residual variance $\widehat{\sigma}^2$ by $\widehat{\phi}$, and using a weighted form of the Hat matrix.

11.6.1.1 Leverage

Hat values, h_i, measuring **leverage** or the potential of an observation to affect the fitted model, are defined as the diagonal elements of the hat matrix H, using the weight matrix W from the final IWLS iteration. This has the same form as in a weighted least squares regression using a fixed W matrix:

$$H = W^{1/2}X(X^{\mathsf{T}}WX)^{-1}X^{\mathsf{T}}W^{1/2} .$$

In contrast to OLS, the weights depend on the y values as well as the X values, so high leverage observations do not necessarily reflect only unusualness in the space of the predictors.

11.6.1.2 Residuals

Several types of residuals can be defined starting from the goodness-of-fit measures discussed in Section 11.1.3. The **raw residual** or **response residual** is simply the difference $y_i - \widehat{\mu}_i$ between the observed response y_i and the estimated mean, $\widehat{\mu} = g^{-1}(\widehat{\eta}_i) = g^{-1}(x_i^{\mathsf{T}}\widehat{\beta})$.

From this, the **Pearson residual** is defined as

$$r_i^P = \frac{y_i - \widehat{\mu}_i}{\sqrt{\widehat{\mathcal{V}}(y_i)}} \tag{11.11}$$

and the **deviance residual** is defined as the signed square root of the contribution of observation i to the deviance in Eqn. (11.4).

$$r_i^D = \text{sign}(y_i - \widehat{\mu}_i)\sqrt{d_i} . \tag{11.12}$$

The Pearson and deviance residuals do not account for dispersion or for differential leverage (which makes their variance smaller), so **standardized residuals** (sometimes called *scaled* residuals) can be calculated as

$$\widetilde{r}_i^P = \frac{r_i^P}{\sqrt{\widehat{\phi}(1 - h_i)}} . \tag{11.13}$$

$$\widetilde{r}_i^D = \frac{r_i^D}{\sqrt{\widehat{\phi}(1 - h_i)}} . \tag{11.14}$$

These have approximate standard normal $\mathcal{N}(0, 1)$ distributions, and will generally have quite similar values (except for small values in $\widehat{\mu}$). Consequently, convenient thresholds like $|\widetilde{r}_i| > 2$ or $|\widetilde{r}_i| > 4$ are useful for identifying unusually large residuals.

Finally, the **studentized residual** (or *deletion* residual) gives the standardized residual that would result from omitting each observation in turn and calculating the change in the deviance. Calculating these exactly would require refitting the model n times, but an approximation is

$$\widetilde{r}_i^S = \text{sign}(y_i - \widehat{\mu}_i)\sqrt{(\widetilde{r}_i^D)^2 + (\widetilde{r}_i^P)^2 h_i/(1 - h_i)} . \tag{11.15}$$

From the theory of classical linear models, these provide formal outlier tests for individual observations (Fox, 2008, Section 11.3) as a *mean-shift* outlier model that dedicates an additional parameter to fit observation i exactly. To correct for multiple testing and a focus on the largest absolute residuals, it is common to apply a Bonferroni adjustment to the p-values of these tests, multiplying them by n.

For a class "glm" object, the function `residuals(object, type)` returns the unstandard-ized residuals for `type="pearson"` or `type="deviance"`.[18] The standardized versions are obtained using `rstandard()`, again with a `type` argument for the Pearson or deviance flavor. `rstudent()` calculates the studentized deletion residuals.

11.6.1.3 Influence

As discussed in Section 7.5 in the context of logistic regression, influence measures attempt to evaluate the effect that an observation exerts on the parameters, fitted values, or goodness-of-fit statistics by comparing a statistic calculated for all the data with the value obtained omitting each observation in turn. Again, approximations are used to estimate these effects without laboriously refitting the model n times.

Overall measures of influence include

- Cook's distance (Eqn. (7.10)), a squared measure of the difference $\widehat{\beta} - \widehat{\beta}_{(-i)}$ in all p coefficients in the model. The approximation used in `cooks.distance()` is

$$C_i = \frac{\widetilde{r}_i h_i}{\widehat{\phi} \, p \, (1 - h_i)} \, .$$

  This follows Williams (1987), but scales the result by the estimated dispersion $\widehat{\phi}$ as an approxi-mate $F_{p,n-p}$ statistic rather than χ_p^2.
- DFFITS, the standardized signed measure of the difference of the fitted value $\widehat{\mu}_i$ using all the data and the value $\widehat{\mu}_{(-i)}$ omitting observation i.

EXAMPLE 11.17: Publications of PhD candidates

For models that inherit methods from the "glm" class (including NB models fit using `glm.nb()`), the simplest initial diagnostic plots are provided by the `plot()` method. Figure 11.33 shows the default *regression quartet* of plots for the negative-binomial model phd.nbin examined in earlier examples. By default, the `id.n=3` most noteworthy observations are labeled with their row names from the original data set.

```
> plot(phd.nbin)
```

The plot of residuals against predicted values in the upper left panel of Figure 11.33 should show no overall systematic trend for a well-fitting model. The smoothed loess curve in red suggests that this is not the case.

Several functions in the car package make these plots more flexibly and with greater control of the details. Figure 11.34 shows the plot of residuals against predicted values two ways. The right panel explains the peculiar pattern of diagonal band of points. These correspond to the different discrete values of the response variable, number of articles published.

```
> library(car)
> residualPlot(phd.nbin, type = "rstandard", col.smooth = "red", id.n = 3)
> residualPlot(phd.nbin, type = "rstandard",
+              groups = PhdPubs$articles, key = FALSE, linear = FALSE,
+              smoother = NULL)
```

Other useful plots show the residuals against each predictor. For a good-fitting model, the average residual should not vary systematically with the predictor. As shown in Figure 11.35, `residualPlot()` draws a lowess smooth, and also computes a curvature test for each of the plots by adding a quadratic term and testing the quadratic to be zero.

---

[18]Other types include raw response residuals (`type="response"`), working residuals (`type="working"`), and par-tial residuals (`type="partial"`).

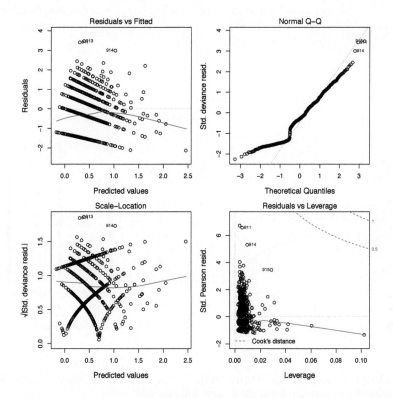

Figure 11.33: Default diagnostic plots for the negative-binomial model fit to the PhdPubs data.

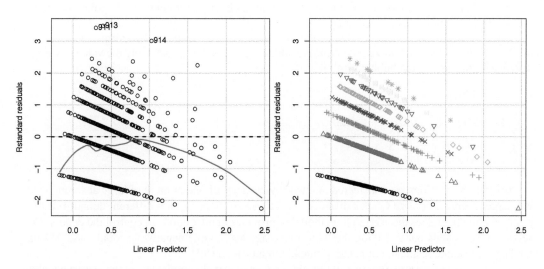

Figure 11.34: Plots of residuals against the linear predictor using residualPlot(). The right panel shows that the diagonal bands correspond to different values of the discrete response.

```
> residualPlot(phd.nbin, "mentor", type = "rstudent",
+              quadratic = TRUE, col.smooth = "red", col.quad = "blue",
+              id.n = 3)
> residualPlot(phd.nbin, "phdprestige", type = "rstudent",
+              quadratic = TRUE, col.smooth = "red", col.quad = "blue",
+              id.n = 3)
```

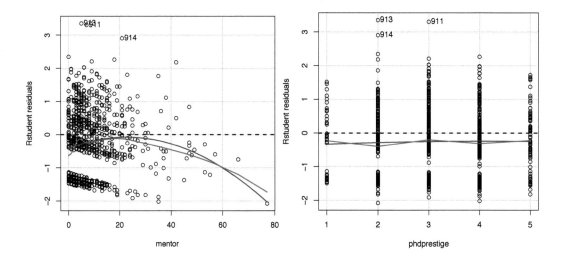

Figure 11.35: Plots of residuals against two predictors in the phd.nbin model. Such plots should show no evidence of a systematic trend for a good-fitting model.

In the plot at the left for number of articles by the student's mentor, the curvature is quite pronounced: at high values of mentor, nearly all of the residuals are negative, these students publishing fewer articles than would be expected. This would indicate a problem in the scale for mentor if there were more observations at the high end; but only about 1.5% points occur for mentor>45, so this can be discounted.

Figure 11.36 gives a better version of the influence plot shown in the lower right panel of Figure 11.33. This plots studentized (deletion) residuals against leverage, showing the value of Cook's distance by the area of the bubble symbol.

```
> influencePlot(phd.nbin)

      StudRes       Hat    CookD
328  -2.0762  0.1023449  0.18325
913   3.3488  0.0036473  0.16652
915   2.1810  0.0287496  0.24345
```

Several observations are considered noteworthy, because of one or more of large absolute residual, large leverage, or large Cook's distance. influencePlot() uses different default rules for point labeling than does the plot() method, but provides many options to control the details. Observation 328 stands out as having the largest leverage and a large negative residual; case 913 has the largest absolute residual, but is less influential than case 915.[19]

The outlierTest() function in car gives a formal test of significance of the largest absolute studentized residuals, with a Bonferroni-adjusted p-value accounting for choosing the largest values

---

[19]The higher case numbers appear in these plots and diagnostics because the data set *PhdPubs* had been sorted by the response, articles.

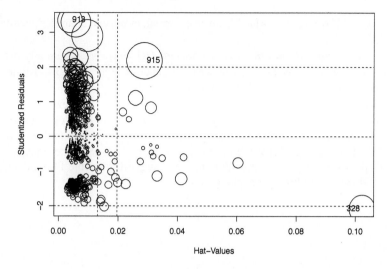

Figure 11.36: Influence plot showing leverage, studentized residuals, and Cook's distances for the negative-binomial model fit to the PhdPubs data. Conventional cutoffs for studentized residuals are shown by dashed horizontal lines at ± 2; vertical lines show 2 and 3 times the average hat-value.

among n such tests. Individually, case 913 is extreme, but it is not at all extreme among $n = 915$ such tests, each using $\alpha = .05$.

```
> outlierTest(phd.nbin)


No Studentized residuals with Bonferonni p < 0.05
Largest |rstudent|:
    rstudent unadjusted p-value Bonferonni p
913  3.3488         0.00084491      0.77309
```

This example started with the negative-binomial model, the best-fitting from the previous examples. It highlighted a few features of the data not seen previously and worth considering, but doesn't seriously challenge the substantive interpretation of the model. This is what we hope for from model diagnostic plots.

$\triangle$

11.6.2 Quantile–quantile and half-normal plots

As we noted above, in theory the standardized and studentized Pearson and deviance residuals have approximate standard normal $\mathcal{N}(0, 1)$ distributions (in large samples) when the fitted model is correct. This suggests a plot of the sorted residuals, $r_{(i)}$, against the corresponding expected values, $z_{(i)}$, an equal-sized sample of size n would have in a normal distribution.[20]

If the distribution of the residuals is approximately normal, the points $(r_{(i)}, z_{(i)})$ should lie along a line with unit slope through the origin; systematic or individual departure from this line signals a potential violation of assumptions. The expected values are typically calculated as $z_{(i)} =$

---

[20]The subscripted notation $r_{(i)}$ (and $z_{(i)}$) denotes an *order statistic*, i.e., the ith largest value in a set arranged in increasing order.

$\Phi^{-1}\{(i-\frac{3}{8})/(n+\frac{1}{4})\}$, where $\Phi^{-1}(\bullet)$ is the inverse normal, or normal quantile function, `qnorm()` in R.

Such plots, called ***normal quantile plots*** or ***normal QQ plots***, are commonly used for GLMs with a quantitative response variable. The upper right panel of Figure 11.33 illustrates the form of such plots produced by `plot()` for a `"glm"` object.

One difficulty with the default plots is that it is hard to tell to what extent the points deviate from the unit line because there is no visual reference for the line or envelope to indicate expected variability about that line. This problem is easily remedied using `qqPlot()` from car.

Figure 11.37 shows the result for the model `phd.nbin`. The envelope lines used here are at the quartiles of the expected normal distribution. They suggest a terrible fit, but, surprisingly, the largest three residuals are within the envelope.

```
> qqPlot(rstudent(phd.nbin), id.n = 3,
+          xlab = "Normal quantiles", ylab = "Studentized residuals")

913 911 914
915 914 913
```

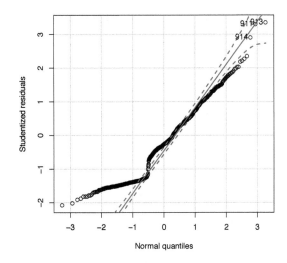

Figure 11.37: Normal QQ plot of the studentized residuals from the NB model for the PhdPubs data. The normal-theory reference line and confidence envelope are misleading here.

For GLMs with discrete responses, such plots are often disappointing, even with a reasonably good-fitting model, because: (a) possible outliers can appear at both the lower and upper ends of the distribution of residuals; (b) the theoretical normal distribution used to derive the envelope may not be well approximated in a given model.

Atkinson (1981, 1987) suggested a more robust and useful version of these QQ plots: half normal plots, with simulated confidence envelopes. The essential ideas are:

- Model departures and outliers are often easier to see for discrete data when the *absolute values* of residuals are plotted, because large positive and negative values are sorted together. This gives the ***half-normal plot***, in which the absolute values of residuals, arranged in increasing order, $|r|_{(i)}$, are plotted against $|z|_{(i)} = \Phi^{-1}\{(n+i-\frac{1}{8})/(2n+\frac{1}{2})\}$. All outliers will then appear in the upper right of such a plot, as points separated from the trend of the remaining cells.

- The normal-theory reference line, $|r|_{(i)} = |z|_{(i)}$ and the normal-theory confidence envelope can be replaced by simulating residuals from the assumed distribution, that need not be normal. The reference line is taken as the mean of S simulations and the envelope with $1 - \alpha$ coverage is taken as the $(\alpha/2, 1 - \alpha/2)$ quantiles of their values.
- Specifically, for a GLM, S sets of random observations $\boldsymbol{y}_j, j = 1, 2, \ldots S$ are generated from the fitted model, each with mean $\widehat{\boldsymbol{\mu}}$, the fitted values under the model and with the *same* distribution. In R, this is readily accomplished using the generic `simulate()` function; the random variation around $\widehat{\mu}$ uses `rnorm()`, `rpois()`, `rnegbin()`, etc., as appropriate for the family of the model.
- The same model is then fit to each simulated $\boldsymbol{y}_j$, giving a new set of residuals for each simulation. Sorting their absolute values then gives the simulation distribution used as reference for the observed residuals.

At the time of writing there is no fully general implementation of these plots in R, but the technique is not too difficult and is sufficiently useful to illustrate here.

EXAMPLE 11.18: Publications of PhD candidates
First, calculate the sorted absolute values of the residuals $|r|_{(i)}$ and their expected normal values, $|z|_{(i)}$. The basic plot will be `plot(expected, observed)`.

```
> observed <- sort(abs(rstudent(phd.nbin)))
> n <- length(observed)
> expected <- qnorm((1:n + n - 1/8) / (2*n + 1/2))
```

Then, use `simulate()` to generate $S = 100$ simulated response vectors around the fitted values in the model. Here this uses the negative-binomial random number generator (`rnegbin()`) with the same dispersion value ($\widehat{\theta} = 2.267$) estimated in the model. The result, called `sims` here, is a data frame of $n = 915$ rows and $S = 100$ columns, named `sim_1`, `sim_2`, `....`

```
> S <- 100
> sims <- simulate(phd.nbin, nsim = S)
> simdat <- cbind(PhdPubs, sims)
```

The next step is computationally intensive, because we have to fit the NB model $S = 100$ times, and a little bit tricky, because we need to use the same model formula as the original, but with the simulated $\boldsymbol{y}$. We first define a function `resids` to do this for a given y, and then use a loop to calculate them all. To save computing time, the coefficients from the `phd.nbin` model are used as starting values.

```
> # calculate residuals for one simulated data set
> resids <- function(y)
+    rstudent(glm.nb(y ~ female + married + kid5 + phdprestige + mentor,
+                data=simdat, start=coef(phd.nbin)))
> # do them all ...
> simres <- matrix(0, nrow(simdat), S)
> for(i in 1:S) {
+        simres[,i] <- sort(abs(resids(dat[,paste("sim", i, sep="_")])))
+ }
```

We can then use `apply()` to compute the summary measures defining the center and limits for the simulated confidence interval.

```
> envelope <- 0.95
> mean <- apply(simres, 1, mean)
> lower <- apply(simres, 1, quantile, prob = (1 - envelope) / 2)
> upper <- apply(simres, 1, quantile, prob = (1 + envelope) / 2)
```

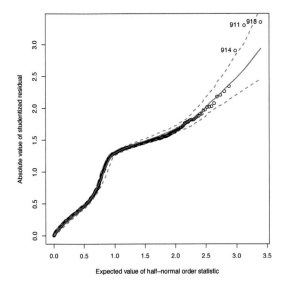

Figure 11.38: Half-normal QQ plot of studentized residuals for the NB model fit to the PhdPubs data. The reference line and confidence envelope reflect the mean and (2.5%, 97.5%) quantiles of the simulation distribution under the negative-binomial model for the same data.

Finally, plot the observed against expected absolute residuals as points, and add the lines for the confidence envelope, producing Figure 11.38.

```
> plot(expected, observed,
+       xlab = "Expected value of half-normal order statistic",
+       ylab = "Absolute value of studentized residual")
> lines(expected, mean, lty = 1, lwd = 2, col = "blue")
> lines(expected, lower, lty = 2, lwd = 2, col = "red")
> lines(expected, upper, lty = 2, lwd = 2, col = "red")
> identify(expected, observed, labels = names(observed), n = 3)
```

The shape of the QQ plot in Figure 11.37 shows a peculiar bend at low values and the half-normal version in Figure 11.38 has a peculiar hump in the middle. What could be the cause?

Figure 11.39 shows two additional plots of the studentized residuals that give a clear answer. The density plot at the left shows a strongly bimodal distribution of the residuals. An additional plot at the right of residuals against the log(response) confirms the guess that the lower mode corresponds to those students who published no articles—excess zeros again!

```
> # examine distribution of residuals
> res <- rstudent(phd.nbin)
> plot(density(res), lwd = 2, col = "blue",
+       main = "Density of studentized residuals")
> rug(res)
>
> # why the bimodality?
> plot(jitter(log(PhdPubs$articles + 1), factor = 1.5), res,
+       xlab = "log (articles + 1)", ylab = "Studentized residual")
```

Now we have something to worry about that *could* affect substantive interpretation or conclusions from this analysis using the NB model, but not accounting for excess zeros. If we believe, following Long (1997), that there is a separate latent class of students who don't publish, it would be sensible to fit a zero-inflated NB model, perhaps with a different subset of predictors for the

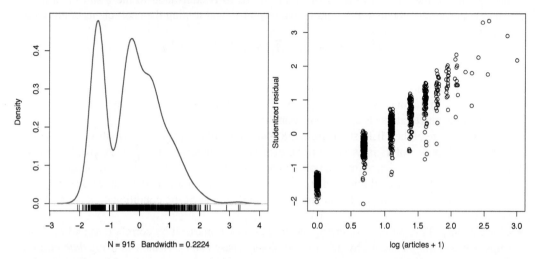

Figure 11.39: Further plots of studentized residuals. Left: density plot; right: residuals against log(articles+1).

zero component. The alternative theory of a "hurdle" to a first publication suggests fitting a hurdle model. We leave these as exercises for the reader.

$\triangle$

11.7 Multivariate response GLM models*

> Far better an approximate answer to the right question, which is often vague, than an exact answer to the wrong question, which can always be made precise.
>
> John W. Tukey (1962), *The future of data analysis*

As noted in Section 10.4, in many studies, there may be several response variables along with one or more explanatory variables, and it is useful to try to model some properties of their joint distribution as well as their separate dependence on the predictors. In the current chapter, the case study (Section 11.5.2) of demand for medical care by the elderly provides a relevant example. There are actually four indicators of medical care, a 2×2 set of (office vs. hospital) place and (physician vs. non-physician) practitioner. That case study analyzed only the office visits by physicians.

This section describes a few steps in this direction. To provide some context, we begin with a capsule overview of classical multivariate response models.

In the case of classical linear models with Gaussian error distributions, the model for a univariate response, $y = X\beta + \epsilon$, with $\epsilon \sim \mathcal{N}(0, \Sigma)$, extends quite readily to the ***multivariate linear model*** (MLM) for q response variables, $Y = \{y_1, y_2, \ldots, y_q\}$. The MLM has the form

$$\underset{(n \times q)}{Y} = \underset{(n \times p)(p \times q)}{X \quad B} + \underset{(n \times q)}{E} \tag{11.16}$$

where Y is a matrix of n observations on q response variables; X is a model matrix with columns for p regressors, typically including an initial column of 1s for the regression constant; B is a matrix of regression coefficients, one column for each response variable; and E is a matrix of errors.

It is important to note that:

- The maximum likelihood estimator of $\boldsymbol{B}$ in the MLM is equivalent to the result of fitting q separate univariate models for the individual responses and joining the coefficients columwise, giving

$$\widehat{\boldsymbol{B}} = \{\widehat{\boldsymbol{\beta}}_1, \widehat{\boldsymbol{\beta}}_2, \ldots, \widehat{\boldsymbol{\beta}}_q\} = (\boldsymbol{X}^{\mathsf{T}}\boldsymbol{X})^{-1}\boldsymbol{X}^{\mathsf{T}}\boldsymbol{Y}\,.$$

- Procedures for statistical inference (hypothesis tests, confidence intervals), however, take account of the correlations among the responses. Multivariate tests can therefore be more powerful than separate univariate tests under some conditions.
- A unique feature of the MLM stems from the assumption of multivariate normality of the errors, so that each row ϵ_i^{T} of $\boldsymbol{E}$ is assumed to be distributed independently, $\epsilon_i^{\mathsf{T}} \sim \mathcal{N}_q(\boldsymbol{0}, \boldsymbol{\Sigma})$, where $\boldsymbol{\Sigma}_{q \times q}$ is the error covariance matrix, constant across observations, like σ^2 in univariate models. Then, the conditional distributions of $\boldsymbol{y}_j \mid \boldsymbol{X}$ are all univariate normal, all bivariate distributions, $\boldsymbol{y}_j, \boldsymbol{y}_k \mid \boldsymbol{X}$ are bivariate normal, and any linear combination of the conditional ys is univariate normal.
- Consequently, all relationships among the ys can be summarized by correlations and relationships between the ys and xs by linear regressions. These can be visualized using *data ellipses* (Friendly et al., 2013) and hypothesis tests in the MLM can be visualized by ellipses using *hypothesis-error plots* (Friendly, 2007, Fox et al., 2009).

This generality of the MLM is lost, however, when we move to multivariate response models in the non-Gaussian case. For binomial responses, Section 10.4 described several approaches toward a multivariate logistic regression model that attempt to separate the marginal dependence of each $\boldsymbol{y}$ on the $\boldsymbol{x}$s from the relationship of the association among the $\boldsymbol{y}$s on the $\boldsymbol{x}$s. The bivariate logistic model for $(\boldsymbol{y}_1, \boldsymbol{y}_2)$, for example, was parameterized (see Eqn. (10.15)) in terms of submodels for a logit for each response, $\eta_1 = \boldsymbol{x}^{\mathsf{T}}\boldsymbol{\beta}_1$, $\eta_2 = \boldsymbol{x}^{\mathsf{T}}\boldsymbol{\beta}_2$ and a submodel for the log odds ratio, $\theta_{12} = \boldsymbol{x}^{\mathsf{T}}\boldsymbol{\beta}_{12}$.

The situation becomes more difficult for multivariate count data responses, because parametric approaches to their joint distribution (e.g., a multivariate Poisson distribution) given a set of explanatory variables are computationally and analytically intractable. Cameron and Trivedi (2013, Chapter 8) provide a detailed description of the problems and some solutions for the bivariate case, including bivariate Poisson, negative-binomial and hurdle models.

Consequently, only a few special cases have been worked out theoretically, and mostly for the bivariate case. For example, King (1989) described a seemingly unrelated bivariate Poisson model for two correlated count variables. This models the separate linear predictors for $\boldsymbol{y}_1$ and $\boldsymbol{y}_2$ as

$$
\begin{aligned}
g(\boldsymbol{\mu}_1) &= \boldsymbol{x}_1^{\mathsf{T}}\boldsymbol{\beta}_1 \\
g(\boldsymbol{\mu}_2) &= \boldsymbol{x}_2^{\mathsf{T}}\boldsymbol{\beta}_2\,,
\end{aligned}
$$

with the covariance between $\boldsymbol{y}_1$ and $\boldsymbol{y}_2$ represented as ξ. As in the MLM, the coefficients have the same point estimates as in equation-by-equation Poisson models. However, there is a gain in efficiency (reduced standard errors) resulting from a bivariate full-information maximum likelihood solution, and efficiency increases with the covariance ξ between the two count variables.

As a result, for lack of a fully general model for multivariate count data, one simple approach is to employ a method for simultaneous estimation of the equation-by-equation coefficients, accepting some loss of efficiency. This allows for hypothesis tests that may not be the most powerful, but provide approximate answers to more interesting questions. We can supplement this with separate analysis of the dependencies among the responses, and how these vary with the explanatory variables.

In R, the VGAM package is the most general available package for analysis of multivariate response GLMs. For multivariate count data, it provides for both Poisson and negative-binomial models. For NB models, the dispersion parameters $\theta_j = \alpha_j^{-1}$ can be allowed to vary with the predictors via a GLM of the form $\log \theta_j = \boldsymbol{x}^{\mathsf{T}}\boldsymbol{\gamma}_j$ or can be constrained to be "intercept-only,"

$\log \theta_j = \gamma_{0j}$, giving separate global dispersion estimates for each response. In the latter case, the resulting coefficients are the same as fitting a separate model for each response using `glm.nb()`.

EXAMPLE 11.19: Demand for medical care

In the examples in Section 11.5.2 we considered a variety of models for the number of office visits to physicians (`visits`) as the primary outcome variable in the study of demand for medical care by the elderly. We noted that other indicators of demand included office visits to non-physicians and hospital visits to both physicians and non-physicians. A more complete analysis of this data would consider all four response indicators together.

A special feature of this example is that the four response variables constitute a 2×2 set of the combinations of *place of visit* (office vs. hospital) and (physician vs. non-physician) *practitioner*. These are all counts, and could be transformed to two binary responses according to place and practitioner. Instead, we treat them individually here.

We start by selecting the variables to consider from the *NMES1988* data, giving a new working data set `nmes2`.

```
> data("NMES1988", package = "AER")
> nmes2 <- NMES1988[, c(1 : 4, 6 : 8, 13, 15, 18)]
> names(nmes2)[1 : 4]        # responses

[1] "visits"   "nvisits"   "ovisits"   "novisits"

> names(nmes2)[-(1 : 4)]    # predictors

[1] "hospital"  "health"     "chronic"   "gender"    "school"
[6] "insurance"
```

11.7.1 Analyzing correlations: HE plots

For purely descriptive purposes, a useful starting point is often an analysis of the $\log(y)$ on the predictor variables using the classical MLM, a rough analog of a multivariate Poisson regression with a log link. Inferential statistics will be biased, but we can use the result to visualize the pairwise linear relations that exist among all responses and all predictors compactly using hypothesis-error (HE) plots (Friendly, 2007).

Zero counts cause problems because the log of zero is undefined, so we add 1 to each y_{ij} in the call to `lm()`. The result is an object of class "mlm".

```
> clog <- function(x) log(x + 1)
> nmes.mlm <- lm(clog(cbind(visits, nvisits, ovisits, novisits)) ~ .,
+                data = nmes2)
```

An HE plot provides a visualization of the covariances of effects for the linear hypothesis (H) for each term in an MLM in relation to error covariances (E) using data ellipsoids in the space of dimension q, the number of response variables. The size of each H ellipsoid in relation to the E ellipsoid indicates the strength of the linear relations between the responses and the individual predictors.[21] The orientation of each H ellipsoid shows the direction of the correlations for that term with the response variables. For 1 degree of freedom terms (a covariate or factor with two levels), the corresponding H ellipsoid collapses to a line.

The heplots (Fox and Friendly, 2014) package contains functions for 2D plots (`heplot()`) of pairs of y variables, 3D plots (`heplot3d()`), and all pairwise plots (`pairs()`). We illustrate this here using `pairs()` for the MLM model, giving the plot shown in Figure 11.40.

---

[21] When the errors, E in Eqn. (11.16), are approximately multivariate normal, the H ellipsoid provides a visual test of significance: the H ellipsoid projects outside the E ellipsoid *if and only if* Roy's test is significant at a chosen α level.

```
> library(heplots)
> vlabels <- c("Physician\noffice visits",
+              "Non-physician\n office visits",
+              "Physician\nhospital visits",
+              "Non-physician\nhospital visits")
> pairs(nmes.mlm, factor.means = "health",
+       fill = TRUE, var.labels = vlabels)
```

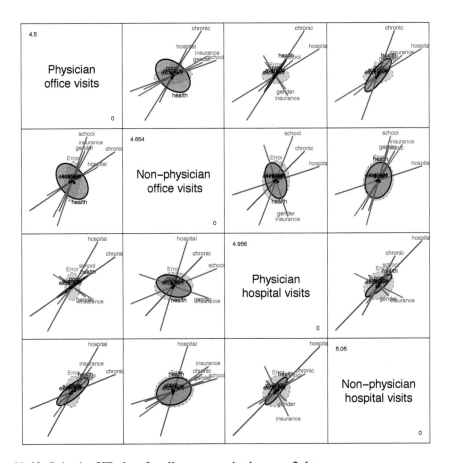

Figure 11.40: Pairwise HE plots for all responses in the nmes2 data.

The top row in Figure 11.40 shows the relationship of physician office visits to the other types of medical services. It can be seen that chronic conditions and hospital stays are positively correlated with both responses, as they also are in all other pairwise plots. Having private health insurance is positively related to some of these outcomes, and negatively to others. Except for difficulties with overlapping labels and the obvious violation of statistical assumptions of the MLM here, such plots give reasonably useful overviews on the relationships among the y and x variables.

11.7.2 Analyzing associations: Odds ratios and fourfold plots

In the analysis below, we first attempt to understand the association among these response variables and how these associations relate to the explanatory variables. It is natural to think of this in terms of the (log) odds ratio of a visit to a physician vs. a non-physician, given that the place is in an office as opposed to a hospital. Following this, we consider some multivariate negative binomial models relating these counts to the explanatory variables.

In order to treat the four response variables as a single response (`visit`), distinguished by `type`, it is necessary to reshape the data from a wide format to a long format with four rows for each input observation.

```
> vars <- colnames(nmes2)[1 : 4]
> nmes.long <- reshape(nmes2,
+    varying = vars,
+    v.names = "visit",
+    timevar = "type",
+    times = vars,
+    direction = "long",
+    new.row.names = 1 : (4 * nrow(nmes2)))
```

Then, the `type` variable can be used to create two new variables, `practitioner` and `place`, corresponding to the distinctions among visits. While we are at it, we create factors for two of the predictors.

```
> nmes.long <- nmes.long[order(nmes.long$id),]
> nmes.long <- transform(nmes.long,
+    practitioner = ifelse(type %in% c("visits", "ovisits"),
+                     "physician", "nonphysician"),
+    place = ifelse(type %in% c("visits", "nvisits"), "office", "hospital"),
+    hospf = cutfac(hospital, c(0 : 2, 8)),
+    chronicf = cutfac(chronic))
```

Then, we can use `xtabs()` to create a frequency table of `practitioner` and `place` classified by any one or more of these factors. For example, the total number of visits of the four types is given by

```
> xtabs(visit ~ practitioner + place, data = nmes.long)

              place
practitioner   hospital office
  nonphysician     2362   7129
  physician        3308  25442
```

From this, we can calculate the odds ratio and visualize the association with a fourfold or mosaic plot. More generally, by including more factors in the call to `xtabs()`, we can calculate and visualize how the *conditional* association varies with these factors. For example, Figure 11.41 shows fourfold plots conditioned by health status. It can be seen that there is a strong positive association, except for those with excellent health: people are more likely to see a physician in an office visit, and a non-physician in a hopsiptal visit. The corresponding log odds ratios are shown numerically using `loddsratio()`.

```
> library(vcdExtra)
> fourfold(xtabs(visit ~ practitioner + place + health, data = nmes.long),
+          mfrow=c(1,3))
> loddsratio(xtabs(visit ~ practitioner + place + health,
+                  data = nmes.long))

log odds ratios for practitioner and place by health

     poor   average excellent
 1.140166  0.972777  0.032266
```

Going further, we can condition by more factors. Figure 11.42 shows the fourfold plots conditioned by the number of chronic conditions (in the rows) and the combinations of gender and private insurance (columns).

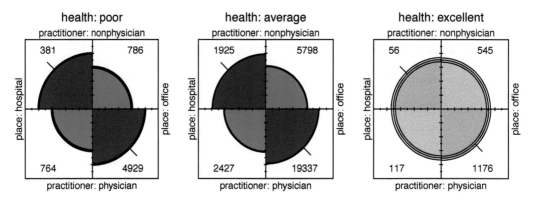

Figure 11.41: Fourfold displays for the association between practitioner and place in the nmes.long data, conditioned on health status.

```
> tab <- xtabs(visit ~ practitioner + place + gender +
+                     insurance + chronicf,
+              data = nmes.long)
> fourfold(tab, mfcol=c(4,4), varnames=FALSE)
```

The systematic patterns seen here are worth exploring further by graphing the log odds ratios directly. The call `as.data.frame(loddsratio(tab))` converts the result of `loddsratio(tab)` to a data frame with factors for these variables, and variables `LOR` and `ASE` containing the estimated log odds ratio ($\hat{\theta}$) and its asymptotic standard error ($\mathrm{ASE}(\hat{\theta})$). Figure 11.43 shows the plot of these values as line graphs with associated ± 1 error bars produced using **ggplot2**.[22]

```
> lodds.df <- as.data.frame(loddsratio(tab))
> library(ggplot2)
> ggplot(lodds.df, aes(x = chronicf, y = LOR,
+                      ymin = LOR - 1.96 * ASE, ymax = LOR + 1.96 * ASE,
+                      group = insurance, color = insurance)) +
+    geom_line(size = 1.2) + geom_point(size = 3) +
+    geom_linerange(size = 1.2) +
+    geom_errorbar(width = 0.2) +
+    geom_hline(yintercept = 0, linetype = "longdash") +
+    geom_hline(yintercept = mean(lodds.df$LOR), linetype = "dotdash") +
+    facet_grid(. ~ gender, labeller = label_both) +
+    labs(x = "Number of chronic conditions",
+         y = "log odds ratio (physician|place)") +
+    theme_bw() + theme(legend.position = c(0.1, 0.9))
```

It can be seen that for those with private insurance, the log odds ratios are uniformly positive, but males and females exhibit a somewhat different pattern over number of chronic conditions. Among those with no private insurance, the log odds ratios generally increase over number of chronic conditions, except for females with 3 or more such conditions.

Beyond this descriptive analysis, you can test hypotheses about the effects of the predictors on the log odds ratios using a simple ANOVA model. Under the null hypothesis, $H_0 : \theta_{ijk...} = 0$, the $\hat{\theta}$ are each distributed normally, $\mathcal{N}(0, \mathrm{ASE}(\hat{\theta}))$, so a weighted ANOVA can be used to test for differences according to the predictors. This analysis gives the results below.

---

[22]A similar plot can be obtained using `cotabplot(~practitioner + place + chronicf + insurance | gender, tab, panel = cotab_loddsratio)`.

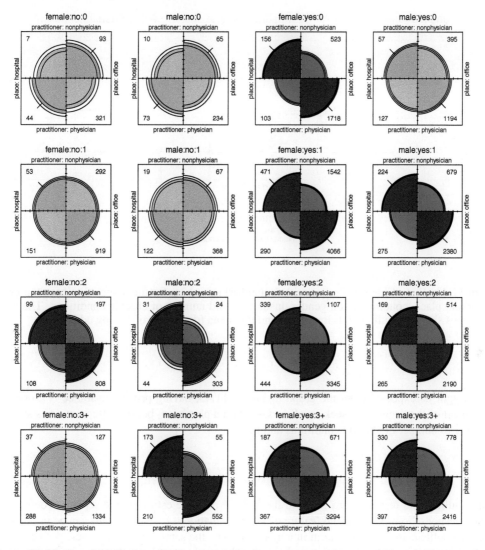

Figure 11.42: Fourfold displays for the association between practitioner and place in the nmes.long data, conditioned on gender, insurance, and number of chronic conditions. Rows are levels of chronic; columns are the combinations of gender and insurance.

```
> lodds.mod <- lm(LOR ~ (gender + insurance + chronicf)^2,
+                 weights = 1 / ASE^2, data = lodds.df)
> anova(lodds.mod)

Analysis of Variance Table

Response: LOR
                    Df Sum Sq Mean Sq F value Pr(>F)
gender               1    0.8     0.8    0.17  0.707
insurance            1    5.3     5.3    1.17  0.358
chronicf             3    4.6     1.5    0.34  0.802
gender:insurance     1   32.5    32.5    7.20  0.075 .
gender:chronicf      3   54.1    18.0    3.99  0.143
insurance:chronicf   3  114.1    38.0    8.43  0.057 .
Residuals            3   13.5     4.5
---
```

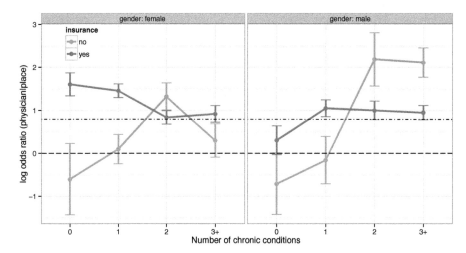

Figure 11.43: Plot of log odds ratios with 1 standard error bars for the association between practitioner and place, conditioned on gender, insurance, and number of chronic conditions. The horizontal lines show the null model (longdash) and the mean (dot–dash) of the log odds ratios.

```
Signif. codes:  0 '***' 0.001 '**' 0.01 '*' 0.05 '.' 0.1 ' ' 1
```

As might be expected from the graph in Figure 11.43, having private insurance is a primary determinant of the decision to seek an office visit with a physician, but this effect interacts slightly according to number of chronic conditions and gender.

$\triangle$

11.7.2.1 Fitting and testing multivariate count data models

With a multivariate response, vglm() in the **VGAM** package estimates the separate coefficients for each response jointly. A special feature of this formulation is that constraints can be imposed to force the coefficients for a given term in a model to to be the same for all responses. A likelihood-ratio test against the unconstrained model can then be used to test for differences in the effects of predictors across the response variables.

This is achieved by formulating the linear predictor as a sum of terms,

$$\eta(\boldsymbol{x}) = \sum_{k=1}^{p} \boldsymbol{H}_k \beta_k \boldsymbol{x}_k \ ,$$

where $\boldsymbol{H}_1, \ldots, \boldsymbol{H}_p$ are *known* full-rank constraint matrices. With no constraints, the $\boldsymbol{H}_k$ are identity matrices $\boldsymbol{I}_q$ for all terms. With vglm(), the constraint matrices for a given model are returned using constraints(), and can be set for a new, restricted model using the constraints argument. To constrain the coefficients for a term k to be equal for all responses, use $\boldsymbol{H}_k = \boldsymbol{1}_q$, a unit vector.

More general Wald tests of hypotheses can be carried out without refitting using linearHypothesis() in the car package. These include (a) joint tests that a subset of predictors for a given response have null effects; (b) across-response tests of equality of coefficients for one or more model terms.

EXAMPLE 11.20: Demand for medical care

In the examples in Section 11.5.2, we described a series of increasingly complex models for physician office visits, including interactions and nonlinear terms. The multivariate case is computationally more intensive, and estimation can break down in complex models. We can illustrate the main ideas here using the multivariate analog of the simple main effects model discussed in Example 11.14.

Using vglm(), the response variables are specified as the matrix form Y using cbind() on the left-hand side of the model formula. The right-hand side, ~ . here specifies all other variables as predictors. family = negbinomial uses the NB model for each y_j, with an intercept-only model for the dispersion parameters by default.

```
> nmes2.nbin <- vglm(cbind(visits, nvisits, ovisits, novisits) ~ .,
+                     data = nmes2, family = negbinomial)
```

The estimated parameters from this model are returned by the coef() method as pairs of columns labeled log(mu), logsize for each response. For example, the parameters for the visits response are in the first two columns, and are the same as those estimated for the model nmes.nbin using glm.nb().

```
> # coefficients for visits
> coef(nmes2.nbin, matrix = TRUE)[,c(1, 2)]

                  loge(mu1)  loge(size1)
(Intercept)        0.929257      0.18781
hospital           0.217772      0.00000
healthpoor         0.305013      0.00000
healthexcellent   -0.341807      0.00000
chronic            0.174916      0.00000
gendermale        -0.126488      0.00000
school             0.026815      0.00000
insuranceyes       0.224402      0.00000

> # theta for visits
> exp(coef(nmes2.nbin, matrix = TRUE)[1, 2])

[1] 1.2066
```

The log(mu) coefficients for all four response variables are shown below.

```
> coef(nmes2.nbin, matrix = TRUE)[,c(1, 3, 5, 7)]

                  loge(mu1)  loge(mu2)  loge(mu3)  loge(mu4)
(Intercept)        0.929257  -0.747798   -1.11284  -1.341793
hospital           0.217772   0.144645    0.41506   0.483883
healthpoor         0.305013  -0.179822    0.16491   0.033509
healthexcellent   -0.341807  -0.038121   -0.42449  -1.006527
chronic            0.174916   0.093430    0.27664   0.243570
gendermale        -0.126488  -0.255508    0.33456   0.052044
school             0.026815   0.068269    0.03559  -0.027475
insuranceyes       0.224402   0.492793   -0.53105   0.484062
```

We notice that the coefficients for hospital and chronic have values with the same signs for all four responses. If it is desired to test the hypothesis that their coefficients are all the same for each of these predictors, first extract the H matrices for the unconstrained model using constraints().

```
> clist <- constraints(nmes2.nbin, type = "term")
> clist$hospital[c(1, 3, 5, 7),]

      [,1]  [,2]  [,3]  [,4]
[1,]     1     0     0     0
```

```
[2,]     0     1     0     0
[3,]     0     0     1     0
[4,]     0     0     0     1
```

Then, reset the constraints for these terms to be unit vectors, forcing them to be all equal.

```
> clist2 <- clist
> clist2$hospital <- cbind(rowSums(clist$hospital))
> clist2$chronic  <- cbind(rowSums(clist$chronic))
> clist2$hospital[c(1, 3, 5, 7), 1, drop = FALSE]

      [,1]
[1,]     1
[2,]     1
[3,]     1
[4,]     1
```

Now, fit the same model as before, but using the constraints in clist2.

```
> nmes2.nbin2 <- vglm(cbind(visits, nvisits, ovisits, novisits) ~ .,
+                     data = nmes2, constraints = clist2,
+                     family = negbinomial(zero = NULL))
```

The coefficients for the constrained model are shown below. As you can see, the coefficients for hospital and chronic have the same estimates for all four responses.

```
> coef(nmes2.nbin2, matrix = TRUE)[,c(1, 3, 5, 7)]

                  loge(mu1)  loge(mu2)  loge(mu3)  loge(mu4)
(Intercept)        0.918002  -0.835090  -0.864251  -1.175650
hospital           0.244655   0.244655   0.244655   0.244655
healthpoor         0.293334  -0.315479   0.366404   0.172403
healthexcellent   -0.334959   0.047830  -0.538294  -1.044784
chronic            0.178563   0.178563   0.178563   0.178563
gendermale        -0.127956  -0.272264   0.330587   0.071228
school             0.026829   0.064401   0.031764  -0.028986
insuranceyes       0.222531   0.477475  -0.529976   0.507129
```

A likelihood-ratio test prefers the reduced model with equal coefficients for these two predictors. The degrees of freedom for this test (6) is the number of constrained parameters in the smaller model.

```
> lrtest(nmes2.nbin, nmes2.nbin2)

Likelihood ratio test

Model 1: cbind(visits, nvisits, ovisits, novisits) ~ .
Model 2: cbind(visits, nvisits, ovisits, novisits) ~ .
    #Df LogLik Df Chisq Pr(>Chisq)
1 35212 -25394
2 35218 -25413  6  39.2    6.4e-07 ***
---
Signif. codes:  0 '***' 0.001 '**' 0.01 '*' 0.05 '.' 0.1 ' ' 1
```

Alternatively, these tests can be performed as tests of linear hypotheses (see Section 11.1.2) on the coefficients B from the original model without refitting. Using linearHypothesis(), a hypothesis matrix L specifying equality of the coefficients for a given predictor can be easily generated using a character vector of the coefficient names.

```
> lh <- paste("hospital:", 1 : 3, " = ", "hospital:", 2 : 4, sep="")
> lh

[1] "hospital:1 = hospital:2" "hospital:2 = hospital:3"
[3] "hospital:3 = hospital:4"
```

Using `lh` as the `linear.hypothesis` argument then gives the following result for the coefficients of `hospital`, rejecting the hypothesis that they are all equal across response variables.

```
> car::linearHypothesis(nmes2.nbin, lh)

Linear hypothesis test

Hypothesis:
hospital:1 - hospital:2 = 0
hospital:2 - hospital:3 = 0
hospital:3 - hospital:4 = 0

Model 1: restricted model
Model 2: cbind(visits, nvisits, ovisits, novisits) ~ .

  Res.Df Df Chisq Pr(>Chisq)
1  35215
2  35212  3  26.4    7.8e-06 ***
---
Signif. codes:  0 '***' 0.001 '**' 0.01 '*' 0.05 '.' 0.1 ' ' 1
```

△

To pursue this analysis further, you could investigate whether any interactions of these effects were interesting and important as in Example 11.14, but now for the multivariate response variables.

To interpret a given model visually, you could use effect plots for the terms predicting each of the responses, as in Example 11.15. The **effects** package cannot handle models fit with **VGAM** directly, but you can use `glm()` or `glm.nb()` to fit the equivalent submodels for each response separately, and then use the `plot(Effect())` methods to display the effects for interesting terms. Figure 11.44 shows one such plot, for the effects of health status on each of the four response variables.

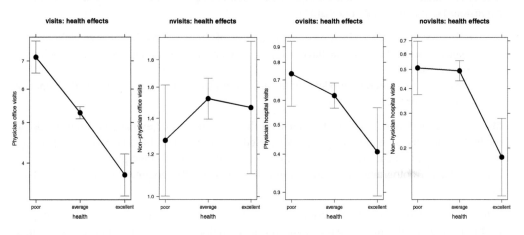

Figure 11.44: Effect plots for the effects of health status on the four response variables in the nmes2 data.

11.8   Chapter summary

- The generalized linear model extends the familiar classical linear models for regression and ANOVA to encompass models for discrete responses and continuous responses for which the assumption of normality of errors is untenable.

- It does this by retaining the idea of a *linear predictor*—a linear function of the regressors, $\eta_i = x^\mathsf{T}\beta$, but then allowing:

  - a *link function*, $g(\bullet)$, connecting the linear predictor η_i to the mean, $\mu_i = \mathcal{E}(y_i)$, of the response variable, so that $g(\mu_i) = \eta_i$. The link function formalizes the more traditional approach of analyzing an ad-hoc transformation of y, such as $\log(y)$, $\sqrt{y}$, y^2, or Box-Cox (Box and Cox, 1964) transformations y^λ to determine an empirical optimal power transformation.
  - a *random component*, specifying the conditional distribution of $y_i \mid x_i$ as any member of the exponential family, including the normal, binomial, Poisson, gamma, and other distributions.

- For the analysis of discrete response variables, and count data in particular, a key feature of the GLM is recognition of a *variance function* for the conditional variance of y_i, not forced to be constant, but rather allowed to depend on the mean μ_i and possibly a dispersion parameter, ϕ.

- From this background, we focus on GLMs for discrete count data response variables that extend considerably the loglinear models for contingency tables treated in Chapter 9. The Poisson distribution with a log link function is an equivalent starting point; however, count data GLMs often exhibit overdispersion in relation to the Poisson assumption that the conditional variance is the same as the mean, $\mathcal{V}(y_i \mid \eta_i) = \mu_i$.

  - One simple approach to this problem is the quasi-Poisson model, which estimates the dispersion parameter ϕ from the data, and uses this to correct standard errors and inferential tests.
  - Another is the wider class of negative-binomial models that allow a more flexible mean-variance function such as $\mathcal{V}(y_i \mid \eta_i) = \mu_i + \alpha\mu_i^2$.

- In practical application, many sets of empirical count data also exhibit a greater prevalence of zero counts than can be fit well using (quasi-) Poisson or negative-binomial models. Two simple extensions beyond the GLM class are

  - zero-inflated models, which posit a latent class of observations that always yield $y_i = 0$ counts, among the rest that have a Poisson or negative-binomial distribution including some zeros;
  - hurdle (or zero-altered) models, with one submodel for the zero counts and a separate submodel for the positive counts.

- Data analysis and visualization of count data therefore requires flexible tools and graphical methods. Some useful exploratory methods include jittered scatterplots and boxplots of $\log(y)$ against predictors enhanced by smoothed curves and trend lines, spine plots, and conditional density plots. Rootograms are quite helpful in visualizing the goodness-of-fit of count data models.

- Effect plots provide a convenient visual display of the high-order terms in a possibly complex GLM. They show the fitted values of the linear predictor $\widehat{\eta}^\star = X^\star\widehat{\beta}$, using a score matrix $X^\star$ that varies the predictors in a given term over their range while holding all other predictors constant. It is important to recognize, however, that like any model summary these show only the fitted effects under a given model, not the data.

- Model diagnostic measures (leverage, residuals, Cook's distance, etc.) and plots of these provide important ancillary information about the adequacy of a given model as a summary of relationships in the data. These help to detect problems of violations of assumptions, unusual or influential observations, or patterns that suggest that an important feature has not been accounted for.

- For multivariate response count data, there is no fully general theory as there is for the MLM with multivariate normality assumed for the errors. Nevertheless, there is a lot one can do to analyze such data combining the ideas of estimation for the separate responses with analysis of dependencies among the responses, conditioned by the explanatory variables.

11.9   Lab exercises

Exercise 11.1   Poole (1989) studied the mating behavior of elephants over 8 years in Amboseli National Park, Kenya. A focal aspect of the study concerned the mating success of males in relation to age, since larger males tend to be more successful in mating. Her data were used by Ramsey and Schafer (2002, Chapter 22) as a case study, and are contained in the **Sleuth2** (Ramsey et al., 2012) package (Ramsey et al., 2012) as *case2201*.

For convenience, rename this to `elephants`, and study the relation between `Age` (at the beginning of the study) and number of successful `Matings` for the 41 adult male elephants observed over the course of this study, ranging in age from 27–52.

```
> data("case2201", package="Sleuth2")
> elephants <- case2201
> str(elephants)

'data.frame': 41 obs. of  2 variables:
 $ Age    : num  27 28 28 28 28 29 29 29 29 29 ...
 $ Matings: num  0 1 1 1 3 0 0 0 2 2 ...
```

(a) Create some exploratory plots of `Matings` against `Age` in the styles illustrated in this chapter. To do this successfully, you will have to account for the fact that `Matings` has a range of only 0–9, and use some smoothing methods to show the trend.
(b) Repeat (a) above, but now plotting `log(Matings+1)` against `Age` to approximate a Poisson regression with a log link and avoid problems with the zero counts.
(c) Fit a linear Poisson regression model for `Matings` against `Age`. Interpret the fitted model *verbally* from a graph of predicted number of matings and/or from the model coefficients. (*Hint*: Using `Age-27` will make the intercept directly interpretable.)
(d) Check for nonlinearity in the relationship by using the term `poly(Age,2)` in a new model. What do you conclude?
(e) Assess whether there is any evidence of overdispersion in these data by fitting analogous quasi-Poisson and negative-binomial models.

Exercise 11.2   The data set *quine* in **MASS** gives data on absenteeism from schools in rural New South Wales, Australia. 146 children were classified by ethnic background (`Eth`), age (`Age`, a factor), `Sex`, and Learner status (`Lrn`), and the number of days absent (`Days`) from school in a particular school year was recorded.

(a) Fit the all main-effects model in the Poisson family and examine the tests of these effects using `summary()` and `car::Anova()`. Are there any terms that should be dropped according to these tests?

(b) Re-fit this model as a quasi-Poisson model. Is there evidence of overdispersion? Test for overdispersion formally, using `dispersiontest()` from **AER**.

(c) Carry out the same significance tests and explain why the results differ from those for the Poisson model.

Exercise 11.3 The data set `AirCrash` in **vcdExtra** was analyzed in Exercise 5.2 and Exercise 6.3 in relation to the `Phase` of the flight and `Cause` of the crash. Additional variables include the number of `Fatalities` and `Year`. How does `Fatalities` depend on the other variables?

(a) Use the methods of this chapter to make some exploratory plots relating fatalities to each of the predictors.

(b) Fit a main effects poisson regression model for `Fatalities`, and make effects plots to visualize the model. Which phases and causes result in the largest number of fatalities?

(c) A linear effect of `Year` might not be appropriate for these data. Try using a natural spline term, `ns(Year, df)` to achieve a better, more adequate model.

(d) Use a model-building tool like `add1()` or **MASS**::`stepAIC()` to investigate whether there are important two-way interactions among the factors and your chosen effect for `Year`.

(e) Visualize and interpret your final model and write a brief summary to answer the question posed.

Exercise 11.4 Male double-crested cormorants use advertising behavior to attract females for breeding. The `Cormorants` data set in **vcdExtra** gives some results from a study by Meagan Mc Rae (2015) on counts of advertising males observed two or three times a week at six stations in a tree-nesting colony for an entire breeding season. The number of advertising birds was counted and these observations were classified by characteristics of the trees and nests. The goal was to determine how this behavior varies temporally over the season and spatially over observation stations, as well as with characteristics of nesting sites. The response variable is `count` and other predictors are shown below. See `help(Cormorants, package="vcdExtra")` for further details.

```
> data("Cormorants", package = "vcdExtra")
> car::some(Cormorants)

        category week station    nest height  density tree_health count
9            Pre    1      C2      no    mid      few        dead     8
48           Pre    1      B2 partial    mid      few     healthy     2
66           Pre    2      C2 partial   high      few     healthy     2
141          Pre    3      B1    full   high      few     healthy     1
143          Pre    3      B2      no    mid      few        dead     1
214    Incubation    5      C3    full   high      few        dead     1
217    Incubation    5      C4      no   high      few        dead    10
219    Incubation    5      C4 partial   high      few        dead     2
319         <NA>   10      C1      no   high     high        dead     1
342         <NA>   13      C2    full    mid moderate        dead     1
```

(a) Using the methods illustrated in this chapter, make some exploratory plots of the number of advertising birds against week in the breeding season, perhaps stratified by another predictor, like tree `height`, nest condition, or observation `station`. To see anything reasonable, you should plot `count` on a log (or square root) scale, jitter the points, and add smoothed curves. The variable `category` breaks the weeks into portions of the breeding season, so adding vertical lines separating those will be helpful for interpretation.

(b) Fit a main-effects Poisson GLM to these data and test the terms using `Anova()` from the **car** package.

(c) Interpret this model using an effects plot.

(d) Investigate whether the effect of `week` should be treated as linear in the model. You could

try using a polynomial term like `poly(week, degree)` or perhaps better, using a natural spline term like `ns(week, df)` from the **splines** package.
(e) Test this model for overdispersion, using either a `quasipoisson` family or `dispersiontest()` in **AER**.

Exercise 11.5 For the *CodParasites* data, recode the `area` variable as an ordered factor as suggested in footnote 13. Test the hypotheses that prevalence and intensity of cod parasites is linearly related to area.

Exercise 11.6 In Example 11.10, we ignored other potential predictors in the *CodParasites* data: `depth`, `weight`, `length`, `sex`, `stage`, and `age`. Use some of the graphical methods shown in this case study to assess whether any of these are related to prevalence and intensity.

Exercise 11.7 The analysis of the *PhdPubs* data in the examples in this chapter were purposely left incomplete, going only as far as the negative binomial model.

(a) Fit the zero-inflated and hurdle models to this data set, considering whether the count component should be Poisson or negative-binomial, and whether the zero model should use all predictors or only a subset. Describe your conclusions from this analysis in a few sentences.
(b) Using the methods illustrated in this chapter, create some graphs summarizing the predicted counts and probabilities of zero counts for one of these models.
(c) For your chosen model, use some of the diagnostic plots of residuals and other measures shown in Section 11.6 to determine if your model solves any of the problems noted in Example 11.17 and Example 11.18, and whether there are any problems that remain.

Exercise 11.8 In Example 11.19 we used a simple analysis of $\log(y + 1)$ for the multivariate responses in the *NMES1988* data using a classical MLM (Eqn. (11.16)) as a rough approximation of a multivariate Poisson model. The HE plot in Figure 11.40 was given as a visual summary, but did not show the data. Examine why the MLM is not appropriate statistically for these data, as follows:

(a) Calculate residuals for the model `nmes.mlm` using

```
> resids <- residuals(nmes.mlm, type="deviance")
```

(b) Make univariate density plots of these residuals to show their univariate distributions. These should be approximately normal under the MLM. What do you conclude?
(c) Make some bivariate plots of these residuals. Under the MLM, each should be bivariate normal with elliptical contours and linear regressions. Add 2D density contours (`kde2d()`, or `geom_density2d()` in **ggplot2**) and some smoothed curve. What do you conclude?

References

Aberdein, J. and Spiegelhalter, D. (2013). Have London's roads become more dangerous for cyclists? *Significance*, 10(6), 46–48.

Adler, D. and Murdoch, D. (2014). *rgl: 3D visualization device system (OpenGL)*. R package version 0.95.1201.

Agresti, A. (1984). *Analysis of Ordinal Categorical Data*. New York: Wiley.

Agresti, A. (1990). *Categorical Data Analysis*. New York: Wiley-Interscience.

Agresti, A. (1996). *An Introduction to Categorical Data Analysis*. New York: Wiley Interscience.

Agresti, A. (2002). *Categorical Data Analysis*. Wiley Series in Probability and Statistics. New York: Wiley-Interscience [John Wiley & Sons], 2nd edn.

Agresti, A. (2007). *An Introduction to Categorical Data Analysis*. New York: Wiley, 2nd edn.

Agresti, A. (2013). *Categorical Data Analysis*. Wiley Series in Probability and Statistics. New York: Wiley-Interscience [John Wiley & Sons], 3rd edn.

Agresti, A. and Winner, L. (1997). Evaluating agreement and disagreement among movie reviewers. *Chance*, 10(2), 10–14.

Aitchison, J. (1986). *The Statistical Analysis of Compositional Data*. London: Chapman and Hall.

Akaike, H. (1973). Information theory and an extension of the maximum likelihood principle. In B. N. Petrov and F. Czaki, eds., *Proceedings of the 2nd International Symposium on Information*. Budapest: Akademiai Kiado.

Andersen, E. B. (1991). *Statistical Analysis of Categorical Data*. Berlin: Springer-Verlag, 2nd edn.

Anderson, E. (1935). The irises of the Gaspé peninsula. *Bulletin of the American Iris Society*, 35, 2–5.

Andrews, D. F. and Herzberg, A. M. (1985). *Data: A Collection of Problems from Many Fields for the Student and Research Worker*. New York, NY: Springer-Verlag.

Antonio, A. L. M. and Crespi, C. M. (2010). Predictors of interobserver agreement in breast imaging using the breast imaging reporting and data system. *Breast Cancer Research and Treatment*, 120(3), 539–546.

Arbuthnot, J. (1710). An argument for divine providence, taken from the constant regularity observ'd in the births of both sexes. *Philosophical Transactions*, 27, 186–190. Published in 1711.

Ashford, J. R. and Sowden, R. D. (1970). Multivariate probit analysis. *Biometrics*, 26, 535–546.

Atkinson, A. C. (1981). Two graphical displays for outlying and influential observations in regression. *Biometrika*, 68, 13–20.

Atkinson, A. C. (1987). *Plots, Transformations and Regression: An Introduction to Graphical Methods of Diagnostic Regression Analysis*. New York: Oxford University Press.

Bangdiwala, S. I. (1985). A graphical test for observer agreement. In *Proceeding of the International Statistics Institute*, vol. 1, (pp. 307–308). Amsterdam: ISI.

Bangdiwala, S. I. (1987). Using SAS software graphical procedures for the observer agreement chart. *Proceedings of the SAS User's Group International Conference*, 12, 1083–1088.

Bartlett, M. S. (1935). Contingency table interactions. *Journal of the Royal Statistical Society, Supplement*, 2, 248–252.

Bates, D., Maechler, M., Bolker, B., and Walker, S. (2014). *lme4: Linear mixed-effects models using Eigen and S4*. R package version 1.1-7.

Becker, M. P. and Clogg, C. C. (1989). Analysis of sets of two-way contingency tables using association models. *Journal of the American Statistical Association*, 84(405), 142–151.

Bentler, P. M. (1980). Multivariate analysis with latent variables: Causal modeling. *Annual Review of Psychology*, 31(1), 419–456.

Benzécri, J.-P. (1977). Sur l'analyse des tableaus binaires associés a une correspondense multiple. *Cahiers de l'Analyse des Données*, 2, 55–71.

Bertin, J. (1981). *Graphics and Graphic Information-processing*. New York: de Gruyter. (trans. W. Berg and P. Scott).

Bertin, J. (1983). *Semiology of Graphics*. Madison, WI: University of Wisconsin Press. (trans. W. Berg).

Bickel, P. J., Hammel, J. W., and O'Connell, J. W. (1975). Sex bias in graduate admissions: Data from Berkeley. *Science*, 187, 398–403.

Birch, M. W. (1963a). An algorithm for the logarithmic series distributions. *Biometrics*, 19, 651–652.

Birch, M. W. (1963b). Maximum likelihood in three-way contingency tables. *Journal of the Royal Statistical Society, Series B*, 25, 220–233.

Bishop, Y. M. M., Fienberg, S. E., and Holland, P. W. (1975). *Discrete Multivariate Analysis: Theory and Practice*. Cambridge, MA: MIT Press.

Bliss, C. I. (1934). The method of probits. *Science*, 79(2037), 38–39.

Böhning, D. (1983). Maximum likelihood estimation of the logarithmic series distribution. *Statistische Hefte (Statistical Papers)*, 24(1), 121–140.

Bollen, K. A. (2002). Latent variables in psychology and the social sciences. *Annual Review of Psychology*, 53(1), 605–634.

von Bortkiewicz, L. (1898). *Das Gesetz der Kleinen Zahlen*. Leipzig: Teubner.

Bouchet-Valat, M. (2015). *logmult: Log-Multiplicative Models, Including Association Models*. R package version 0.6.1.

Box, G. E. P. (1979). Some problems of statistics and everyday life. *Journal of the American Statistical Association*, 74(365), 1–4.

Box, G. E. P. and Cox, D. R. (1964). An analysis of transformations (with discussion). *Journal of the Royal Statistical Society, Series B*, 26, 211–252.

Box, G. E. P. and Draper, N. R. (1987). *Empirical Model Building and Response Surfaces*. New York, NY: John Wiley & Sons.

Bradu, D. and Gabriel, K. R. (1978). The biplot as a diagnostic tool for models of two-way tables. *Technometrics*, 20, 47–68.

Brinton, W. C. (1939). *Graphic Presentation*. New York, NY: Brinton Associates.

Brockmann, H. J. (1996). Satellite male groups in horseshoe crabs, *Limulus polyphemus*. *Ethology*, 102(1), 1–21.

Brown, P. J., Stone, J., and Ord-Smith, C. (1983). Toxaemic signs during pregnancy. *Journal of the Royal Statistical Society, Series C (Applied Statistics)*, 32, 69–72.

Brunswick, A. F. (1971). Adolescent health, sex, and fertility. *American Journal of Public Health*, 61(4), 711–729.

Burt, C. (1950). The factorial analysis of qualitative data. *British Journal of Statistical Psychology*, 3, 166–185.

Cameron, A. C. and Trivedi, P. K. (1990). Regression-based tests for overdispersion in the poisson model. *Journal of Econometrics*, 46, 347–364.

Cameron, A. C. and Trivedi, P. K. (1998). *Regression analysis of count data*. Econometric society monographs. Cambridge (U.K.), New York: Cambridge University Press.

Cameron, A. C. and Trivedi, P. K. (2013). *Regression analysis of count data*. Econometric society monographs. Cambridge (U.K.), New York: Cambridge University Press, 2nd edn.

Carlyle, T. (1840). *Chartism*. London: J. Fraser.

Caussinus, H. (1966). Contribution à l'analyse statistique des tableaux de corrélation. *Annales de la Faculté des Sciences de l'Université de Toulouse*, 39 (année 1965), 77–183.

Chambers, J. M., Cleveland, W. S., Kleiner, B., and Tukey, P. A. (1983). *Graphical Methods for Data Analysis*. Belmont, CA: Wadsworth.

Chang, W. and Wickham, H. (2015). *ggvis: Interactive Grammar of Graphics*. R package version 0.4.1.

Christensen, R. (1997). *Log-Linear Models and Logistic Regression*. New York, NY: Springer, 2nd edn.

Cicchetti, D. V. and Allison, T. (1971). A new procedure for assessing reliability of scoring EEG sleep recordings. *American Journal of EEG Technology*, 11, 101–109.

Cleveland, W. S. (1993a). A model for studying display methods of statistical graphics. *Journal of Computational and Graphical Statistics*, 2, 323–343.

Cleveland, W. S. (1993b). *Visualizing Data*. Summit, NJ: Hobart Press.

Cleveland, W. S., McGill, M. E., and McGill, R. (1988). The shape parameter of a two-variable graph. *Journal of the American Statistical Association*, 83, 289–300.

Cleveland, W. S. and McGill, R. (1984). Graphical perception: Theory, experimentation and application to the development of graphical methods. *Journal of the American Statistical Association*, 79, 531–554.

Cleveland, W. S. and McGill, R. (1985). Graphical perception and graphical methods for analyzing scientific data. *Science*, 229, 828–833.

Cohen, A. (1980). On the graphical display of the significant components in a two-way contingency table. *Communications in Statistics—Theory and Methods*, A9, 1025–1041.

Cohen, J. (1960). A coefficient of agreement for nominal scales. *Educational and Psychological Measurement*, 20, 37–46.

Cohen, J. (1968). Weighted kappa: Nominal scale agreement with provision for scaled diasgreement or partial credit. *Psychological Bulletin*, 70, 213–220.

Cook, R. D. (1993). Exploring partial residual plots. *Technometrics*, 35(4), 351–362.

Cook, R. D. and Weisberg, S. (1999). *Applied Regression Including Computing and Grapics*. New York: Wiley.

Cragg, J. G. (1971). Some statistical models for limited dependent variables with application to the demand for durable goods. *Econometrica*, 39, 829–844.

Croissant, Y. (2013). *mlogit: multinomial logit model*. R package version 0.2-4.

de la Cruz Rot, M. (2005). Improving the presentation of results of logistic regression with r. *Bulletin of the Ecological Society of America*, 86, 41–48.

Dahl, D. B. (2014). *xtable: Export tables to LaTeX or HTML*. R package version 1.7-4.

Dalal, S., Fowlkes, E. B., and Hoadley, B. (1989). Risk analysis of the space shuttle: Pre-*Challenger* prediction of failure. *Journal of the American Statistical Association*, 84(408), 945–957.

Dalgaard, P. (2008). *Introductory Statistics with R*. Springer, 2nd edn.

Deb, P. and Trivedi, P. K. (1997). Demand for medical care by the elderly: A finite mixture approach. *Journal of Applied Econometrics*, 12, 313–336.

Dragulescu, A. A. (2014). *xlsx: Read, write, format Excel 2007 and Excel 97/2000/XP/2003 files*. R package version 0.5.7.

Edwards, A. W. F. (1958). An analysis of geissler's data on the human sex ratio. *Annals of Human Genetics*, 23(1), 6–15.

Emerson, J. W. (1998). Mosaic displays in S-PLUS: A general implementation and a case study. *Statistical Graphics and Computing Newsletter*, 9(1), 17–23.

Emerson, J. W. and Green, W. A. (2014). *gpairs: The Generalized Pairs Plot*. R package version 1.2.

Emerson, J. W., Green, W. A., Schloerke, B., Crowley, J., Cook, D., Hofmann, H., and Wickham, H. (2013). The generalized pairs plot. *Journal of Computational and Graphical Statistics*, 22(1), 79–91.

Evers, M. and Namboordiri, N. K. (1977). A Monte Carlo assessment of the stability of log-linear estimates in small samples. In *Proceedings of the Social Statistics Section*. Alexandria, VA: American Statistical Association.

Feynman, R. P. (1988). *What Do You Care What Other People Think? Further Adventures of a Curious Character*. New York: W. W. Norton.

Fienberg, S. E. (1975). Perspective Canada as a social report. *Social Indicators Research*, 2, 153–174.

Fienberg, S. E. (1980). *The Analysis of Cross-Classified Categorical Data*. Cambridge, MA: MIT Press, 2nd edn.

Fienberg, S. E. and Rinaldo, A. (2007). Three centuries of categorical data analysis: Log-linear models and maximum likelihood estimation. *Journal of Statistical Planning and Inference*, 137(11), 3430–3445.

Finney, D. J. (1947). *Probit analysis*. Cambridge, England: Cambridge University Press.

Firth, D. (2003). Overcoming the reference category problem in the presentation of statistical models. *Sociological Methodology*, 33, 1–18.

Firth, D. and Menezes, R. X. d. (2004). Quasi-variances. *Biometrika*, 91, 65–80.

Fisher, R. A. (1925). *Statistical Methods for Research Workers*. London: Oliver & Boyd.

Fisher, R. A. (1936a). Has Mendel's work been rediscovered? *Annals of Science*, 1, 115–137.

Fisher, R. A. (1936b). The use of multiple measurements in taxonomic problems. *Annals of Eugenics*, 8, 379–388.

Fisher, R. A. (1940). The precision of discriminant functions. *Annals of Eugenics*, 10, 422–429.

Fisher, R. A., Corbet, A. S., and Williams, C. B. (1943). The relation between the number of species and the number of individuals. *Journal of Animal Ecology*, 12, 42.

Fleiss, J. L. (1973). *Statistical Methods for Rates and Proportions*. New York: John Wiley and Sons.

Fleiss, J. L. and Cohen, J. (1972). The equivalence of weighted kappa and the intraclass correlation coefficient as measures of reliability. *Educational and Psychological Measurement*, 33, 613–619.

Fleiss, J. L., Cohen, J., and Everitt, B. S. (1969). Large sample standard errors of kappa and weighted kappa. *Psychological Bulletin*, 72, 332–327.

Fowlkes, E. B. (1987). Some diagnostics for binary logistic regression via smoothing. *Biometrika*, 74(3), 503–5152.

Fox, J. (1987). Effect displays for generalized linear models. In C. C. Clogg, ed., *Sociological Methodology, 1987*, (pp. 347–361). San Francisco: Jossey-Bass.

Fox, J. (2003). Effect displays in R for generalized linear models. *Journal of Statistical Software*, 8(15), 1–27.

Fox, J. (2008). *Applied Regression Analysis and Generalized Linear Models*. Thousand Oaks, CA: SAGE Publications, 2nd edn.

Fox, J. (2015). Appendices to *Applied Regression Analysis, Generalized Linear Models, and Related Methods*, third edition. Online document. Available at `http://socserv.socsci.mcmaster.ca/jfox/Books/Applied-Regression-3E/Appendices.pdf`.

Fox, J. and Andersen, R. (2006). Effect displays for multinomial and proportional-odds logit models. *Sociological Methodology*, 36, 225–255.

Fox, J. and Friendly, M. (2014). *heplots: Visualizing Hypothesis Tests in Multivariate Linear Models*. R package version 1.0-12.

Fox, J., Friendly, M., and Monette, G. (2009). Visualizing hypothesis tests in multivariate linear models: The *heplots* package for R. *Computational Statistics*, 24(2), 233–246. (Published online: May 15, 2008).

Fox, J. and Weisberg, S. (2011a). *An R Companion to Applied Regression*. Thousand Oaks CA: SAGE Publications, 2nd edn.

Fox, J. and Weisberg, S. (2011b). *An R Companion to Applied Regression*. Thousand Oaks CA: SAGE Publications, 2nd edn.

Fox, J. and Weisberg, S. (2015a). *car: Companion to Applied Regression*. R package version 2.0-25/r421.

Fox, J. and Weisberg, S. (2015b). Visualizing fit and lack of fit in complex regression models: Effect plots with partial residuals. submitted, *Journal of Computational and Graphical Statistics*.

Fox, J., Weisberg, S., Friendly, M., and Hong, J. (2015). *effects: Effect Displays for Linear, Generalized Linear, and Other Models*. R package version 3.0-4/r200.

Friendly, M. (1991). *SAS System for Statistical Graphics*. Cary, NC: SAS Institute, 1st edn.

Friendly, M. (1992). Mosaic displays for loglinear models. In *ASA, Proceedings of the Statistical Graphics Section*, (pp. 61–68). Alexandria, VA.

Friendly, M. (1994a). A fourfold display for 2 by 2 by K tables. Tech. Rep. 217, York University, Psychology Dept.

Friendly, M. (1994b). Mosaic displays for multi-way contingency tables. *Journal of the American Statistical Association*, 89, 190–200.

Friendly, M. (1994c). SAS/IML graphics for fourfold displays. *Observations*, 3(4), 47–56.

Friendly, M. (1995). Conceptual and visual models for categorical data. *The American Statistician*, 49, 153–160.

Friendly, M. (1997). Conceptual models for visualizing contingency table data. In M. Greenacre and J. Blasius, eds., *Visualization of Categorical Data*, chap. 2, (pp. 17–35). San Diego, CA: Academic Press.

Friendly, M. (1999a). Extending mosaic displays: Marginal, conditional, and partial views of categorical data. *Journal of Computational and Graphical Statistics*, 8(3), 373–395.

Friendly, M. (1999b). Extending mosaic displays: Marginal, conditional, and partial views of categorical data. *Journal of Computational and Graphical Statistics*, 8(3), 373–395.

Friendly, M. (2000). *Visualizing Categorical Data*. Cary, NC: SAS Institute.

Friendly, M. (2002). Corrgrams: Exploratory displays for correlation matrices. *The American Statistician*, 56(4), 316–324.

Friendly, M. (2003). Visions of the past, present and future of statistical graphics: An ideo-graphic view. American Psychological Association. Toronto, ON, URL: http://datavis.ca/papers/apa-2x2.pdf.

Friendly, M. (2007). HE plots for multivariate general linear models. *Journal of Computational and Graphical Statistics*, 16(2), 421–444.

Friendly, M. (2013). Comment on the generalized pairs plot. *Journal of Computational and Graphical Statistics*, 22(1), 290–291.

Friendly, M. (2014a). *HistData: Data sets from the history of statistics and data visualization.* R package version 0.7-5.

Friendly, M. (2014b). *Lahman: Sean Lahman's Baseball Database.* R package version 3.0-1.

Friendly, M. (2015). *vcdExtra: vcd Extensions and Additions.* R package version 0.6-7.

Friendly, M. and Denis, D. (2005). The early origins and development of the scatterplot. *Journal of the History of the Behavioral Sciences*, 41(2), 103–130.

Friendly, M. and Kwan, E. (2003). Effect ordering for data displays. *Computational Statistics and Data Analysis*, 43(4), 509–539.

Friendly, M. and Kwan, E. (2011). Comment (graph people versus table people). *Journal of Computational and Graphical Statistics*, 20(1), 18–27.

Friendly, M., Monette, G., and Fox, J. (2013). Elliptical insights: Understanding statistical methods through elliptical geometry. *Statistical Science*, 28(1), 1–39.

Gabriel, K. R. (1971). The biplot graphic display of matrices with application to principal components analysis. *Biometrics*, 58(3), 453–467.

Gabriel, K. R. (1980). Biplot. In N. L. Johnson and S. Kotz, eds., *Encyclopedia of Statistical Sciences*, vol. 1, (pp. 263–271). New York: John Wiley and Sons.

Gabriel, K. R. (1981). Biplot display of multivariate matrices for inspection of data and diagnosis. In V. Barnett, ed., *Interpreting Multivariate Data*, chap. 8, (pp. 147–173). London: John Wiley and Sons.

Gabriel, K. R., Galindo, M. P., and Vincente-Villardón, J. L. (1997). Use of biplots to diagnose independence models in three-way contingency tables. In M. Greenacre and J. Blasius, eds., *Visualization of Categorical Data*, chap. 27, (pp. 391–404). San Diego, CA: Academic Press.

Gabriel, K. R. and Odoroff, C. L. (1990). Biplots in biomedical research. *Statistics in Medicine*, 9, 469–485.

Galton, F. (1886). Regression towards mediocrity in hereditary stature. *Journal of the Anthropological Institute*, 15, 246–263.

Gart, J. J. and Zweiful, J. R. (1967). On the bias of various estimators of the logit and its variance with applications to quantal bioassay. *Biometrika*, 54, 181–187.

Geissler, A. (1889). Beitrage zur frage des geschlechts verhaltnisses der geborenen. *Z. K. Sachsischen Statistischen Bureaus*, 35(1), n.p.

Gesmann, M. and de Castillo, D. (2015). *googleVis: R Interface to Google Charts*. R package version 0.5.8.

Gifi, A. (1981). *Nonlinear Multivariate Analysis*. The Netherlands: Department of Data Theory, University of Leiden.

Glass, D. V. (1954). *Social Mobility in Britain*. Glencoe, IL: The Free Press.

Goodman, L. A. (1970). The multivariate analysis of qualitative data: Interactions among multiple classifications. *Journal of the American Statistical Association*, 65, 226–256.

Goodman, L. A. (1971). The analysis of multidimensional contingency tables: Stepwise procedures and direct estimates for building models for multiple classifications. *Technometrics*, 13, 33–61.

Goodman, L. A. (1972). Some multiplicative models for the analysis of cross classified data. In *Proceedings of the sixth Berkeley Symposium on Mathematical Statistics and Probability*, (pp. 649–696). Berkeley, CA: University of California.

Goodman, L. A. (1973). The analysis of multidimensional contingency tables when some variables are posterior to others: A modified path analysis approach. *Biometrika*, 60, 179–192.

Goodman, L. A. (1978). *Analyzing Qualitative Categorical Data: Log-Linear Models and Latent-Strucutre Analysis*. Cambridge, MA: Abt Books.

Goodman, L. A. (1979). Simple models for the analysis of association in cross-classifications having ordered categories. *Journal of the American Statistical Association*, 74, 537–552.

Goodman, L. A. (1981). Association models and canonical correlation in the analysis of cross-classifications having ordered categories. *Journal of the American Statistical Association*, 76(374), 320–334.

Goodman, L. A. (1983). The analysis of dependence in cross-classifications having ordered categories, using log-linear models for frequencies and log-linear models for odds. *Biometrics*, 39, 149–160.

Goodman, L. A. (1985). The analysis of cross-classified data having ordered and/or unordered categories: Association models, correlation models, and asymmetry models for contingency tables with or without missing entries. *Annals of Statistics*, 13(1), 10–69.

Goodman, L. A. (1986). Some useful extensions of the usual correspondence analysis approach and the usual log-linear models approach in the analysis of contingency tables. *International Statistical Review*, 54(3), 243–309. With a discussion and reply by the author.

Gower, J., Lubbe, S., and Roux, N. (2011). *Understanding Biplots*. Wiley.

Gower, J. C. and Hand, D. J. (1996). *Biplots*. London: Chapman & Hall.

Grayson, D. K. (1990). Donner party deaths: A demographic assessment. *Journal of Anthropological Research*, 46(3), 223–242.

Greenacre, M. (1984). *Theory and Applications of Correspondence Analysis*. London: Academic Press.

Greenacre, M. (1989). The Carroll-Green-Schaffer scaling in correspondence analysis: A theoretical and empirical appraisal. *Journal of Marketing Research*, 26, 358–365.

Greenacre, M. (1990). Some limitations of multiple correspondence analysis. *Computational Statistics Quarterly*, 3, 249–256.

Greenacre, M. (1994). Multiple and joint correspondence analysis. In M. J. Greenacre and B. Jörg, eds., *Correspondence Analysis in the Social Sciences*. London: Academic Press.

Greenacre, M. (1997). Diagnostics for joint displays in correspondence analysis. In J. Blasius and M. Greenacre, eds., *Visualization of Categorical Data*, (pp. 221–238). Academic Press.

Greenacre, M. (2007). *Correspondence analysis in practice*. Boca Raton: Chapman & Hall/CRC.

Greenacre, M. (2013). Contribution biplots. *Journal of Computational and Graphical Statistics*, 22(1), 107–122.

Greenacre, M. and Hastie, T. J. (1987). The geometric interpretation of correspondence analysis. *Journal of the American Statistical Association*, 82, 437–447.

Greenacre, M. and Nenadic, O. (2014). *ca: Simple, Multiple and Joint Correspondence Analysis*. R package version 0.58.

Greenwood, M. and Yule, G. U. (1920). An inquiry into the nature of frequency distributions of multiple happenings, with particular reference to the occurrence of multiple attacks of disease or repeated accidents. *Journal of the Royal Statistical Society, Series A*, 83, 255–279.

Haberman, S. J. (1972). Statistical algorithms: Algorithm AS 51: Log-linear fit for contingency tables. *Applied Statistics*, 21(2), 218–225.

Haberman, S. J. (1973). The analysis of residuals in cross-classified tables. *Biometrics*, 29, 205–220.

Haberman, S. J. (1974). *The Analysis of Frequency Data*. Chicago: University of Chicago Press.

Haberman, S. J. (1979). *The Analysis of Qualitative Data: New Developments*, vol. II. New York: Academic Press.

Haldane, J. B. S. (1955). The estimation and significance of the logarithm of a ratio of frequencies. *Annals of Human Genetics*, 20, 309–311.

Hamilton, N. (2014). *ggtern: An extension to ggplot2, for the creation of ternary diagrams*. R package version 1.0.3.2.

Harrell, Jr, F. E. (2001). *Regression Modeling Strategies: With Applications to Linear Models, Logistic Regression, and Survival Analysis*. Graduate Texts in Mathematics. New York: Springer.

Harrell, Jr., F. E. (2015). *rms: Regression Modeling Strategies*. R package version 4.3-0.

Hartigan, J. A. and Kleiner, B. (1981). Mosaics for contingency tables. In W. F. Eddy, ed., *Computer Science and Statistics: Proceedings of the 13th Symposium on the Interface*, (pp. 268–273). New York, NY: Springer-Verlag.

Hartigan, J. A. and Kleiner, B. (1984). A mosaic of television ratings. *The American Statistician*, 38, 32–35.

Hastie, T. J. and Tibshirani, R. J. (1990). *Generalized Additive Models*. London: Chapman & Hall.

Hauser, R. M. (1979). Some exploratory methods for modeling mobility tables and other cross-classified data. In K. F. Schuessler, ed., *Sociological Methodology 1980*, (pp. 413–458). San Francisco: Jossey-Bass.

Hedeker, D. (2005). Generalized linear mixed models. In *Encyclopedia of Statistics in Behavioral Science*. John Wiley & Sons, Ltd.

van der Heijden, P. G. M., de Falguerolles, A., and de Leeuw, J. (1989). A combined approach to contingency table analysis using correspondence analysis and log-linear analysis. *Applied Statistics*, 38(2), 249–292.

van der Heijden, P. G. M. and de Leeuw, J. (1985). Correspondence analysis used complementary to loglinear analysis. *Psychometrika*, 50, 429–447.

Hemmingsen, W., Jansen, P. A., and Mackenzie, K. (2005). Crabs, leeches and trypanosomes: an unholy trinity? *Marine Pollution Bulletin*, 50(3), 336–339.

Heuer, J. (1979). *Selbstmord Bei Kinder Und Jugendlichen*. Stuttgard: Ernst Klett Verlag. [Suicide by children and youth.].

Hilbe, J. (2011). *Negative Binomial Regression*. Cambridge University Press, 2nd edn.

Hilbe, J. M. (2014). *Modeling Count Data*. New York, NY: Cambridge University Press.

Hoaglin, D. C. (1980). A poissonness plot. *The American Statistician*, 34, 146–149.

Hoaglin, D. C. and Tukey, J. W. (1985). Checking the shape of discrete distributions. In D. C. Hoaglin, F. Mosteller, and J. W. Tukey, eds., *Exploring Data Tables, Trends and Shapes*, chap. 9. New York: John Wiley and Sons.

Hocking, T. D. (2013). *directlabels: Direct labels for multicolor plots in lattice or ggplot2*. R package version 2013.6.15.

Hofmann, H. (2000). Exploring categorical data: Interactive mosaic plots. *Metrika*, 51(1), 11–26.

Hofmann, H. (2001). Generalized odds ratios for visual modeling. *Journal of Computational and Graphical Statistics*, 10(4), 628–640.

Hofmann, H. and Theus, M. (2005). Interactive graphics for visualizing conditional distributions. Unpublished manuscript.

Holm, S. (1979). A simple sequentially rejective multiple test procedure. *Scandinavian Journal of Statistics*, 6(2), 65–70.

Hosmer, Jr, D. W., Lemeshow, S., and Sturdivant, R. X. (2013). *Applied Logistic Regression*. New York: John Wiley and Sons, 3rd edn.

Hothorn, T., Zeileis, A., Farebrother, R. W., and Cummins, C. (2014). *lmtest: Testing Linear Regression Models*. R package version 0.9-33.

Hout, M., Duncan, O. D., and Sobel, M. E. (1987). Association and heterogeneity: Structural models of similarities and differences. *Sociological Methodology*, 17, 145–184.

Hummel, J. (1996). Linked bar charts: Analyzing categorical data graphically. *Computational Statistics*, 11, 23–33.

Hurley, C. B. (2004). Clustering visualizations of multidimensional data. *Journal of Computational and Graphical Statistics*, 13, 788–806.

Husson, F., Josse, J., Le, S., and Mazet, J. (2015). *FactoMineR: Multivariate Exploratory Data Analysis and Data Mining*. R package version 1.29.

Ihaka, R., Murrell, P., Hornik, K., Fisher, J. C., and Zeileis, A. (2015). *colorspace: Color Space Manipulation*. R package version 1.2-6.

Immer, F. R., Hayes, H., and Powers, L. R. (1934). Statistical determination of barley varietial adaptation. *Journal of the American Society of Agronomy*, 26, 403–419.

Jackman, S., Tahk, A., Zeileis, A., Maimone, C., and Fearon, J. (2015). *pscl: Political Science Computational Laboratory, Stanford University*. R package version 1.4.9.

Jansen, J. (1990). On the statistical analysis of ordinal data when extravariation is present. *Journal of the Royal Statistical Society. Series C (Applied Statistics)*, 39(1), 75–84.

Jinkinson, R. A. and Slater, M. (1981). Critical discussion of a graphical method for identifying discrete distributions. *The Statistician*, 30, 239–248.

Johnson, K. (1996). *Unfortunate Emigrants: Narratives of the Donner Party*. Logan, UT: Utah State University Press.

Johnson, N. L., Kotz, S., and Kemp, A. W. (1992). *Univariate Discrete Distributions*. New York, NY: John Wiley and Sons, 2nd edn.

Kemp, A. W. and Kemp, C. D. (1991). Weldon's dice data revisited. *The American Statistician*, 45, 216–222.

Kendall, M. G. and Stuart, A. (1961). *The Advanced Theory of Statistics*, vol. 2. London: Griffin.

Kendall, M. G. and Stuart, A. (1963). *The Advanced Theory of Statistics*, vol. 1. London: Griffin.

King, G. (1989). A seemingly unrelated Poisson regression model. *Sociological Methods and Research*, 17(3), 235–255.

Kleiber, C. and Zeileis, A. (2008). *Applied Econometrics with R*. New York: Springer-Verlag. ISBN 978-0-387-77316-2.

Kleiber, C. and Zeileis, A. (2014). Visualizing count data regressions using rootograms. Working papers, Faculty of Economics and Statistics, University of Innsbruck.

Kleiber, C. and Zeileis, A. (2015). *AER: Applied Econometrics with R*. R package version 1.2-3.

Koch, G. and Edwards, S. (1988). Clinical efficiency trials with categorical data. In K. E. Peace, ed., *Biopharmaceutical Statistics for Drug Development*, (pp. 403–451). New York: Marcel Dekker.

Kosambi, D. D. (1949). Characteristic properties of series distributions. *Proceedings of the National Institute of Science of India*, 15, 109–113.

Kosslyn, S. M. (1985). Graphics and human information processing: A review of five books. *Journal of the American Statistical Association*, 80, 499–512.

Kosslyn, S. M. (1989). Understanding charts and graphs. *Applied Cognitive Psychology*, 3, 185–225.

Kundel, H. L. and Polansky, M. (2003). Measurement of observer agreement. *Radiology*, 228(2), 303–308.

Labby, Z. (2009). Weldon's dice, automated. *Chance*, 22(4), 6–13.

Lambert, D. (1992). Zero-inflated poisson regression, with an application to defects in manufacturing. *Technometrics*, 34, 1–14.

Landis, J. R. and Koch, G. G. (1977). The measurement of observer agreement for categorical data. *Biometrics*, 33, 159–174.

Landis, R. J., Heyman, E. R., and Koch, G. G. (1978). Average partial association in three-way contingency tables: A review and discussion of alternative tests,. *International Statistical Review*, 46, 237–254.

Landwehr, J. M., Pregibon, D., and Shoemaker, A. C. (1984). Graphical methods for assessing logistic regression models. *Journal of the American Statistical Association*, 79, 61–71.

Lang, J. B. and Agresti, A. (1994). Simultaneously modeling joint and marginal distributions of multivariate categorical responses. *Journal of the American Statistical Association*, 89(426), 625–632.

Larsen, W. A. and McCleary, S. J. (1972). The use of partial residual plots in regression analysis. *Technometrics*, 14, 781–790.

Lavine, M. (1991). Problems in extrapolation illustrated with space shuttle O-ring data. *Journal of the American Statistical Association*, 86, 912–922.

Lawrance, A. J. (1995). Deletion influence and masking in regression. *Journal of the Royal Statistical Society. Series B (Methodological)*, 57(1), 181–189.

Lazarsfeld, P. F. (1950). The logical and mathematical foundation of latent structure analysis. In S. A. Stouffer, L. Guttmann, E. A. Suchman, P. F. Lazarsfeld, S. A. Star, and J. A. Clausen, eds., *Studies in Social Psychology in World War II, vol. IV, Measurement and Prediction*, (pp. 362–412). Princeton, NJ: Princeton University Press.

Lazarsfeld, P. F. (1954). A conceptual introduction to latent structure analysis. In P. F. Lazarsfeld, ed., *Mathematical Thinking in the Social Sciences*, (pp. 349–387). Glencoe, IL: Free Press.

Lebart, L., Morineau, A., and Warwick, K. M. (1984). *Multivariate Descriptive Statistical Analysis: Correspondence Analysis and Related Techniques for Large Matrices*. New York: John Wiley and Sons.

Lee, A. J. (1997). Modelling scores in the Premier League: Is Manchester United really the best? *Chance*, 10(1), 15–19.

Leifeld, P. (2013). texreg: Conversion of statistical model output in R to LaTeX and HTML tables. *Journal of Statistical Software*, 55(8), 1–24.

Leifeld, P. (2014). *texreg: Conversion of R regression output to LaTeX or HTML tables*. R package version 1.34.

Lemeshow, S., Avrunin, D., and Pastides, J. S. (1988). Predicting the outcome of intensive care unit patients. *Journal of the American Statistical Association*, 83, 348–356.

Lenth, R. V. (2014). *rsm: Response-surface analysis*. R package version 2.07.

Lenth, R. V. and Hervé, M. (2015). *lsmeans: Least-Squares Means*. R package version 2.16.

Lewandowsky, S. and Spence, I. (1989). The perception of statistical graphs. *Sociological Methods & Research*, 18, 200–242.

Lindsey, J. K. (1995). *Modelling Frequency and Count Data*. Oxford, UK: Oxford University Press.

Lindsey, J. K. and Altham, P. M. E. (1998). Analysis of the human sex ratio by using overdispersion models. *Journal of the Royal Statistical Society: Series C (Applied Statistics)*, 47(1), 149–157.

Lindsey, J. K. and Mersch, G. (1992). Fitting and comparing probability distributions with log linear models. *Computational Statistics and Data Analysis*, 13, 373–384.

Linzer, D. and Lewis., J. (2014). *poLCA: Polytomous variable Latent Class Analysis*. R package version 1.4.1.

Long, J. S. (1990). The origins of sex differences in science. *Social Forces*, 68(4), 1297–1316.

Long, J. S. (1997). *Regression Models for Categorical and Limited Dependent Variables*. Thousand Oaks, CA: SAGE Publications.

Lumley, T. and Zeileis, A. (2015). *sandwich: Robust Covariance Matrix Estimators*. R package version 2.3-3.

Maindonald, J. and Braun, J. (2007). *Data Analysis and Graphics Using R*. Cambridge: Cambridge University Press, 2nd edn.

Mc Rae, M. (2015). *Spatial, Habitat and Frequency Changes in Double-crested Cormorant Advertising Display in a Tree-nesting Colony*. Masters project, environmental studies, York University.

McCullagh, P. (1980). Regression models for ordinal data. *Journal of the Royal Statistical Society*, 42, 109–142.

McCullagh, P. and Nelder, J. A. (1989). *Generalized Linear Models*. London: Chapman and Hall.

McCulloch, C. E. and Neuhaus, J. M. (2005). Generalized linear mixed models. In *Encyclopedia of Biostatistics*. John Wiley & Sons, Ltd.

Mendenhall, W. and Sincich, T. (2003). *A Second Course in Statistics: Regression Analysis*. Prentice Hall / Pearson Education.

Merkle, E. C. and You, D. (2014). *nonnest2: Tests of Non-nested Models*. R package version 0.2.

Mersey, L. (1912). Report on the loss of the "Titanic" (S. S.). Parliamentary command paper 6352.

Meyer, D., Zeileis, A., and Hornik, K. (2006). The strucplot framework: Visualizing multi-way contingency tables with vcd. *Journal of Statistical Software*, 17(3), 1–48.

Meyer, D., Zeileis, A., and Hornik, K. (2015). *vcd: Visualizing Categorical Data*. R package version 1.3-3.

Milan, L. and Whittaker, J. (1995). Application of the parametric bootstrap to models that incorporate a singular value decomposition. *Journal of the Royal Statistical Society. Series C (Applied Statistics)*, 44(1), 31–49.

Mirai Solutions GmbH (2015). *XLConnect: Excel Connector for R*. R package version 0.2-11.

Mosteller, F. and Wallace, D. L. (1963). Inference in an authorship problem. *Journal of the American Statistical Association*, 58(302), 275–309.

Mosteller, F. and Wallace, D. L. (1984). *Applied Bayesian and Classical Inference: The Case of the Federalist Papers*. New York, NY: Springer-Verlag.

Mullahy, J. (1986). Specification and testing of some modified count data models. *Journal of Econometrics*, 33, 341–365.

Murrell, P. (2011). *R Graphics*. Boca Raton, FL: Chapman & Hall/CRC.

Nelder, J. A. and Wedderburn, R. W. M. (1972). Generalized linear models. *Journal of the Royal Statistical Society, Series A*, 135, 370–384.

Neter, J., Wasserman, W., and Kutner, M. H. (1990). *Applied Linear Statistical Models : Regression, Analysis of Variance, and Experimental Designs*. Homewood, IL: R. D. Irwin, Inc., 3rd edn.

Noack, A. (1950). A class of random variables with discrete distributions. *Annals of Mathematical Statistics*, 21, 127–132.

Ord, J. K. (1967). Graphical methods for a class of discrete distributions. *Journal of the Royal Statistical Society, Series A*, 130, 232–238.

Pareto, V. (1971). *Manuale di economia politica ("Manual of political economy")*. New York: A.M. Kelley. Translated by Ann S. Schwier. Edited by Ann S. Schwier and Alfred N. Page.

Pearson, K. (1900). On the criterion that a given system of deviations from the probable in the case of a correlated system of variables is such that it can be reasonably supposed to have arisen by random sampling. *Philosophical Magazine*, 50(5th Series), 157–175.

Peterson, B. and Harrell, Jr, F. E. (1990). Partial proportional odds models for ordinal response variables. *Applied Statistics*, 39, 205–217.

Pilhoefer, A. (2014). *extracat: Categorical Data Analysis and Visualization*. R package version 1.7-1.

Pinheiro, J., Bates, D., and R-core (2015). *nlme: Linear and Nonlinear Mixed Effects Models*. R package version 3.1-120.

Poole, J. H. (1989). Mate guarding, reproductive success and female choice in African elephants. *Animal Behavior*, 37, 842–849.

Powers, D. A. and Xie, Y. (2008). *Statistical Methods for Categorical Data Analysis*. Bingley, UK: Emerald, 2nd edn.

Pregibon, D. (1981). Logistic regression diagnostics. *Annals of Statistics*, 9, 705–724.

R Core Team (2015). *foreign: Read Data Stored by Minitab, S, SAS, SPSS, Stata, Systat, Weka, dBase, ...* R package version 0.8-63.

Raftery, A. E. (1986). Choosing models for cross-classifications. *American Sociological Review*, 51, 146–146.

Ramsey, F. L. and Schafer, D. W. (2002). *The Statistical Sleuth: A Course in Methods of Data Analysis*. Belmont, CA: Duxbury, 2nd edn.

Ramsey, F. L., Schafer, D. W., Sifneos, J., and Turlach, B. A. (2012). *Sleuth2: Data sets from Ramsey and Schafer's* Statistical Sleuth *(2nd ed)*. R package version 1.0-7.

Revelle, W. (2015). *psych: Procedures for Psychological, Psychometric, and Personality Research*. R package version 1.5.1.

Riedwyl, H. and Schüpbach, M. (1983). Siebdiagramme: Graphische darstellung von kontingenztafeln. Tech. Rep. 12, Institute for Mathematical Statistics, University of Bern, Bern, Switzerland.

Riedwyl, H. and Schüpbach, M. (1994). Parquet diagram to plot contingency tables. In F. Faulbaum, ed., *Softstat '93: Advances In Statistical Software*, (pp. 293–299). New York: Gustav Fischer.

Ripley, B. (2015a). *MASS: Support Functions and Datasets for Venables and Ripley's MASS*. R package version 7.3-40.

Ripley, B. (2015b). *nnet: Feed-Forward Neural Networks and Multinomial Log-Linear Models*. R package version 7.3-9.

Roberto, C., Giordano, S., Cazzaro, M., and Lang, J. (2014). *hmmm: hierarchical multinomial marginal models*. R package version 1.0-3.

le Roux, N. and Lubbe, S. (2013). *UBbipl: Understanding Biplots: Data Sets And Functions*. R package version 3.0.4.

RStudio, Inc. (2011). *manipulate: Interactive Plots for RStudio*. R package version 0.98.977.

RStudio, Inc. (2015). *shiny: Web Application Framework for R*. R package version 0.11.1.

Sarkar, D. (2015). *lattice: Lattice Graphics*. R package version 0.20-31.

Schloerke, B., Crowley, J., Cook, D., Hofmann, H., Wickham, H., Briatte, F., Marbach, M., and Thoen, E. (2014). *GGally: Extension to ggplot2*. R package version 0.5.0.

Schwartz, G. (1978). Estimating the dimensions of a model. *Annals of Statistics*, 6, 461–464.

Searle, S. R., Speed, F. M., and Milliken, G. A. (1980). Population marginal means in the linear model: An alternative to least squares means. *The American Statistician*, 34(4), 216–221.

Shneiderman, B. (1992). Tree visualization with treemaps: A 2-D space-filling approach. *ACM Transactions on Graphics*, 11(1), 92–99.

Shrout, P. E. and Fleiss, J. L. (1979). Intraclass correlations: Uses in assessing rater reliability. *Psychological Bulletin*, 86, 420–428.

Simpson, E. H. (1951). The interpretation of interaction in contingency tables. *Journal of the Royal Statistical Society, Series B*, 30, 238–241.

Skellam, J. G. (1948). A probability distribution derived from the binomial distribution by regarding the probability of success as variable between the sets of trials. *Journal of the Royal Statistical Society. Series B (Methodological)*, 10(2), 257–261.

Snee, R. D. (1974). Graphical display of two-way contingency tables. *The American Statistician*, 28, 9–12.

Snow, G. (2013). *TeachingDemos: Demonstrations for teaching and learning*. R package version 2.9.

Spence, I. (1990). Visual psychophysics of simple graphical elements. *Journal of Experimental Psychology: Human Perception and Performance*, 16, 683–692.

Spence, I. and Lewandowsky, S. (1990). Graphical perception. In J. Fox and J. S. Long, eds., *Modern Methods of Data Analysis*, chap. 1, (pp. 13–57). SAGE Publications.

Srole, L., Langner, T. S., Michael, S. T., Kirkpatrick, P., Opler, M. K., and Rennie, T. A. C. (1978). *Mental Health in the Metropolis: The Midtown Manhattan Study*. New York: NYU Press.

Stokes, M. E., Davis, C. S., and Koch, G. G. (2000). *Categorical Data Analysis Using the SAS System*. Cary, NC: SAS Institute, 2nd edn.

Stubben, C., Milligan, B., and Nantel, P. (2012). *popbio: Construction and analysis of matrix population models*. R package version 2.4.

Temple Lang, D., Swayne, D., Wickham, H., and Lawrence, M. (2014). *rggobi: Interface between R and GGobi*. R package version 2.1.20.

Theus, M. and Lauer, S. R. W. (1999). Visualizing loglinear models. *Journal of Computational and Graphical Statistics*, 8(3), 396–412.

Thornes, B. and Collard, J. (1979). *Who Divorces?* London: Routledge & Kegan.

Thurstone, L. L. (1927). A law of comparitive judgment. *Psychological Review*, 34, 278–286.

Tufte, E. (2006). *Beautiful Evidence*. Cheshire, CT: Graphics Press.

Tufte, E. R. (1983). *The Visual Display of Quantitative Information*. Cheshire, CT: Graphics Press.

Tufte, E. R. (1990). *Envisioning Information*. Cheshire, CT: Graphics Press.

Tufte, E. R. (1997). *Visual Explanations: Images and Quantities, Evidence and Narrative*. Cheshire, CT: Graphics Press.

Tukey, J. W. (1962). The future of data analysis. *Annals of Mathematical Statistics*, 33, 1–67 and 81.

Tukey, J. W. (1977). *Exploratory Data Analysis*. Reading, MA: Addison Wesley.

Tukey, J. W. (1990). Data-based graphics: Visual display in the decades to come. *Statistical Science*, 5(3), 327–339.

Tukey, J. W. (1993). Graphic comparisons of several linked aspects: Alternative and suggested principles. *Journal of Computational and Graphical Statistics*, 2(1), 1–33.

Turner, H. and Firth, D. (2014). *gnm: Generalized Nonlinear Models*. R package version 1.0-7.

Upton, G. J. G. (1976). The diagrammatic representation of three-party contests. *Political Studies*, 24, 448–454.

Upton, G. J. G. (1994). Picturing the 1992 British general election. *Journal of the Royal Statistical Society, Series A*, 157(Part 2), 231–252.

Urbanek, S. and Wichtrey, T. (2013). *iplots: iPlots - interactive graphics for R*. R package version 1.1-7.

Vaidyanathan, R. (2013). *rCharts: Interactive Charts using Javascript Visualization Libraries*. R package version 0.4.5.

Von Eye, A. and Mun, E. (2006). *Analyzing Rater Agreement: Manifest Variable Methods*. New York: Psychology Press, Taylor & Francis.

Vuong, Q. H. (1989). Likelihood ratio tests for model selection and non-nested hypotheses. *Econometrica*, 57(2), pp. 307–333.

Wainer, H. (1996). Using trilinear plots for NAEP state data. *Journal of Educational Measurement*, 33(1), 41–55.

Wand, M. (2015). *KernSmooth: Functions for Kernel Smoothing Supporting Wand & Jones (1995)*. R package version 2.23-14.

Wang, P. C. (1985). Adding a variable in generalized linear models. *Technometrics*, 27, 273–276.

Warnes, G. R., Bolker, B., Gorjanc, G., Grothendieck, G., Korosec, A., Lumley, T., MacQueen, D., Magnusson, A., Rogers, J., and others (2014). *gdata: Various R programming tools for data manipulation.* R package version 2.13.3.

Wei, T. (2013). *corrplot: Visualization of a correlation matrix.* R package version 0.73.

Wickham, H. (2009). *ggplot2: Elegant Graphics for Data Analysis.* Springer New York.

Wickham, H. (2014a). *plyr: Tools for splitting, applying and combining data.* R package version 1.8.1.

Wickham, H. (2014b). *reshape2: Flexibly Reshape Data: A Reboot of the Reshape Package.* R package version 1.4.1.

Wickham, H. and Chang, W. (2015). *ggplot2: An Implementation of the Grammar of Graphics.* R package version 1.0.1.

Wilkinson, L. (2005). *The Grammar of Graphics.* New York: Springer, 2nd edn.

Williams, D. A. (1987). Generalized linear model diagnostics using the deviance and single case deletions. *Applied Statistics*, 36, 181–191.

Wimmer, G. and Altmann, G. (1999). *Thesaurus of univariate discrete probability distributions.* Essen: Stamm.

Wong, R. S.-K. (2001). Multidimensional association models: A multilinear approach. *Sociological Methods and Research*, 30(2), 197–240.

Wong, R. S.-K. (2010). *Association Models.* Quantitative Applications in the Social Sciences. Los Angeles: SAGE Publications.

Wood, S. (2015). *mgcv: Mixed GAM Computation Vehicle with GCV/AIC/REML Smoothness Estimation.* R package version 1.8-6.

Wood, S. N. (2004). Stable and efficient multiple smoothing parameter estimation for generalized additive models. *Journal of the American Statistical Association*, 99(467), 673–686.

Wood, S. N. (2006). *Generalized Additive Models: An Introduction with R.* Chapman and Hall/CRC Press.

Woolf, B. (1995). On estimating the relation between blood group and disease. *Annals of Human Genetics*, 19, 251–253.

Wright, K. (2013). Revisiting Immer's barley data. *The American Statistician*, 67(3), 129–133.

Wright, K. (2015). *agridat: Agricultural Datasets.* R package version 1.11.

Xie, Y. (2014). *animation: A gallery of animations in statistics and utilities to create animations.* R package version 2.3.

Xie, Y. (2015). *knitr: A General-Purpose Package for Dynamic Report Generation in R.* R package version 1.9.

Yee, T. W. (2015). *VGAM: Vector Generalized Linear and Additive Models.* R package version 0.9-7.

Yip, K. C. and Yau, K. K. (2005). On modeling claim frequency data in general insurance with extra zeros. *Insurance: Mathematics and Economics*, 36(2), 153–163.

Zeileis, A. (2004). Econometric computing with hc and hac covariance matrix estimators. *Journal of Statistical Software*, 11(10), 1–17.

Zeileis, A. (2006). Object-oriented computation of sandwich estimators. *Journal of Statistical Software*, 16(9), 1–16.

Zeileis, A., Hornik, K., and Murrell, P. (2009). Escaping RGBland: Selecting colors for statistical graphics. *Computational Statistics & Data Analysis*, 53, 3259–3270.

Zeileis, A. and Kleiber, C. (2014). *countreg: Count Data Regression*. R package version 0.1-2/r88.

Zeileis, A., Kleiber, C., and Jackman, S. (2008). Regression models for count data in R. *Journal of Statistical Software*, 27(8).

Zeileis, A., Meyer, D., and Hornik, K. (2007). Residual-based shadings for visualizing (conditional) independence. *Journal of Computational and Graphical Statistics*, 16(3), 507–525.

Zelterman, D. (1999). *Models for Discrete Data*. New York: Oxford University Press.

Colophon

This book was produced using R version 3.2.1 (2015-06-18) and knitr (1.11). Writing, editing and compositing was done using RStudio. Hence, we can be assured that the code examples produced the output in the text.

The principal R package versions used in examples and illustrations are listed below. At the time of writing, most of these were current on CRAN repositories (e.g., http://cran.us. r-project.org/) but some development versions are indicated in the "source" column. "R-Forge" refers to the development platform (https://r-forge.r-project.org) used by many package authors to prepare and test new versions. By the time you read this, most of these should be current on CRAN.

package	version	date	source
AER	1.2-4	2015-06-06	CRAN
ca	0.58	2014-12-31	CRAN
car	2.1-0	2015-09-03	CRAN
colorspace	1.2-6	2015-03-11	CRAN
corrplot	0.73	2013-10-15	CRAN
countreg	0.1-3	2015-04-18	R-Forge
directlabels	2013.6.15	2013-07-23	CRAN
effects	3.0-5	2015-09-10	R-Forge
ggparallel	0.1.2	2015-08-21	CRAN
ggplot2	1.0.1	2015-03-17	CRAN
ggtern	1.0.6.1	2015-10-12	CRAN
gmodels	2.16.2	2015-07-22	CRAN
gnm	1.0-8	2015-04-22	CRAN
gpairs	1.2	2014-03-09	CRAN
heplots	1.0-16	2015-07-13	CRAN
Lahman	4.0-1	2015-09-15	CRAN
lattice	0.20-33	2015-07-14	CRAN
lmtest	0.9-34	2015-06-06	CRAN
logmult	0.6.2	2015-04-22	CRAN
MASS	7.3-45	2015-11-10	CRAN
mgcv	1.8-9	2015-10-30	CRAN
nnet	7.3-11	2015-08-30	CRAN
plyr	1.8.3	2015-06-12	CRAN
pscl	1.4.9	2015-03-29	CRAN
RColorBrewer	1.1-2	2014-12-07	CRAN
reshape2	1.4.1	2014-12-06	CRAN
rms	4.4-0	2015-09-28	CRAN
rsm	2.7-4	2015-10-07	CRAN
sandwich	2.3-4	2015-09-24	CRAN
vcd	1.4-2	2015-10-18	R-Forge
vcdExtra	0.6-11	2015-09-17	CRAN
VGAM	1.0-0	2015-10-29	CRAN
xtable	1.8-0	2015-11-02	CRAN

To prepare your R installation for running the examples in this book, you can use the following commands to install these packages.

```
> packages <- c("AER", "ca", "car", "colorspace", "corrplot", "countreg",
+     "directlabels", "effects", "ggparallel", "ggplot2", "ggtern",
+     "gmodels", "gnm", "gpairs", "heplots", "Lahman", "lattice", "lmtest",
+     "logmult", "MASS", "mgcv", "nnet", "plyr", "pscl", "RColorBrewer",
+     "reshape2", "rms", "rsm", "sandwich", "splines", "vcd", "vcdExtra",
+     "VGAM", "xtable")
> install.packages(packages)
> # if countreg is not yet on CRAN:
> install.packages("countreg", repos = "http://R-Forge.R-project.org")
```

Author Index

Example Index

Subject Index